HANDBOOK OF NATURALLY OCCURRING COMPOUNDS WITH ANTIOXIDANT ACTIVITY IN PLANTS

Handbook of Naturally Occurring Compounds with Antioxidant Activity in Plants

Rosa Martha Pérez Gutiérrez

Nova Science Publishers, Inc.
New York

LIBRARY OF CONGRESS CATALOGING-IN-PUBLICATION DATA

Pérez Gutiérrez, Rosa Martha.
Handbook of naturally occurring compounds with antioxidant activity in plants / Rosa Martha Perez Gutierrez.
p. ; cm.
Includes bibliographical references and index.
ISBN 1-59454-820-X
1. Antioxidants--Therapeutic use--Handbooks, manuals, etc. 2. Plants, Useful--Handbooks, manuals, etc. I. Title.
[DNLM: 1. Antioxidants--therapeutic use. 2. Antioxidants—isolation & purification. 3. Plant Preparations--pharmacology. QV 325 P438h 2006]
RB170.G882 2006
615'.321--dc22 2005031794

Published by Nova Science Publishers, Inc. ✦ New York

With Love to Julio, Julito, Alexandra and Alex

CONTENTS

PROLOGUE

Lipid oxidation is a highly deteriorative process in foods, as it leads to unacceptable properties for the customer and a loss in nutritional value. Hence, the presence of antioxidants in foods is essential for their quality and safety. Consumers are now including phytonutrients in their diet with the notion that antioxidant compounds may reduce the incidence of cancer and aging in humans.

This book presents 847 compounds isolated and identified from plants that present antioxidant activity. These substances have been classified by chemical groups and each provides the most relevant information of its pharmacological activity, action mechanism, chemical structure, spectroscopic date and other properties. Chemical structures have been drawn to indicate the stereochemistry. In this handbook the summary of the scientific information of plants that present biological activity and the compounds responsible for this activity is presented, which introduces the reader to the study of medicinal plants and also providing bibliographic references, where a detailed study of its chemistry and pharmacology can be found. This dictionary will be of great help for pre and post-graduate students, as well as professors, research of the medical industry, who work in some way with isolates bioactive compounds obtained from plants.

INTRODUCTION

Oxidation is well known to be a major cause of food and materials degradation. The process of oxidation in the human body damages cell membranes and other structures. When oxygen is metabolised, it creates 'free radicals' which steal electrons from other molecules, causing damage. The body can cope with some free radicals and needs them to function effectively. However, oxidation can be accelerated by stress, cigarette smoking, alcohol, sunlight, pollution and other factors. An overload of free radicals has been linked to certain diseases, including heart disease and some cancers. Antioxidants are found in certain foods that neutralise free radicals. These include vitamins A, C and E, and the minerals copper, zinc and selenium. Other compounds, such as the phytochemicals in plants and zoochemicals from animal products, are believed to have greater antioxidant effects than either vitamins or minerals. The effect of free radicals some of the degenerative conditions caused by free radicals include: deterioration of the eye lens, which contributes to blindness, inflammation of the joints (arthritis), damage to nerve cells in the brain, which contributes to conditions such as Parkinson's or Alzheimer's disease, acceleration of the ageing process, increased risk of coronary heart disease, since free radicals encourage low density lipoprotein (LDL) cholesterol to adhere to artery walls and certain cancers, triggered by damaged cell DNA. Aging also may be the sum of the deleterious free-radical reactions which occur continuously throughout cells and tissues. In this context, natural antioxidants are receiving increasing attention. They can be an alternative to the use of synthetic compounds in food and pharmaceutical technology or serve as lead compounds for the development of new drugs with the prospect of improving the treatment of various disorders.

Lipid peroxidation has many deleterious effects on membrane structure and function, since it disrupts membrane order and generates many potentially cytoloxic products unsaturated aldehydes and lipid hydroperoxides that are highly toxic *in vivo* and *in vitro*. These compounds are able to inactivate enzymes, modify biomolecules by the covalent binding of their fragment, and initiate free radical-mediated modification of proteins and lipid peroxidation. Cells and living organisms have developed several defensive mechanisms to protect them from reactive oxygen species. It has been demonstrated in many *in vitro* experiments, that enzymes and secondary compounds of vascular plants protect against oxidative damage by inhibiting the formation of free radicals and/or trapping reactive oxygen species, including peroxyl radicals. Among chain-breaking antioxidants, are several lignans and flavonoids, whose antioxidant activity has been related to the presence of reactive phenolic groups. Also, phenolic compounds, vitamin E, and synthetic antioxidants,2-

*ter*tbutyl-4-methoxyphenol (BRA) and 2,6-di.*tert*-butyl-4-methoxyphenol (BRT) are used to Protect cosmetics, drugs and foods from oxidative degradation.

Natural compounds with antioxidative properties which have been reported from plant, microbial or animal sources are reviewed. Antioxidants have been found within most natural substance classes. Flavonoids are by far the most comprehensively investigated natural products that can stabilize highly reactive and potentially harmful free radicals and are considered to reduce the incidence of chronic diseases. Reactivity towards various radicals has been studied under many experimental conditions and structure-activity relationships have been established. Tannins, coumarins and, more recently, xanthones have also been investigated in some detail. Studies on alkaloids have concentrated so far mostly on benzylisoquinoline alkaloids. At the same time, particular interest has been provided by some plant species or genera normally consumed as food or spices. Curcuminoids isolated from various Curcuma have shown potent antioxidant properties. Phenolic diterpenes from *Rosmarinus officinalis* possess very promising activity.The exceptional stability of sesame oil from *Sesamum indicum* led to the isolation of sesamol, together with several bisfuranyllignans. The search for new natural antioxidants is not confined to higher plants. A great variety of metabolites including in particular phenazine and carbazole derivatives have been isolated from *Streptomycetes* or fungal strains. The antioxidant phenolic compounds are reported to remove free radicals and protect the structural integrity of cells and tissues.

Chapter 1

ALKALOIDS AND NITROGEN COMPOUNDS

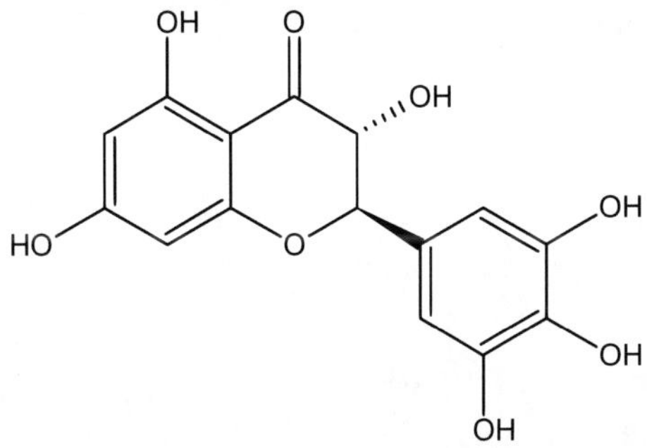

1
Ampelopsin

Found in leaves of *Rhododendron cinnabarium* (Ericaceae) showed antioxiant activity (Harbone and Baxter, 1993).

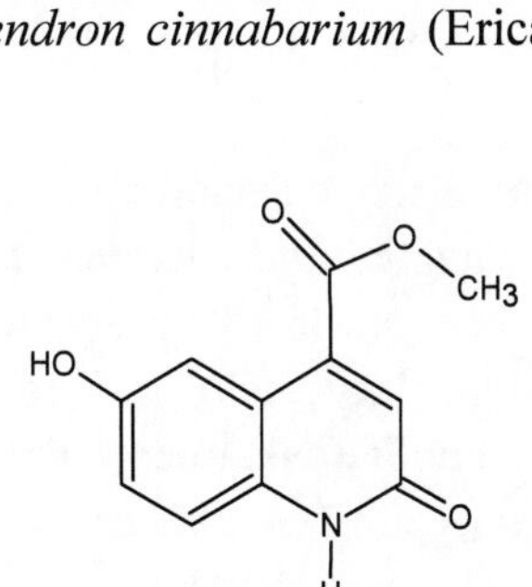

2
4-Carbomethoxy-6-hydroxy-2-quinolone

The aleurone layer of *Oryza sativa* L. (Gramineae) ev. Heugjinmi yielded a new quinolone alkaloid, 4-carbomethoxy-6-hydroxy-2-quinolone, showing moderate antioxidative activity in a 1,1-diphenyl-2-picrylhydrazyl free-radical scavenging assay. (Chung et al., 2001).

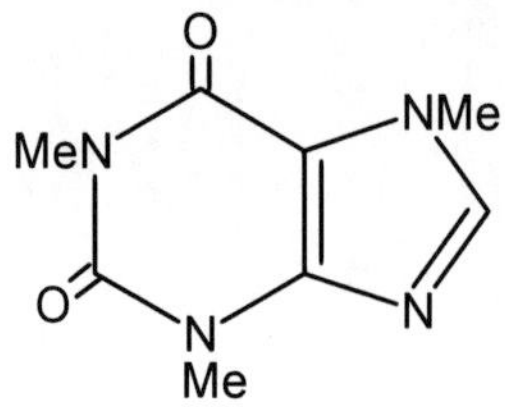

3
Caffeine

Caffeine, from the leaves of tea (*Thea sinensis,*) and coffee (*Caffea Arabica*), was shown to have antioxidative activity (in a linoleic acid oxidation test) comparable to that of BHA and BHT (Larson, 1988).

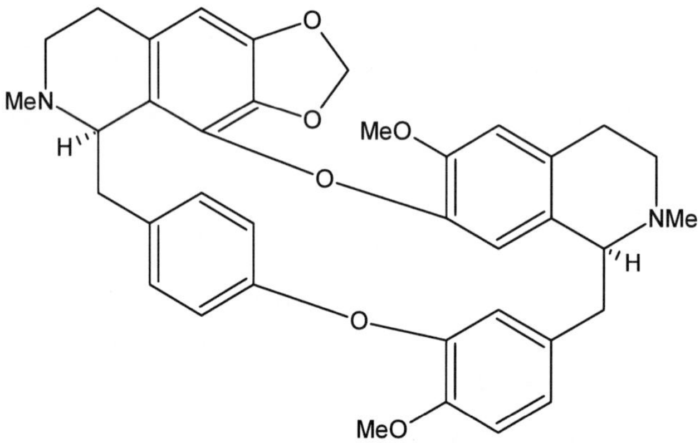

4
Cepharanthine

Increasingly, evidence from a variety of sources is indicating that the basic nitrogen compounds of higher plants include many representatives that are potent inhibitors of various oxidative processes, both *in vivo* and *in vitro*. Lipid peroxidation induced by radioactive cobalt irradiation (in soybean lecithin liposomes) or ferrous ion (in rat liver mitochondria) was inhibited by the bisbenzylisoquinoline alkaloid cepharanthine (*Stephania cepharantha, Stephania erecta*, from the roots of *Stephania sasakii* and *Stephania epigeae*, Menispermaceae). A concentration of 2.5 x 10^{-5} M prevented 50% of the mitochondrial lipid peroxidation which occurred in its absence in 20 min (Larson, 1988).

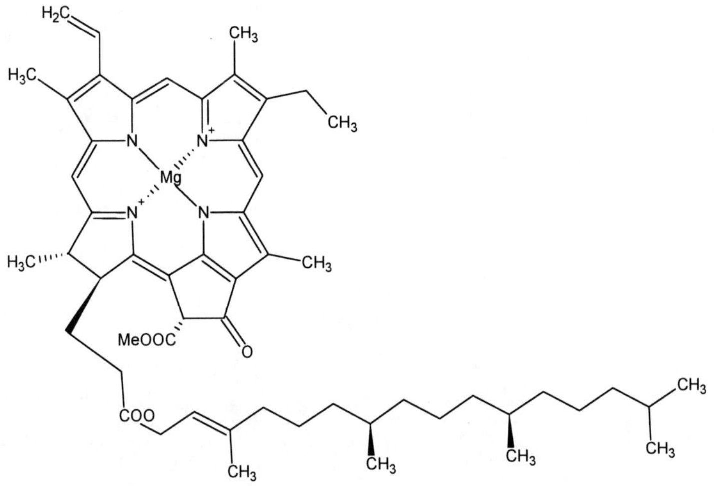

5
Chlorophyll a

Although both chlorophyll promote the oxidation of lipids in the light, they are inhibitors of auto oxidation under dark conditions. At 30 °C, 2 x 10^{-5} M chlorophyll a was superior by a factor of ca two to BHT. The compound appear to be unreactive toward lipid hydroperoxides, but do react with peroxy radicals; electron spin resonance data indicate the presence of a tetrapyrrole radical cation (Larson, 1988).

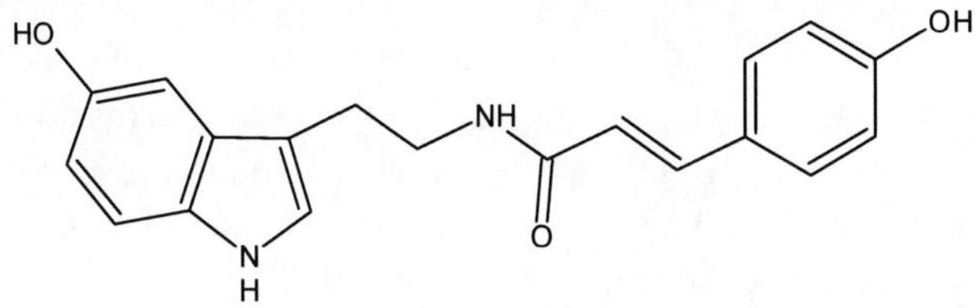

6
N-(*p*-Coumaroyl)serotonín

The antioxidative serotonin derivative was isolated from an ethanol extract of Japanese barnyard millet *Echinochloa utilis* (Graminae) grains. Exhibited a strong antioxidant activity almost equivalent to that of butylated hydroxy-anisole at the same concentration (w/v), (Watanabe, 1999).

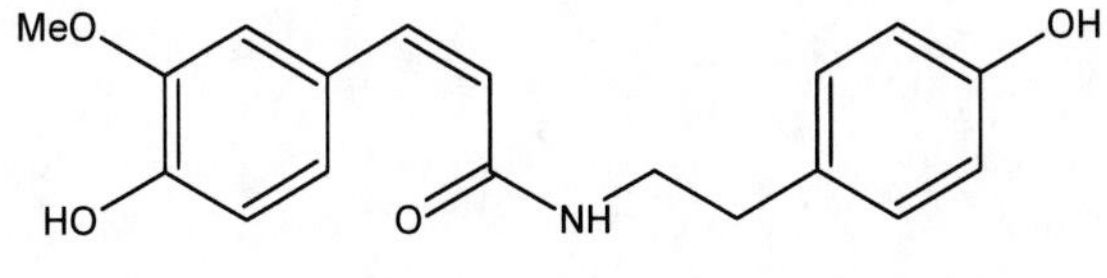

7
N-Formylnornuciferin

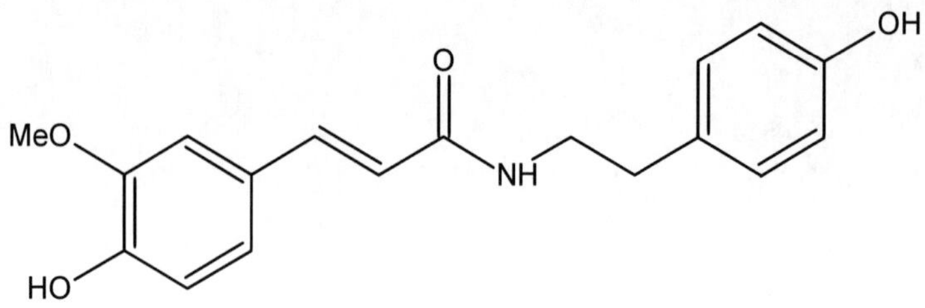

8
N-*cis*-Feruloyltyramine

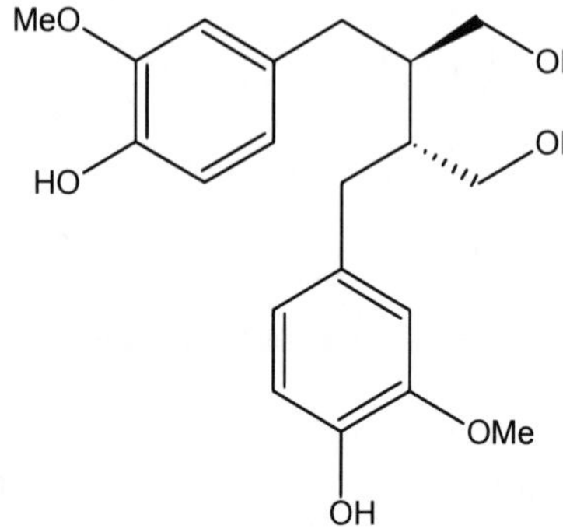

9
N-*trans*-Feruloyltyramine

These alkaloids were isolated from *Tinospora crisp* a Miers (Menispermaceae), exhibiting antioxidant and radical scavenging properties towards β-carotene and 2,2-diphenyl-l-picrylhydrazyl (DPPH) radical when tested in dilution assays on the reduction of the DPPH radical. These 3 compounds proved to be more active than the synthetic antioxidant butylhydroxytoluene (BHT), (Cavinlet al., 1998).

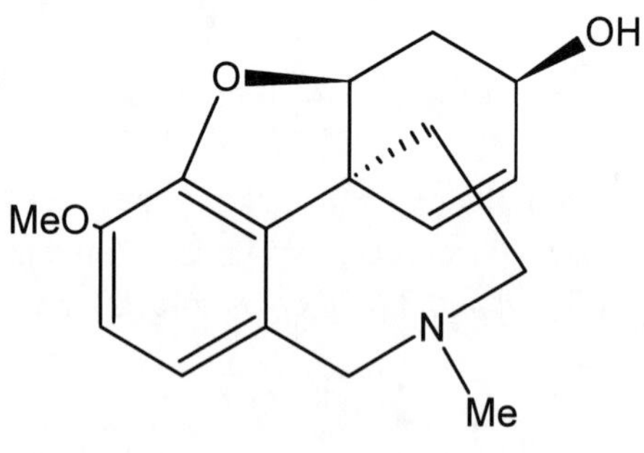

10
Galantamine

Alkaloid from *Sarothamnus scoparius* (Leguminosae), showed antioxidant activity (Vedavanam et al., 1999). This compound was a scavenger of reactive oxygen species (ROS).

The antioxidant activity depended on the enolic OH group in the molecule. The presence of a quaternary nitrogen in the compound increased the strength of the scavenging effect.

11
Glycoprotein

An active glycoprotein was isolated from *Aloe vera* gel (Liliaceae) and showed a radical scavenging activity against superoxide anion generated by the xanthine-xanthine oxidase system as well as inhibition of cyclooxygenase-2 and reduction of thromboxane A_2 synthase level *in vitro* (Yagi et al., 2003).

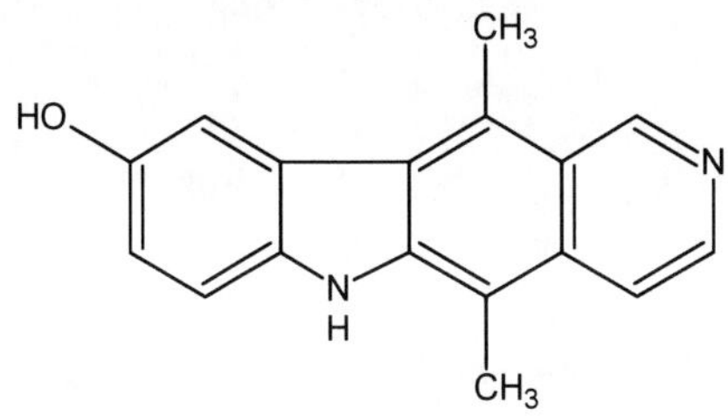

12
9-Hydroxyellipticine

Alkaloid from the stem bark of *Strychnos dinklagei* (Strychnaceae), was shown to be a remarkably effective inhibitor of several *in vitro* peroxidative systems. The compound completely inhibited peroxidation in rat liver microsomes at less than 1 nmol/mg protein (Larson, 1988).

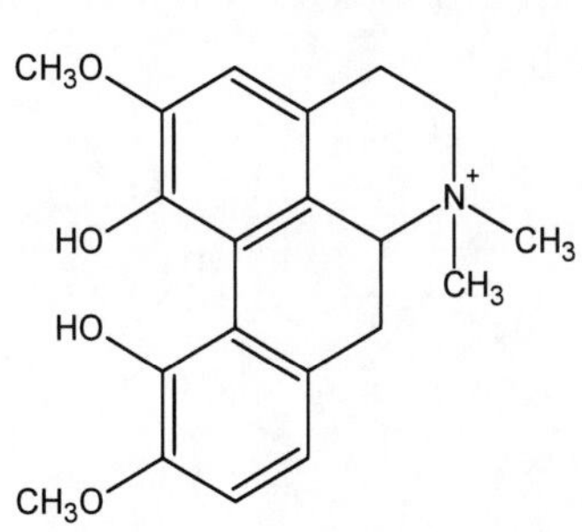

13
Magnoflorine

The methanol extract from the seed of *Xanthoxylum piperitum* (Rutaceae) showed strong antioxidative activity against linoleic acid by the ferric thiocyanate and thiobarbituric acid (TBA) methods. Japanese pepper contained 3.9 and 2.9 mg/100 g of dry weight (dw) of tocopherols in the pericarp and seed, respectively. Magnoflorine was present only in the seed, and not in the pericarp. Magnoflorine exhibited antioxidative activity against linoleic acid and radical-scavenging activity against the DPPH radical (Hisatomi et al., 2000).

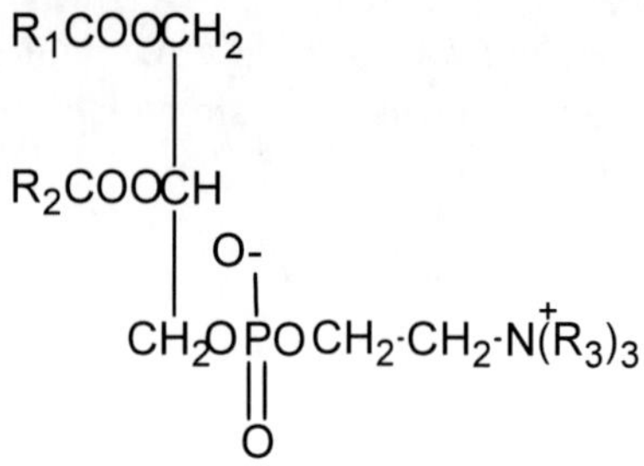

14
$R_1 = R_2$ = long-chain fatty acids: R_3 = H
Phosphatidylethanolamine

15
$R_1 = R_2$ = long-chain fatty acids: R_3 = Me
Phosphatidylcholine

Some complex lipids, such as phospholipids, have antioxidant activity. It is possible that these compounds act by chelating trace metals that may promote oxidation. Complex lipids may contain some constituents with much higher activity than others. As an example, fractionation of a crude soybean lecithin indicated that an alcohol-soluble fraction was far superior to an alcohol-insoluble fraction at preventing the thermal oxidation of lard (Larson, 1988). The complex lipid, phosphatidylethanolamine was shown to be a very effective synergist of flavonoid antioxidation in a test in which the induction period for lard auto oxidation was measured. It was similarly shown that phosphatidylcholine as well as phosphatidylethanolamine synergized the antioxidant effect of α-tocopherol. The presence of nitrogen together with phosphorus, in these complex lipids presumably increases their metal-complexing ability (Larson, 1988).

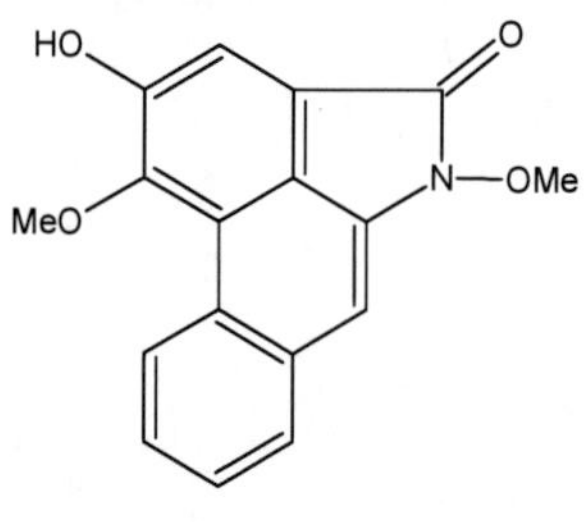

16
Piperlactam S

This alkaloid was isolated from *Piper kadsura* (Choisy) OUI, (Piperaceae). Piperlactam S (1 to 20, μM) concentration-dependently prevented the copper-catalyzed oxidative modification of human low-density lipoproteins (LDL). Fe^{2+}-induced oxidative modification of cell membrane was also significantly attenuated by piperlactam S as measured by thiobarbituric acid-reactive substances (TBARS). Furthermore, piperlactam S effectively minimized the loss of cell viability induced by Fenton's reagent ($H_2O_2/FeSO_4$) in cultured endothelial cells and significantly reversed $H_2O_2/FeSO_4^-$ induced impairment of endothelium-dependent relaxation to acetylcholine in rat aorta. Since the oxidative modification of LDL plays an important role in the genesis of atherosclerosis, this compound may help to reduce the risk of atherosclerosis (Tsai et al., 2002).

17
Protein 33 kDa TI

Trypsin inhibitors (TIs), root storage proteins, were purified from sweet potato (*Ipomoea batatas* L. Convolvulaceae) roots by trypsin affinity.This purified 33 kDa TI had scavenging activity against 1,1-diphenyl-2-picrylhydrazyl (DPPH) radical. There was positive correlation between scavenging effects against DPPH (2 to 22%) and amounts of 33 kDa TI (1.92 to 46 pmol). Using electron paramagnetic resonance (EPR) spectrometry for hydroxyl radical detection, it was found that 33 kDa TI could capture hydroxyl radical, and the intensities of EPR signal were significantly decreased from 1.5 to 6 pmol of 33 kDa TI compared to those of the controls (Hou et al., 2001).

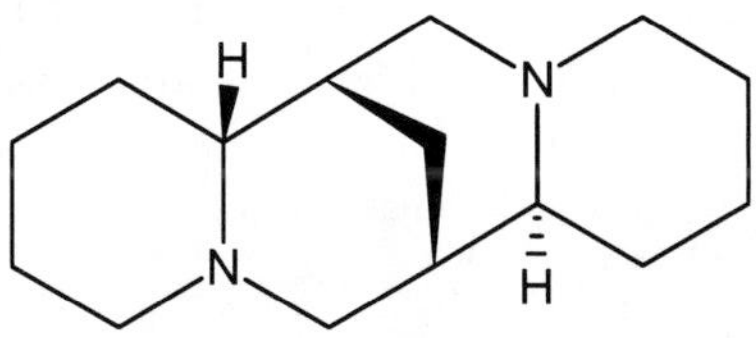

18
Sparteine

Alkaloids of the quinolizidine type, for example sparteine, have been found to be stored principally in the epidermal, cells from several genera of the family Leguminosae, subfamily Faboideae, notably *Lupinus pusillus* (Leguminosae) the concentrations in these tissues were up to 20 mM. It is unlikely that these alkaloids are acting as UV light filtering agents, because their absorption in the solar UV range would be minimal. Showed antioxidant activity (Larso, 1988).

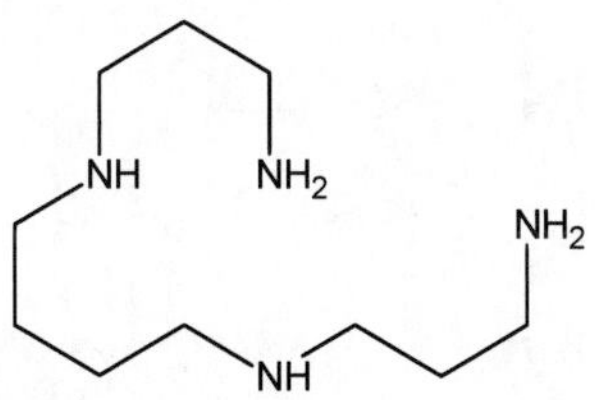

19
Spermine

Polyamines such as spermine are related to simple acyclic alkaloids found in legumes. This compound was shown in electron spin resonance experiments to scavenge the superoxide radical at rather high spermine concentrations (0.01-0.03 M). This finding may have some bearing on the observed effects of polyamines as membrane-stabilizing substances (Larson, 1988).

$(CH_3)_3N$

20
Trimethylamine

Tertiary amines such as trimethylamine (Widespread in higher plants), have been shown to be very effective quenchers of peroxy radicals (Larson, 1988). The radicals derived from the amines have very large termination rate constants, contributing to the efficient cessation of radical chain reactions.

21
YTM mucilage

The mucilages of Yam (*Dioscorea batatas*, Dioscoreaceae), tuber has been to be a mannan-protein complex. This mucilage might play roles as antiradicals and antioxidant (Hou, et al., 2002).

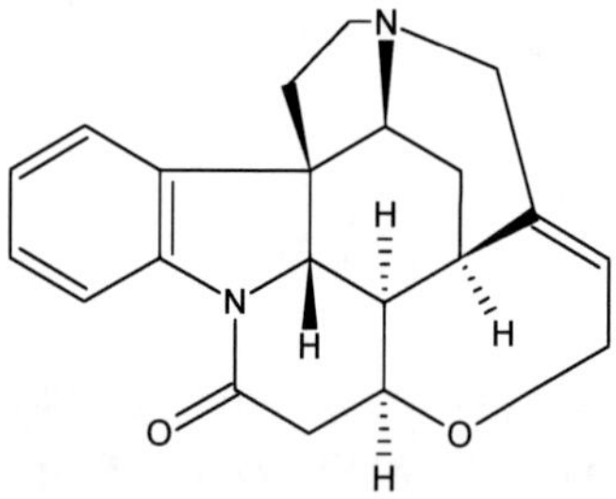

22
Strychnine

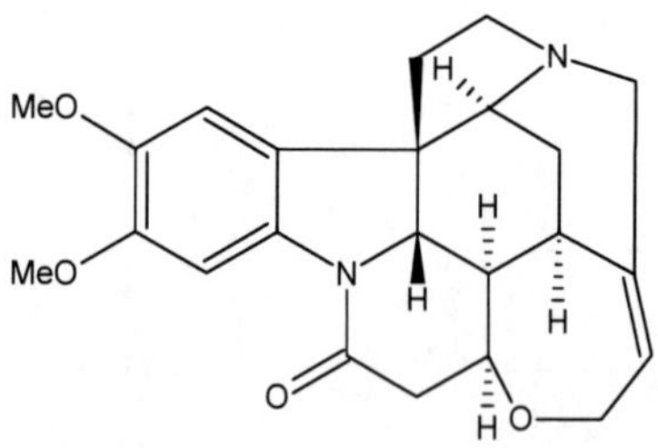

23
Brucine

Several alkaloids of various structural types have been found to be potent inhibitors of $^1O_2$. Both alkaloids were isolated from *Strychnos nux-vomica* and many other *Strychnos* spp. (Strychnaceae). Particularly effective are indole alkaloids such as strychnine and brucine that have a basic nitrogen atom in a rigid, cage-like structure. Such alkaloids appear to be strictly physical quenchers, and are not destroyed chemically by the process of quenching. Thus, in principle, they could inactivate many molecules of singlet oxygen per molecule of alkaloid (Larson, 1988).

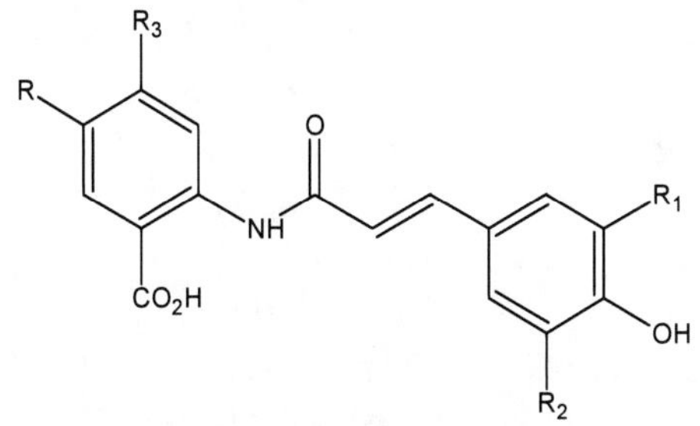

24
R = H, R_1 = OH, R_2 = H, R_3 = H
N-[3,4-Dihydroxy-(E)-cinnamoyl]anthranilic acid

25
R = H, R_1 = OMe, R_2 = H, R_3 = H
N-[4-Hydroxy-3-methoxy-(E)-cinnamoyl]anthranilic acid

26
R = H, R_1 = OMe, R_2 = OMe, R_3 = H
N-[4-Hydroxy-3,5-dimethoxy-(E)-cinnamoyl]anthranilic acid

27
R = OH, R_1 = H, R_2 = H, R_3 = H
N-[4'-Hydroxy-(E)-cinnamoyl]-5-hydroxyanthranilic acid

28
R = OH, R_1 = OH, R_2 = H, R_3 = H
N-[3',4'-Dihydroxy-(E)-cinnamoyl]-5-hydroxyanthranilic acid

29
R = OH, R_1 = OMe, R_2 = H, R_3 = H
N-[4'-Didroxy-3'-methoxy-(E)-cinnarnoyl]-5-hydroxyanthranilic acid

30
R = OH, R_1 = OMe, R_2 = OMe, R_3 = H
N-[4'-Didroxy-3',5'-dimethoxy-(E)-cinnamoyl]-5-hydroxyanthranilic acid

Several avenanthramides, amides of anthranilic acid were isolated from *Avena sativa* L (Graminae). As assessed by the reactivity toward 1,1-diphenyl-2-pilcrylhydrazyl (DPPH). All avenanthramides inhibited azo-initiated peroxidation of linoleic acid. N-[3,4-Dihydroxy-(E)-

cinnamoyl]anthranilic acid and N-[4-hidroxy)-3.5-dimethoxy-(E)-cinnamoyl]anthranilic acid were initially the most effective compounds (Bratt et al., 2003).

N-[3,4-Dihydroxy-(E)-cinnamoyl]anthranilic acid: IR Vmax cm^{-1} 3369, 1636, 1221; UV (MeOH)λ max 339 nm; mp 221-230 °C; Anal. Ca]cd for $C_{16}H_{13}NO_5$. $0.5H_2O$; C 62.34, H, 4.57, N, 4.54 %; Found: C. 61.84: H, 4.47; N, 4.23 %; ^{1}HNMR (DMSO-d_6): δ 11.24 (s, COOH,1H), 9.53 (s, OH, 1H), 9.13 (s, OH, 1H), 8.58 (dd, J = 8.4, 1.1 Hz, 1H), 7.99 (ddd, J = 7.9, 1.7, 0.4 Hz, 1H), 7.59 (ddd, J = 7.4, 7.4, 1.7 Hz, IH), 7.44 (d. J = 15.5 Hz, 1H), 7.15 (ddd, J = 7.9, 7.3, 1.2 Hz, 1H), 7.08 (d, J = 2.1 Hz, 1H), 7.00 (ddd, J = 8.1, 2.2, 0.4 Hz, 1H), 6.77 (d, J = 8.2 Hz, 1H), 6.50 (d, J= 15.5 Hz, 1H); ^{13}C NMR (DMSO-d_6): δ 169.5, 164.1, 148.0, 145.6, 141.9, 134.0, 131.1, 125.9, 122.6, 121.2, 120.3, 118.4, 116.6, 115.8, 114.6.

N-[4-Hydroxy)-3.5-dimethoxy-(E)-cinnamoyl]anthranilic acid: IR Vmax, cm^{-1} 3401, 1609, 1268; UV (MeOH) λ max 232.3 nm; mp 199- 200 °C; Anal. Calcd for $C_{18}H_{17}NO_6$. 0.H2O: C 59.83, H. 5.30, N, 3.88; Found: C 59.54, H 5.20, N, 3.45. ^{1}HNMR (DMSO-d_6): δ 11.29 (s, COOH, 1H), 8.90 (s, OH, 1 H), 8.63 (ddd, J = 8.5, 1.2, 0.4 Hz, 1H), 8.0 (ddd, J = 8.0, 1.7, 0.4 Hz, 1H), 7.60 ,(ddd, J = 7.7, 7.2, 1.8 Hz, 1H), 7.53 (d, J = 15.4 Hz, 1H), 7.15 (ddd, J = 7.9, 7.2, 1.2 Hz, 1H), 7.04 (s, 2H), 6.76 (d, J = 15.4 Hz, 1H), 3.82 (s, 6H); ^{13}C NMR (DMSO-d_6): δ 169.4, 164.2, 148.0, 142.4, 141.1, 137.9, 134.0, 131.1, 124.7, 122.6, 120.3, 119.2, 116.5, 106.5, 56.1.

N-[4'-Hydroxy.-3'.5'-dimethoxy-(E)-cinnamoyl]-5-hydroxyanthranilic acid: IRvmax cm^{-1} 3371, 1608,1221; UV(MeOH) λ max 226,3 nm; mp 206 °C; Anal. Calcd for $C_{18}H_{17}NO_7$.0.5 H_2O. C 58.69, H 4.93, N. 3.80%; Found; C 58.44; H 4.87, N3.76% ; ^{1}HNMR (DMSO-d6): δ 10,84 (s, COOH, 1H), 9.59 (s, OH, 1H), 8.86 (s, OH, 1H), 8.38 (d, J = 9.0 Hz, 1H), 7.47 (d, J = 15.4 Hz, 1H), 7.37 (d, J = 3.0 Hz, 1H), 7.02 (dd. J = 9.0, 3.0 Hz, 1H), 7.00 (s, 2H), 6.72 (d. J = 15.4 Hz, 1H), 3.81 (s, 6H); ^{13}C NMR (DMSO): δ 169.1, 163.7, 152.4, 148.0, 141.5, 137.7, 133.0, 124.9, 122.4, 120.9, 119.5, 118.3, 116.5, 106.0, 56.1.

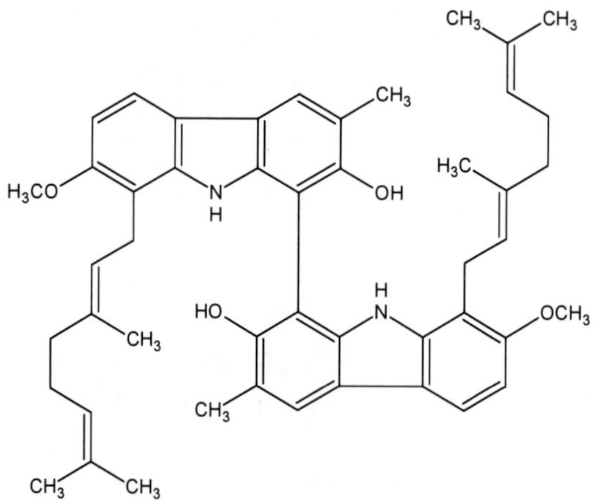

31

Bismurrayafoline E

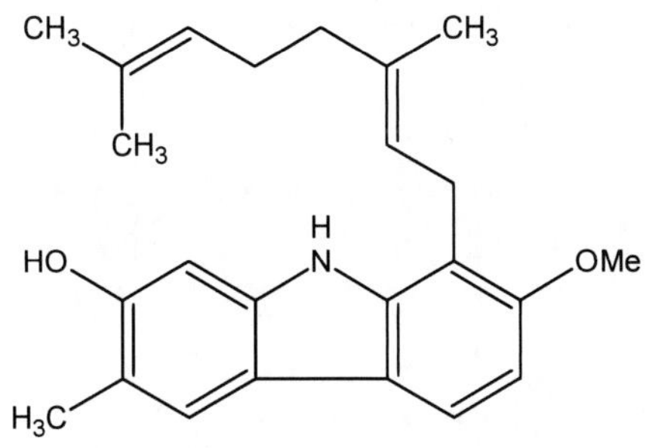

32
Euchrestine B

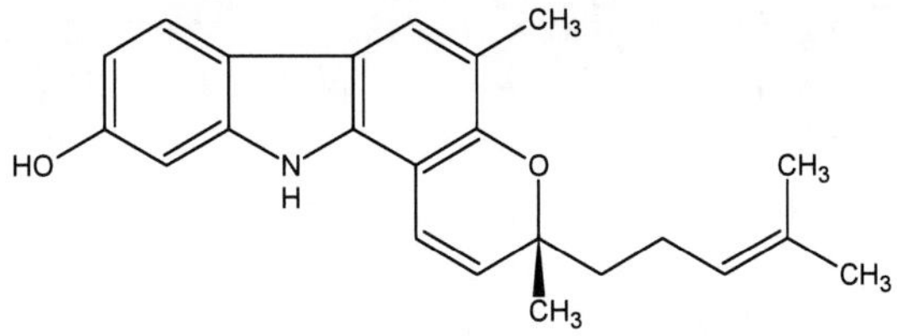

33
Mahanine

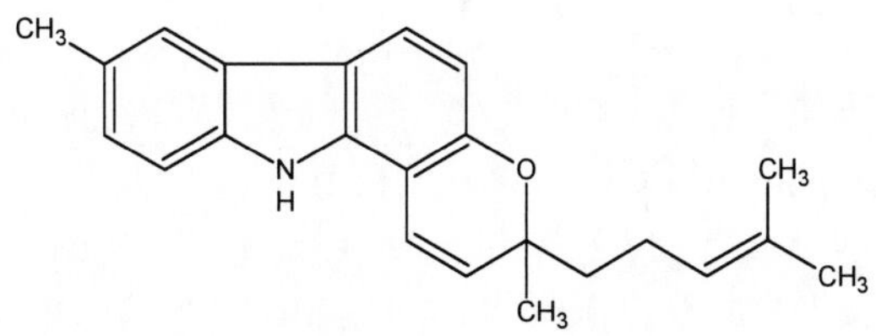

34
Mahanimbicine

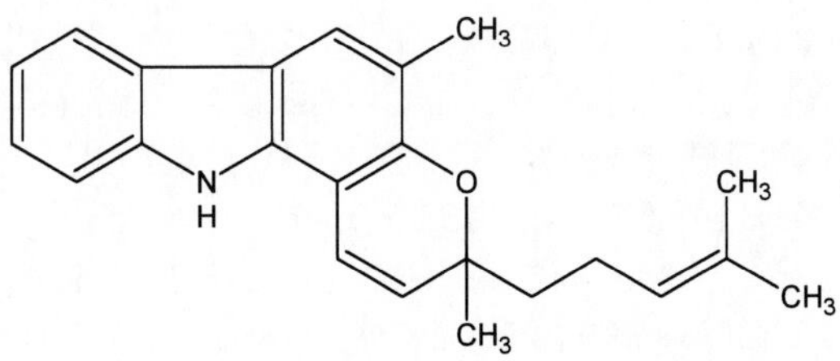

35
Mahanimbine

Five carbazole alkaloids were isolated from *Murraya koenigii* (Rutaceae) leaves. The OSI value of carbazoles at 110°C decreased in the order euchrestine B and mahanine > α-tocopherol > BHT > bismurrayafoline E > mahanimbicine > mahanimbine > control. The DPPH radical scavenging activity for these carbazoles was in the order ascorbic acid >

bismurrayafoline E > euchrestine B, mahanine and α-tocopherol > BHT > mahanimbicine > mahanimbine (Tachibana et al., 2001).

Euchrestine B: UV. λmax (MeOH) nm (log ε): 211.8 (4.55), 237.8 (4.68), 264.6 (4.41),310.4 (4.17), 317.2 (4.16), 330.8 (3.88); EIMS 70 eV *m/z* (rel int.): 363 $[M]^+$ (100), 294 (20), 280 (26), 264 (14), 263 (22), 240 (39), $C_{24}H_{29}NO_2$; ^1H NMR: δ 1.57 (3H, br s, 7'-CH_3), 1.62 (3H, br s, 7'-CH_3), 1.87 (3H, d, J = 1.0 Hz, 3'-CH_3), 2.03-2.12 (4H, m, H-4', H-5'), 2.38 (3H, s, 3-CH_3), 3.62 (2H, d, J = 6.8 Hz, H-l '), 3.90 (3H, s, OCH_3), 4.78 (lH, br s, OH), 5.07 (lH, m, H-6'), 5.32 (lH, qt, J = 1.0, 6.8 Hz, H-2'), 6.80 (lH, s, H-l), 6.83 (lH, d, J = 8.5 Hz, H-6), 7.67 (lH, s, H-4), 7.71 (lH, d, J = 8.5 Hz, H-5), 7.74 (lH, br s, NH); ^{13}C NMR: δ 16.2 (3- CH_3), 16.4 (3'- CH_3), 17.7 (7'- CH_3), 23.8 (C-1'), 25.7 (C-7'- CH_3), 26.6 (C-5'), 39.7 (C-4'), 56.7 (OCH_3), 96.7 (C-1), 104.6 (C-6), 111.2 (C-8), 115.8 (C-3), 117.1 (C-5), 117.8 (C-4a), 117.9 (C-4b), 121.1 (C-4), 122.4 (C-2'), 124.0 (C-6'), 131.7 (C-7'), 136.5 (C-3'), 139.6 (C-9a), 140.3 (C-8a), 152.1 (C-2), 154.7 (C-7).

Bismurrayafoline E: UV λmax (CH_2Cl_2) nm (log ε): 238.2 (5.02), 312.0 (4.62), 334.4 (4.4l); EIMS 70 eV m/z (rel int.): 724 $[M]^+$ (100), 479 (3), 362 (3); HR-EIMS 70 eV: found: *m/z* 724.4277 $[M]^+$, $C_{48}H_{56}N_2O_4$ requires 724.4237;^1H NMR: δ 1.25 (6H, d, J = 1.2 Hz, 3'-CH_3), 1.42 (6H, br s, 7'-CH_3), 1.56 (6H, br s, 7'-CH_3), 1.43-1.52 (4H, m, H-4'), 1.58-1.64 (4H, m, H-5'), 2.48 (6H, s, 3-CH_3), 3.40 (2H, dd, J = 7.3,15.5 Hz, H-l'), 3.47 (2H, dd, J = 7.3, 15.5 Hz, H-l'), 3.89 (6H, s, 7-OCH_3), 4.78 (2H, m, H-6'), 5.08 (2H, qt, J = 1.2, 7.3 Hz, H-2'), 5.19 (2H, s, OH), 6.85 (2H, d, J = 8.5, H-6), 7.57 (2H, br s, NH), 7.77 (2H, d, J = 8.5 Hz, H-5), 7.85 (2H, s, H-4); ^{13}C NMR: δ 15.2 (3'-CH_3), 16.7 (3-CH_3), 17.5 (7'-CH_3), 23.7 (1'), 25.6 (7'-CH_3), 26.3 (5'), 39.1 (4'), 56.7 (OCH_3), 99.4 (1), 104.9 (6), 111.0 (8), 117.3 (4a), 117.3 (5), 117.4 (3), 118.2 (4b), 121.7 (2'), 122.3 (4), 124.1 (6'), 131.1 (7'), 136.5 (3'), 137.8 (9a), 140.5 (8a), 150.0 (2), 154.8 (7).

Mahanine: UV λmax (MeOH) nm (log ε): 211.0 (4.60), 240.6 (4.66), 285.4 (4.40, sh), 295.2 (4.56), 325.2 (3.90), 342.2 (3.92), 356.6 (3.86); EIMS 70 eV *m/z* (rel int.): 347 $[M]^+$ (100), 332 (19), 304 (8), 278 (15),264 (75), $C_{23}H_{25}NO_2$; $[\alpha]_D25$ = +7.6° (CHC1$_3$; c 5.38); ^{1}HNMR: δ 1.43 (3H, s, 3-CH_3), 1.57 (3H, br s, 4'-CH_3), 1.65 (3H, br s, 4'- CH_3), 1.72-1.78 (2H, m, H-l'), 2.11-2.20 (2H, m, H-2'), 2.31 (3H, br s, 5-CH_3), 5.01 (lH, br s, OH), 5.11 (lH, m, H-3'), 5.63 (lH, d, J = 9.8 Hz, H-2), 6.57 (lH, d, J = 9.8 Hz, H-l), 6.68 (lH, dd, J = 2.2, 8.3 Hz, H-8), 6.77 (lH, d, J = 2.2 Hz, H-10), 7.54 (lH, s, H-6), 7.71 (lH, d, J = 8.3 Hz, H-7), 7.73 (lH, br s, NH); ^{13}C NMR: δ 16.0 (5-CH_3), 17.6 (4'-CH_3), 22.7 (2'), 25.7 (3-CH_3), 25.7 (4'-CH_3), 40.7 (C-1'), 78.0 (C-3), 97.0 (C-10), 104.3 (C-11b), 108.3 (C-8), 116.7 (C-6a), 117.5 (C-1), 118.0 (C-7a), 118.3 (C-5), 120.0 (C-7), 120.4 (6), 124.2 (3'), 128.7 (C-2), 131.7 (C-4'), 134.7 (C-11a), 140.7 (C-l0a), 148.9 (C-4a), 153.5 (C-9).

Mahanimbicine:UV λmax (MeOH) nm (log ε): 238.2 (4.61), 280.0 (4.29,sh), 289.6 (4.56), 334.2 (3.78), 353.2 (3.75); EIMS 70 eV *m/z* (rel int.): 331 $[M]^+$, 316 (6), 248 (100), 210 (3), $C_{23}H_{25}NO$; $[\alpha]_D25$ = +58.1° ($CHCl_3$; c 0.26); ^1H NMR: δ 1.45 (3H, s, 3-CH_3), 1.57, 1.65 (each 3H, br s, 4'-CH_3), 1.69-1.82 (2H, m, H-l'), 2.13-2.19 (2H, m, H-2'), 2.50 (3H, s, 8-CH_3), 5.10 (lH, m, H-3'), 5.66 (lH, d, J = 9.8 Hz, H-2), 6.65 (1H, d, J = 9.8 Hz, H-l), 6.71 (lH, d, J = 8.5 Hz, H-5), 7.14 (lH, dd, J = 1.7, 8.3 Hz, H-9), 7.28 (lH, br, d, J = 8.3 Hz, H-I0), 7.72 (lH, d, J = 1.7 Hz, H-7), 7.74 (lH, d, J = 8.5 Hz, H-6), 7.85 (lH, s, NH); ^{13}C NMR: δ 17.6, (4'-CH_3), 21.5 (8-CH_3), 22.7 (C-2'), 25.7 (4'-CH_3), 26.0 (3-CH_3), 40.8 (C-l'), 78.3 (C-3), 104.5 (C-11b), 109.4 (C-5), 110.1 (C-10), 117.3 (C-l), 117.4 (C-6a), 119.5 (C-7), 120.4 (C-6), 124.1

(C-3'), 124.1 (C-7a), 125.7 (C-9), 128.7 (C-2), 129.0 (C-8), 131.7 (C-4'), 136.6 (C-11a), 137.7 (C-10a), 151.7 (C-4a).

Mahanimbine: UV λmax (MeOH) nm (log ε): 223.0 (4.56sh), 238.0 (4.67),287.6 (4.61), 329.4 (3.88), 342.6 (3.88); ElMS 70 eV *m/z* (rel int.): 331 $[M]^+$ (24), 248 (100), 210 (5), $C_{23}H_{25}NO$; $[\alpha]_D25$ = +30.0° ($CHCl_3$; c 0.69); ^1H NMR: δ 1.45 (3H, s, 3-CH_3), 1.58, 1.66 (each 3H, br s, 4'- CH_3), 1.73-1.79 (2H, m, H-l '), 2.12-2.22 (2H, m, H-2'), 2.33 (3H, s, 5-CH_3), 5.11 (lH, m, H-3'), 5.66 (lH, d, J = 9.8 Hz, H-2), 6.65 (lH, d, J = 9.8 Hz, H-l), 7.17 (lH, br t, J = 8.1 Hz, H-8), 7.30 (lH, br t, J = 8.1 Hz, H-9), 7.37 (lH, br d, J = 8.1 Hz, H-10), 7.66 (lH, s, H-6), 7.87 (lH, br s, NH), 7.91 (lH, br d, J = 7.8 Hz, H-7); ^{13}C NMR: δ 16.1 (5-CH_3), 17.6, (4'-CH_3), 22.7 (C-2'), 25.7 (4'-CH_3), 25.8 (3-CH_3), 40.8 (C-l'), 78.1 (C-3), 104.2 (C-11b), 110.3 (C-10), 116.6 (C-6a), 117.5(C-l), 118.4 (C-5), 119.3 (C-7), 119.5 (C-8), 121.2 (C-6), 123.9 (C-7a), 124.2 (C-3'), 124.2 (C-9), 128.5 (C-2), 131.7 (C-4'), 134.8 (C-11a), 139.4 (C-10a), 149.9 (C-4a).

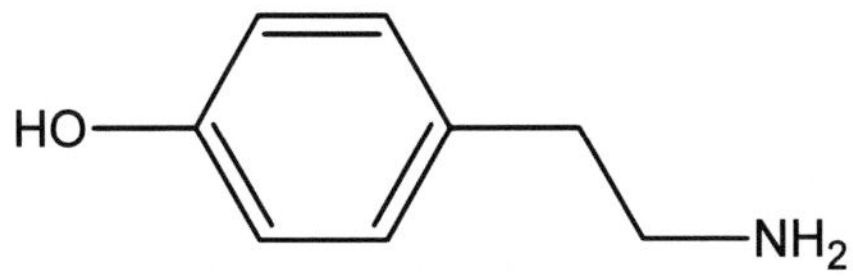

36
Tyramine

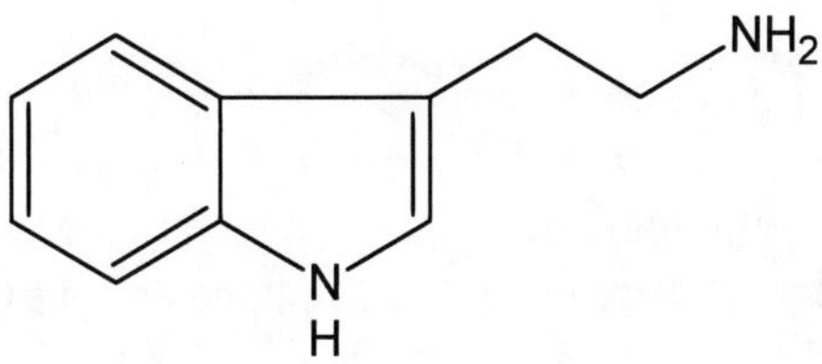

37
Tryptamine

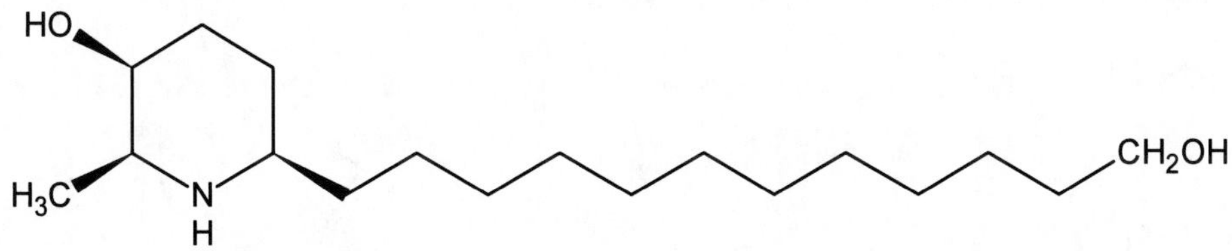

38
2β-Methyl-3-β-hydroxy-6-β-piperidinododecanol

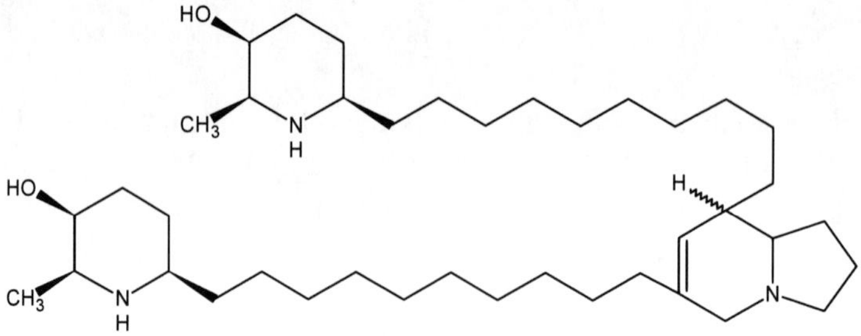

39
Juliprosopine

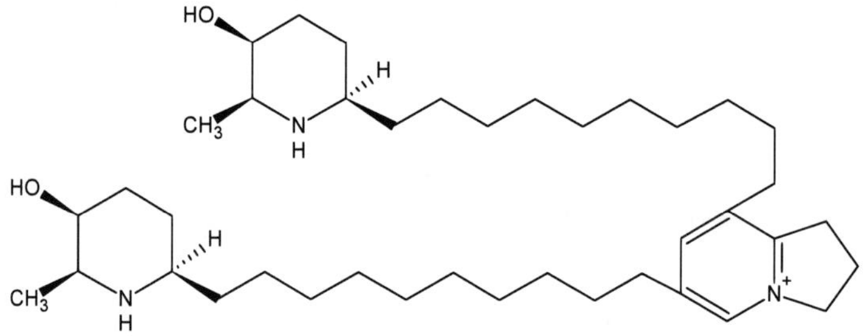

40
Juliprosine

These compounds were isolated and characterize from the aerial parts of Argentinian collections of *Prosopis alpataco*, *Prosopis argentina*, *Prosopis chilensis*, *Prosopis ilexuosa* and *Prosopis pugionata* (Leguminosae). The phenethylamine showed effect in the DNA binding assay suggesting that activity in, such compounds might be related to the absence of the *p*-hydroxy group. At 0.5 mg/ml the tryptamine displayed a stronger effect (28%) but higher activity was observed in the piperidine derivatives 2-β-methyl-3-β-hydroxy-6-β-piperidinedodecanol, juliprosopine and juliprosine with 50, 47 and 54%, respectively (Tapia et al., 2000). Piperidine alkaloids with related structures have been shown to display DNA-damaging activity as well as antitumoral effects (Gunatilaka et al., 1998).

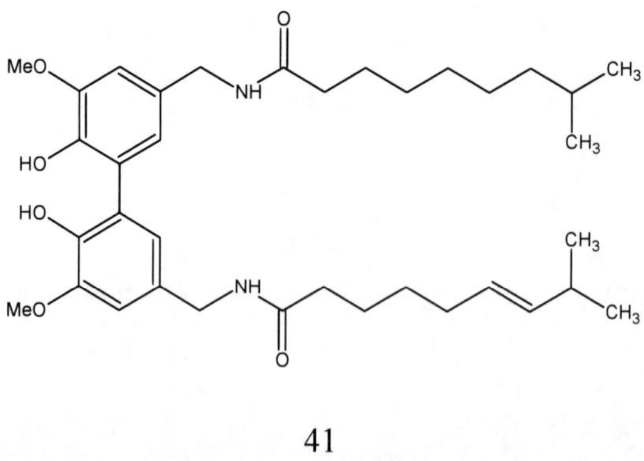

41
6",7"-Dihydro-5',5'"-dicapsaicin

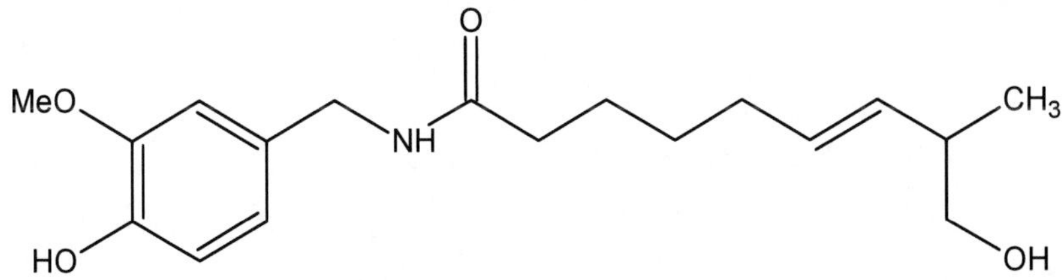

42
ω-Hydroxycapsaicin

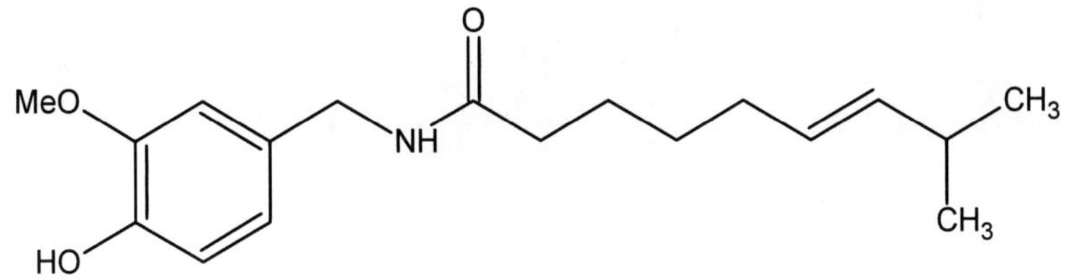

43
Capsaicin

These capsaicin derivative, were isolated from the fruit of *Capsicum annuum* (Solanaceae). ω-Hydroxycapsaicin showed almost the same antioxidant activity as capsaicin, but did not have a pungent taste. Both compounds inhibited lipid peroxidation in a concentration-dependent manner. However, 6",7"-dihydro-5',5'''-dicapsaicin was about 25 times more potent than α-tocopherol; that is, the concentration of their IC_{50} is about 10 μM, while that of α-tocopherol was 250 μM. In addition, the IC_{50} of it compound was almost the same as that of capsaicin (Ochi et al., 2003). Capsaicin is a pungent principle of various *Capsicum* spp. (Solanaceae) is an irritant used for desensitisation of sensory neurones. Used as a selective probe for studying neurogenic inflammation and the role of nociceptors in human physiol. Used as a counterirritant and topical analgesic for some skin conditions. Shows anticancer and antiinflammatory properties (Martindale, 1989).

6",7"-Dihydro-5',5'''-dicapsaicin: Light yellow oil; IR: (KBr) vmax 3412, 2928, 1645, 1544, 1460 cm^{-1}; HRFABMS: *m/z* 611.4091 $[M + H]^+$ (calcd for $C_{36}H_{54}N_2O_6$, 611.4060); ^{1}HNMR ($CDCl_3$): δ 0.85 (6H, d, J = 6.6 Hz, H-9′′, H-10′′), 0,95 (6H, d, J = 6.6 Hz, H-9, H-10), 3.91 (6H, s, OMe x 2), 4.39 (4H, d, J = 5.6 Hz, H-7', H-7′′′), 5.34 (lH, dt, J = 15.5, 6.0 Hz, H-6), 5.40 (lH, dd, J = 15.5, 5.7 Hz, H- 7), 6.88 (2H, brs, H-6', H-6′′′), 6.90 (2H, brs, H-2', H-2′′′); ^{13}C NMR: δ 173.1 (C-1), 37.0 (C-2), 25.4 (C-3), 29.4 (C-4), 32.4 (C-5), 126.7 (C-6), 138.3 (C-7), 31.1 (C-8), 22.8 (C-9), 22.8 (C-10), 130.5 (C-1′), 110.4 (C-2′), 147.5 (C-3′), 142.3 (C-4′), 124.2 (C-5′), 122.9 (C-6′), (C-7′), 56.4 (OMe), 170.3 (C-1′′), 36.9 (C-2′′), 25.4 (C-3′′), 29.5 (C-4′′), 26.0 (C-5′′), 27.4 (C-6′′), 39.1 (C-7′′), 28.1 (C-8′′), 22.8 (C-9′′), 22.8 (C-10′′), 130.5 (C-1′′′), 110.4 (C-2′′′), 147.5 (C-3′′′), 142.3 (C-4′′′), 124.3 (C-5′′′), 122.9 (C-6′′′), 43.7 (C-7′′′), 56.4 (OMe).

ω-Hydroxycapsaicin: light yellow oil; $[\alpha]_D25$ = +3.5° ($CHCl_3$; c 0.28); IR: (KBr) vmax 3339, 2929, 1647, 1517,1462 cm^{-1}; FABMS: *m/z* 322.2; ^{1}HNMR ($CDCl_3$): δ 0.97 (3H, d, J = 6.5 Hz, H-10), 3.89 (3H, s, OMe), 4.36 (2H, d, J = 5.5 Hz, H-7'), 5.49 (lH, m, H-6), 5.27 (lH, dd, J = 15.5, 7.8, H-7), 6.77 (lH, d, J = 8.0 Hz, H-6'), 6.82 (lH, s, H-2'), 6.87 (lH, d, J = 8.0 Hz, H-5'); ^{13}C NMR: δ 172.9 (C-1), 36.7 (C-2), 25.2 (C-3), 29.0 (C-4), 32.2 (C-5), 131.5 (C-6), 133.2 (C-7), 99.9 (C-8), 67.5 (C-9), 16.8 (C-10), 130.6 (C-1′), 110.9 (C-2′), 146.9 (C-3′), 145.3 (C-4′), 114.6 (C-5′), 121.0 (C-6′), 43.7 (C-7′), 56.1 (OMe).

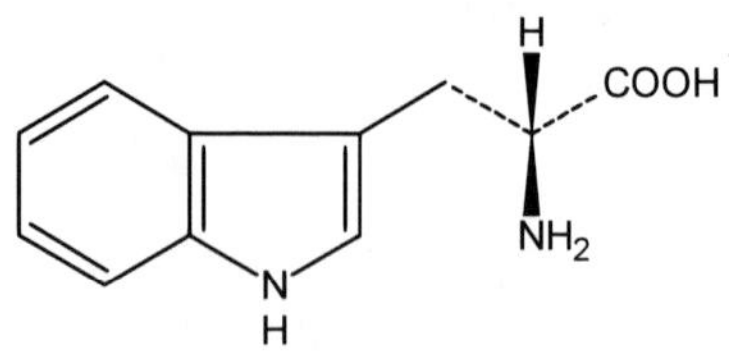

44
Tryptophan

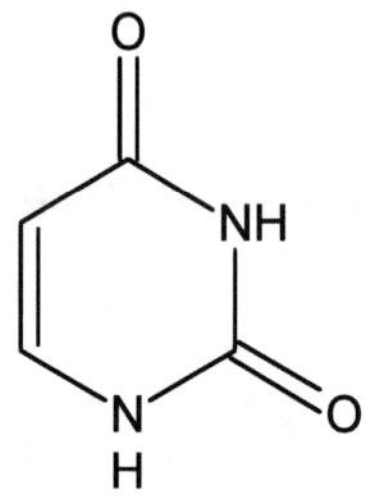

45
Uracil

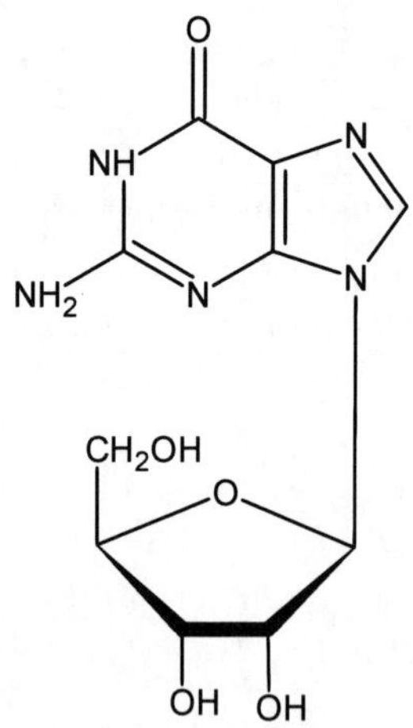

46
Guanosine

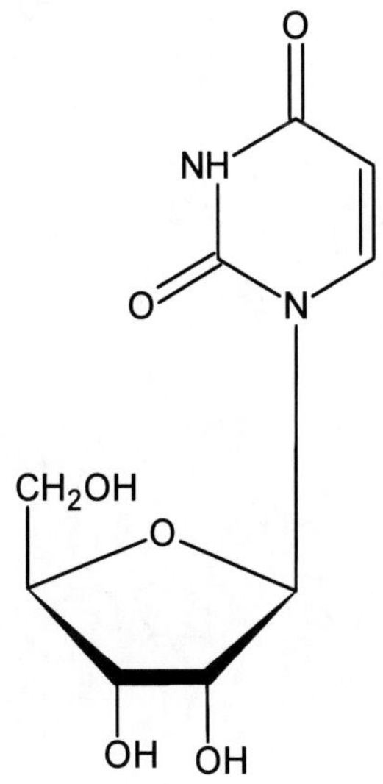

47
Uridine

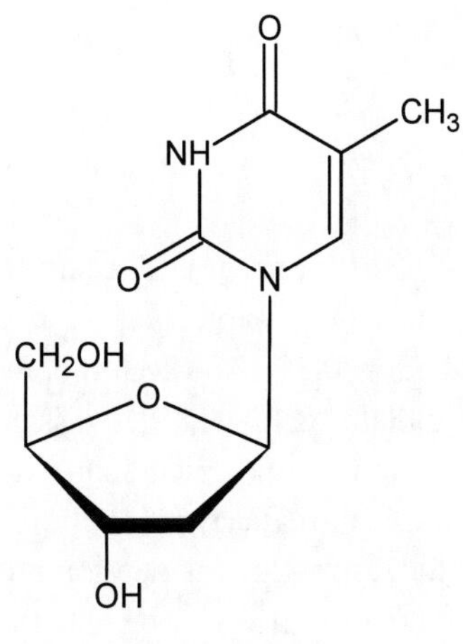

48
Thymidine

These compounds were isolated from the leaves of *Peucedanum japonicum* Thunb. (Umbelliferae). Showed antioxidant activity on the test of radical scavenging on 1, 1-diphenyl-2-picrylhydrazyl radical and for inhibition of oxidation of liposome induced by 2,2'-azobis{2-amidinopropane)dihydrochloride (Hisamoto et al., 2003).

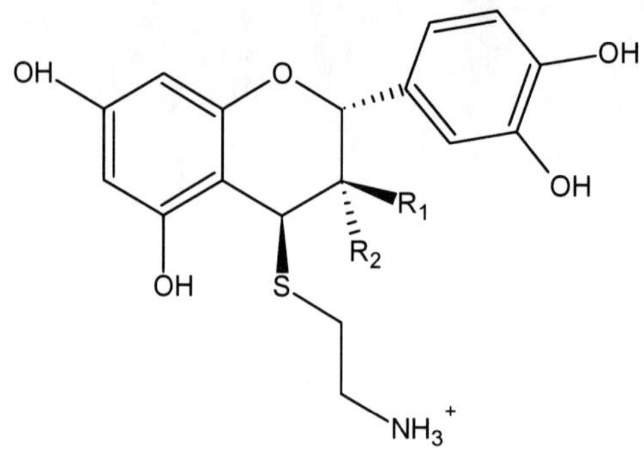

49
R_1=H, R_2=OH,
4β-(2-Aminoethylthio)epicatechin

50
R_1=H, R_2=Gal
4β-(2-Aminoethylthio)epicatechin-3-O-gallate

51
R_1=OH, R_2=H
4β-(2-Aminoethylthio)catechin

Found in the grape *Vitis vinifera* (Vitaceae). The antiradical efficacies of compounds were compared with those of Trolox and (-)-epicatechin. The ED_{50} values obtained were as follows: Trolox, 0.26; (-)-epicatechin, 0.19; 4β-(2-aminoethylthio)epicatechin 0.11; 4β-(2-aminoethylthio)epicatechin 3-O-gallate 0.11 (DPPH assay). The introduction of a 2-aminomethylthio group onto position 4 of epicatechin and epicatechin gallate yielded compounds with improved antiradical efficacy in solution. The antiradical activity of epicatechins is exerted through the formation of semiquinone free radicals and quinones, triggered by the abstraction of one proton and one electron from some of the phenolic hydroxyl groups. The amino group and/or the sulfur atom of the new derivatives might modulate the reactivity of the molecules. The oxidation of the thio-ether to sulfoxide and/or sulfone might contribute to this antioxidant effect (Torres and Bobet, 2001).

4β-(2-Aminoethylthio)epicatechin: ES-MS: positive ions, found 366.1029, calculated for $C_{17}H_{20}NO_6S$ $[M + H]^+$ 366.1011, 365.0943 (69% intensity) calculated for $C_{17}H_{19}NO_6S$ $[M]^+$ 365.0933; ^{1}HNMR [(CD_3)2-CO, 300 MHz]: δ 2.9-3.8 (4H, m, S-CH_2-CH_2-N), 4.02-4.07 (lH, 2m, 3-H), 4.08-4.18 (2H, m, S- CH_2-CH_2-NH_2), 4.10-4.22 (lH, m, d, J = 2.1 Hz, 4-H), 5.12-5.23 (lH, 2s,2-H), 5.89 (lH, d, J = 2.4 Hz, 6-H), 6.07-6.09 (lH, 2d, J = 2.4 Hz each, 8-H), 6.83 (2H, m, 5'-H, 6'-H), 7.10 (lH, d, J = 2.1 Hz, 2'H); ^{1}HNMR (D_2O, 300 MHz): δ 2.64-3.26 (4H, 3m, S-CH_2-CH_2-N), 3.77 (lH, bs, 3-H), 3.91 (lH, bs, 4-H), 5.11 (lH:-bs, 2-:-H), 5.84 (lH, bd, 6-H), 5.89 (lH, bd, 8-H), 6.72 (2H, bs, 5'-H, 6'-H), 6.82 (lH, bs, 2'-H).

4β-(2-Aminoethylthio)epicatechin 3-O-Gallate: ES-MS: positive ions, *m/z* 518.2 $[M + H]^+$; HR-FAB-MS, positive ions, found 518.1129, calculated for $C_{24}H_{24}NO_{10}S$ $[M + H]^+$

518.1121; ^{1}HNMR [$(CD_3)_2$-CO, 300 MHz]: δ 3.03-3.86 (4H, m, S- CH_2-CH_2-N), 4.15-4.28 (lH, 2d, J = 2 Hz each, 4-H), 4.17-4.37 (2H, m, S-CH_2-CH_2-NH_2), 5.21-5.31 (lH, 2m, 3-H), 5.36-5.47 (lH, 2bs, 2-H), 6.01-6.02 (lH, 2d, J = 2.4 Hz each, 6-H), 6.08-6.11 (lH, 2d, J = 2.4 Hz each, 8-H), 6.75-6.80 (lH, 2d, J = 8 Hz each, 5'-H), 6.84-6.90 (lH, m, J = 8, 2 Hz, 6'-H), 6.98-7.01 (2H, m, galloyl H), 7.14-7.17 (lH, 2d J = 2 Hz each, 2'-H); ^{1}HNMR (D_2O, 300 MHz): δ 2.71-3.27 (4H, 3m, S-CH_2-CH_2-N), 3.96 (lH, bd, 4-H), 5.04-5.11 (lH, 2bs, 3-H), 5.30 (lH, bs, 2-H), 5.90 (2H, bs, 6-H, 8-H), 6.60-6.68 (2H, m, 5'-H, 6'-H), 6.70- 6.73 (2H, m, galloyl H), 6.82 (lH, bs, 2'-H).

4β-(2-Aminoethylthio)catechin: ES-MS: positive ions *m/z*: 366.3 $[M + H]^+$; HR-FAB-MS, positive ions, found 365.0935, calculated for $C_{17}H_{19}NO_6S$ (M]+ 365.0933; ^{1}HNMR [$(CD_3)_2$-CO, 300 MHz]: δ 2.92-3.70 (4H, m, S- CH_2-CH_2-N), 4.01-4.21 (lH, 2d, J = 9.6 Hz each, 3-H; 2H, m, S-CH_2-CH_2-NH_2), 4.29-4.51 (lH, 2d, J = 4.2 Hz each, 4-H), 4.82-5.02(lH, 2d, J = 9 Hz each, 2-H), 5.80 (lH, m, 6-H), 6.09 (lH, m, 8-H), 6.80 (2H, m, 5'-H, 6'-H), 6.94-6.96 (1H. m, 2'-H); ^{1}HNMR (D_2O, 300 MHz):7 δ 2.7-4.2 (5H, several multiplets, S-CH_2-CH_2-N, 3-H), 4.73-4.76 (lH, d, J = 9.3 Hz, 2-H), 5.70-5.71 (lH, bd, 6-H), 5.91-5.92 (lH, bd, 8-H), 6.75 (2H, bs, 5'-H and 6'-H), 6.82 (lH, bs, 2'-H).

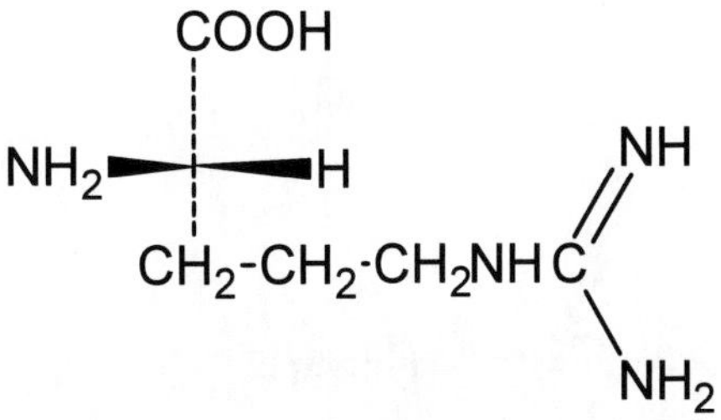

52
Arginine

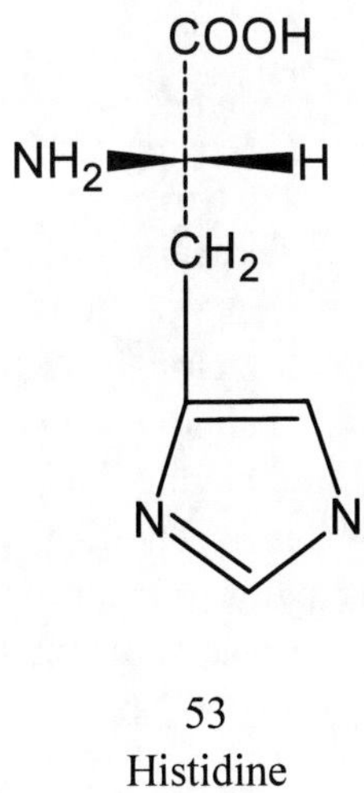

53
Histidine

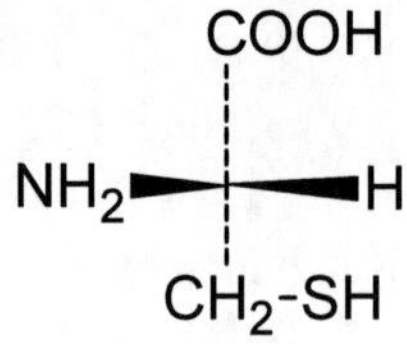

54
Cysteine

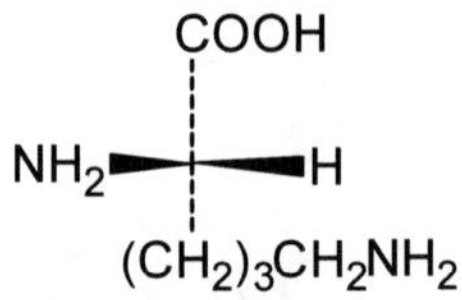

55
Lysine

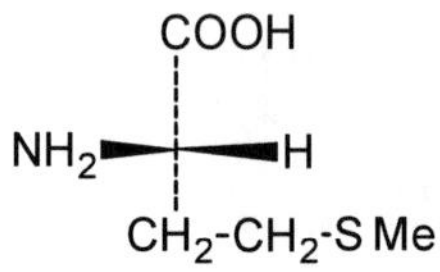

56
Methionine

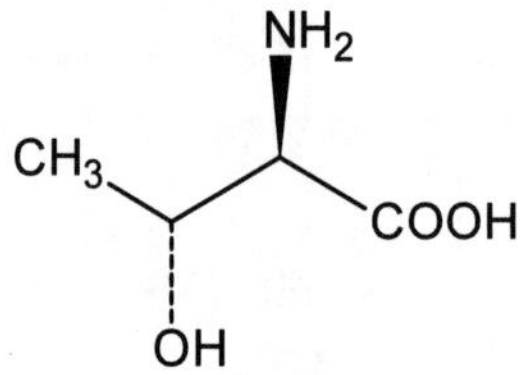

57
Threonine

Many amino acids have been tested for their antioxidant activity, especially in food-based systems. Among the amino acids for which antioxidant activity has been claimed are arginine histidine, cysteine, tryptophan, lysine, methionine, and threonine. These amino acids may exhibit antioxidant potential under some conditions of temperature, pH, or oxygen concentration but have no effect or actually promote oxidation in others. The alanine and histidine to inhibit the oxidation of linoleic acid at pH 9.5 and to promote it at pH 7.5 (Larson, 1988).

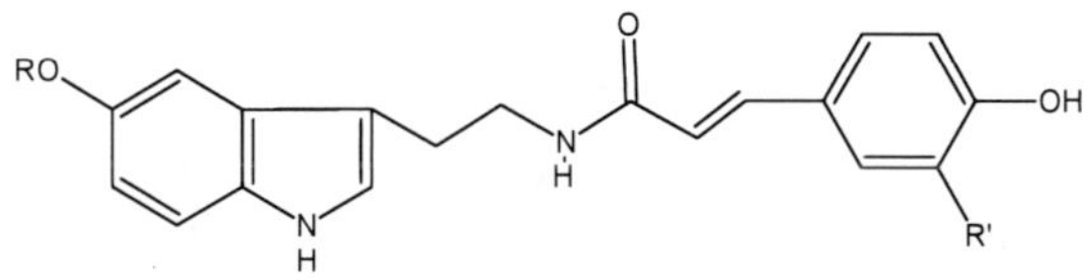

58
R'=OCH_3 R=H
N-[2-(5-Hydroxy-lH-indol-3-yl)ethyl]ferulamide

59
R'=H R=H
N-[2-(5-Hydroxy-lH-indol-3-yl)ethyl]-*p*-coumaramide

60
R'=H R=glc
N-[2-[5-β-O-Glucosyloxy)-1 H-indol-3-yl)ethyl]-*p*-coumaramide

61
R'=OCH_3 R=glc
N-[2-[5-(β-O-Glucosyloxy)-1H-indol-3-yl)ethyl]ferulamide

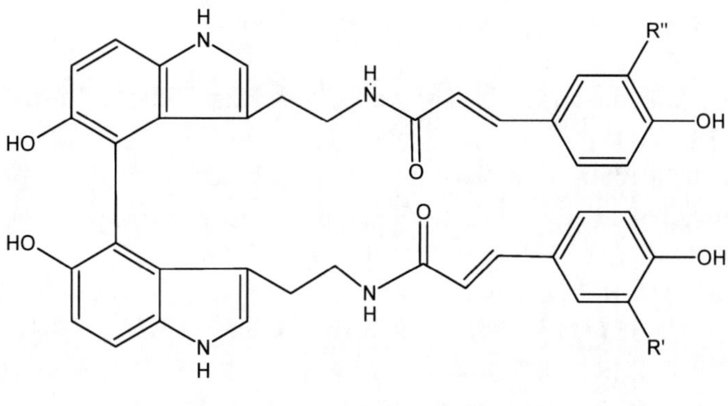

62
R'=H R''=H
N,N'-[2,2'-(5,5'-Dihydroxy-4,4'-bi-lH-indol-3,3'-yl)diethyl]-di-*p*-coumaramide

63
R'=OCH_3 R''=H
N-[2-[3'-[2-(*p*-Coumaramido)ethyl]-5,5'-dihydroxy-4,4'-bi-lH-indol-3-yl]
ethyl]ferulamide

64
R'= R''= OCH_3
N,N'-[2,2'-(5,5'-Dihydroxy-4,4'-bi-lH-indol-3,3'-yl)diethyl]-diferulamide

Seven antioxidative serotonin derivatives were isolated from safflower (*Carthamus tinctorius* L. Compositae) oil cake. The compounds were found to have relatively strong antioxidative activity in ferric thiocyanate and the α-α-diphenyl-β-picrylhydrazyl (DPPH) methods (Zhang et al., 1997).

N-[2-(5-Hydroxy-lH-indol-3-yl)ethyl]ferulamide: Colorless needles, mp 115-117°C. IR (KBr, cm^{-1}): 3400,1650,1600; FAB-MS *m/z*: 353 $[M+H]^+$; ^{1}HNMR(CD_3COCD_3): δ 7.09 (d, J= 1.8 Hz, H-2), 7.03 (d, J = 2.4 Hz, H-4), 6.71 (dd, J = 2.4, 8.5 Hz, H-6), 7.19 (d, J = 8.5 Hz, H-7), 2.91 (m, H-10), 3.59 (m, H-11), 7.15 (d, J = 1.8 Hz, H-2´´), 6.83 (d, J = 8.5 Hz, H-5´´), 7.05 (dd, J = 1.8, 8.5 Hz, H-6´´), 7.47 (d, J = 15 Hz, H-7´´), 6.52 (d, J = 15 Hz, H-8´´), 3.87 (s, OMe), 9.75 (s, NH), 17.26 (t, CONH); ^{13}CNMR(CD_3COCD_3): δ 124.0 (C-2), 112.6 (C-3), 103.5 (C-4), 151.6 (C-5), 112.3 (C-6), 112.5 (C-7), 132.5 (C-8), 129.4 (C-9), 26.4 (C-10), 40.8 (C-11), 128.3 (C-1´´), 111.3 (C-2´´), 148.6 (C-3´´), 149.1 (C-4´´), 116.1 (C-5´´), 122.6 (C-6´´), 140.3 (C-7´´), 120.2 (C-8´´), 166.5 (CO), 56.2 (OMe).

N-[2-(5-Hydroxy-lH-indol-3-yl)ethyl]-*p*-coumaramide: Colorless needles, mp 195-196°C. IR (KBr, cm^{-1}): 3400,1650, 1600; FAB-MS *m/z*: 323 $[M+H]^+$; ^{1}HNMR(CD_3COCD_3): δ 7.09 (d, J = 2.4 Hz, H-2), 7.05 (d, J =2.4 Hz, H-4), 6.73 (dd, J = 2.4, 8.5 Hz, H-6), 7.20 (d, J = 8.5 Hz, H-7), 2.94 (m, H-10), 3.63 (m, H-11), 7.40 (ddd, J = 1.8, 3.0, 8.5 Hz, H-2´´), 6.86 (ddd, J = 1.8, 3.0, 8.5 Hz, H-3´´), 6.86 (ddd, J = 1.8, 3.0, 8.5 Hz, H-5´´), 7.40 (tt, J = 1.8, 3.0, 8.5 Hz, H-6´´), 7.52 (d, J = 15 Hz, H-7´´), 6.52 (d, J = 15 Hz, H-8´´), 9.78 (s, NH), 7.43 (t, CONH); ^{13}CNMR(CD_3COCD_3): δ 124.1 (C-2), 112.4 (C-3), 103.5 (C-4), 151.5 (C-5), 112.4 (C-6), 112.5 (C-7), 132.2 (C-8), 129.3 (C-9), 26.4 (C-10), 40.8 (C-11), 127.7 (C-1´´), 130.2 (C-2´´), 116.6 (C-3´´), 159.8 (C-4´´), 116.6 (C-5´´), 130.2 (C-6´´), 140.4 (C-7´´), 119.6 (C-8´´), 165.1 (CO).

N,N'-[2,2'-(5,5'-Dihydroxy-4,4'-bi-lH-indol-3,3'-yl)diethyl]-di-*p*-coumaramide: Colorless powder, mp 180-182 °C; UV λmax (MeOH): nm (ε): 305 (34700), 294 (35000), 221 (43000). IR (KBr, cm^{-1}): 3400,1650, 1600. HR-FAB-MS: *m/z*: Calcd for $C_{38}H_{34}N_4O_6$: 642.2580 $[M+H]^+$, Found: 642.2582. ^{1}HNMR (DMSO-d_6): δ 6.89 (d, J = 1.8 Hz, H-2), 6.73 (d, J = 8.5 Hz, H-6), 7.15 (d, J = 8.5 Hz, H-7), 2.06 (m, H-10), 2.78 (m, H-11), 7.30 (d, J = 8.5 Hz, H-2´´), 6.74 (d, J = 8.5 Hz, H-3´´), 6.74 (d, J = 8.5 Hz, H-5´´), 7.30 (d, J = 8.5 Hz, H-6´´), 7.21 (d, J = 15 Hz, H-7´´), 6.35 (d, J = 15 Hz, H-8´´),10.4 (s, NH), 7.36 (t, CONH), 7.75 (s, 5-OH), 9.80 (br, 4´´-OH); ^{13}CNMR (DMSO-d6): δ 123.2 (C-2), 112.5 (C-3), 113.7 (C-4), 147.6 (C-5), 110.9 (C-6), 110.3 (C-7), 131.0 (C-8), 127.1 (C-9), 25.0 (C-10), 39.1 (C-11), 125.8 (C-1´´), 129.0 (C-2´´), 115.6 (C-3´´), 158.6 (C-4´´), 115.6 (C-5´´), 129.0 (C-6´´), 138.2 (C-7´´), 118.7 (C-8´´), 165.1 (CO).

N-[2-[3'-[2- (*p*-Coumaramido) ethyl]-5,5'-dihydroxy-4,4'-bi-l-H-indol-3-yl]ethyl]-ferulami-de: Colorless powder, mp 179-181°C. UV: λmax (MeOH) nm (ε): 309 (52700), 292 (51000), 221 (70500); IR (KBr, cm^{-1}): 3400, 1650, 1600. HR-FAB-MS: *m/z* Calcd for $C_{39}H_{36}N_4O_7$: 672.2584 $[M+H]^+$, Found: 672.2572; ^{1}HNMR (DMSO-d_6): δ 6.89 (d, J = 1.8 Hz, H-2), 6.73 (d, J = 8.5 Hz, H-6), 7.15 (d, J = 8.5 Hz, H-7), 2.06 (m, H-10), 2.78 (m, H-11), 7.03 (d, J = 1.8 Hz, H-2´´), 6.74 (d, J = 8.0 Hz, H-5´´), 6.91 (dd, J = 1.8, 8.0 Hz, H-6´´), 7.21 (d, J = 15 Hz, H-7´´), 6.36 (d, J = 15 Hz, H-8´´), 7.30 (d, J = 8.5 Hz, H-2´´´, 6´´´), 6.74 (d, J = 8.5 Hz, H-3´´´,5´´´), 7.21 (d, J = 15 Hz, H-7´´´), 6.40 (d, J = 15 Hz, H-8´´´), 3.73 (s, OMe), 10.2 (s, NH), 7.38 (t, CONH), 7.75 (s, 5-OH), 9.40 (br, 4´´-OH), 9.80 (br, 4´´´OH); ^{13}CNMR (DMSO-d_6): δ 123.1 (C-2), 112.5 (C-3), 113.7 (C-4), 147.6 (C-5), 110.9 (C-6), 110.2 (C-7), 131.0 (C-8), 127.1 (C-9), 25.0 (C-10), 39.0 (C-11), 125.8 (C-1´´), 110.6 (C-2´´), 147.6 (C-

3′′), 148.0 (C-4′′), 115.4 (C-5′′), 121.3 (C-6′′), 138.1 (C-7′′), 118.8 (C-8′′), 126.3 (C-1′′′), 128.9 (C-2′′′y 6′′′), 115.6 (C-3′′′ y 5 ′′′), 158.6 (C-4′′′), 138.4 (C-7′′′), 119.1 (C-8′′′), 165.1 (CO), 55.3 (OMe).

N,N' - [2,2' - (5,5'- Dihydroxy - 4,4' – bi – lH-indol-3,3'-yl)diethyl]-diferulamide: Colorless powder, mp 158-160 °C. UV: λmax (MeOH) nm (ε): 317 (38800), 289 (35400); IR (KBr, cm^{-1}): 3400,1650,1600; FAB-MS *m/z*: 703 $[M+H]^+$; ^{1}HNMR (DMSO-d_6): δ 6.90 (d, J = 1.8 Hz, H-2), 6.74 (d, J = 8.5 Hz, H-6), 7.14 (d, J = 8.5 Hz, H-7), 2.07 (m, H-10), 2.80 (m, H-11), 7.02 (d, J = 1.8 Hz, H-2′′), 6.74 (d, J = 7.93 Hz, H-5′′), 6.90 (dd, J = 1.8, 7.9 Hz, H-6′′), 7.21 (d, J = 15 Hz, H-7′′), 6.40 (d, J = 15 Hz, H-8′′), 3.72 (s, OMe), 10.4 (s, NH), 7.38 (t, CONH), 7.75 (s, 5-OH), 9.40 (br, 4′′-OH); ^{13}CNMR (DMSO-d_6): δ 123.9 (C-2), 113.2 (C-3), 114.0 (C-4), 148.4 (C-5), 111.6 (C-6), 111.0 (C-7), 131.7 (C-8), 127.9 (C-9), 25.7 (C-10), 40.7 (C-11), 125.0 (C-1′′), 111.3 (C-2′′), 148.4 (C-3′′), 148.8 (C-4′′), 116.2 (C-5′′), 122.0 (C-6′′), 139.2 (C-7′′), 129.8 (C-8′′), 165.8 (CO) 56.1 (OMe).

N-[2-[5-β-O-Glucosyloxy)-1H-indol-3-yl)ethyl]-*p*-coumaramide: Colorless needles, mp 240-242 °C. $[\alpha]_D$ -14.9° (c= 0.72, MeOH). IR (KBr, cm^{-1}): 3400, 1650. FAB-MS *m/z*: 485 $[M+H]^+$; ^{1}HNMR (DMSO-d_6): δ 7.13 (d, J = 2.4 Hz, H-2), 7.26 (d, J = 1.8 Hz, H-4), 6.87 (d, J = 1.8, 8.5 Hz, H-6), 7.23 (d,) J = 8.5 Hz, H-7), 2.96 (m, H-10), 3.40 (m, H-11), 7.39 (d, J = 8.5 Hz, H-2′′), 6.78 (d, J = 8.5 Hz, H-3′′), 6.78 (d, J = 8.5 Hz, H-5′′), 7.39 (d, J = 8.5 Hz, H-6′′), 7.33 (d, J = 15 Hz, H-7′′), 6.42 (d, J = 15 Hz, H-8′′), 10.6 (s, OMe), 8.09 (t, NH), 9.86 (br, 4′′-OH), 4.76 (d, J = 7.93 Hz, glc-1), 3.26 (m, glc-2), 3.31 (m, glc-3), 3.19 (m, glc-4), 3.31 (m, glc-5), 3.46 (m, glc-6), 3.76 (m); ^{13}CNMR (DMSO-d_6): δ 123.4 (C-2), 111.4 (C-3), 105.1 (C-4), 151.0 (C-5), 112.5 (C-6), 111.5 (C-7), 132.3 (C-8), 127.3 (C-9), 26.2 (C-10), 39.3 (C-11), 125.8 (C-1′′), 129.1 (C-2′′), 115.7 (C-3′′), 158.6 (C-4′′), 115.7 (C-5′′), 129.1 (C-6′′), 138.5 (C-7′′), 118.6 (C-8′′), 165.4 (CO), 102.3 (glc-1), 73.4 (glc-2), 76.8 (glc-3), 71.9 (glc-4), 76.7 (glc-5), 60.8 (glc-6).

N-[2-[5-(β-O-Glucosyloxy)-1H-indol-3-yl)ethyl]ferulamide: Colorless needles, mp 143-145°C. $[\alpha]_D$ -8.76° (c = 1.25, MeOH). IR (KBr, cm^{-1}); 3400, 1650. FAB-MS *m/z*: 485 $[M+H]^+$; ^{1}HNMR (DMSO-d_6): δ 7.13 (d, J = 1.8 Hz, H-2), 7.26 (d, J = 1.8 Hz, H-4), 6.86 (d, J = 1.8, 8.5 Hz, H-6), 7.23 (d, J = 8.5 Hz, H-7), 2.83 (m, H-10), 3.40 (m, H-11), 7.11 (d, J = 1.8 Hz, H-2′′), 6.78 (d, J = 7.93 Hz, H-5′′), 6.98 (dd, J = 1.8, 7.93 Hz, H-6′′), 7.33 (d, J = 15 Hz, H-7′′), 6.45 (d, J = 15 Hz, H-8′′), 3.80 (s, OMe), 10.6 (s, NH), 8.04 (t, CONH), 8.28 (br, 4′′-OH), 4.76 (d, J = 7.32 Hz, glc-1), 3.26 (m, glc-2), 3.31 (m, glc-3), 3.19 (m, glc-4), 3.31 (m, glc-5), 3.46 (m, glc-6), 3.74 (m); ^{13}CNMR (DMSO-d_6): δ 123.1 (C-2), 111.2 (C-3), 104.7 (C-4), 150.6 (C-5), 112.1 (C-6), 111.0 (C-7), 131.8 (C-8), 127.0 (C-9), 24.8 (C-10), 40.0 (C-11), 125.9 (C-1′′), 110.2 (C-2′′), 147.3 (C-3′′), 147.7 (C-4′′), 115.1 (C-5′′), 121.0 (C-6′′), 138.5 (C-7′′), 118.6 (C-8′′), 165.4 (CO), 55.0 (OMe), 101.9 (glc-1), 73.0 (glc-2), 76.3 (glc-3), 69.5 (glc-4), 76.5 (glc-5), 60.4 (glc-6).

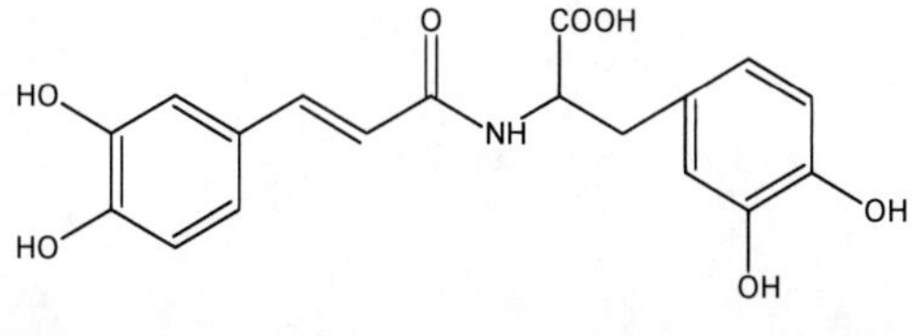

65

N-(3',4'-Dihydroxy-*trans*-cinnamoyl)-3-(3,4-dihydroxyphenyl)-L-alanine

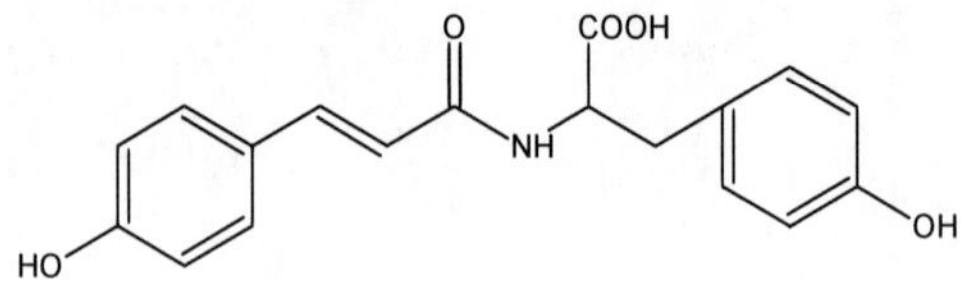

66
N-(4'-Hydroxy-*trans*-cinnamoyl)-3-(4-hydroxyphenyl)-L-alanine

Found in *Theobroma cacao* (Esterculaceae) showed antioxidant activity (Sanbongi et al., 1998).

N-(3',4'-Dihydroxy-*trans* -cinnamoyl)-3-(3,4-dihydroxyphenyl)-L-alanine: FD-MS, *m/z* 360 (M +H), 382 (M +Na); ^{1}HNMR (in DMSO-d_6): δ 8.62, 8.67, 9.07, 9.31 (4H, br, OH), 8.75 (lH, d, J = 8.0 Hz, NH), 7.20 (lH, d, J = 15.7 Hz, H8), 6.93 (lH, d, J = 1.8 Hz, H2), 6.82 (lH, dd, J = 8.0, 1.8 Hz, H6), 6.74 (lH, d, J = 8.1 Hz, H5), 6.62 (1H, d, J = 1.8 Hz, H2'), 6.61 (lH, d, J = 7.8 Hz, H5´), 6.57 (lH, d, J = 7.9, 1.8 Hz, H6'), 6.41 (lH, d, J = 15.7 Hz, H7), 4.48 (lH, m, H8'), 2.90 (H1, dd, J = 13.7, 4.9 Hz, H7'), 2.72 (H1, dd, J = 13.7, 9.1 Hz, H7'); ^{13}C NMR: δ 128.1 (C-1), 115.8 (C-2), 144.9 (C-3), 147.2 (C-4), 117.9 (C-5), 122.8 (C-6), 142.6 (C-7), 117.6(C-8), 169.5 (C-9), 130.1 (C-1´), 117.7(C-2´), 144.7 (C-3´), 143.7 (C-4´), 117.0 (C-5´), 122.4 (C-6´), 36.9 (C-7´), 55.3 (C-8´), 176.0 (C-9´).

N-(4'-Hydroxy-*trans*-cinnamoyl)-3-(4-hydroxyphenyl)-L-alanine: FD-MS, *m/z* 328 (M + H), 360 (M + Na); ^{1}HNMR (in DMSO-d_6): δ 9.19, 8.84 (4H, br, OH), 8.18 (1H, d, J = 8.2 Hz, NH), 7.37 (2H, d, J = 8.6 Hz, H2', 6'), 7.27 (1H, d, J = 15.6 Hz, H8), 7.02 (2H, d, J = 8.6 Hz, H2,6), 6.78 (2H, d, J = 8.6 Hz, H3',5'), 6.64 (2H, d, J = 8.2 Hz, H3,5), 6.48 (1H, d, J = 15.6 Hz, H7), 4.46 (1H, m, H8'), 2.97 (1H, dd, J = 13.7, 4.7 Hz, H7'), 2.79 (H1, dd, J = 13.7,9.4 Hz, H7'); ^{13}C NMR: δ 127.7 (C-1), 130.7 (C-2), 116.7 (C-3), 160.6 (C-4), 116.7 (C-5), 130.7 (C-6), 142.4 (C-7), 117.9 (C-8), 169.0 (C-9), 129.1 (C-1´), 131.3 (C-2´), 116.2 (C-3).

Chapter 2

CAROTENOIDS

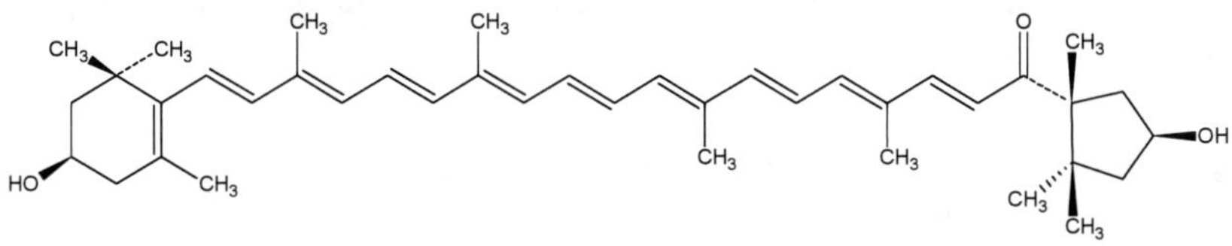

67
Capsanthin

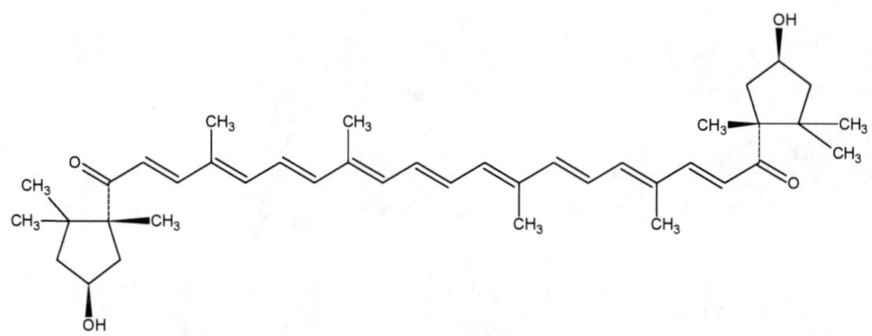

68
Capsorubin

Capsanthin suppressed hydroperoxide formation as well as α-carotene, lutein, and zeaxanthin. Interestingly, capsanthin decomposed more slowly than the other carotenoids, and the radical scavenging effect of capsanthin was found to last longer. Also, the capsanthin esterified partially and/or totally with fatty acids (mono- and/or diesterified capsanthin), isolated from paprika *Capsicum annuum* (Solanaceae). Suppressed oxidation of methyllinoleate in a similar manner as nonesterified cansanthin. This finding suggests that the radical scavenging ability of capsanthin was notinfluenced by esterification (Matsufuji et al., 1998).

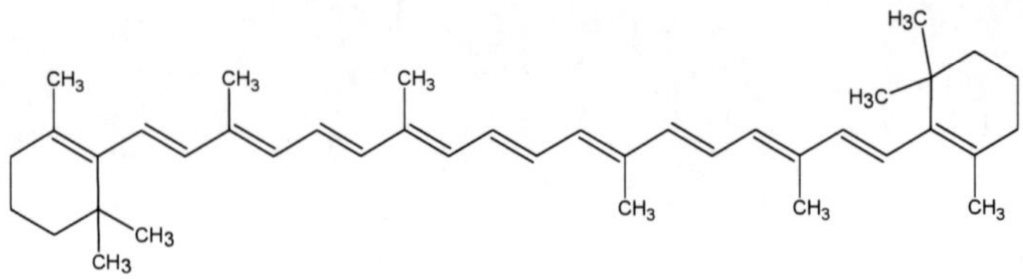

69

β-Carotene

Singlet oxygen is very powerfully quenched by β-carotene, with a rate constant near that of diffusion control, exceeding 101° M^{-1} sec^{-1}. This rate constant exceeds that for the reaction of 10^2 with most biologically important unsaturated fatty acids by four to five orders of magnitude, thus allowing a relatively low concentration of β-carotene to effectively protect membrane lipids from reactions of $^1O_2$ leading to peroxidation. The second-order rate constant for the reaction of β-carotene with $Cl_3COO^{\cdot}$ was 1.5 x 10^9 M^{-1} sec^{-1}, about a tenth of the diffusion controlled rate. It was 20-100 times more reactive than 25-diphenylfuran and 1,4-diazabicyclooctane. At higher oxygen concentrations β-carotene may be converted to radical species that have some chain-carrying ability and may actually promotote oxidation. The authors suggest that β-carotene and related compounds may be concentrated in cellular regions that are exposed to low partial pressures of oxygen (Larson, 1988).

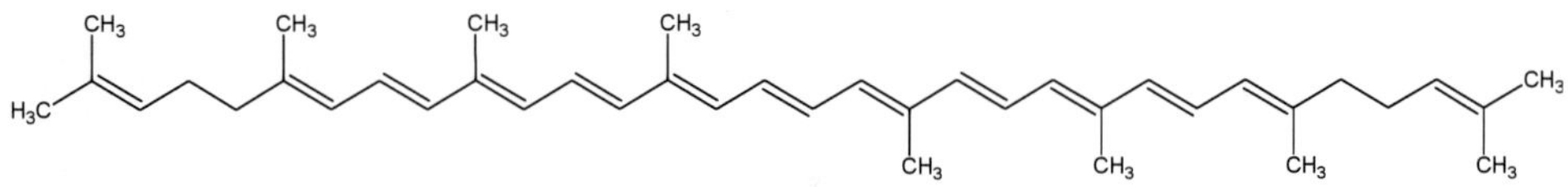

70

Lycopene

Found in *Lycopersicum esculentum* (Solanaceae). Showed antioxidant activity (Martinez-Valverde et al., 2002).

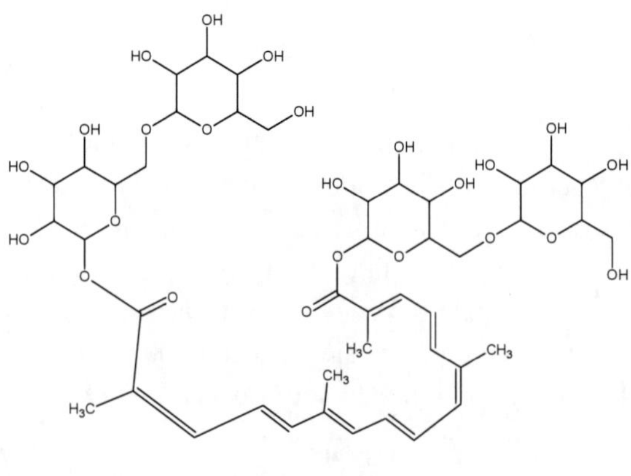

71

trans-Crocin

Crocin, a water soluble carotenoid is found in the fruits of *Gardenia jasminoides* Ellis and in the stigmas of saffron (*Crocus sativus* Linne), has an antioxidant activity at concentrations up to 40 ppm. The antioxidant property of crocin as evaluated by the thiocyanate method was better than with the thiobarbituric acid method. The adduct between the linoleic acid radical and crocin was detected by LC-MS. When crocin reacted with oxygen in the presence of $FeSO_4$, intermediates such as monohydroperoxides and dihydroperoxides of crocin were formed and detected by LC-MS (Pham et al., 2000).

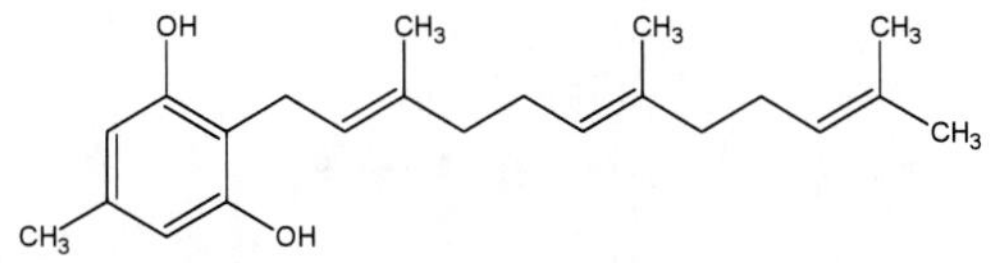

72
Grifolin

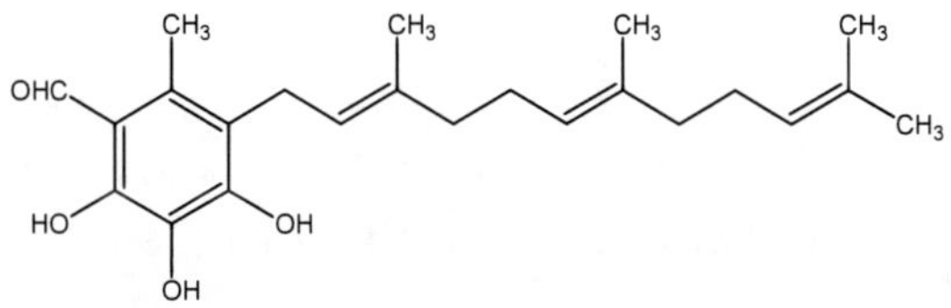

73
1-Formyl-3-hydroxyneogrifolin

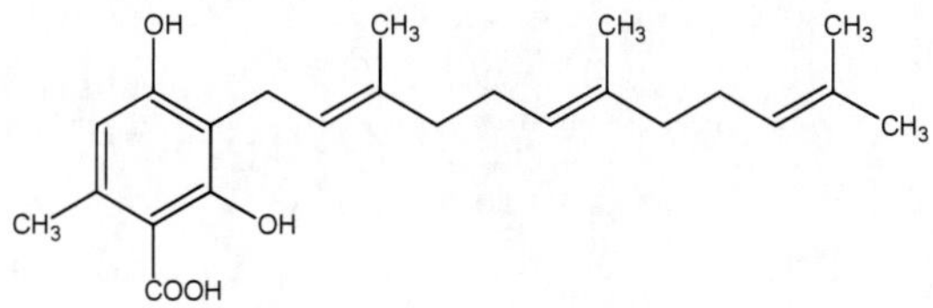

74
Grifolic acid

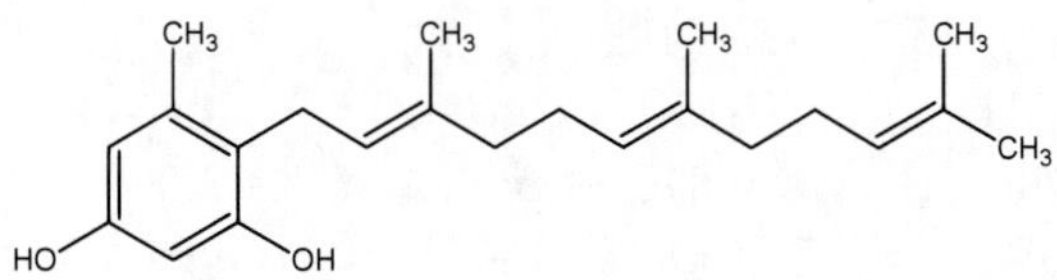

75
Neogrifolin

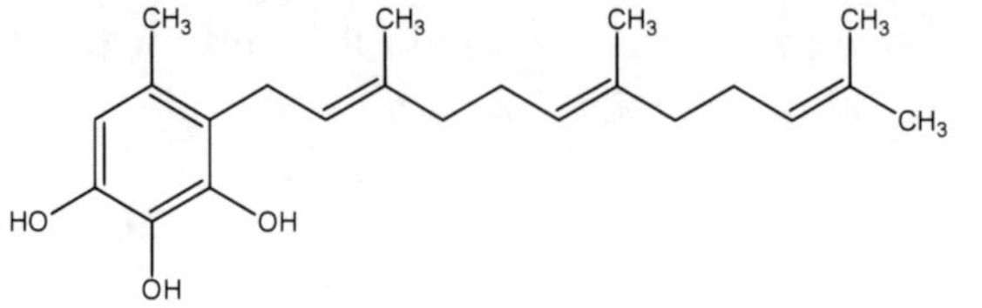

76

3-Hydroxyneogrifolin

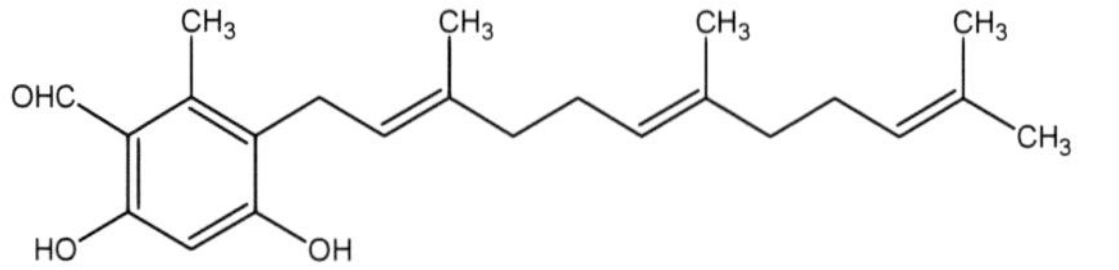

77

1-Formylneogrifolin

From the Japanese *Albatrellus ovinus* belonging to Scutigeraceae were isolated five neogrifolin derivatives. 1-Formyl-3-hydroxyneogrifolin showed more potent antioxidative activity properties than either α-tocopherol or BHA (Nukata et al., 1996, Nukataa et al., 2002).

Grifolin: ^{1}HNMR ($CDCl_3$): δ 6.2 (s, H-1), 6.2 (H-3), 2.18 (s, H-8), 3.37 (d, J = 7.0 Hz, H-1´), 5.26 (t, J = 7.0 Hz, H-2´), 1.90-2.1 (m, H- 4´), 1.9-2.1 (m, H-5´), 5.06 (m, H-6'), 1-9-2.1 (m, H-8´), 1-9-2.1 (m, H-9´), 5.06 (m, H-10´), 1.66 (s, H-12´), 1.57 (s, H-13´), 1.57 (s, H-14´), 1.80 (s, H-15´); ^{13}C NMR: δ 109.3 (C-1), 137.2 (C-2), 109.3 (C-3), 154.6 (C-4), 111.0 (C-5), 154.6 (C-6), 21.0 (C-8), 22.3 (C-1´), 122.0 (C-2´), 138.5 (C-3´), 39.7 (C-4´), 26.7 (C-5´), 123.7 (C-6´), 135.4 (C-7´), 39.7 (C-8´), 26.7 (C-9´), 124.5 (C-10´), 131.0 (C-11´), 25.7 (C-12´), 17.7 (C-13´), 16.1 (C-14´), 16.0 (C-15´).

Neogrifolin: ^{1}HNMR ($CDCl_3$): δ 6.23 (d, J = 3.0 Hz, H-1), 6.18 (d, J = 3.0 Hz, H-3), 2.21 (s, H-8), 3.27 (d, J = 7.0 Hz, H-1´), 5.11 (t, J = 7.0 Hz, H-2´), 1.84-2.04 (m, H- 4´), 1.84-2.04 (m, H-5´), 5.04 (m, H-6'), 1.84-2.04 (m, H-8´), 1.84-2.04 (m, H-9´), 5.04 (m, H-10´), 1.66 (s, H-12´), 1.58 (s H-13´), 1.58 (s, H-14´), 1.80 (s, H-15´); ^{13}C NMR: δ 109.3 (C-1), 137.2 (C-2), 109.3 (C-3), 154.6 (C-4), 111.0 (C-5), 154.6 (C-6), 21.0 (C-8), 22.3 (C-1´), 122.0 (C-2´), 138.5 (C-3´), 39.7 (C-4´), 26.7 (C-5´), 123.7 (C-6´), 135.4 (C-7´), 39.7 (C-8´), 26.7 (C-9´), 124.5 (C-10´), 131.0 (C-11´), 25.7 (C-12´), 17.7 (C-13´), 16.1 (C-14´), 16.0 (C-15´).

3-Hydroxyneogrifolin: ^{1}HNMR ($CDCl_3$): δ 6.33 (d, J = 3.0 Hz, H-1), 2.17 (s, H-8), 3.30 (d, J =7.1 Hz, H-1´), 5.17 (tq, J = 7.1, 0.8 Hz, H-2´), 2.04 (m, H- 4´), 2.10 (m, H-5´), 5.07 (m, H-6'), 1.96 (m, H-8´), 2.05 (m, H-9´), 5.08 (m, H-10´), 1.67 (d, J = 1.1 Hz, H-12´), 1.59 (d, J = 0.5 Hz, H-13´), 1.58 (d, J = 0.5 Hz, H-14´), 1.79 (d, J = 0.8 Hz, H-15´), 6.2 (s, OH); ^{13}C NMR: δ 109.3 (C-1), 141.6 (C-2), 129.9 (C-3), 142.7 (C-4), 118.1 (C-5), 127.8 (C-6), 19.4 (C-8), 25.7 (C-1´), 122.1 (C-2´), 137.7 (C-3´), 39.7 (C-4´), 26.3 (C-5´), 123.7 (C-6´), 135.5 (C-7´), 39.7 (C-8´), 26.7 (C-9´), 124.3 (C-10´), 131.4 (C-11´), 25.6 (C-12´), 17.7 (C-13´), 16.0 (C-14´), 16.2 (C-15´).

1-Formylneogrifolin: ^{1}HNMR ($CDCl_3$): δ 6.24 (s, H-3), 10.16 (s, H-7), 2.48 (s, H-8), 3.34 (d, J =6.6 Hz-H-1´), 5.06 (tq, J = 6.6, 1.1 Hz, H-2´), 2.02 (m, H- 4´), 2.10 (m, H-5´), 5.07 (m,

H-6'), 1.95 (m, H-8′), 2.03 (m, H-9′), 5.07 (m, H-10′), 1.68 (d, J = 1.1 Hz, H-12′), 1.58 (d, J= 0.5 Hz, H-13′), 1.58 (d, J = 0.5 Hz, H-14′), 1.79 (d, J = 1.1 Hz, H-15′), 6.24 (s, OH); ^{13}C NMR: δ 113.8 (C-1), 164.4 (C-2), 101.3 (C-3), 162.4 (C-4), 119.4 (C-5), 142.0 (C-6), 193.6 (C-7), 13.8 (C-8), 24.6 (C-1′), 121.5 (C-2′), 137.5 (C-3′), 39.7 (C-4′), 26.3 (C-5′), 123.6 (C-6′), 135.4 (C-7′), 39.7 (C-8′), 26.7 (C-9′), 124.3 (C-10′), 131.3 (C-11′), 25.7 (C-12′), 17.7 (C-13′), 16.0 (C-14′), 16.3 (C-15′).

1-Formyl-3-hydroxyneogrifolin: ^{1}HNMR ($CDCl_3$): δ 10.16 (s, H-7), 2.43 (s, H-8), 3.35 (d, J = 6.3 Hz, H-1′), 5.04 (tq, J = 6.3, 0.8 Hz, H-2′), 2.0 (m, H- 4′), 2.07 (m, H-5′), 5.07 (m, H-6'), 1.94 (m, H-8′), 2.01 (m, H-9′), 5.07 (m, H-10′), 1.67 (d, J = 1.1 Hz, H-12′), 1.58 (d, J = 0.5 Hz, H-13′), 1.57 (d, J = 0.5 Hz, H-14′), 1.77 (d, J = 0.8 Hz, H-15′), 5.7 (s, OH) ; ^{13}C NMR: δ 112.9 (C-1), 149.7 (C-2), 128.4 (C-3), 149.5 (C-4), 119.9 (C-5), 133.1 (C-6), 194.2 (C-7), 12.9 (C-8), 24.4 (C-1′), 121.8 (C-2′), 135.7 (C-3′), 39.7 (C-4′), 26.3 (C-5′), 124.0 (C-6′), 135.0 (C-7′), 39.7 (C-8′), 26.7 (C-9′), 124.3 (C-10′), 131.3 (C-11′), 25.7 (C-12′), 17.6 (C-13′), 16.0 (C-14′), 16.2 (C-15′).

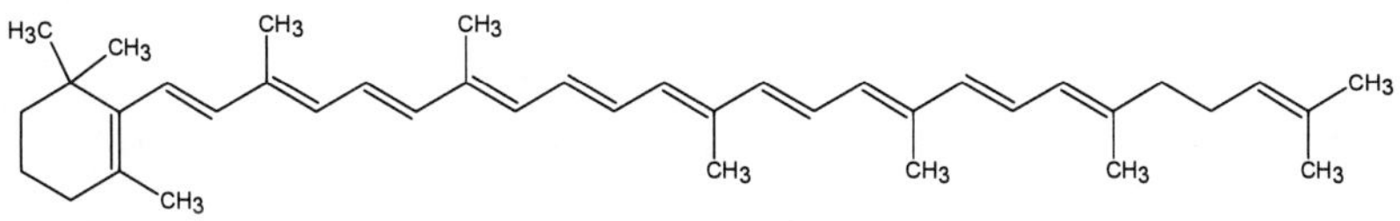

78

γ-Carotene

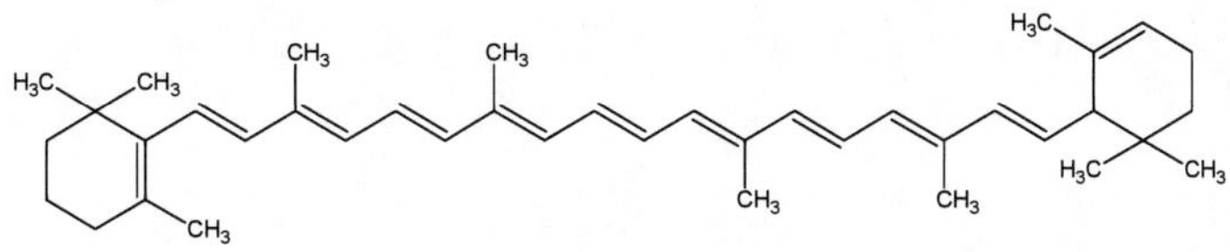

79

α-Carotene

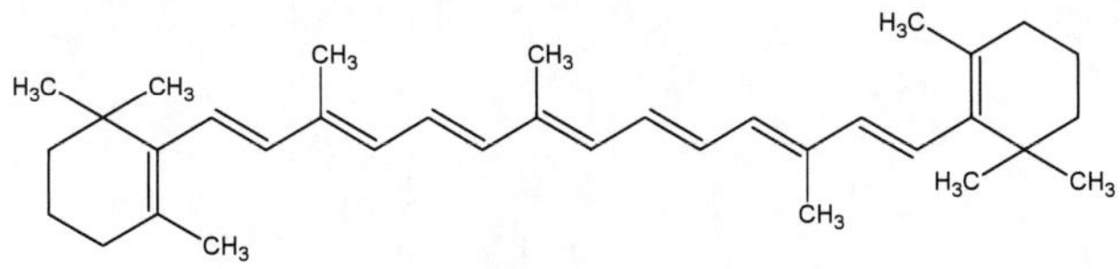

80

Septapreno-β-carotene

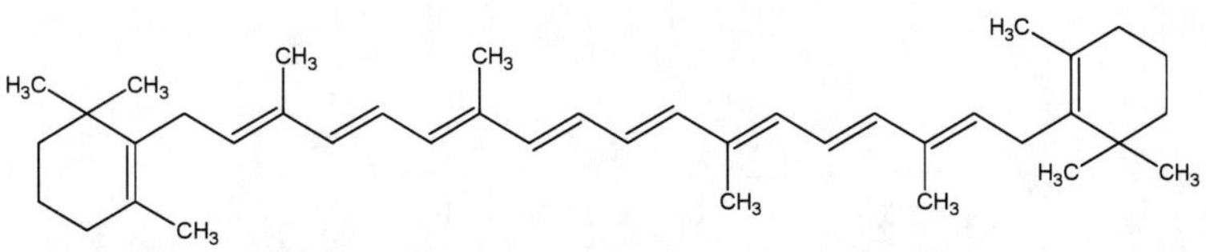

81

7,7' Dihydro-β-carotene

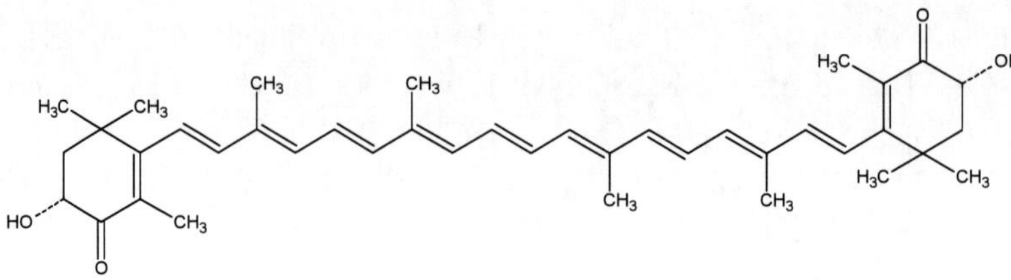

82
Astaxanthin

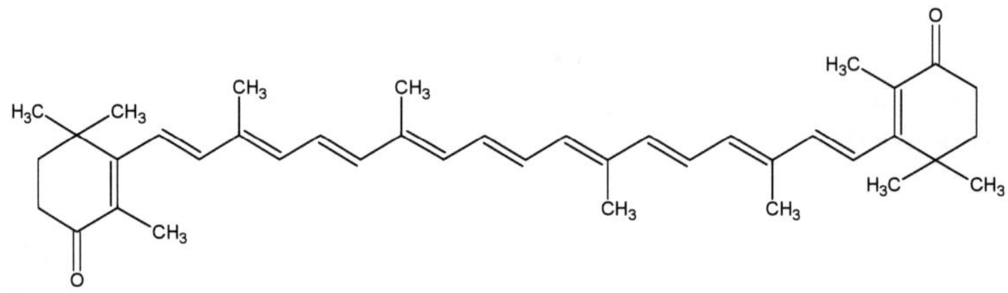

83
Canthaxanthin

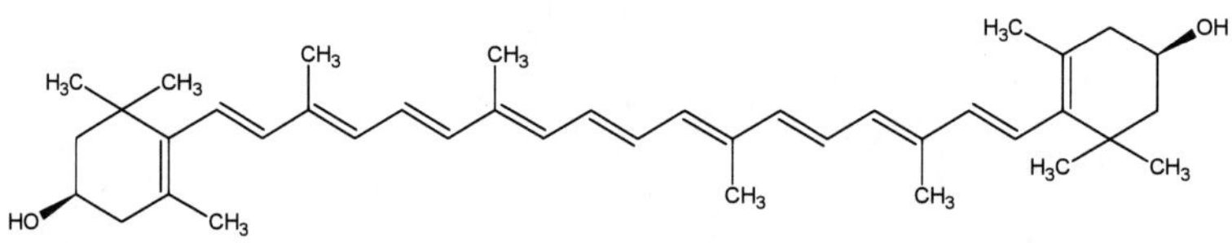

84
Zeaxanthin

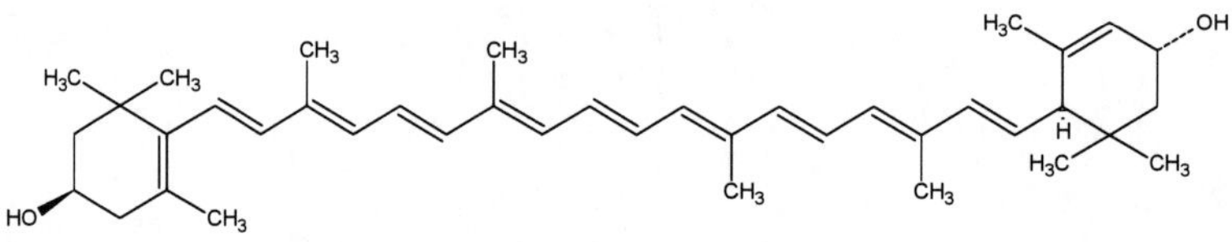

85
Lutein

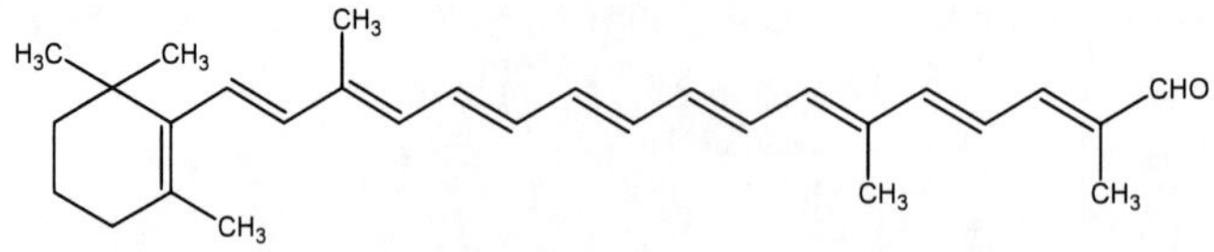

86
β-Apo-8'-carotenal

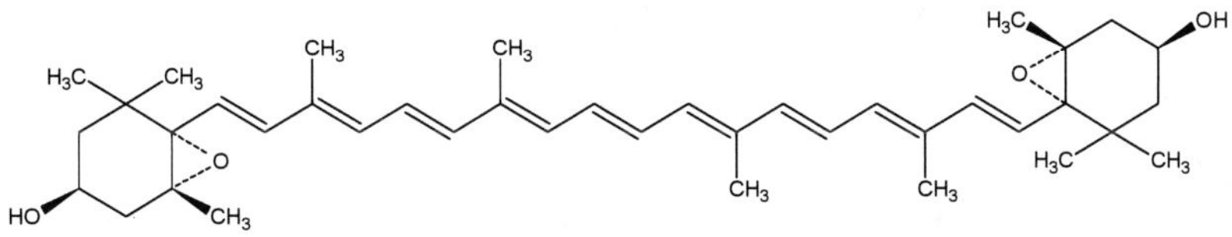

87
Violaxanthin

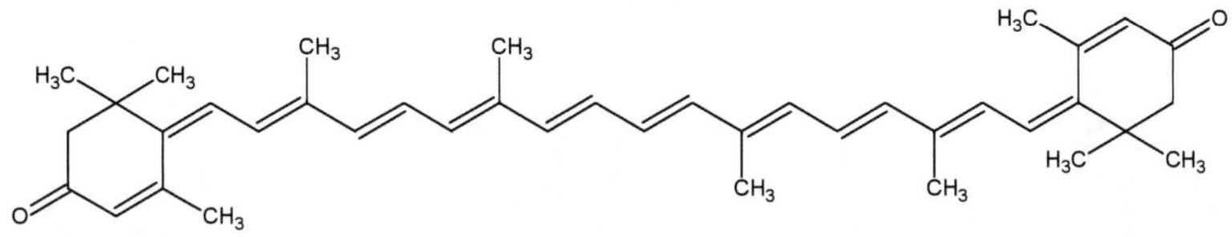

88
Rhodoxanthin

Carotenoids are abundant in many fruits and vegetables and they play diverse roles in photobiology, photochemistry and medicine. Carotenoids have the ability to act as chain-breaking anti-oxidants and thus protect cells and organisms against photo-oxidation. These carotenoids possess ability to quench singlet molecular oxygen, and interaction with a range of free radicals. Showed mechanisms associated with the anti- and pro-oxidant behaviour, they can produced interactions with other anti-oxidants. As well as cancer prevention, the potential anti-oxidant, properties of carotenoids may help to inhibit the onset of other diseasecs that are believed to be initiated by free radicals. These include atherosclerosis, cataracts, age-related macular degeneration and multiple sclerosis (Edge et al., 1997). Several mechanisms have been suggested for carotenoid protection in biological systems, which include the deactivation of electronically activated species such as singlet oxygen ($^1O_2$) and the deactivation of reactive chemical species, such as peroxyl or alkoxyl radicals, that can be generated within cells and might otherwise initiate harmful oxidative reactions. This antioxidant activity is related to the chemical structures of the carotenoids and it has been reported to be particularly effective at low O_2 tension (Palozza and Krinsky, 1992).

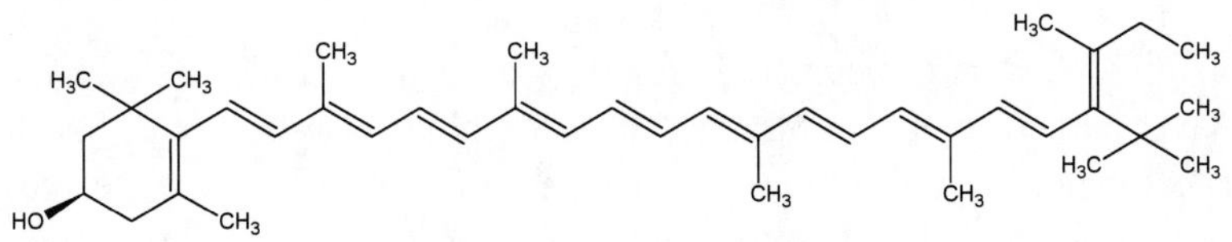

89
β-Cryptoxanthin

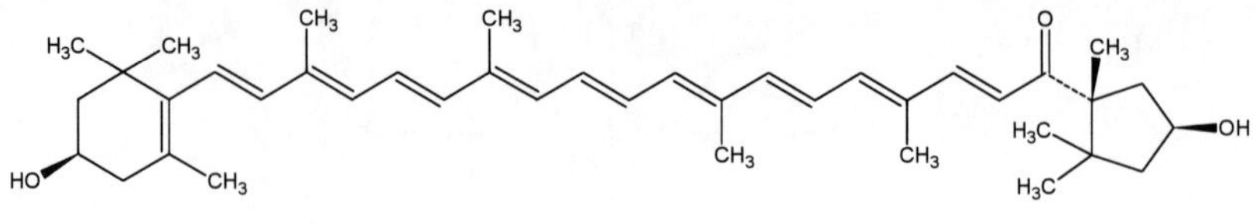

90
3-Deoxy-cryptocapsin

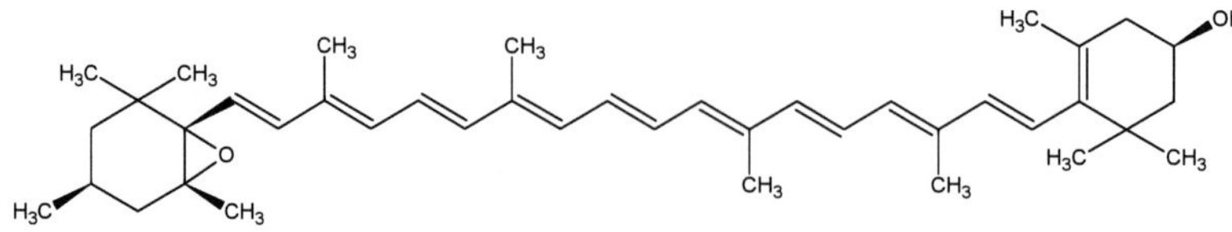

91
Antheraxanthin

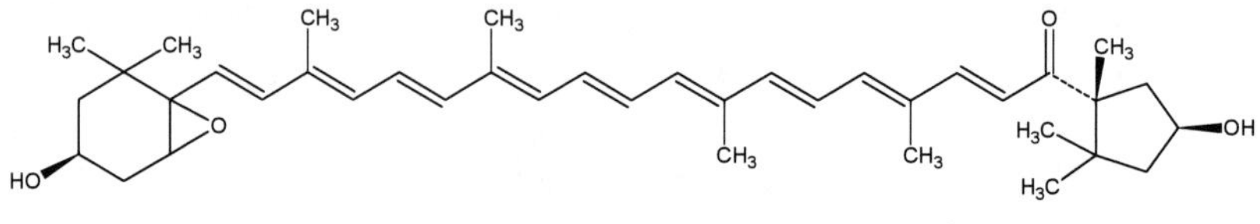

92
Capsanthin-5,6-epoxide

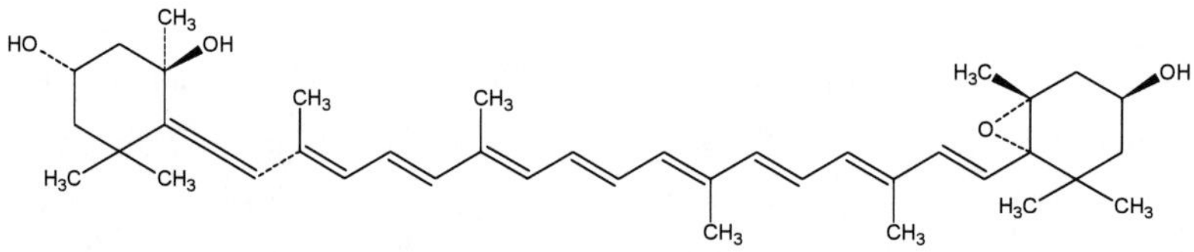

93
Neoxanthin

Found in fully green and maure red fruits of the Yolo Wonder a variety of *Capsicum annuum* (Solanaceae). The disappearance of chlorophyll was followed by an increased synthesis of carotenoids.Capsanthin was the main carotenoid. It appeared as diester, monoester and free forms, while cryptocapsin was not esterified. These compounds showed antioxidant activity (Camaraand Moneger, 1978).

Chapter 3

COUMARINS

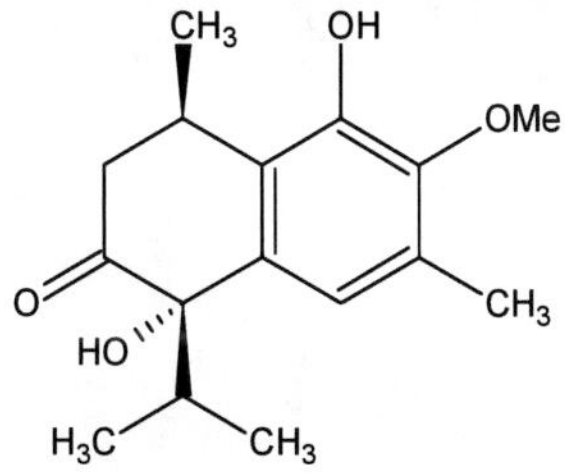

94
Mansonone N

Found in *Mansonia gagei* Drumm (Sterculiaceae). Mansonone N exhibit radical scavenging properties in the DPPH, at 2.5 μg. Quercetin used as a reference was five times more active (Tiew et al., 2003).

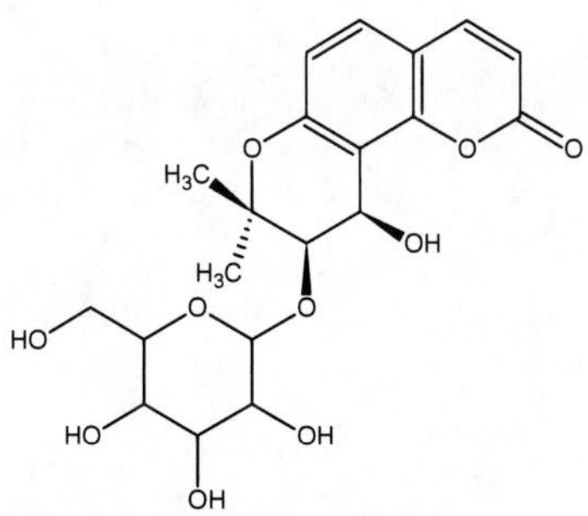

95
Praeroside II

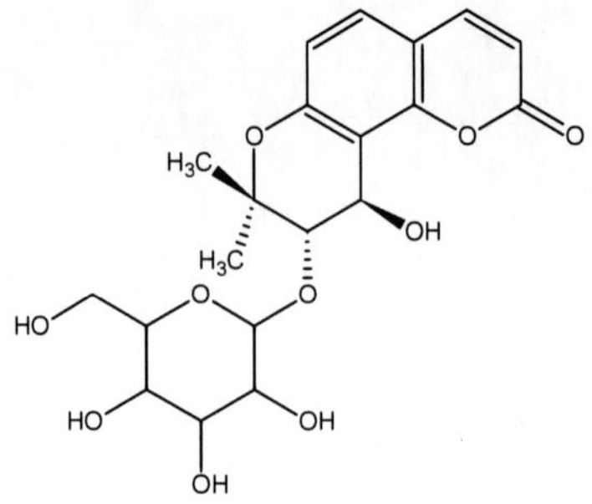

96
Praeroside III

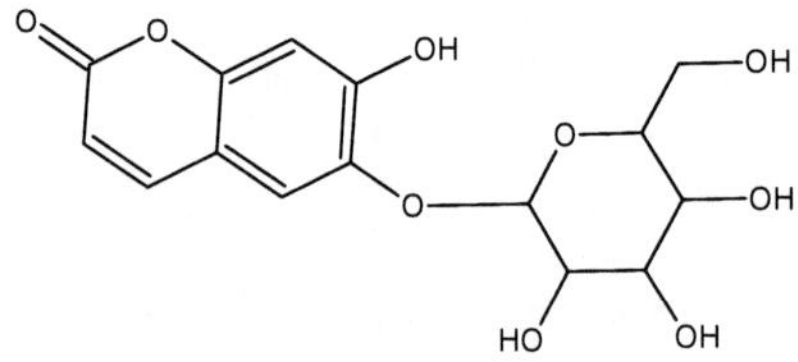

97
Esculin

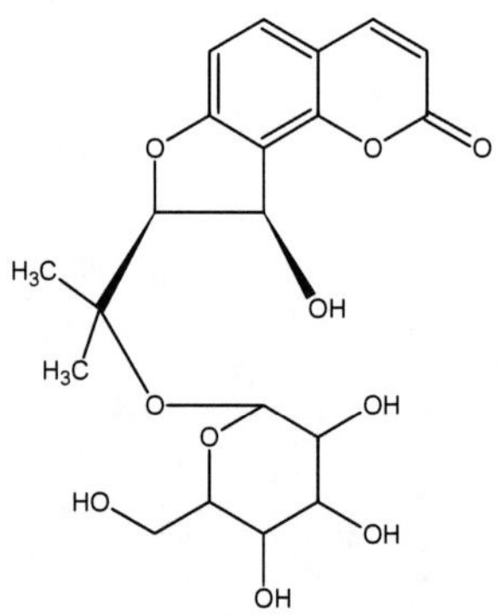

98
Apterin

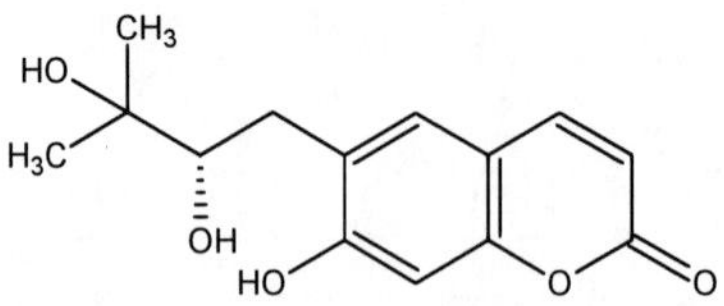

99
(R)-Peucedanol

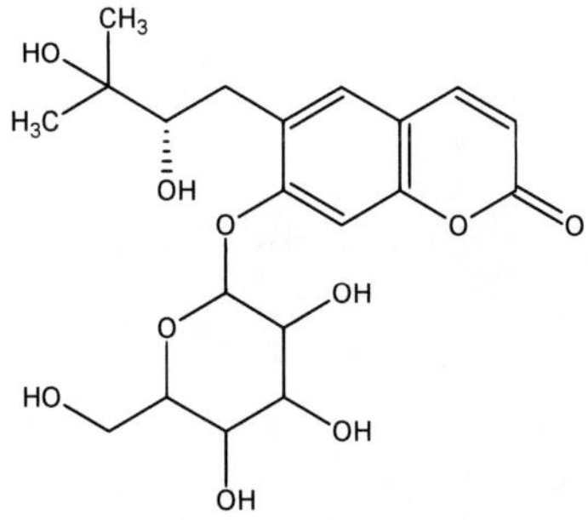

100
(R)-Peucedanol 7-O-β-D-glucopyranoside

All isolated compounds from *Peucedanum japonicum* Thunb. (Umbelliferae) showed antioxidant effect for free radical scavenging on 1,1-diphenyl-2-picrylhydrazyl radical and for inhibition of oxidation of liposome induced by 2,2'-azobis(2-amidinopropane)dihydrochloride (Hisamoto et al., 2003).

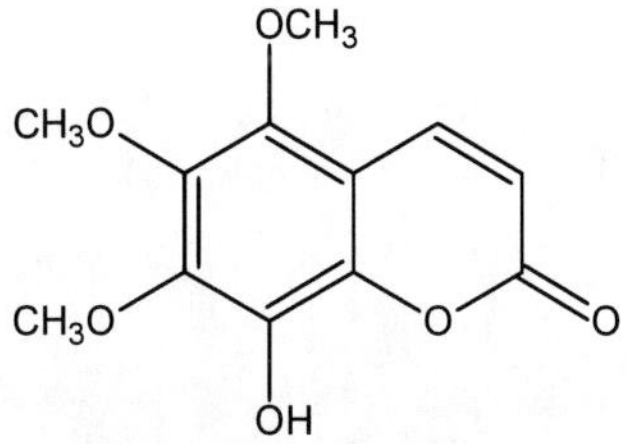

101
8-Hidroxy -5,6,7-trimetoxycoumarin

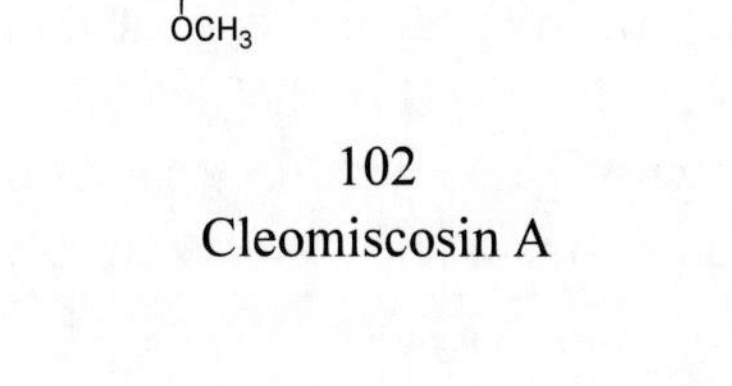

102
Cleomiscosin A

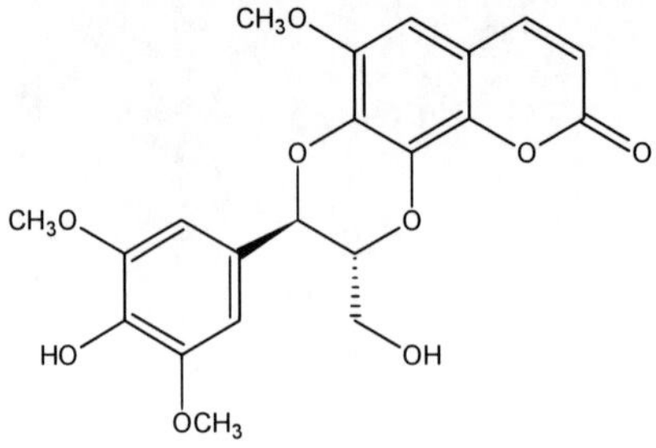

103
Cleomiscosin C

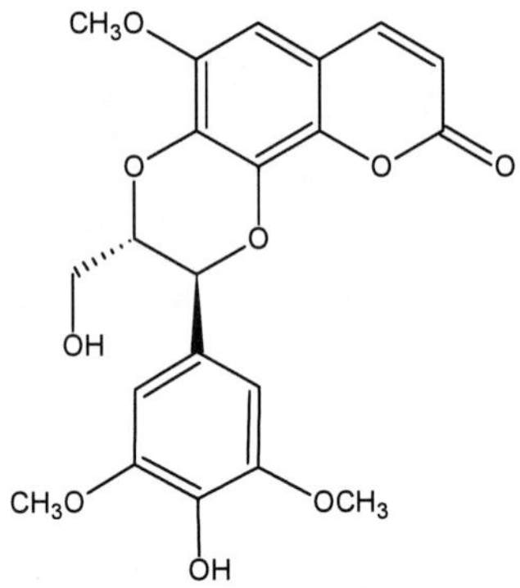

104
Cleomiscosin D

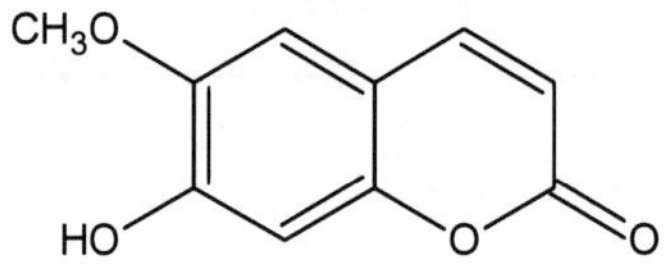

105
Scopoletin

Found in *Hibiscus syriacus* (Malvaceae). The 8-hydroxy-5,6,7-trimethoxycoumarin and scopoletin inhibited monoamine oxidase in a dose-dependent fashion with IC_{50} values of 44.5 and 19.4 μg/mL, respectively. cleomiscosins did not show activity up to 70 μg/mL. Cleomiscosin C inhibited lipid peroxidation with IC_{50} values of 1.4 μg/mL. This activity was comparable to vitamin E (0.8 μg/mL), which was used as a control. Cleomiscosins D was about 8 times less active (IC_{50} 5.5 μg/mL) than cleomiscosins C, its regioisomer with the alternative linkage orientation of the benzodioxan moiety. Cleomiscosins A was about 13 times less active (IC_{50} 9.0 μg/mL) than cleomiscosins C, from which it differs only in the lack of a methoxyl group at C-3'. In addition, 8-hydroxy-5,6,7-trimethoxycoumarin and scopoletin did not show significant antioxidative activity. In the microsomal assay the linkage orientation of the benzo-dioxan on the coumarin core and the substituents on the phenyl group are important determinants for lipid peroxidation inhibitory activity (Yun et al., 2001).

8-Hydroxy-5,6,7-trimethoxycoumarin. ^{1}HNMR: δ 6.33 (lH, d, J = 9.3 Hz, H-3), 7.94 (lH, d, J = 9.3 Hz, H-4), 6.91 (lH, s, H-5), 6.74 (2H, d, H-2′, 6′), 4.73(1H, d, J = 8.1 Hz, H-7′), 4.47 (lH, dq, J = 8.1, 6.3 Hz, H-8′), 1.17 (3H, d, J = 6.3 Hz, H-9′), 3.78 (6-OMe), 3.77 (6H, s, 3′-5′OMe), 8.58 (1H, s, 4′-OH).

Scopoletin. UV: [neutral] λmax 229 (ε 13800), 252 (ε 9120); 260 (sh) (8710), 297 (9550), 344 (13500) (MeOH) (Derep) [neutral] max 230, 255, 295, 345, (MeOH) (Berdy) [neutral] λmax 228 (ε 23000); 254 (ε 7600); 297 (ε 8700); 345 (ε 20500) (EtOH) (Berdy).

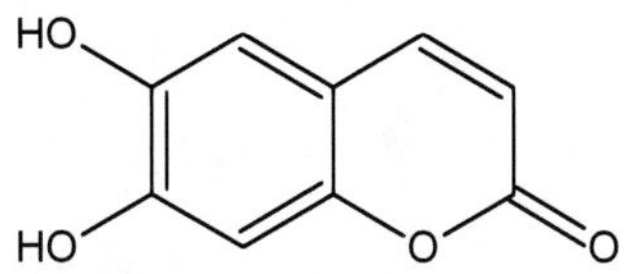

106
Esculetin

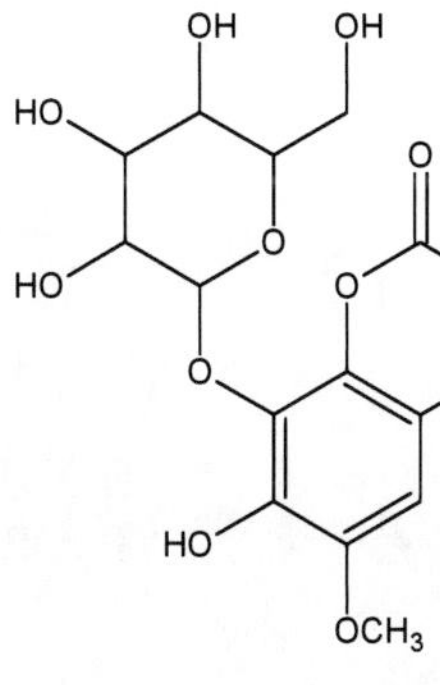

107
Fraxin

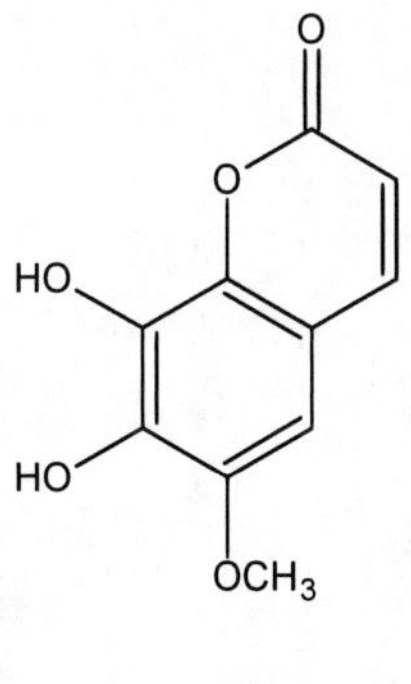

108
Fraxetin

Fraxinus omus L. (Oleaceae) is rich in hydroxycoumarins. They occur free or as glucosides, these compounds are the main components of the bark. The antioxidant activity of fraxetin and esculetin was higher than that of the corresponding glucosides fraxin and esculin is comparable to that of other well-known antioxidants such as caffeic acid. The data also demonstrate that fraxetin is a stronger antioxidant than esculetin, especially in the system of lower oxidizability (TGL). (Kostova, 2001).

Chapter 4

FLAVONOIDS

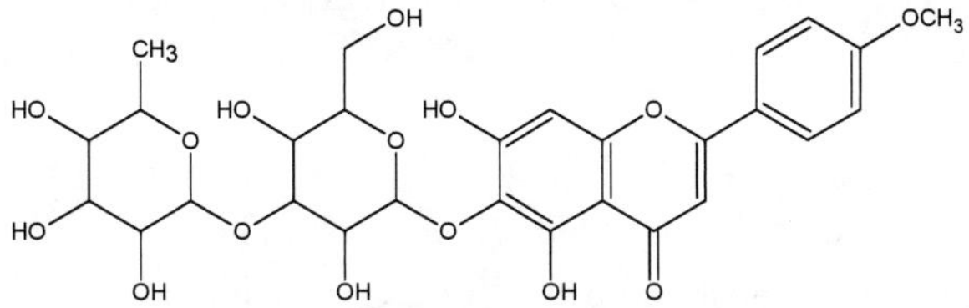

109

Acacetin 6-C-[α-L-rhamnopyranosyl-(1→3)-β-D-glucopyranoside]

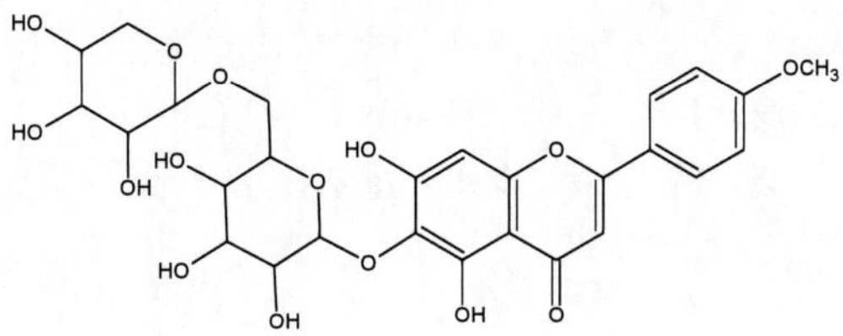

110

Acacetin 6-C-[β-D-xylopyranosyl-(1→6)-
β-D-glucopyranoside]

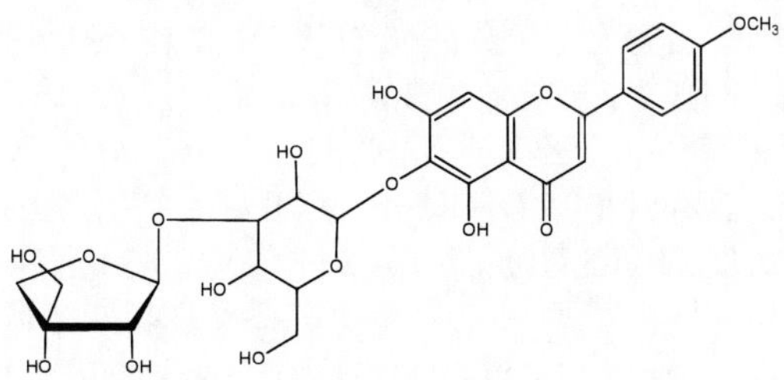

111

Acacetin 6-C-[β-D-apiofuranosyl- (1→3)-β-D-glucopyranoside]

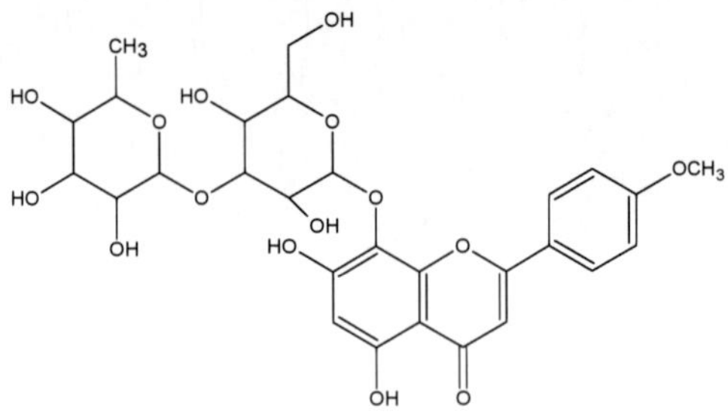

112

Acacetin 8-C-[α-L-rhamnopyranosyl-(1→3)-β-D-glucopyranoside]

Fractionation of a methanolic extract of the leaves of *Anthurium versicolor* has resulted in the isolation of four flavone glycosides, showed antioxidant activity (Aquino et al., 2001).

Acacetin 6-C-[α-L-rhamnopyranosyl-(1→3)-β-D-glucopyranoside: Amorphous powder; mp 210-212 °C; $[\alpha]_D$ -79.5° (c 1, MeOH); UV (MeOH): λmax (log ε) 329 (4.40), 273 (4.30) nm; FABMS: *m/z* 591 [M -H]⁻, 445 [(M-H)-146]⁻, 429 [(M-H)-162]⁻; anal. C 56.82%, H 5.409% calcd for $C_{28}H_{32}O_{14}$, C 56.76%, H 5.44%; ^{1}HNMR: δ 3.91 (3H, s, OMe), 6.50 (1H, s, H-8), 6.63 (1H, s, H-3), 7.10 (2H, d, J = 8.0 Hz, H-3' and H-5), 7.93 (2H, d, J = 8.0 Hz, H-2' and H-6); ^{13}C NMR: δ 183.8 (C-4), 165.6 (C-2 and C-7), 164.4 (C-4'), 162.1 (C-5), 158.7 (C-9), 129.2 (C-6' and C-2), 124.4 (C-1'), 115.6 (C-3' and C-5), 109.2 (C-6), 105.0 (C-10), 104.3 (C-3), 95.5 (C-8), 56.1 (OMe); ^{1}H and ^{13}C NMR for the sugar chain: δ 4.93 (d, J = 7.5 Hz, Glc-l"), 4.31 (dd, J = 8.5, 7.5 Hz, Glc-2"), 3.62 (t, J = 8.5 Hz, Glc-3"), 3.56 (t, J = 8.5 Hz, Glc-4"), 3.46 (m, Glc-5"), 3.78 (dd, J = 12.0, 4.5 Hz, Glc-6"), 5.22 (d, J = 1.5 Hz, Rha-l'"), 3.99 (dd, J = 1.5, 2.5 Hz, Rha-2'"), 3.77 (dd, J =2.5, 9.5 Hz, Rha-3'"), 3.43 (t, J =9.5 Hz, Rha-4'"), 4.07 (m, Rha-5'"), 1.31 (d, J = 6.5 Hz, Rha-6'").75.3 (Glc-l"), 72.8 (Glc-2"), 87.1 (Glc-3"), 70.4 (Glc-4"), 82.6 (Glc-5"), 62.8 (Glc-6"), 103.0 (Rha-l'"), 72.2 (Rha-2'"), 72.4 (Rha-3'"), 74.0 (Rha-4'"), 10.1 (Rha-5'"), 17.9 (Rha-6'").

Acacetin 6-C-[β-D-xylopyranosyl-(l→6)-β-D-glucopyranoside]: Amorphous powder; mp 200-201 °C; $[\alpha]_D$ -58.2° (c 0.5, MeOH); UV (MeOH): λmax (log ε) 336 (4.44), 276 (4.30) nm; FABMS: *m/z* 577 lM -H]⁻; 445 l(M -H) -132], 429 l(M -H) -148]⁻; anal. C 56.10%, H 5.190% calcd for $C_{27}H_{30}O_4$, C 56.03%, H 5.230%; ^{1}H and ^{13}C NMR data for the aglycon were almost superimposable on those of compound acacetin 6-C-[α-L-rhamnopyranosyl-(1→3)-β-D-glucopyranoside. ^{1}H and ^{13}C NMR for the sugar chain: δ 4.92 (d, J = 7.5 Hz, Glc-l"), 4.16 (dd, J = 9.0, 7.5 Hz, Glc-2"), 3.52 (t, J = 9.0 Hz, Glc-3"), 3.48 (t, J = 9.0 Hz, Glc-4"), 3.60 (m, Glc-5"), 3.79 (dd, J =12.0, 4.5 Hz, Glc-6"), 3.99 (dd, J = 12.0, 3.5 Hz, Xy-l'"), 4.28 (d, J = 7.5, Xy-2'"), 3.22 (dd, J = 7.5, 9.0 Hz, Xy-3'"), 3.31 (t, J = 9.0 Hz, Xy-4'"), 3.49 (m, Xy-5'"), 3.20 (dd, J = 10.5, 2.0 Hz, Xy-6'"); 75.2 (Glc-l"), 73.0 (Glc-2"), 79.9 (Glc-3"), 71.0 (Glc-4"), 81.5 (Glc-5), 70.3 (Glc-6), 105.3 (Xy-l'"), 74.8 (Xy-2'"), 77.6 (Xy-3'"), 71.4 (Xy-4'"), 66.9 (Xy-5'").

Acacetin 6-C-[β-D-apiofuranosyl-(1→3)-β-D-glucopyranoside]: Amorphous powder; mp 208-210 °C; $[\alpha]_D$ -63.1° (c 0.2, MeOH); UV (MeOH): λmax 328 (4.38), 272 (4.32) nm; FABMS *m/z* 577 [M –H]⁻; 445 [(M -H) -132)-, 429 [(M -H) -148]⁻; anal. C 56.050%, H 5.20%, calcd for $C_{27}H_{30}O_4$, C 56.03% H 5.23%; ^{1}H and ^{13}C NMR data for the aglycon were almost superimposable on those of compound Acacetin 6-C-[α-L-rhamnopyranosyl-(1→3)-β-D-glucopyranoside. ^{1}H and ^{13}C NMR for the sugar chain: δ 4.95 (d, J =7.5 Hz, Glc-l"), 4.37

(dd, J = 9.0,7.5 Hz, Glc-2"), 3.62 (t, J = 9.0 Hz, Glc-3"), 3:55 (t, J = 9.0 Hz, Glc-4"), 3.45 (m, Glc-5"), 3.77 (dd, J = 12.0, 4.5 Hz, Glc-6"), 3.92 (dd, J = 12.0, 3.5 Hz, Ap-l'''), 5.31 (d, J = 2.0 Hz, Ap-2'''), 4.03 (d, J = 2.0 Hz, Ap-3'''), 4.18 (d, J = 10.0 Hz, Ap-4'''), 3.64 (br s Ap-5'''); 75.2 (Glc-l"), 72.4 (Glc-2"), 87.8 (Glc-3"), 70.2 (Glc-4"), 82.8 (Glc-5"), 62.5 (Glc-6"), 112.2 (Ap-l'''), 77.5 (Ap-2'''), 80.5 (Ap-3'''), 74.9 (Ap-4'''), 64.9 (Ap-5''').

Acacetin 8-C-[α-L-rhamnopyranosyl-(1→3)-β-D-glucopyranoside: Amorphous powder; mp 207-209 °C; $[\alpha]_D$ -90.4° (c 0.2, MeOH); UV (MeOH): λmax 331 (4.42),270 (4.30) nm; FABMS: *m/z* 591 $[M-H]^-$; 445 $[(M\text{-}H)-146]^-$; anal. C 56.75% H 5.40% calcd for $C_{28}H_{32}O_{14}$, C 56.76% H 5.44%; ^1H and ^{13}C NMR: δ 3.90 (3H, s, OMe), 6.37 (lH, s, H-6), 6.61 (lH, s, H-3), 7.12 (2H, d, J = 8.0 Hz, H-3' and H-5), 7.99 (2H, d, J = 8.0 Hz, H-2' and H-6); ^{13}C NMR δ 183.6 (C-4), 165.9 (C-7), 165.2 (C-2), 164.4 (C-4), 162.1 (C-5), 157.5 (C-9), 129.3 (C-6' and C-2´), 124.4 (C-1'),115.8 (C-3' and C-5), 105.1 (C-8), 104.8 (C-10), 104.3 (C-3), 99.2 (C-6), 56.2 (OMe). ^1H and ^{13}C NMR data for the sugar chain were almost superimposable on those of compound aacetin 6-C-[α-L-rhamnopyranosyl-(1→3)-β-D-glucopyranoside.

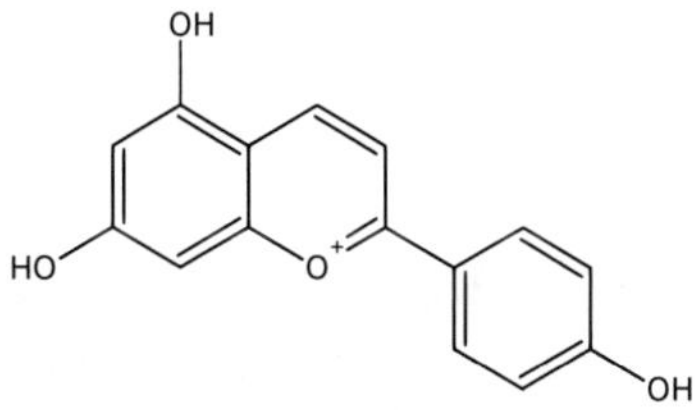

113
Apigeninidin

Found in *Glycine max* (Leguminosae). The apigeninidin shown a significant capacity to limit lipid peroxidation. It can be due at that the antioxidant activity of the pigment depends on its posibility of accessing the membrane location where radicals are generated. Thus, the antioxidant capacity of apiginінidin could be related not only to the pigment structural features but to the membrane characteristics as well. Also apigeninidin shows scavenger activity against ascorbyl radicals *in vitro* (Boverisa et al., 2001).

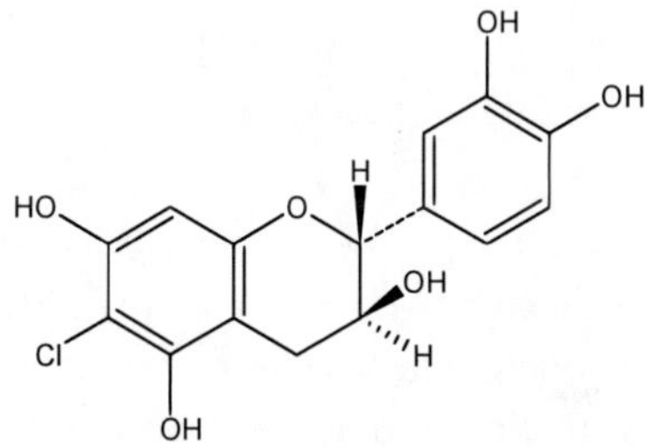

114
6-Chlorocatechin

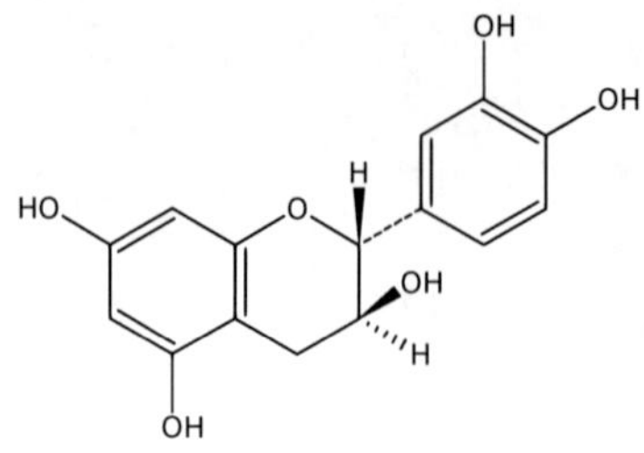

115
Catechin

The compounds have been isolated from the roots of *Rumex patiemia* L. (Poligonaceae). Showed cytotoxic effects and radical scavenging properties (Demirezera et al., 2001). Catechin showed suppression of carrageenan-induced rat paw edema as well as TPA-induced rat ear edema (tetradecanoylphorbol acetate). Isolated from the bark of *Ceiba pentandra* L. Gaether (Bombacaceae), (Noreen et al., 1988).

6-Chlorocatechin: Amorphous: $[\alpha]_D$ = + 50° (c = 0.16. DMSO); Rf 0.36 ($CHCl_3/CH_3OH/H_2O$. 80:20:2). ESIMS: *m/z* 649 (35), 647 (100) $[2M-H]^+$, 325 (24), 323 (70) $[M-H]^+$, 287 (5) $[M-H-Cl]^+$; UV: λmax (MeOH) nm: 208, 233sh, 281; IR νmax (KBr): cm^{-1}: 3381 (OH), 2922 (CH), 1623 (C=C, arom); ^{1}HNMR (300 MHz; CD_3OD): δ 4.75 (d, J = 6.7 Hz, H2), 4.02 (ddd, J = 7.2, 6.7, 5.2 Hz, H-3), 2.76 (dd, J = 16.3, 5.2 Hz, H-4α), 2.55 (dd, J = 16.3, 7.2 Hz, H-4β), 6.07 (s, H-8), 6.81 (d, J = 1.8 Hz, H-1´), 6.75 (d, J = 8.1 Hz, H-5´), 6.69 (dd, J = 8.1, 1.8 Hz, H-6´); ^{13}C NMR (75.5 MHz; CD_3OD): δ 82.90 (C-2), 68.28 (C-3), 27.89 (C-4α), 102.02 (C-4a), 155.50 (C-5), 100.18 (C-6), 153.40 (C-7), 96.62 (C-8), 152.09 (C-8a), 131.88 (C-1´), 114.86 (C-2´), 146.21 (C-3´), 146.25 (C-4´), 116.12 (C-5´), 119.56 (C-6´).

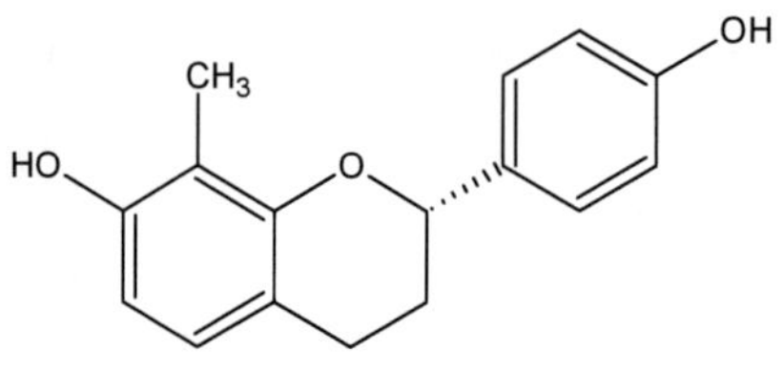

116
(-)-7,4'-Dihydroxy-8-methylflavan

Found in *Narcissus pseudonarcisus* L. (Amaryllidaceae) and *Pancratium littorale* (Amaryllidaceae). The radical scavenging activity of (-)-7,4-dihydroxy-8-methylflavan was found to be weaker than that of quercetin, known as a very efficient antioxidant agent, but equipotent to that of BHT, a synthetic antioxidant widely used in food technology (Ioset, et al., 2001).

(-)-7,4'-Dihydroxy-8-methylflavan: $C_{16}H_{16}0_3$; White amorphous powder; UV: λmax (MeOH): 229, 278 nm; $[\alpha]_D$ -49.2 ($CHCl_3$; c 0.12): TSP-MS *m/z* (rel. int.):257 $[M + H]^+$ (100), 200 (62), 183 (72); DCI-MS *m/z* (rel. int.): 257 $[M + H]^+$ (38), 211 (18), 207 (100); ^{1}H-NMR (500 MHz. $CDCl_3$): δ 7.30 (2H, dd, J = 6.8, 1.9 Hz,H-2', H-6'), 6.85 (2H, dd, J = 6.4, 2.0 Hz, H-3',H-5'), 6.79 (1H, d, J= 7.8 Hz. H-5), 6.38(lH, d, J = 8.3 Hz, H-6), 5.00 (lH, dd, J=

10.3, 2.5 Hz, H-2), 2.89 (lH, m, H-4a), 2.73 (lH, m, H-4b), 2.14 (lH, m, H-3a), 2.12 (3H, s, CH,-8), 1.99 (lH, m, H-3b); 13C-NMR (125 MHz, $CDCl_3$): δ 155.0 (C-4'), 153.6 (C-9), 152.6 (C-7), 134.5 (C-1′), 127.2 (C-2', C-6'), 126.5 (C-5), 115.2 (C-3', C-5'), 113.9 (C-10), 111.4 (C-8), 107.2 (C-6), 77.2 (C-2), 30.0 (C-3), 24.8 (C-4), 8.1 (CH_3-8).

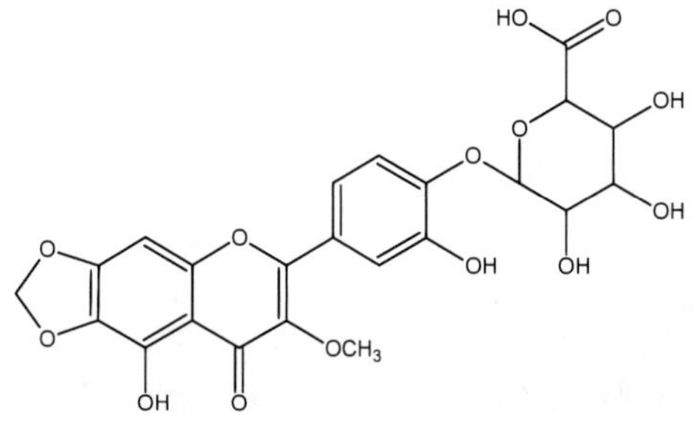

117
6-(3,4-Dihydroxy-phenyl)-9-hydroxy-7-methoxy-(1,3]dioxolo[4,5-g]chromen-8-one 4'-β-glucuronid

This glucurinated flavonoid (GF) was isolated from spinach (*Spinaceae oleracea* L., Chenopodiaceae). The free radical scavenging properties was found to be superior to vitamin E homologue Trolox in inhibiting reactive oxygen species (ROS) formation in the auto oxidation system of linoleic acid and in fibroblasts exposed to metal oxidation. Also was found to inhibit the ESR signal intensity of DMPO-Oc radical formation during the riboflavin photodynamic reaction. 10 mM GF caused approximately 90% inhibition in the intensity of the electron spin resonance (ESR) signal. Using the Fenton reaction. GF was found to inhibit DMPO-OH radical formation. A concentration of 2 mM GF caused a 70 % inhibition in the intensity of the DMPO-OH radical ESR signal. The natural antioxidants derived from spinach may directly affect the scavenging of ROS and, as a consequence may be considered as effective sources for combating oxidative damage (Bergman et al., 2003).

6-(3,4-Dihydroxy-phenyl)-9-hydroxy-7-methoxy-(1,3]dioxolo[4,5-g]chromen-8-one 4'-β-glucuronid: ^{1}HNMR: δ 3.81 (s, 3-OMe), 6.10 (OCH_2O), 6.68 (s, H-8), 7.64 (d, J = 2.2 Hz, H-2′), 7.25(d, J = 8.8 Hz, H-5′), 7.59 (dd, J = 8.8, 2.2 Hz, H-6′), 5.03 (d, J = 7.7 Hz, H-1′′), 3.60 (dd, J = 9.3, 7.7 Hz, H-2′′), 3.55 (t, J = 9.1 Hz, H-3′′),3.64 (dd, J = 9.6, 8.9 Hz, H-4′′), 4.03 (d, J = 9.6 Hz, H-5′′), 4.03 (d, J = 9.6 Hz, H-6′′); ^{13}C NMR: δ 157.4 (C-2), 140.2(C-3), 60.6 (3-OMe), 180.6 (C-4), 108.9 (C-4a), 142.4 (C-5), 131.2 (C-6), 104.3 (OCH_2O), 156.1 (C-7), 90.4 (C-8), 153.9 (C-8a), 126.6 (C-1′), 117.3 (C-2′), 148.3 (C-3′), 148.9 (C-4′), 117.8 (C-5′), 121.9 (C-6′), 103.5 (C-1′′), 74.5 (C-2′′), 76.9 (C-3′′), 73.0 (C-4′′), 76.7 (C-5′′), 172.3 (C-6′′), (Bergmana et al., 2001).

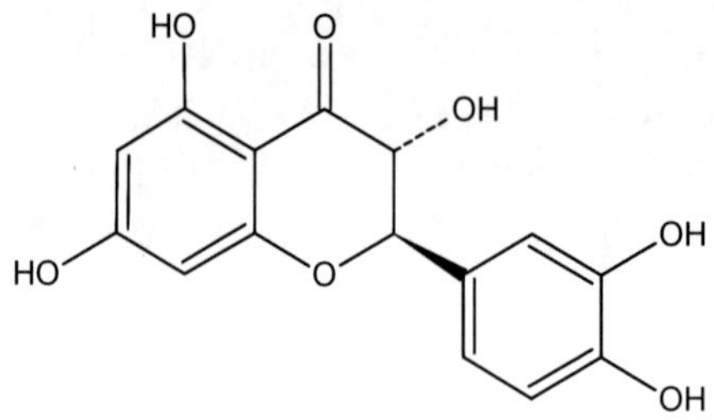

118
Dihydroquercetin

Dihydroquercetin (DHQ) also named taxifolin was found in the larch wood *Larix dahurica* and *L. gmelinni* (Pinaceae), which possesses high antioxidant activity toether with stability against autoxidatio, exhibits protection of capillaries, reduces oedema, hepatoprotective (antitoxic activity) as well as hypolipidaemic and diuretic properties (Kolhir et al., 1998). In model in vitro experiments DHQ repressed the formation of lipid peroxides which resulted in diminution of the amount of DHQ by 38%-61% relative to control values. Also showed dose dependent inhibition of the processes of lipid peroxidation in rat liver homogenate: however at a concentration of 100 mg/kg this effect was 10% smaller than that shown by DHQ. The evaluation of antioxidative and hepatoprotective properties of on a homogenate of livers from rats that had hepatitis elicited by tetracycline, showed levels of MDA 3.4 times higher than analogous levels in livers of healthy animals. Under the influence of DHQ (100 mg/kg). the level of MDA was lowered by 37%, compared with animals not treated by a protective agent (Kolhir et al., 1996).

2R,3R-Dihydroquercetin: UV (MeOH): λmax 207, 290 nm; CD (c = 0.010% in MeOH): $[\theta]_{333}$ + 1.3 x 10^4; $[\theta]_{296}$ -5.2 x 10^4; $[\theta]_{253}$ + 0.8 x 10^4; ESIMS (negative mode): pseudo molecular ion at *m/z* 303 $[M(304) -H^+]^-$, MS/MS of *m/z* 303 at *m/z* 285 $[M -H^+ -H_2O]^-$; ^{1}HNMR (300 MHz,CD_3OD): δ 4.51 (1H, d, J = 11.5 Hz, H3), 4.90 (1H, d, J = 11.5 Hz, H2), 5.88 (1H, d, J = 2 Hz, H8), 5.92 (1H, d, J = 2 Hz, H6), 6.80 (1H, d, J = 8 Hz, H5'), 6.84 (1H, dd, J = 8, 2 Hz, H6'), 6.94 (1H, d, J = 2 Hz, H2'). ^{13}CNMR (75.5 MHz, CD_3OD): δ 73.7 (C3), 85.1 (C2), 96.3 (C6), 97.3 (C8), 102.0 (C10), 115.9 (C2'), 116.0 (C5'), 120.9 (C6'), 129.9 (C1 '), 145.7 (C3'), 146.7 (C4'), 164.9 (C9), 165.7 (C5), 169.9 (C7), 198.1 (C4).

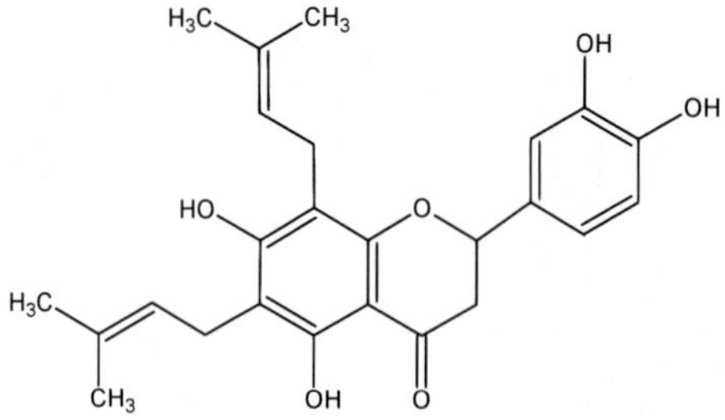

119
6,8-Diprenyleriodictyol

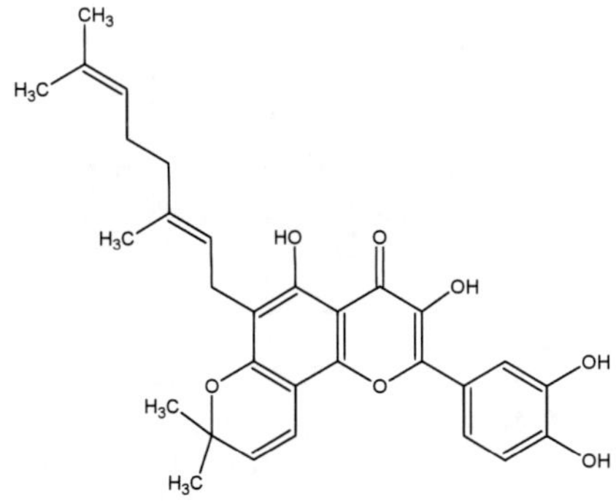

120
Dorsmanin C

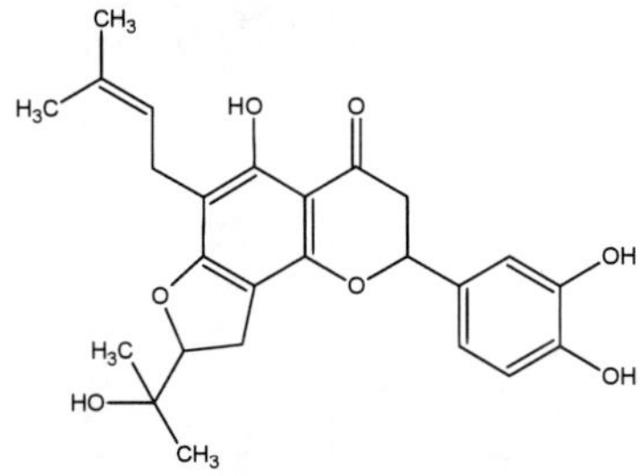

121
Dorsmanin F

These compounds were isolated from *Dorstenia mannii* (Moraceae). The prenylated flavonoids were found to be potent scavengers of the stable free radicall,l-diphenyl-2-picrylhydrazyl (DPPH), and are more potent than butylated hydroxy toluene (BHT), a common antioxidant used as a food additive. The prenylated flavonoids also inhibited CU^{2+}-mediated oxidation of human low density lipoprotein (LDL). Dose-response studies indicated that the prenylated flavonoids were effective inhibitors of lipoprotein oxidation with IC_{50} values < 1 μM and had similar inhibitory potency compared to quercetin, but was not directly related to Cu binding. Unlike quercetin, they did not show any pro-oxidant activity at high doses in the CU^{27+}-mediated lipoprotein oxidation system (Dufalla et al., 2003).

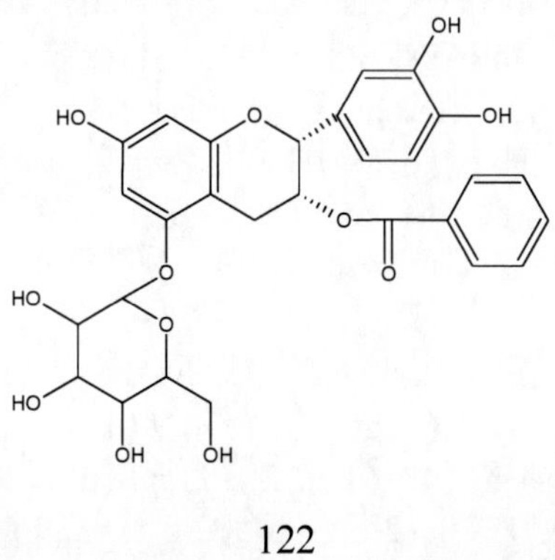

122
(-)Epicatechin-5-O-β-D-glucosyl-3-benzoate

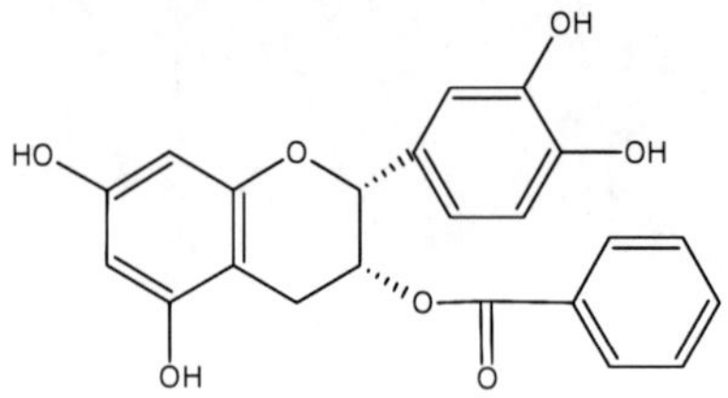

123
(-)-Epicatechin-3-benzoate

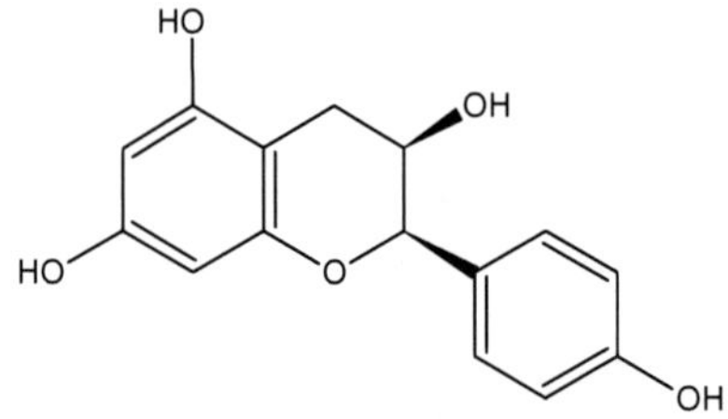

124
(-)Epiafzelechin

These compounds were isolated from an EtOAc extract of *Celastrus orbiculatus* Thunb. (Celastraceae) aerial parts that exhibited significant antioxidant effect in a 1,1-diphenyl-2-picrylhydrazyl (DPPH) free radical assay. (-)-Epicatechin-5-O-β-D-glucosyl-3-benzoate, and (-)-epicatechin-3-benzoate with IC_{50} values of 25 and 17 μg/ mL, respectively, exhibited moderate activities as free-radical scavengers in the DPPH assay, in comparison with reference antioxidants such as 2(3)-tert-butyl-4-hydroxy-anisole (IC_{50} 4.5 μg/mL), vitamin E (IC_{50} 6.2 μg/mL), and (-)-epigallocatechin-3-O-gallate (IC_{50} 5.6 μg/mL). (-)Epiafzelechin with IC_{50} value of 7.5 μg/ mL, exhibited more potent antioxidant activity than other compounds (Hwang et al., 2001).

(-)-Epicatechin-5-O-β-D-glucosyl-3-benzoate: Bright pink needles from 1:1 mixture of CH_2Cl_2 and MeOH; $FeCl_3$, dark-blue; mp 191-192 ^oC; $[\alpha]_D$-95° (c 0.3, MeOH): CD MeOH Δε (λ, nm), -1.83 (275), -5.50 (236); UV (MeOH): λmax (log ε) 213 (4.67), 229 (4.47), 283 (3.78) nm; IR (KBr): νmax 3400, 1700, 1620, 1600, 1520, 1450, 1360, 1280 cm^{-1}, FABMS (positive-ion mode) *m/z* 557 $[M + H]^+$; HRFABMS (positive-ion mode) calcd for $C_{28}H_{29}O_{12}$ 557.1659, found 557.1653; ^{1}HNMR (DMSO-d_6, 300 MHz): δ 9.32 (lH, br s, OH-7), 7.79 (2H, br d, J= 7.5 Hz, H-2′′, H-6′′), 7.61 (lH, br t, J= 7.5 Hz, H-4''), 7.47 (2H, t, J = 7.5 Hz, H-3′′, H-5′′), 6.92 (lH, d, J = 1.9 Hz, H-2′), 6.73 (lH, dd, J = 8.2, 1.9 Hz, H-6′), 6.63 (lH, d, J = 8.2 Hz, H-5′), 6.17 (lH, d, J = 2.1 Hz, H-6), 6.06 (lH, d, J = 2.1 Hz, H-8), 5.51 (lH, m, H-3), 5.09 (lH,br s, H-2), 4.71 (lH, d, J = 6.9 Hz, anomeric H), 3.69 (lH, dd, J = 10.7, 5.3 Hz, glc H-6), 3.50 (lH, m, glc H-6), 3.10-3.30 (4H, glc H-2, -3, -4, -5), 2.99 (2H, m, H-4); ^{13}C NMR (DMSO-d_6, 300 MHz): δ 25.5 (t, C-4), 60.6 (t, glc C-6), 68.7 (d, C-3), 69.5 (d, glc C-4), 73.2 (d, glc C-2), 76.3 (d, glc C-3), 76.5 (d, glc C-5), 77.0 (d, C.2), 95.5 (d, C-6), 96.6 (d, C-8), 99.5 (s, C-10), 100.5 (d, glc C-1), 114.0 (d, C-2′), 115.1 (d, C-5'), 117.2 (d, C-6′), 128.8 (2d, C-3′′, C-5′′), 129.0 (s, C-1′), 129.1 (2d, C-2′′, C-6′′), 129.5 (s, C-1′′), 133.3 (d, C-4′′), 144.7 (s, C-3'), 144.8 (s, C-4′), 155.1 (s, C-7), 156.5 (s, C-5), 156.8 (s, C-9), 164.8 (s, COO).

(-)-Epicatechin-3-benzoate: Light brown needles from 1:1 mixture of CH_2Cl_2 and MeOH; $FeCl_3$, dark-blue; mp 168-170 °C; $[\alpha]_D$ -100° (c 0.6, MeOH): CD MeOH Δε (λ, nm), -0.88 (281), -5.69 (237); UV (MeOH): λmax (log ε) 212 (4.53), 230 (4.35), 284 (3.57) nm; IR (KBr) νmax 3400, 17l0, 1610, 1600, 1520,1450, 1360, 1280 cm^{-1}, ESIMS (negative-ion mode) *m/z* 393 $[M -H]^+$; ^{1}HNMR (DMSO-d_6, 300 MHz): δ 7. 79 (2H, br d, J = 7.5 Hz, H-2′′, H-6′′), 7.61 (lH, br t, J = 7.5 Hz, H-4′′), 7.47 (2H, t, J = 7.5 Hz, H-3′′, H-5′′), 6.92 (lH, d, J = 1.8 Hz, H-2′), 6.73 (lH, dd, J = 8.1, 1.8 Hz, H-6′), 6.64 (lH, d, J = 8.1 Hz, H-5′), 5.93 (lH, d, J = 2.4 Hz, H-6), 5.85 (lH, d, J = 2.4 Hz, H-8), 5.44 (lH, m, H-3), 5.10 (lH, br s, H-2), 2.99 (2H, dd, J = 16.8,4.5 Hz, H-4α.), 2.76 (lH, dd, J = 16.8, 2.0 Hz, H-4β); ^{13}C NMR (DMSO-d_6, 300 MHz): δ 25.5 (t, C-4), 69.2 (d, C-3), 76.2 (d, C-2), 94.2 (d, C-6), 95.6 (d, C-8), 97.0 (s, C-10), 114.0 (d, C-2′), 115.1 (d, C-5′), 117.2 (d, C-6′), 128.8 (dd, C-3′′, C-5′′), 129.0 (s, C-1′), 129.2 (dd, C-2′′, C-6′′),129.6 (s, C-1′′), 133.3 (d, C-4′′), 144.8 (s, C-3′), 144.9 (s, C-4′), 155.5 (s, C-7), 156.5 (s, C-5), 156.9 (s, C-9), 165.0 (s, COO).

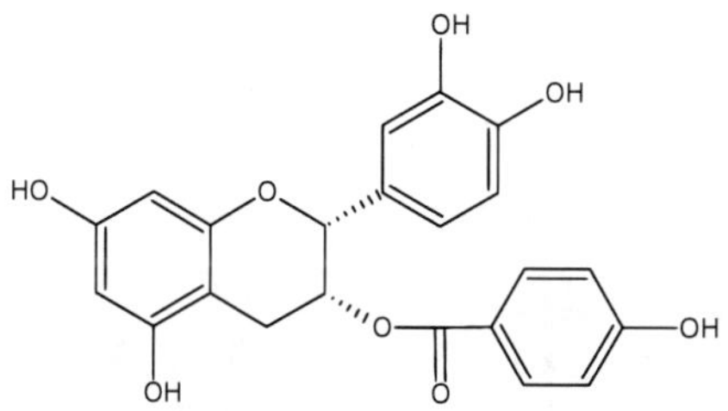

125
(-)-Epicatechin 3-O-*p*-hydroxybenzoate

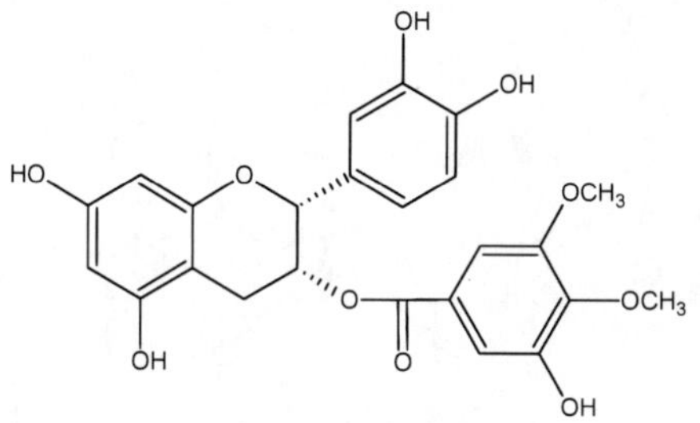

126
(-)-Epicatechin 3-O-(3,4-di-O-methyl)gallate

Two catechins were isolated from ethanol extracts of buckwheat (*Fagopyrum esculentum* Moench, Polygonaceae) groats. The antioxidant activity of these catechins was higher than that of rutin (Watanabe, 1998).

(-)-Epicatechin 3-O-*p*-hydroxybenzoate: ^{1}HNMR: δ 5.05 (s, H-2), 5.36 (m, H-3), 2.67-2.98 (m, H-4), 5.92 (d, J = 2 Hz, H-6), 5.84 (d, J = 2 Hz, H-8), 6.90 (s, H-2′), 6.64 (d, J = 8 Hz, H-5′), 6.71 (d, J = 8 Hz, H-6′), 7.64 (d, J = 9 Hz, H-2′′), 6.78 (d, J = 9 Hz, H-3′′), 6.78 (d, J = 9 Hz, H-5′′), 7.64 (d, J = 9 Hz, H6′′); ^{13}CNMR δ 77.4 (C-2), 69.5 (C-3), 25.9 (C-4), 156.5 (C-5), 96.1 (C-6), 156.5 (C-7), 95.2 (C-8), 156.1 (C-9), 98.5 (C-10), 130.5 (C-1′),

114.5 (C-2´), 144.6 (C-3´), 144.7 (C-4´), 115.6 (C-5´), 118.6 (C-6´), 166.6 (COO), 121.2 (C-1´´), 132.1 (C-2´´), 115.7 (C-3´´), 162.0 (C-4´´), 115.7 (C-5´´), 132.1 (C-6´´).

(-)-Epicatechin 3-O-(3,4-di-O-methyl)gallate: ^{1}HNMR: δ 5.08 (s, H-2), 5.37 (m, H-3), 2.72-3.00 (m, H-4), 5.85 (d, J = 2 Hz, H-6), 5.94 (d, J = 2 Hz, H-8), 6.91 (s, H-2´), 6.67 (d, J = 8 Hz, H-5´), 6.75 (d, J = 8 Hz, H-6´), 6.87 (d, J = 2 Hz, H-2´´), 6.99 (d, J = 2 Hz, H-5´´), 3.7,0 (s, H-6´´), 3.75 (s, OMe); ^{13}CNMR: δ 77.3 (C-2), 70.3 (C-3), 25.7 (C-4), 156.6 (C-5), 96.1 (C-6), 156.6 (C-7), 95.2 (C-8), 156.1 (C-9), 98.4 (C-10), 130.5 (C-1´), 114.3 (C-2´), 144.8 (C-3´), 144.7 (C-4´), 115.7 (C-5´), 118.4 (C-6´), 166.3 (COO), 125.6 (C-1´´), 105.6 (C-2´´), 153.3 (C-3´´), 141.2 (C-4´´), 150.3 (C-5´´), 111.1 (C-6´´), 56.2 (OMe).

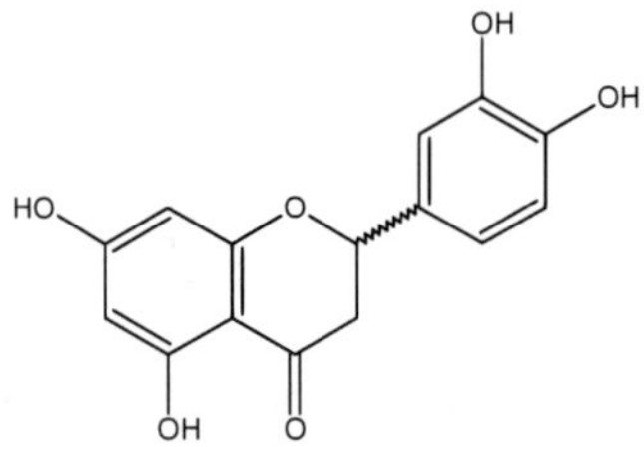

127
Eriodicytol

A flavonoid, eriodicytol was isolated as antioxidative components from the leaves of *Thymus vulgaris* (Labiatae). This compounds inhibited superoxide anion production in the xanthine/xanthine oxidase system. Mitochondrial and microsomal lipid peroxidation induced by Fe(I11)-ADP/NADH or Fe(IlI)-ADP/NADPH were also inhibited by these. This phenolic compound was shown to be effective to protect biological systems against various oxidative stresses (Haraguchi et al., 1996).

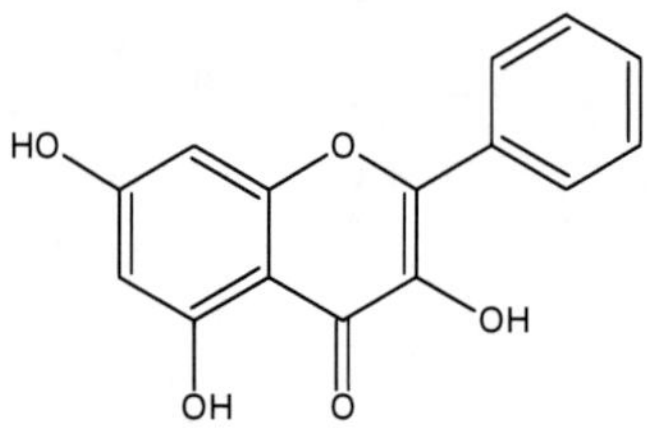

128
Galangin

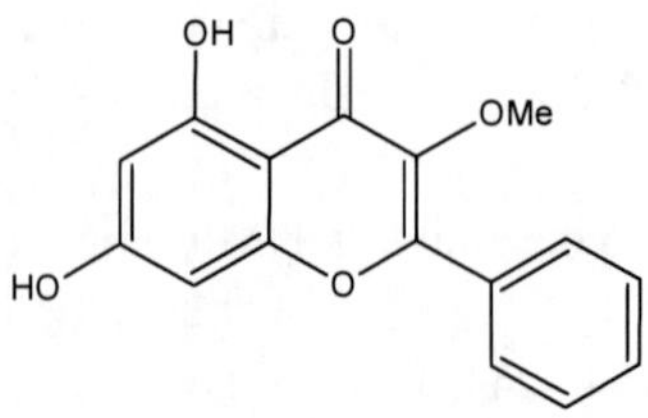

129
Galangin 3-O-methyl ether

Found in the root of *Alpinia officinarum* (Zingiberaceae). May be are two possible mechanisms by which to scavenge peroxynitrite: nitration and electron donation. Phenolic compounds can inhibit peroxynitrite-mediated nitration of tyrosine either by acting as alternative substrates for nitration (Choi et al., 2002).

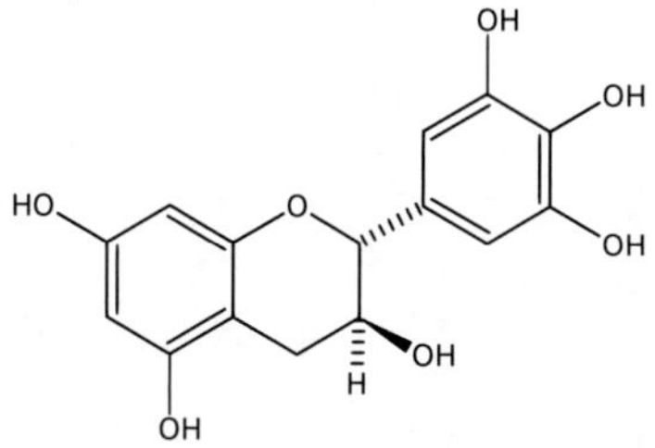

130
(+) Gallocatechin

Found in the fruit of *Chrysophyllum cainito* (Sapotaceae) together with nine known polyphenolic antioxidants, (+)-catechin, (-)-epicatechin, (+)-gallocatechin, (-)-epigallocatechin, quercetin, quercitrin, isoquercitrin, myricitrin, and gallic acid, have been identified from the fruits. All compounds showed the highest antioxidant activity (IC_{50} 40 ,μM) in the DPPH assay (Luo et al., 2002) .

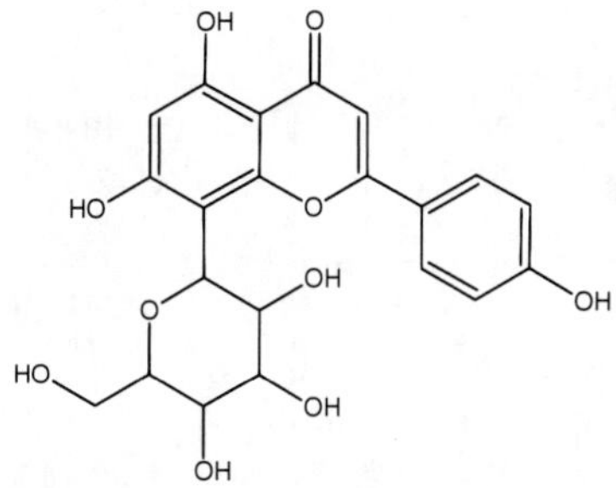

131
8-O-Glucosylvitexin

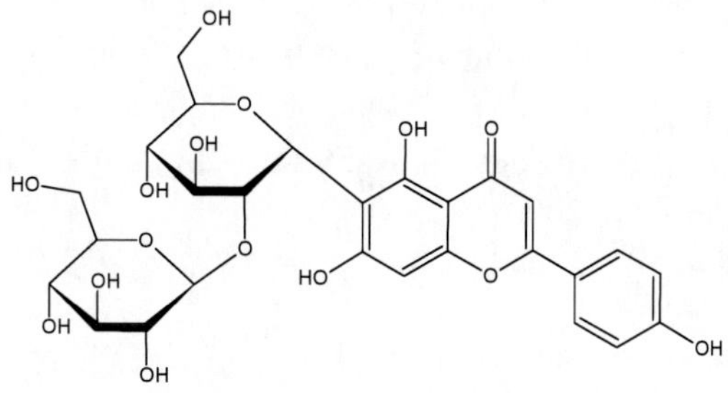

132
2"(3")-O-Glycosylisovitexin

One active component from Barley leaves (*Hordeum vulgare* L. Var. nudum Hook, Graminae) was identified as 2"(3")-O-glycosylisovitexin. Its antioxidative activity was almost equivalent to that of a-tocopherol in a lipid peroxidation system at the level of 100 μg/l.5 mg of ethyllinoleate (Osawa et al., 1992, Kitta et al., 1992). 2-O-Glucosylvitexin it is a major isoflavonoid antioxidant in young green barley leaves of *Hordeum vulgare* (Markham and Mitchell, 2003).

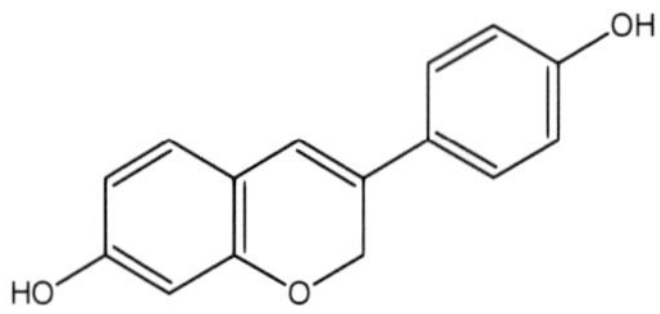

133
Haginin E

Found in the stems of *Lespedeza homoloba* (Leguminosae) showed antioxidant activity against lipid peroxidation in the rat brain homogenate test. 3,9-Dihydroxypterocarp-6a-en and lespedezol A_2 showed significant antiallergic activity in allergic (type I) mice (Miyasea et al., 1999).

Haginin E: Amorphous powder, UV: λmax nm (log ε) 245.5 (4.20), 333 (4.39). FAB-MS: *m/z* 240 $[M]^+$; ^{1}HNMR (upper moiety): δ 5.07 (d, J = 1.5 Hz, H-2), 6.74 (br,s, H-4), 6.94 (d, J = 8 Hz, H-5), 6.41 (dd, J = 8.0, 2.0 Hz, H-6), 6.33 (d, J = 2Hz, H-8), 7.36 (d, J = 8.5 Hz, H-2′), 6.87 (d, J = 8.5 Hz, H-3′), 6.87 (d, J = 8.5 Hz, H-5′), 7.36 (d, J = 8.5 Hz, H-6′); ^{13}C NMR: δ 67.7 (C-2), 128.9 (C-3), 118.3 (C-4), 116.5 (C-4a), 128.4 (C-5), 109.5 (C-6), 159.1 (C-7), 103.4 (C-8), 155.3 (C-8a), 129.3 (C-1′), 126.6 (C-2′), 116.4(C-3′), 158.1(C-4′), 116.4 (C-5′), 126.6 (C-6′).

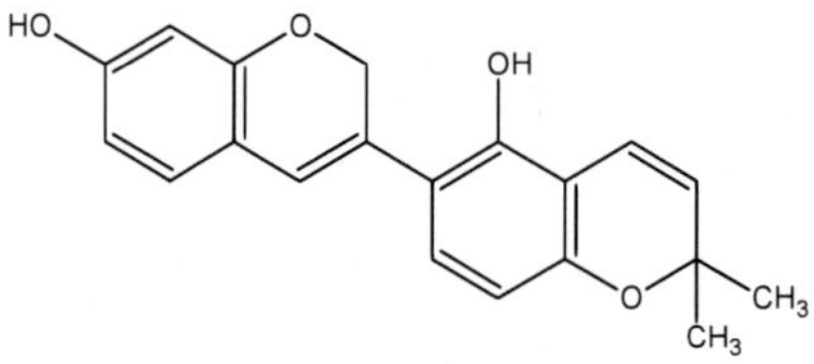

134
Isoflav-3-ene, glabrene

Found in *Glycyrrhiza glabra* (Leguminosae), have an antioxidant activity three time to that of vitamin E (Harbone et al., 2000).

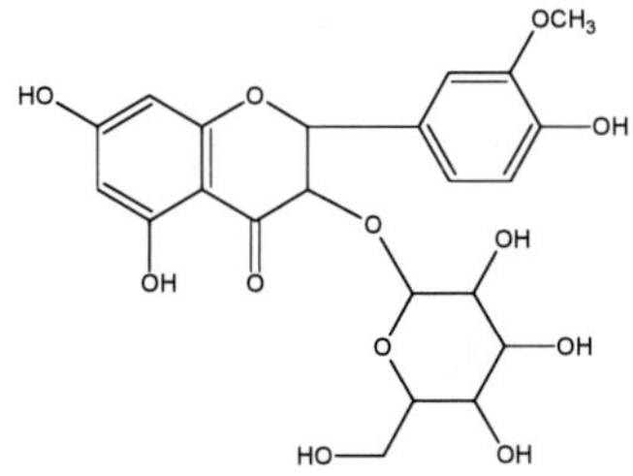

135
Isorhamnetin-3-O-β-D-glucopyranoside

Major antioxidative substance in leaves of Atsumi-kabu (Red Turnip, *Brassica campestris* L. Cruciferaceae) was Isorhamnetin-3-O-β-D-glucopyranoside (Igarashi et al., 1990).

Isorhamnetin-3-O-β-D-glucopyranoside: An amorphous yellow powder, FAB-MS *m/z* 479 $(M+H)^+$, 501 $(M+Na)^+$, 317 $(M+H\ -162)^+$; λnm: (MeOH) 253, 263 sh, 298 sh, 352; (MeOH- $AlC1_3$) 269, 299 sh, 363 sh, 398; (MeOH-$AlC1_3$-HCl) 266, 297sh, 352, 398; (MeOH-NaOAc) 268, 320, 362. Acid hydrolysis and enzymatic hydrolysis with β-glucosidase afforded glucose. ^{1}HNMR (200 MHz, DMSO-d_6): δ 3.85 (3H, s, OCH_3), 5.58 (1H, d, J = 7 Hz, anomeric proton of glucose), 6.22 (1H, d, J = 2Hz, 6-H), 6.48 (1H, d, J = 2Hz, 8-H), 6.99 (1H,d, J= 8 Hz, 5'-H), 7.50 (1H, dd, J = 2, 8 Hz, 6'-H), 7.96 (1 H, d, J = 2 Hz, 2'-H); ^{13}CNMR (50 MHz, DMSO-d_6): δ 178.0 (C-4),164.7 (C-7), 161.8 (C-5), 156.94 (C-2 or C-9), 156.87 (C-2 or C-9), 149.9 (C-3'), 147.4 (C-4'), 133.4 (C-3), 122.5 (C-6´), 121.5 (C-1´), 115.6 (C-5´), 113.9 (C-2´), 104.4 (C-10), 101.1 (C-1´´), 99.1 (C-6), 94.1 (C-8), 77.7 (C-5´´), 76.7 (C-3´´),74.6 (C-2´´), 70.1 (C-4´´), 60.8 (C-6"), 55.9 (OCH_3).

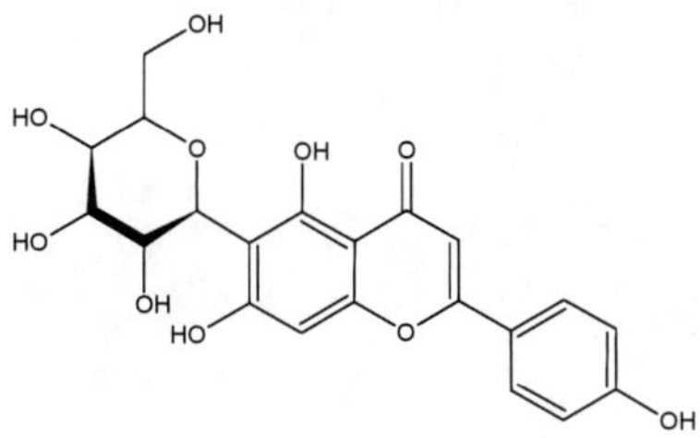

136
Isovitexin

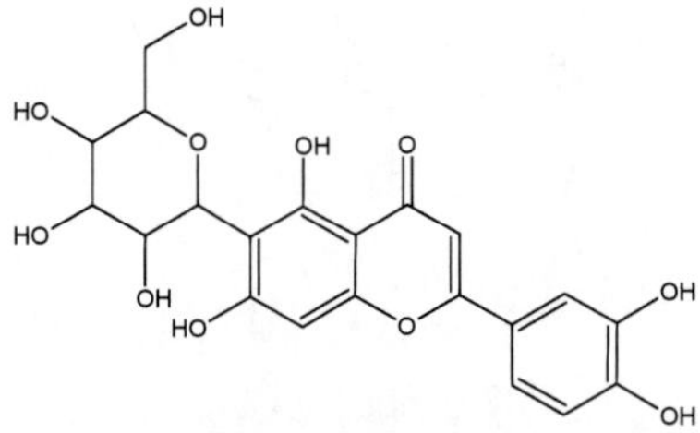

137
Isoorientin

C-glycosylflavones, especially isovitexin and isoorientin, were found to be antioxidative constituents of the fruits of *Boreava orientalis* (Cruciferae). The activity of isoorientin was stronger than that of isovitexin. (Sakushimao et al., 2000).

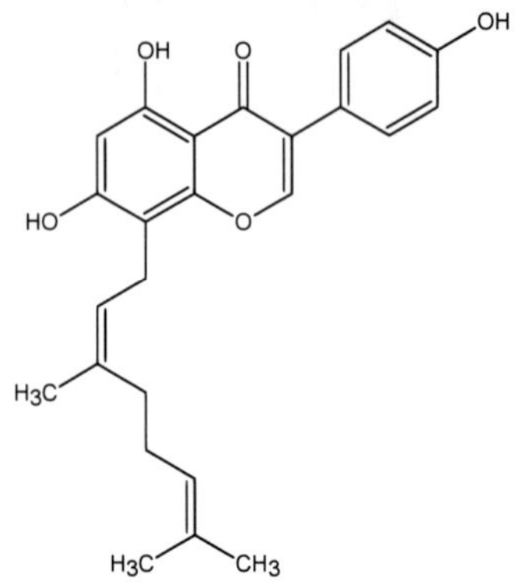

138
Lespedezol E_1

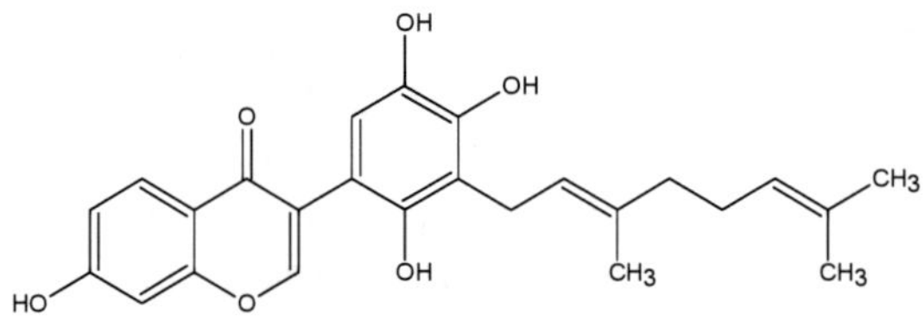

139
Lespedezol E_2

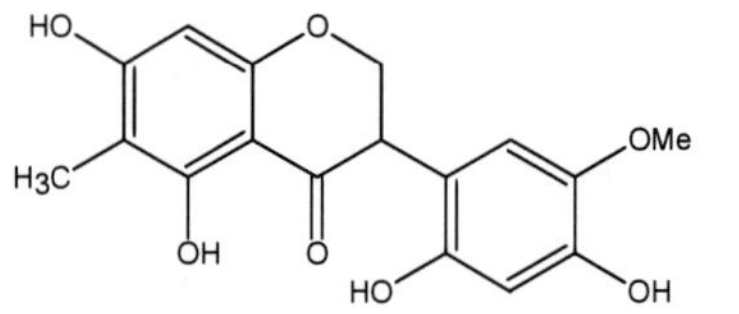

140
Lespedezol E

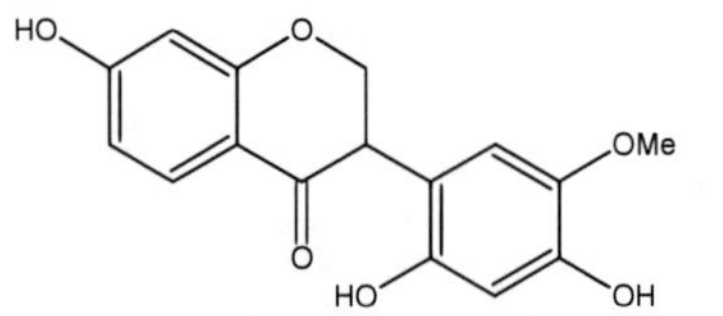

141
Lespedezol D

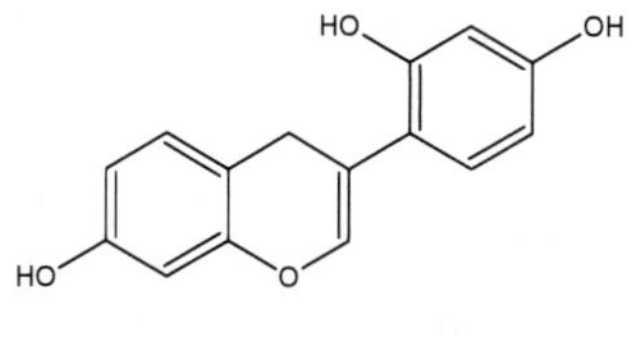

142

Lespedezol F_1

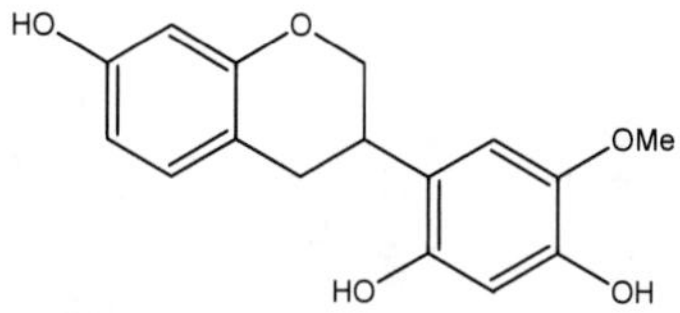

143

Lespedezol G_1

The stems of *Lespedeza homoloba* (Leguminosae) possess phenolic compounds. These compounds had strong antioxidative activity against lipid peroxidation in the rat brain homogenate test. 3,9-Dihydroxy-pterocarp-6a-en and lespedezol A_2 showed significant antiallergic activity in allergic (type I) mice (Miyasea et al., 1999, Miyase et al., 1999a).

Lespedezol E_1: Amorphous powder, λmax nm (log ε): 264.5 (4.30), 341 (4.01). FAB-MS *m/z:* 407 [M+H]; ^{1}HNMR: δ 8.24 (s, H-2), 6.37 (s, H-6), 7.47 (d, J = 8.5 Hz, H-2′), 6.90 (d, J = 8.5 Hz, H-3′), 7.47 (d, J = 8.5 Hz, H-5′) 6.90 (d, J = 8.5, Hz, H-6′); Side chain: 3.46 (br, d, J = 7 Hz, H-1), 5.27 (m, H-2), 1.98 (m, H-4), 2.03 (m, H-5), 5.05 (m, H-6), 1.58 (br, s, H-8), 1.54 (br, s, H-9), 1.82 (br, s, H-10); ^{13}C NMR: δ 154.3 (C-2), 123.2 (C-3), 182.0 (C-4), 106.3 (C-4a), 156.4 (C-5), 99.5 (C-6), 162.2(C-7), 107.3 (C-8), 161.5 (C-8a), 123.7 (C-1′), 131.1 (C-2′), 116.0 (C-3′), 158.4 (C-4′), 116.0 (C-5′), 131.1 (C-6′); Side chain: 22.0 (C-1), 123.1 (C-2), 135.7 (C-3), 40.4 (C-4), 27.3 (C-5), 125.1 (C-6), 131.6 (C-7), 25.8 (C-8), 17.7 (C-9), 16.2 (C-10).

Lespedezol E_2: Amorphous powder, λmax nm (log ε): 260 (4.29), 288 (4.26). FAB-MS m/z: 422 [M]+. 423 [M+H]; ^{1}HNMR: δ 8.21 (s, H-2), 8.15 (d, J = 9 Hz, H-5), 7.09 (dd, J = 9.0, 2.0 Hz, H-6), 6.99 (d, J = 2.0 Hz, H-8), 6.65 (s, H-6′); Side chain: 3.48 (br, d, J = 7 Hz, H-1), 5.37 (m, H-2), 1.97 (m, H-4), 2.08 (m, H-5), 5.11 (m, H-6), 1.62 (d, J = 1.0 Hz, H-8), 1.57 (br, s, H-9), 1.81 (d, J = 1.0 Hz, H-10); ^{13}C NMR: δ 156.3 (C-2), 111.7 (C-3), 179.0 (C-4), 117.3 (C-4a), 128.7 (C-5), 116.7 (C-6), 164.1 (C-7), 103.0 (C-8), 158.8 (C-8a), 125.6 (C-1′), 149.6 (C-2′), 119.1 (C-3′), 146.2 (C-4′), 138.9 (C-5′), 114.3 (C-6′); Side chain: 23.9 (C-1), 124.2 (C-2), 134.7 (C-3), 40.6 (C-4), 27.5 (C-5), 125.6 (C-6), 131.6 (C-7), 25.8 (C-8), 17.7 (C-9), 16.3 (C-10).

Lespedol D: Amorphous powder, λmax nm (log ε): 226 sh (4.23), 278 (4.13), 315 sh (3.92); $[\alpha]_D$ 0° (MeOH: c 0.71); CD no Cotton effect (MeOH; c 0.016); FAB-MS *m/z:* 302 $[M]^+$; ^{1}HNMR: δ 4.54 (dd, J = 11, 5.0 Hz, H-2), 4.16 (dd, J = 10, 5.0 Hz, H-3), 7.77 (d, J = 8.5 Hz, H-5), 6.58 (dd, J = 8.5, 2.0 Hz, H-6), 6.41 (d, J = 2.0 Hz, H-8), 6.46 (s, H-2′), 6.78 (s, H-6′), 3.71 (s, OMe); ^{13}C NMR: δ 71.7 (C-2), 47.8 (C-3), 191.6 (C-4), 115.5 (C-4a), 130.1 (C-5), 111.3 (C-6), 165.2 (C-7), 103.5 (C-8), 164.7 (C-8a), 113.5 (C-1′), 150.8 (C-2′), 104.6 (C-3′), 147.8 (C-4′), 141.9 (C-5′), 115.2 (C-6′), 57.4 (s, OMe), 7.1 (Me).

Lespedol E: Amorphous powder, λmax nm (log ε): 226.5 sh (4.27), 295 (4.26), 330 sh (3.85). $[\alpha]_D$ 0° (MeOH: c1.21). CD no Cotton effect (MeOH; c 0.0103); FAB-MS m/z: 333 $[M+H]^+$; ^{1}HNMR: δ 4.43 (dd, J = 10.5, 5.5 Hz, H-2), 4.25 (dd, J = 10.5, 5.5 Hz, H-3), 6.03 (br, s, H-8), 6.47 (s, H-3′), 6.79 (s, H-6′), 3.72 (s, OMe), 1.97 (br, s, Me), 12.62 (s, C-OH); ^{13}C NMR: δ 71.0 (C-2), 47.4 (C-3), 198.1 (C-4), 103.4 (C-4a), 162.8 (C-5), 103.9 (C-6), 164.8 (C-7), 195.2 (C-8), 162.0 (C-8a), 113.1 (C-1′), 150.8 (C-2′), 104.5 (C-3′), 148.0 (C-4′), 142.0 (C-5′), 115.0 (C-6′), 57.4 (s, OMe), 7.1 (Me).

Lespedezol F_I: Amorphous powder, λmax nm (log ε): 272 (3.85). FAB-MS *m/z*: 256 $[M]^+$. ^{1}HNMR: δ 6.87 (dd, J = 1.5, 1.5, Hz, , H-2), 3.61 (br, s, H-4), 6.94 (br, d, J = 8.0 Hz, H-5), 6.53 (dd, J = 8.0, 2.5 HZ, H-6), 6.38 (d, J = 2.5 Hz, H-8), 6.45 (d, J = 2.5 Hz, H-3′); 6.37 (dd, J = 8.0, 2.5 Hz, H-5′), 7.01 (d, J = 8.0 Hz, H-6′); ^{13}C NMR: δ 139.2 (C-2), 112.4 (C-3), 27.5 (C-4), 111.9 (C-4a), 130.6 (C-5), 111.6 (C-6), 157.6 (C-7), 103.4 (C-8), 152.2 (C-8a), 117.6 (C-1′), 156.9 (C-2′), 104.0 (C-3′), 158.4 (C-4′), 107.8 (C-5′), 130.2 (C-6′).

Lespedezol G_1 : Amorphous powder, λmax nm (log ε): 221 sh (4.13), 287 (3.89). $[\alpha]_D$ -16.7° (MeOH; c 0.98). $[\theta]_{233}$ -10,030, [ɵ]:188 + 1000 (MeOH; c 0.0402). FAB-MS *m/z*: 288 $[M]^+$. ^{1}H NMR: δ 4.00 (dd, J = 10, 10 Hz, H-2), 3.48 (m, H-3), 2.79 (ddd, J = 15.5, 5.0, 2.0 Hz, H-4), 6.89 (br, d, J = 8.0, 2.5 HZ, H-5), 6.36 (dd, J = 8.0, 2.5 Hz, H-6), 6.25 (d, J = 2.5 Hz, H-8); 6.48 (s, H-3′), 6.78 (s, H-6′), 3.74 (s, OMe); ^{13}C NMR: δ 70.6 (C-2), 32.9 (C-3), 31.3 (C-4), 114.4 (C-4a), 131.0 (C-5), 108.8 (C-6), 157.5 (C-7), 103.7 (C-8), 156.2 (C-8a), 118.8 (C-1′), 150.2 (C-2′), 104.3 (C-3′), 146.9 (C-4′), 142.0 (C-5′), 113.1 (C-6′), 57.6 (OMe).

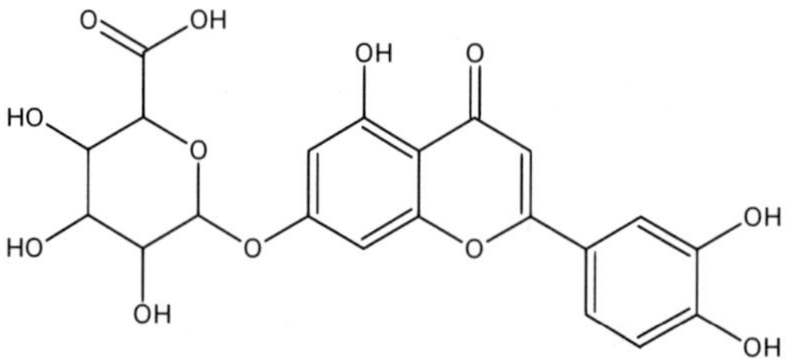

144
Luteolin 7-O-β-glucoronoside

Flavonoids luteolin-7-O-glucoside and luteolin7-O-β-glucoronoside occurring in *Vernonia amygdalina* (Compositae) leaves together with luteolin. The most abundant compound was luteolin 7-O-β-glucuronoside. The antioxidant activity of the flavones was determined by measuring the coupled oxidation of β-carotene and linoleic acid. It was shown that luteolin was a significantly more potent antioxidant than the synthetic antioxidant butylated hydroxytoluene (BHT) at the same, concentration (15 mg/L). The two glycosides showed similar activities but significantly lower activities than luteolin or BHT (Igile et al., 1994).

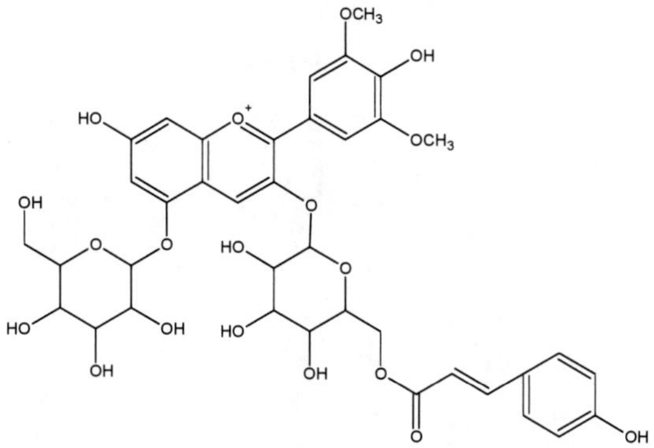

145
Malvidin 3-O-(6-O-*p*-coumaroylglucoside)- 5-glucoside

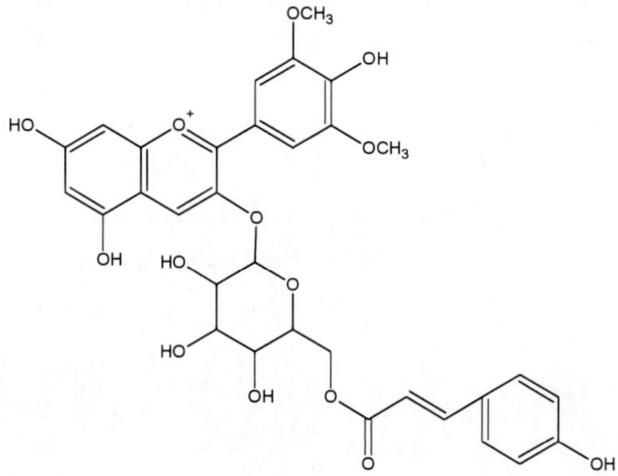

146
Malvidin 3-O-(6-O-*p*- coumaroyl)-glucoside

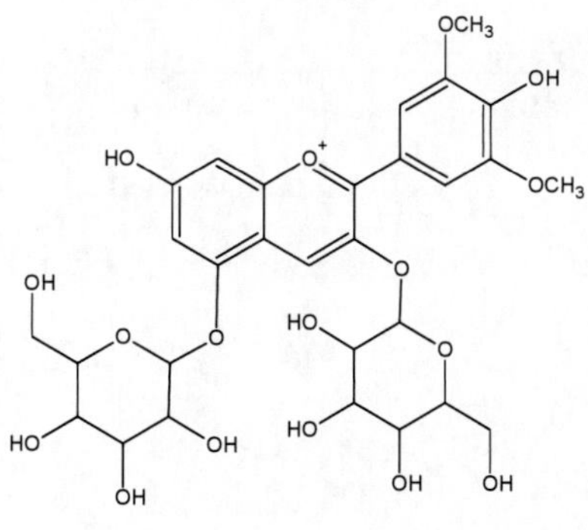

147
Malvidin 3,5-diglucoside

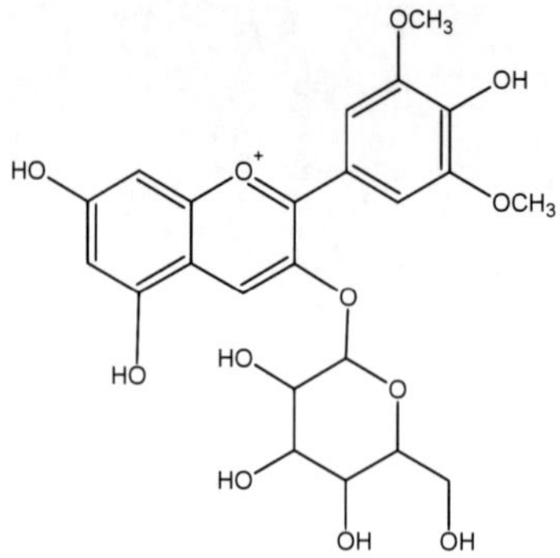

148
Malvidin 3-glucoside

These anthocyanins were isolated from in *Vitis* sp (Muscat Bailey A Grape, Vitaceae). Malvidin 3-O-(6-O-*p*-coumaroylglucosido)-5-glucoside showed only 2.6 nmol/mg of linoleic acid malonaldehyde production. *p*-Coumaric acid attached to malvidin 3-glucoside and malvidin 3,5-diglucoside was found to play an important role in preventing malonaldehyde production from linoleic acid. The antioxidative activity of an acylated anthocyanin was compared with other chemicals. Malvidin 3-O-(6-O-*p*-coumaroylglucosido)-5-glucoside proved to be 2 times more effective than commercially available antioxidants such as (+)-catechin and α-tocopherol (Tamura, and Yamagami, 1994).

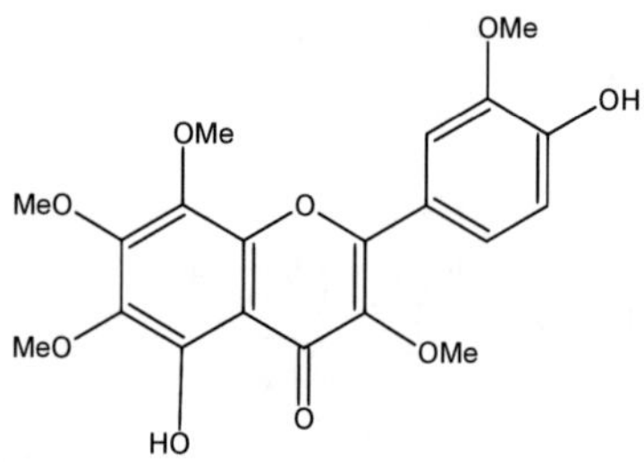

149
3'-Methoxycalicopterin

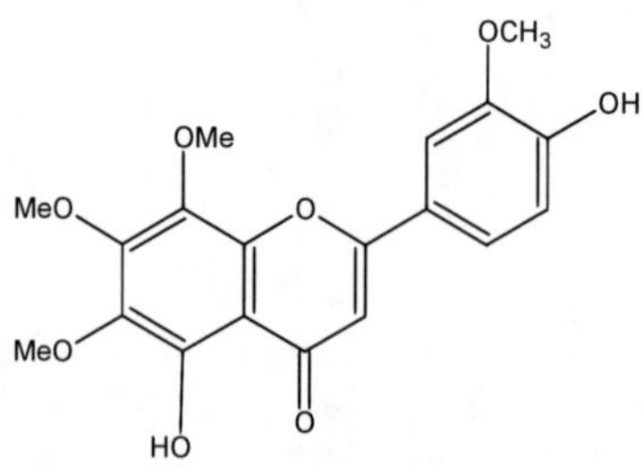

150
7'-Methylsudanchitin

The flavonoids were isolated from *Baccharis incarum* (Compositae). These flavonoids are considerably less active than the lignans or quercetin. Differences observed in their antioxidant capacities assessed from uminescence and TBAR determinations, indicate that visible luminescence is not an appropriate method for assessing he antioxidant capacities of flavonoids or additives absorbing at wavelengths close to those of emission of the excited tates involved (Faure et al., 1990).

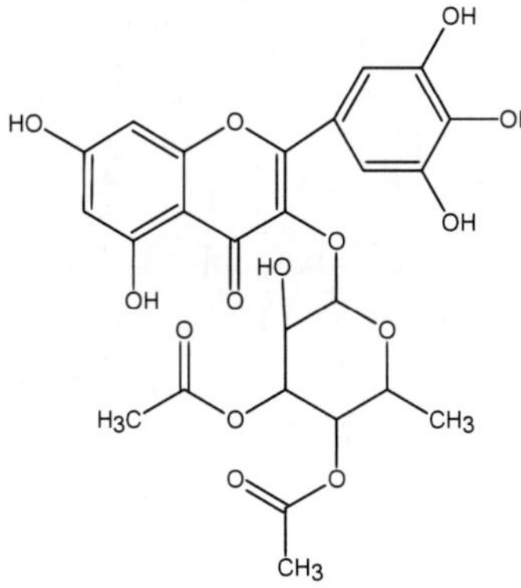

151
Myricetin 3-(3",4"-diacetylrhamnoside)

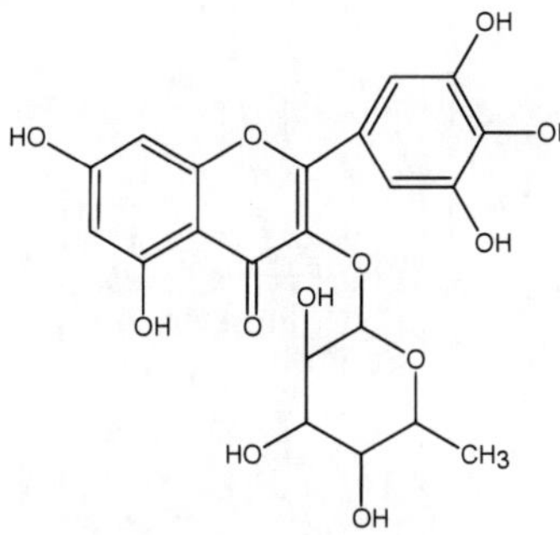

152
Myricetin 3-rhamnoside

Myricetin 3-(3",4"-diacetylrhamnoside), a novel flavonol glycoside together with myricetin 3-rhamnoside, myricetin 7-rhamnoside, myricetin 3-xyloside, myricetin 3-arabinoside, quercetin-3-galactoside, 3'-O-methylquercetin-3-glucoside, myricetin, quercetin, kaempferol and gallic acid have been isolated from a methanolic extract of *Myrsine africana* L. leaves (Manguro et al., 1996). These compounds showed antioxidant activity.

Myricetin 3-(3",4"-diacetylrhamnoside): UV: λmax, nm: 352, 302, 258; (+ $AlCl_3$) 438, 316, 270; (+$AlCl_3$/HCl) 400, 310, 268; (+ NaOMe) 396, 266, degenerate with time; (+ NaOAc) 376, 318, 272; (+ NaOAc/H_3BO_3) 374, 300, 260. IRνcm^{-1}: 3500-3200 (OH), 1720 (C=O. ester), 1650 (C=O, α, β-unsaturated); MS (70eV): *m/z* (%) 318 (100), 153 (40), 136 (25), 109 (8); ^{1}HNMR (CD_3OD): δ 6.80 (s. H-2'. 6'), 6.25 (d, J = 2. 3 Hz, H-8). 6.10 (d, J = 2.3Hz, H-6), 5.27 (d, J = 1.4Hz. H-1"). 5.05 (dd, J = 8.30, 3.8 Hz, H-3"), 4.90 (m, H-4"), 3.40

(m. H-5"), 1.94 (s. OAc-3"), 1.90 (s, OAc-4"), 0.70 (d, J = 6.30 Hz, CH3-6"); ^{13}C NMR (CD_3OD): δ 159.90 (C-2), 135.70 (C-3), 179.40 (C-4), 163.30 (C-5), 98.80 (C-6), 165.90 (C-7), 94.80 (C-8), 158.60 (C-9), 105.60 (C-10), 121.80 (C-1´), 109.40 (C-2´), 149.10 (C-3´), 137.10 (C-4´), 147.10 (C-5´), 109.40 (C-6 ´), 102.50 (C-1´´), 69.30 (C-2´´), 71.90 (C-3´´), 73.20 (C-4´´), 69.60 (C-5´´), 17.60 (C-6´´), 172.30 (OAc), 171.90 (OAc), 20.87 (OAc-Me), 20.77 (OAc-Me).

Myricetin 3-rhamnoside: Pale yellow powder. UV: λmax, nm: 358, 300, 258; (+ $AlC1_3$) 428, 312, 272; (+ $AlC1_3$/HC1) 404, 310, 270; (+NaOMe) 392, 324, 272; (+NaOAc) 366, 318, 270; (+NaOAc/H_3BO_3) 374, 300, 260; IR (nujol) νcm^{-1}: 3500-3000 (OH), 1660 (C=O, α, β-unsaturated); MS (70 eV): *m/z* (%) 318 (100), 244 (20), 153(75), 136 (42), 128; ^{1}HNMR (CD_3OD) + one drop DMSO-d_6): δ: 12.67 (s, D_2O exchangeable, 5-OH), 10.24 (s, D_2O exchangeable, 7-OH), 8.60 (br s, D_2O exchangeable, 3',5'-OH), 8.20 (s, D_2O exchangeable 4'-OH), 7.04 (s, H-2', 6'), 6.36 (d, J = 2.1 Hz, H-8), 6.24 (d, J = 2.1 Hz, H-6), 5.44 (d, J = 0.8 Hz, H-l "), 4.24 (m, H-2"), 3.85 (dd, J = 9.2 Hz, H-3"), 3.35 (dd, J = 9.3 Hz, H-4"), 3.50 (m, H-5"), 0.98 (d, J = 6.0 Hz, CH3-6"); ^{13}C NMR (DMSO-d_6): δ 157.80 (C-2), 134.56 (C-3), 178.15 (C-4), 161.65 (C-5), 98.88 (C-6), 164.53 (C-7), 93.73 (C-8), 156.75 (C-9), 104.28 (C-10), 119.89 (C-1´), 108.15 (C-2´), 146.08 (C-3´), 136.74 (C-4´), 146.08 (C-5´), 108.15 (C-6´), 102.17 (C-1´´), 70.17 (C-2´´), 70.65 (C-3´´), 11.44 (C-4´´), 70.71 (C-5´´), 17.56 (C-6´´).

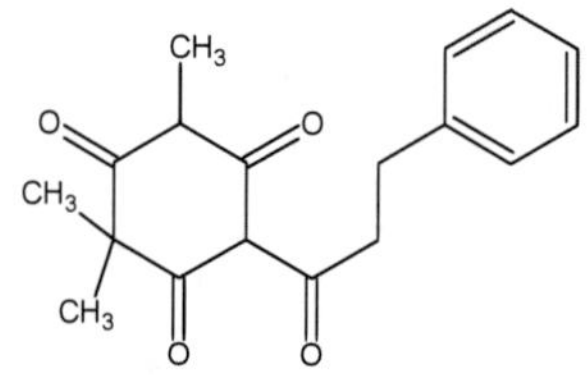

153
Myrigalon A

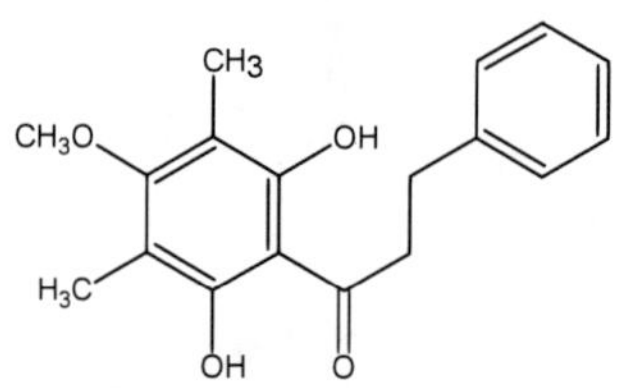

154
Myrigalon B

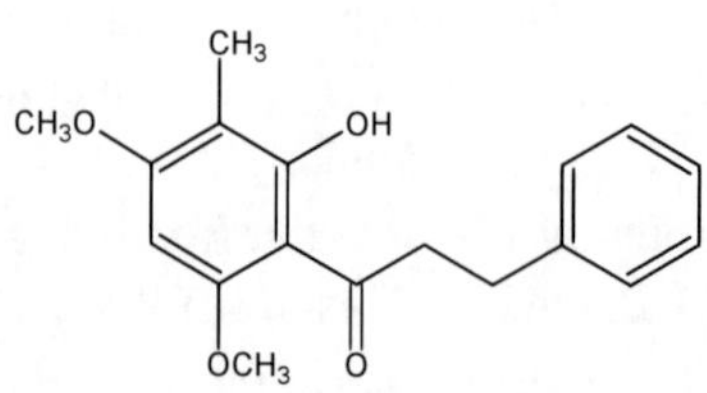

155
Myrigalon E

From diethyl ether extract of the fruit exudate of *Myrica gale* L. (Myricaceae) C-methylated dihydro-chalcones isolated from it. The main antioxidant of the extract, myrigalon B inhibited lipid peroxidation in hepatocytes with an IC_{50} value of 23 ± 1μM, whereas in mitochondria the value was 5.2 ± 1μM, Other myrigalones were considerably less active myrigalon A and E as antioxidants (Mathiesen et al., 1995).

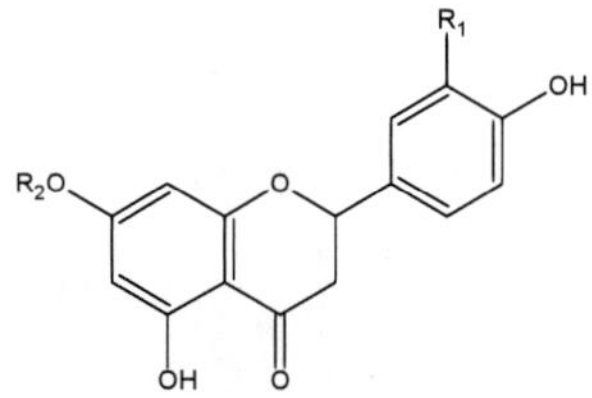

156
R_1 = OH, R_2 = neohesperidose Neoeriocitrin

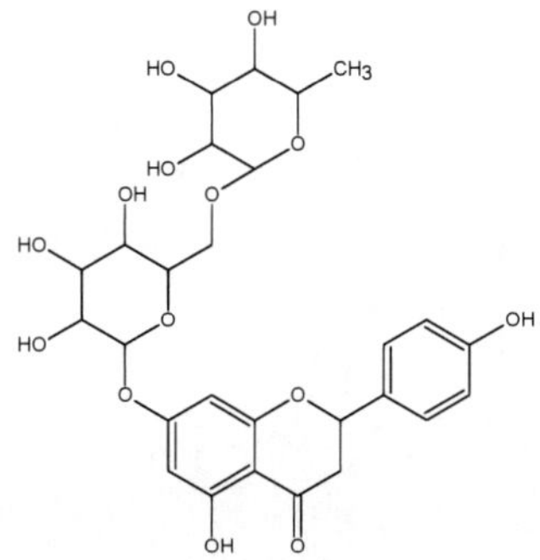

157
Narirutin

Together with eriocitrin, hesperetin, hesperedin, neohesperidin, naringenin and naringin were isolated from the peels and seeds of lemon (*Citrus limon* Femminello Comune), bergamot (*C. bergania* Fantastico), and sour orange (*C. aurantium,* unknown cultivar) and one sample of sweet orange (*C. sinensis* Biondo Commune) seed (Bocco et al., 1998). These compounds showed antioxidant effect.

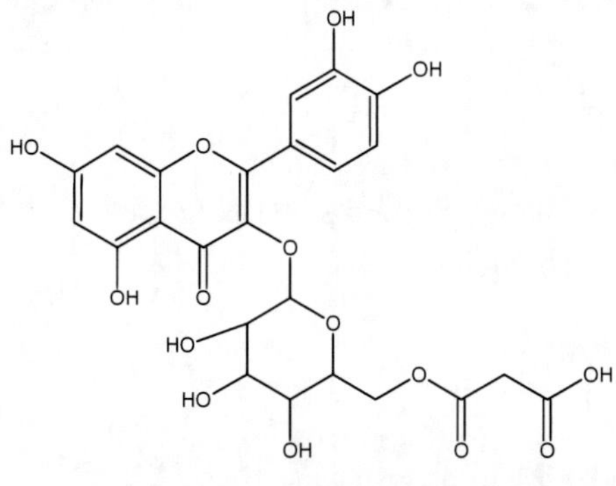

158
Quercetin-3-(6-malonylglucoside)

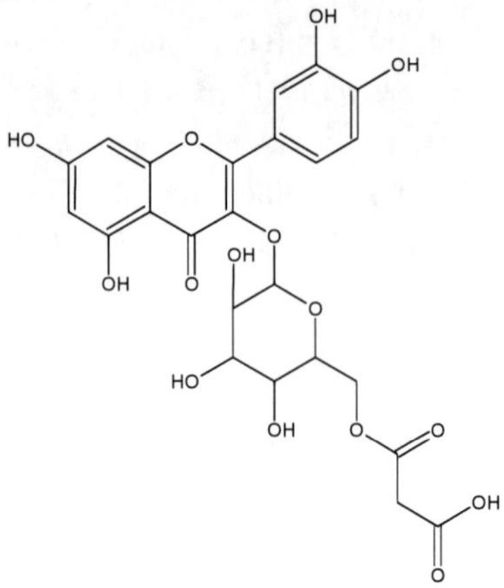

159
Quercetin-3-(6-malonylgalactoside)

Found in *Corchorus olitorius* L. (Tilaceae). Among the antioxidants isolated quercetin glycosides, 3-galactoside and 3-glucoside were slightly more active compared with 3-(6-malonylglucoside) containing 12% 3-(6-malonylgalacto- side), (Azuma et al., 1999).

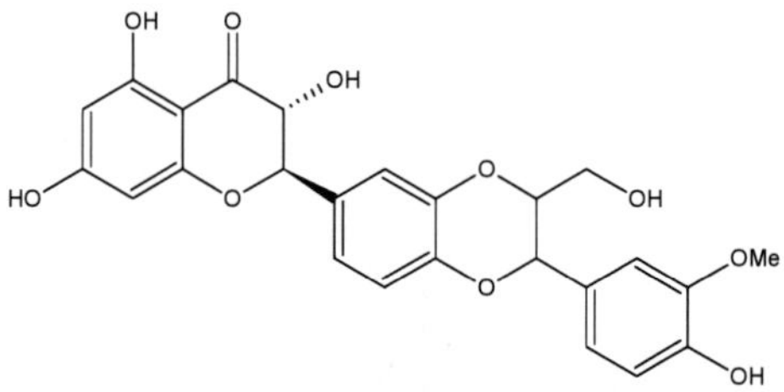

160
Silymarin

Found in Spanish peanuts *Arachis hypogea* (Leguminosae), shown to have high antioxidant activity. The complex flavonol-lignan derivative silymarin was effective in inhibiting lipid peroxidation in several different microsomal or mitochondrial preparations. Silymarin at 5 x 10^{-5} M inhibited 76% of the perchromate-initiated peroxidation of rat liver mitochondria (Larson 1988).

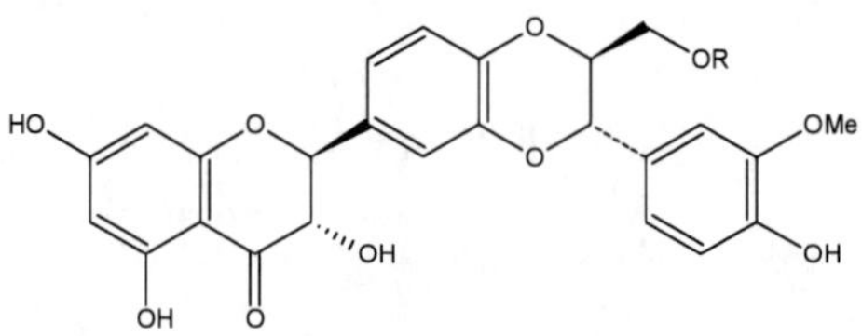

161
R=H
Silybin

162
R=β-D-galactosyl
Silybin galactoside

163
R=β-D-glucosyl
Silybin glucoside

164
R=β-D-lactosyl
Silybin lactoside

165
R=β-D-maltosyl
Silybin maltoside

These flavolignan were isolated from *Silybum marianum* L. (Compositae). In both DPPH and TEAC assays silybin glycosides were found to be significantly more efficient free radical scavengers than silybin. The test compounds showed weak scavenging activity towards the stable DPPH radical compared with β-tocopherol or ascorbic acid (Kosina et al., 2002).

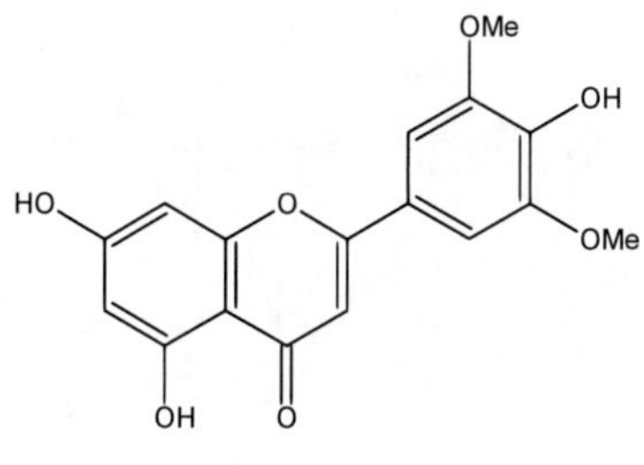

166
Tricin

Two flavonoids, were isolated from an ethanol extract of Japanese barnyard millet (*Echinochloa utilis*) Their structures were established to be luteolin, and tricin. Although the antioxidant activity of luteolin was lower it was nearly equal to that of quercetin, whereas the activity of tricin was lower than that of luteolin (Watanabe, 1999).

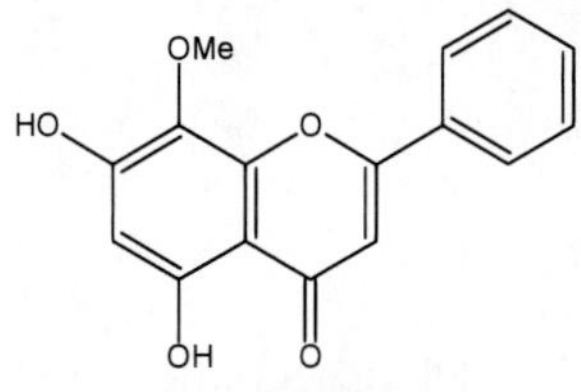

167
Wogonin

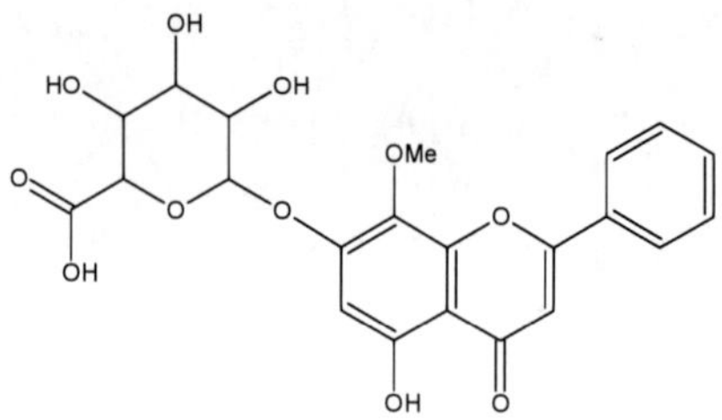

168
Wogonoside

The three major flavonoid components from *Sideritis javalambrensis* (Labiatae) wogonin (5,7-dihydroxy-8-methoxyflavone), baicalein (5,6,7-trihydroxyflavone) and its 7-glucuronide (baicalin) showed antioxidant activity. Baicalin was the most active compound with the highest (72%) inhibition of oxidation and represented 75% of the flavone fraction in the extract. Thus, the presence of a glucuronide moiety at C-7 seems to significantly increase antioxidant activity (Harbone et al., 2000). Wogonoside was isolated from *Scutellaria baicalensis* Georgi wogonin and wogonoside showed significant effects only on NADPH-induced lipid peroxidation (Gao et al., 1999).

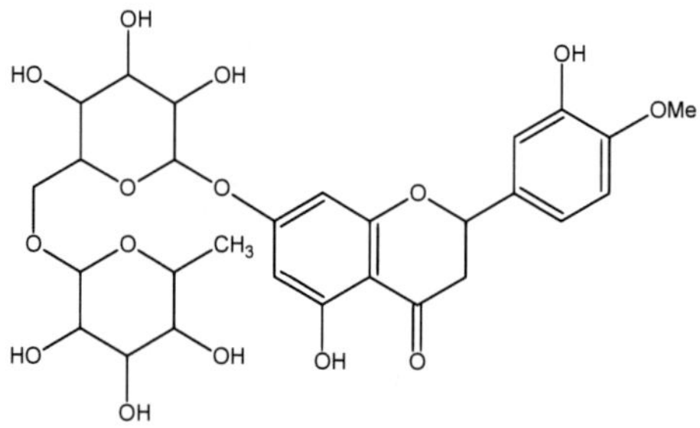

169
Hesperidin

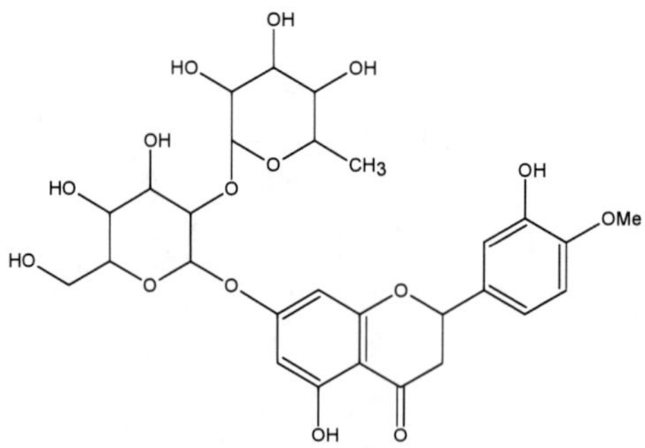

170
Neohesperidin

Hesperidin is isolated in large amounts from the discarded rinds of the ordinary orange *Citrus aurantium L. Citrus sinensis, C. uncia* and other species of the genus Citrus (family Rutaceae). The presence of hesperetin-7-β-neohesperidoside (neohesperidin) was recently reported in *C. humilis* and *C. cornigera* (Zhao et al., 1999). Hesperidin has been found to reduce superoxide in electron transfer plus concerted proton transfer reaction *in vitro*. This activity was also exhibited in liver homogenates for hydroperoxide-induced chemiluminescence (Fraga et al., 1987). Hesperidin was found to be inactive or only moderately active in comparison with other flavonoids and nonflavonoidal antioxidants (Garg et al., 2001), also found not to inhibit the liberation of reactive oxygen species from stimulated neutrophils (Limasset et al., 1993). Thus hesperidin appears not to be a particularly active antioxidant in comparison with most other flavonoids (Grag et al., 2001).

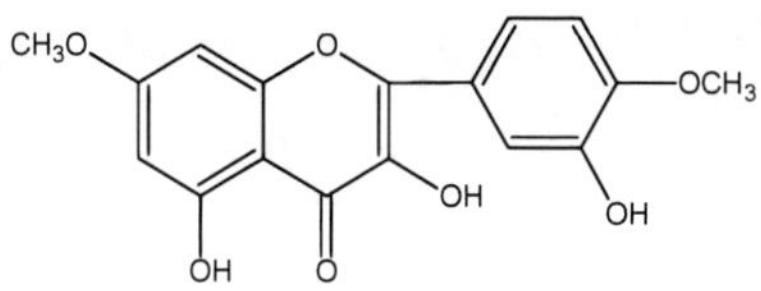

171
7,4′-Dimethylquercetin

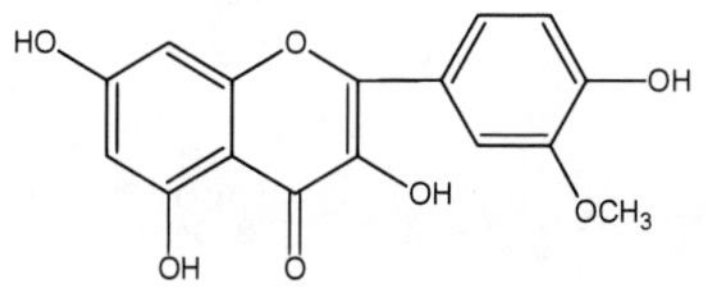

172
3′-Methylquercetin

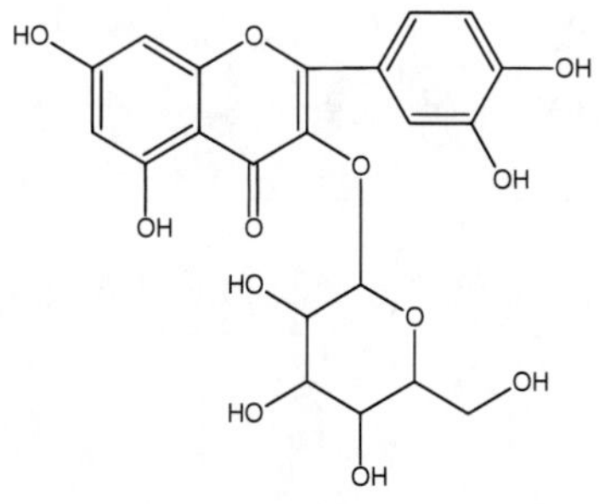

173
Isoquercetrin

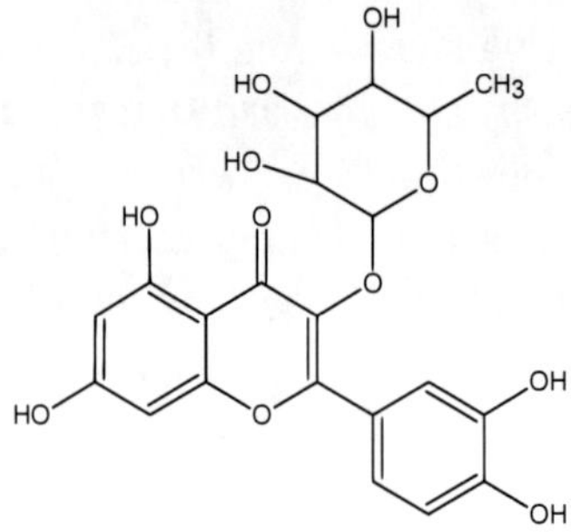

174
Quercitrin

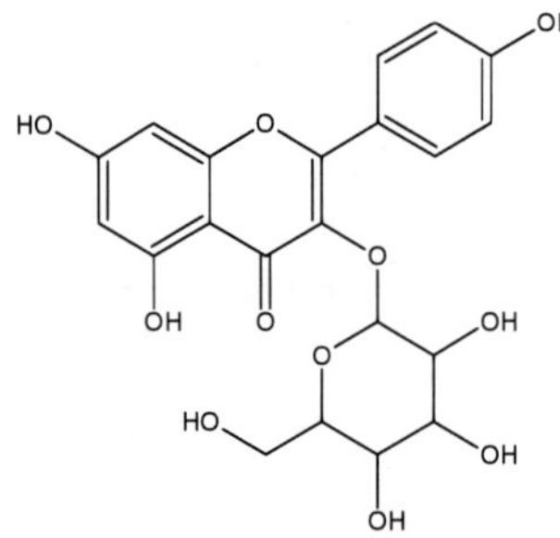

175
Kaempferol-3-glucoside

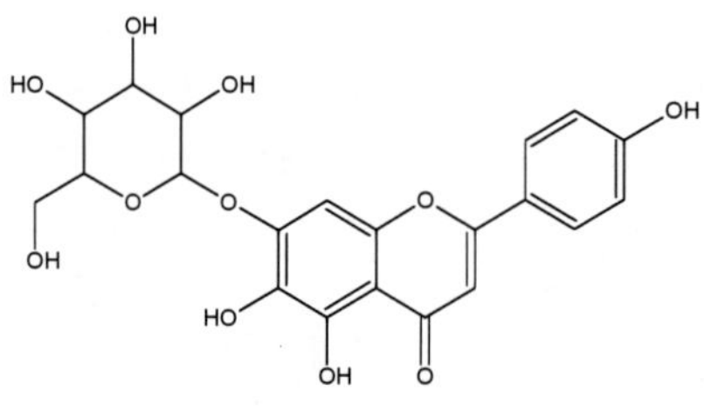

176
Scutillarein 7-O-β-D-glucopyranoside

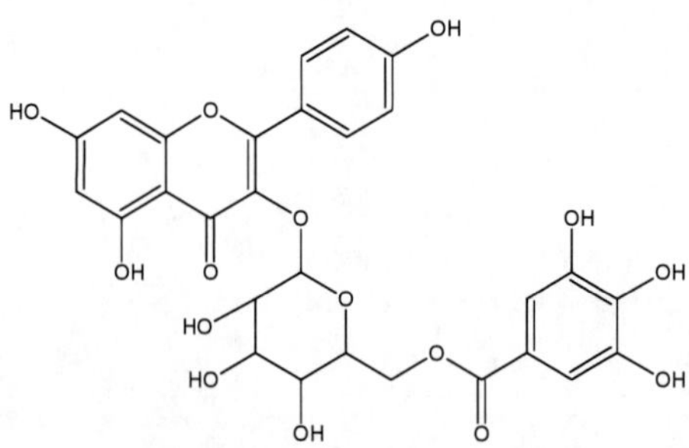

177
Galloyl kaempferol-3-glucoside

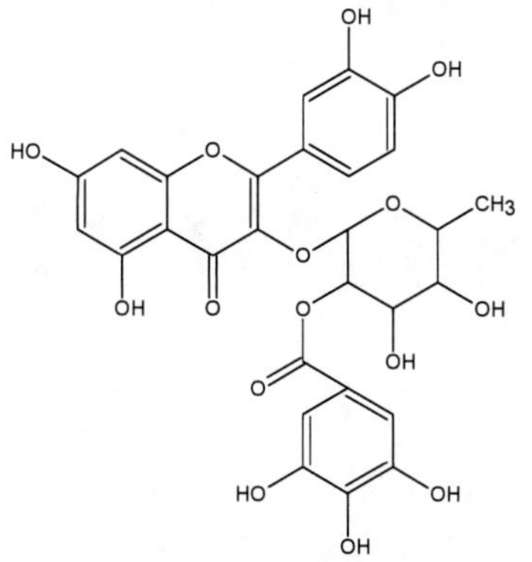

178
2′′-O-(3,4,5-Trihydroxybenzoyl)quercitrin

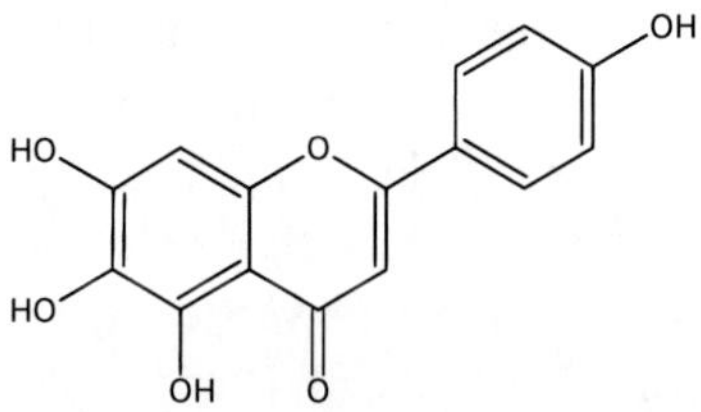

179
Scutillarein

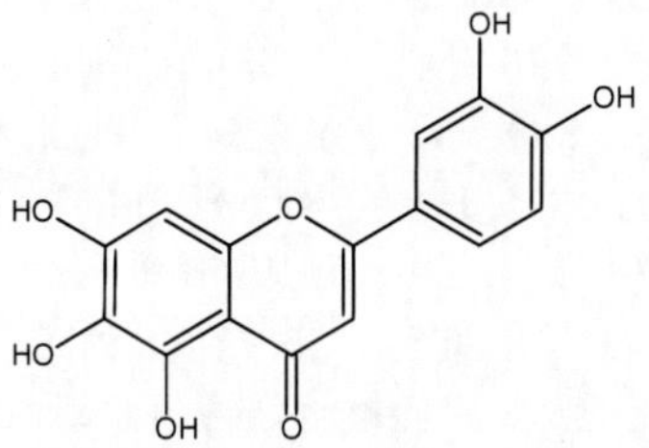

180
6-Hydroxyluteolin; 3',4',5,6,7-pentahydroxyflavone

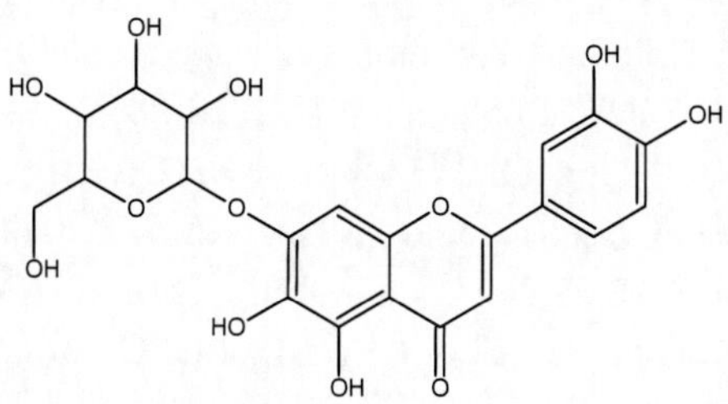

181
6-Hydroxyluteolin 7-O-β-D-glucopyranoside

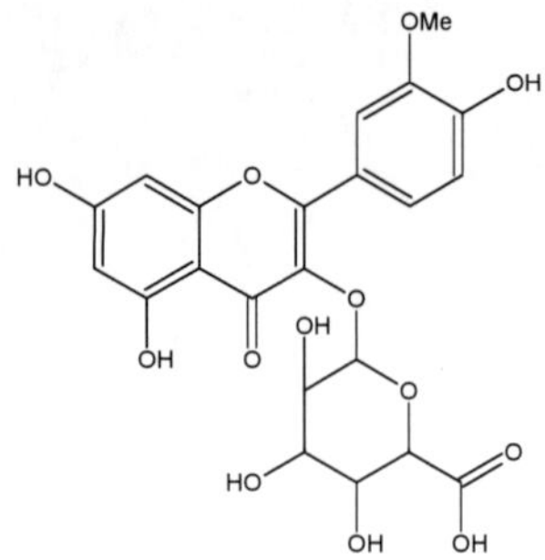

182
Quercetin 3-O-β-D-glucuronide

Antioxidative flavonoids were isolated from the leaves of *Polygonum hydropiper* (Polygonaceae) a culinary herb. The antioxidative activities measured by a ferric thiocyanate method were in the order isoquercetin > 7,4′-dimethylquercetin > quercetin > 3′-methylquercetin. (Haraguchi et al., 1992). Showed strong anti-oxidative activity when were measurement of the Trolox equivalent antioxidant capacity (TEAC) values, against ABTS (2.2'-azinobis(3-ethyl-benzo-thiazoline-6-sulphonic acid) radicals and phenyl-tert-butyl nitrone (PBN) azo initiator (AI). The most powerful of the antioxidants was 2"-O-(3,4,5-trihydroxybenzoyl) quercitrin (galloyl quercitrin), (Peng et al., 2003).

7,4′-Dimethylquercetin: Yellow needles; mp 236-240 °C; UV λmax (MeOH): 207.0 (log ε = 4.76), 254.0 (4.59), 265.0 (4.59), 265.0 (sh), 371.0 (4.60) nm; UV λmax (MeOH-$AlCl_3$): 205.4, 264.4, 430.2 nm; UV λmax (MeOH-NaOAc): 207.4, 254.6, 372.8 nm; UV λmax (MeOH-$AlCl_3$/HC1): 206.2, 263,6, 361.8, 429.9 nm; UV λmax (MeOH-H_3BO_3-NaOAc) 210.6, 254.6, 371.2 nm; MS *m/z* 330 (M^+, 100%),287 (M -CO-CH3, 43%),151, 149, 135; ^{1}HNMR (DMSO-d6): δ 3.86 (6H, s, OCH_3 X2), 6.34 (1 H, d, J = 2 Hz, C_6H), 6.78 (1 H, d, J = 2 Hz, C_8H), 6.90 (1 H, d, J = 8 Hz, C5, H), 7.78 (1 H, dd, J = 8,2 Hz, $C_{6'}H$), 7.79 (1 H,d, J= 2 Hz, $C_{2'}H$), 9.53 (1 H,s, OH), 9.76 (1 H, s, OH), 12.46 (1 H, s, C_5OH); ^{13}C NMR (DMSO-d6): δ 175.9 (s, C4), 164.9 (s, C7), 160.4 (s, C5), 156.9 (s, C9), 148.9 (s,C2), 147.4 (s, C3'), 147.0 (s, C4'), 136.1 (s, C3), 121.9 (d, C6'), 121.9 (s, C1'), 115.5 (d, C5'), 111.8 (d, C2'), 104.0 (s, C10), 97.4 (d, C6), 91.9 (d, C8), 56.0 (q, CH_3O), 55.9 (q, CH_3O).

3′-Methylquercetin: Yellow powder, mp 305 °C; UV λmax (MeOH): 204.4 (log ε = 4.61), 254.4 (4.26), 370.8 (3.68) nm; UV λmax (MeOH-$AlCl_3$): 206.4, 264.2, 361.0, 429.6 nm; UV λmax (MeOH-NaOAc) 202.0, 256.2, 273.2, 321.2, 380.4 nm; UV λmax (MeOH-$AlCl_3$/HC1): 206.4, 263.0, 357.8, 428.4 nm; UV λmax (MeOH- H_3BO_3-NaOAc) 213.2, 254.8, 373.4 nm; MS *m/z* 316 (M^+, 12.9%), 314 (M- 2,40.8%), 287 (M-CO, -H, 16.2%),135; ^{1}HNMR (DMSO-d_6): δ 3.18 (3 H, s, OH), 3.85 (3 H, s, CH_3O), 6.18(1 H, d, J = 2 Hz, C_6H), 6.46 (1 H, d, J = 2 Hz, C_8H), 6.95 (1 H, d, J = 8 Hz, $C_5{'}H$), 7.70 (1 H, dd, J = 8, 2 Hz, $C_{6'}H$), 7.75 (1 H, d, J = 2 Hz, $C_{2'}H$), 12.62 (1 H, s, C_5OH); ^{13}C NMR (DMSO-d_6): δ 175.4 (s, C4), 163.7 (s, C7), 160.2 (s, C5), 155.8 (s, C2 or C9), 148.4 (s, C2 or C9),147 .0 (s, C3'), 146.2 (s, C4'), 135.4 (s, C3), 121.7 (d, C6'), 121.4 (8, C-l'), 115.2 (d, C5'), 111.3 (d. C2'), 102.6 (s, C10), 97.9 (d, C6), 93.4 (d, C8), 55.6 (q,CH_3O).

Isoquercetin: Pale yellow needles; mp 242-243 °C; UV λmax (MeOH): 206.8 (log ε = 4.56), 256.0 (4.26), 359.6 (4.18) nm; UV λmax (MeOH-$AlC1_3l$): 209.4, 274.4, 427.4 nm; UV λmax (MeOH-NaOAc) 207.2, 257.8, 268.0, 368.4 nm; UV λmax (MeOH-AlCl3/HCl): 209.0, 268.8, 298.4, 366.2, 402.4 nm; UV λmax (MeOH- H_3BO_3-NaOAc): 213.8, 261.6, 300.4, 379.0 nm; ^{1}HNMR (DMSO-d_6): δ 3.0-3.6 (sugar CH), 5.37 (1 H, d, J = 7 Hz, anomeric H), 6.17 (1H, d, J = 2 Hz, C_6H), 6.35 (1 H, d, J = 2 Hz, C_8H), 6.80 (1 H, d, J = 8 Hz, $C_{5'}$H), 7.60 (1 H, d, J = 2 Hz, $C_{2'}$H), 7.80 (1 H, dd, J = 8, 2 Hz, $C_{6'}$H), 12.5 (1 H, s, C_5OH).

Quercitrin (3-O-α-L-rhamnopyranosyloxy-3',4',5,7-tetrahydroxyflavone): MS: Negative ES-MS *m/z* (rel. int.): 447 ([M-H]$^-$,(100); UV λmax 345.9 nm; ^{1}HNMR (CD_3OD): δ 7.38 (d, J= 2. 1 Hz, H-2'), 7.35 (dd, $J_{6'-5'}$ = 8.3 Hz, H-6'), 6.95 (d, H-5'), 6.41 (d, J_{8-6} = 2.2 Hz, H-8), 6.24 (d, H-6), 5.40 (d, $J_{1''-2''}$ = 1.7 Hz, H-1"), 4.26 (dd, $J_{2''-3''}$ = 3.4 Hz, H-2"), 3.79 (dd, $J_{3''-4''}$ = 9.3 Hz, H-3"), 3.46 (dq, $J_{4''-5''}$ = 9.6, $J_{5''-6''}$ = 6.0 Hz, H-5"), 3.37 (dd, H-4"), 0.98 (d, H-6").

Kaempfero-3-glucoside-(3-O-β-D-glucopyranosyloxy-4',5,7-trihydroflavone): MS: Negative ES-MS, *m/z* (rel. int.): 447 ([M-H]$^-$,100); UV λmax 267.0 and 356.2 nm; ^{1}HNMR (CD_3OD): δ 7.91 (dAA' , $J_{2'-3'}$ + $J_{2'-5'}$ = 8.9 Hz, H-2'/H-6'), 6.96 (d, BB', H-3'/H-5'), 6.43 (d, J_{8-6} = 2.1 Hz, H-8), 6.25 (d, H-6), 5.27 (d, $J_{1''-2''}$ = 7.6 Hz, H-1"), 4.01 (dd, $J_{6''A-5''}$ = 2.2, $J_{6''A-6''B}$ = 12.1 Hz, H-6"A), 3.77 (dd, $J_{6''B-5''}$ = 6.1 Hz, H-6"B), 3.65-3.44 (m, H-2"/H-3"/H-5"), 3.39 (dd, $J_{4''-3''}$ = 9.5 Hz, $J_{4''-5''}$ = 9.5 Hz, H-4").

Scutillarein 7-O-β-D-glucopyranoside: UV λmax 267.0 and 338.7 nm; Negative ES-MS, *m/z* (rel. int.): 447 ([M-H]$^-$,100); ^{1}HNMR (CD_3OD): δ 7.92 (dAA', $J_{2'-3'}$ + $J_{2'-5'}$ = 8.8 Hz, H-2' H-6'), 7.06 (s, H-8), 6.97 (d, BB', H-3' /H-5'), 6.68 (s, H-3), 5.50 (d, $J_{1''-2''}$ = 7.3 Hz, H-1"), 4.01 (dd, $J_{6''A-5''}$ = 2.1 Hz, $J_{6''A-6''B}$ = 12.0 Hz, H-6"A), 3.77 (dd, $J_{6''B-5''}$ = 6.0 Hz, H-6"B), 3.66-3.55 (m, H-2"/H-3" /H-5"), 3.46 (dd, $J_{4''-3''}$ = 9.2, $J_{4''-5''}$ = 9.2 Hz, H-4"); ^{13}C (DMSO): δ 182.3 (s, C-4), 164.1 (s, C-2), 161.2 (s, C-4'), 151.3 (s, C-9). 149.1 (s, C-7), 146.4 (s, C-5). 130.4 (s, C-6), 128.4 (d, C-2'/C-6'), 121.3 (s, C-1'), 115.9 (d, C-3'/C-5'), 105.9 (s, C-10), 102.5 (d, C-3), 101.1 (d, C-1"), 94.2 (d, C-8), 77.3 (d, C-5"), 75.7 (d, C-3"), 73.1 (d, C-2"), 69.6 (d, C-4"). 60.6 (t, C-6");

Galloyl kaempfero - l3- glucoside - (6′′-O-(3,4,5- trihydroxybenzoyl)-3-O-β-D-glucopyranosyl oxy-3',4',5,7-tetra-hydroxy flavone): MS: Neg. ES-MS. *m/z* (rel. int.): 615 ([M-H]$^-$. (100); UV λmax 263.4 and 353.1 nm; ^{1}HNMR (CD_3OD): δ 7.57 [m, H-2′/H-6' second order system], 6.98 (s. H-2′′/H-6′′′), 6.76 [d (second order), $J_{5'-6'}$ = 8.3 Hz, H-5'], 6.38 (d, J $_{8-6}$ = 2.1 Hz, H-8), 6.22 (d, H-6), 5.24 (d, J $_{1''-2''}$ = 7.4 Hz, H-1′′), 4.38 (dd, A, $J_{6'A-5''}$ = 5.0′, $J_{6''A-6''B}$ = 12.0 Hz, H-6′′], 4.31 [dd, $J_{6''B-5''}$ = 2.0 Hz, H-6′′B), 3.65-3.45 (m, H-2′′-H-5′′).

Scutillarein: MS: Negative ES-MS. *m/z* (rel. int.): 285 ([M-H]$^-$, UV: λmax 282.6 and 335.8 nm; ^{1}HNMR (CD_3OD): δ 7.89 (dAA', $J_{2'-3'}$ + $J_{2'-5'}$ = 8.9 Hz, H-2'/H-6'), 6.97 [d, BB', H-3'/H-5'], 6.63-6.62 [s x 2, H-3/H-8].

6-Hydroxyluteolin: 3',4',5,6,7–pentahydroxyflavone: MS: Negative ES-MS, *m/z* (rel. int.): 301 ([M-H]$^-$,100); UV λmax 268.2 (double- hump) and 364.4 nm; ^{1}HNMR (CD_3OD): δ 7.42 (dd, $J_{6'-2'}$ = 2.2 Hz, $J_{6'-5'}$ = 8.9 Hz, H-6'), 7.41 (d, H-2'), 6.94 (d, H-5'), 6.61 (s, H-3), 6.58 (s, H-8);. ^{13}C (CD_3OD) δ 183.9 (s, C-4), 166.0 (s, C-2), 154.3, 151.7 (s x 2, C-9, C-7), 150.5 (s, C-4'), 147.5 (s, C-5), 146.7 (s, C-3'), 130.3 (s, C-6), 123.5 (s, C-l'), 119.9 (d, C-6'), 116.5 (d, C-5'), 113.8 (d, C-2'), 105.2 (s, C-10), 103.1 (d, C-3), 94.6 (d, C-8).

6-Hydroxyluteolin 7-O-β-D-glucopyranoside: MS: Negative ES-MS, *m/z* (rel. int.): 463 ([M-H]$^-$, 100); UV λmax 213.7,281.3 and 342.3 nm; ^{1}HNMR (CD_3OD): δ 7.45, 7.44 [m (second order system), H-2' /H-6'], 7.04 (s, H-8), 6.94 ([d, $J_{5'-6'}$ = 8.2 Hz, H-5'), 6.62 (s, H-3), 5.10 (d, $J_{1''-2''}$ = 7.4 Hz, H-l"), 4.01 (dd, $J_{6''A-5''}$ = 2.2, $J_{6''/A-6'B}$ = 12.0 Hz, H-6"A), 3.63 (dd, $J_{6''B-}$

$_{5''}$ = 6.0 Hz, H- 6"B), 3.65-3.55 (m, H-2" /H-3" /H-5"), 3.47 (dd, $J_{4''-3''}$ = 9.1 $J_{4''-5''}$ = 9.1 Hz, H-4").

Quercetin 3-O-β-D-glucuronide: UV λmax 256.3 and 356.2 nm; MS: Negative ES-MS, *m/z* (rel. int.): 491 ($[M-H]^-$ 100); ^{1}HNMR(CD_3OD): δ 7.63, 7.64 [m (second order system), H-2′/H-6'], 6.89 [m (second order), H-5'], 6.43 (d, J $_{8,6}$ = 2.0 Hz, H-8,), 6.25 (d, H-6), 5.26 (d, $J_{1''-2''}$ = 7.6 Hz, H-1"), 3.80 (d, $J_{5''-4''}$ = 9.7 Hz, H-5"), 3.70 (s, 3'-OMe), 3.6l (dd, H-4", J4′′-3" = 9.0 Hz), 3.56 (dd, H-2", $J_{2''-3''}$ = 9.1 Hz), 3.49 (dd, H-3").

Galloyl quercitrin (2′′-O-3,4,5-trihydroxybenzoyl quercitrin): MS: Negative ES-MS, *m/z* (rel. int.): 599 ($[M-H]^-$, 100); UV λmax 264.4 and 348.0 nm; ^{1}HNMR (CD_3OD): δ 7.40 (bs, H-2'), 7.39 (dd, $J_{6'-2'}$ = 2.2 Hz, $J_{6'-5'}$ = 8.1 Hz, H-6'), 7.11 (s, H-2'"/H-6'"), 6.42 (d, J_{8-6} = 2.1 Hz, H-8), 6.24 (d, H-6), 5.67 (dd, $J_{2''-1''}$ = 1.7, $J_{2''-3''}$ = 3.4 Hz, H-2"), 5.55 (d, H-1"), 4.05 (m, H-5"), 3.51 (m, H-3" /H-4"), 1.07 (d, $J_{6''-5''}$ = 5.7 Hz, H-6").

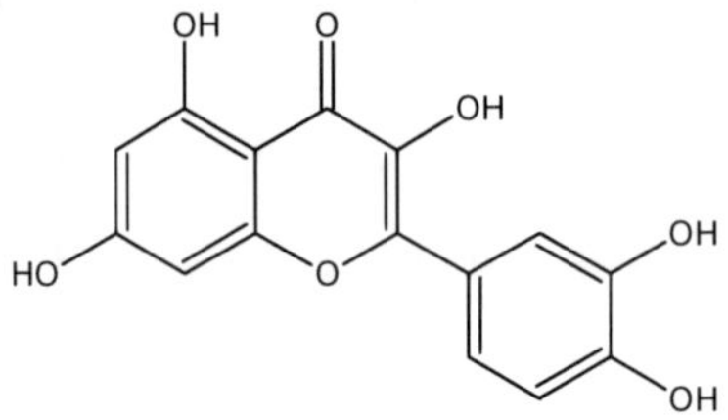

183
Quercetin

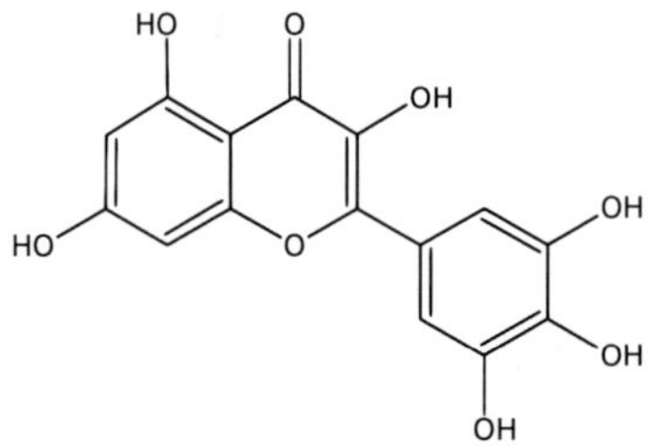

184
Myricetin

Flavonoids widespread in the plant world, occurring in seeds, flowers and stems. Quercetin and myricetin were isolated from *Peganum Harmala* (Rutaceae). The plant-derived phenolic compounds quercetin and myricetin are powerful inhibitors of iron-induced lipid peroxidation in rat liver microsomes, at low micromolar concentrations (IC50 ≤ 1,5 μM). At 100 μM, the three phenolic compounds enhanced OH formation up to eight-fold. The hydroxyl radical generation was inhibited by catalase and superoxide dismutase, suggesting a mechanism in which the phenols oxidize to produce superoxide radical, which then assists ·OH generation from H_2O_2 in the presence of Fe^{3+}-EDTA. At concentrations up to 75 μM, quercetin and myricetin also accelerate bleomycin-dependent DNA damage in the presence of Fe^{3+}, possibly by reducing the Fe^{3+}-bleomycin-DNA complex to the Fe^{2+} form (Laughton et al., 1989).

Quercetin: Yellow powder; mp 310 °C; HPLC online UV-vis spectra: retention time at 20.2 ± 0.1 min; absorption maxima at 252.7 and 369.6 nm; UV λmax (MeOH) 204.1 (log ε = 4.56); 255.3 (4.11), 291.8 (sh), 370.0 (3.98) nm; MS: Negative ES-MS, *m/z* (rel. int.): 301 ($[M-H]^-$, (l00); ^{1}HNMR (CD_3OD): δ 7.77 [d, H-2', J(2'-6') 2.1 Hz], 7.67 [dd, H-6', J(6'-5') 8.5 Hz], 6.92 [d, H-5'], 6.43 [d, H-8, J(8-6) 2.0 Hz], 6.22 [d, H-6].

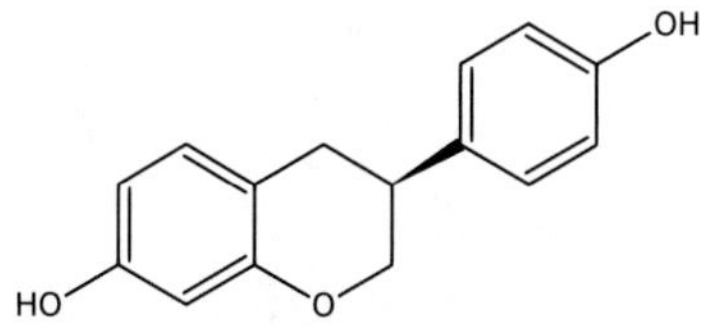

185
Equol

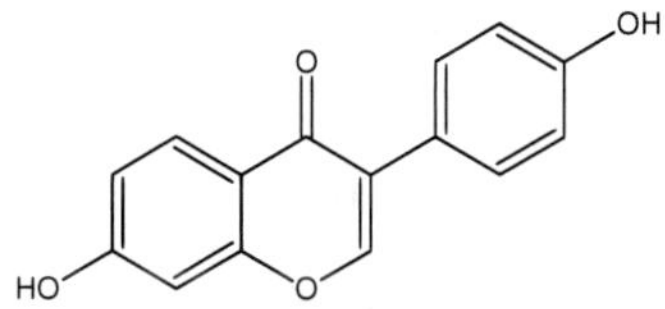

186
Daidzein

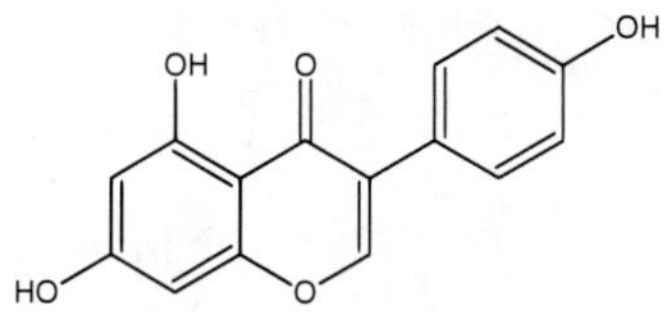

187
Genistin

Found in Soyabean. These compounds were found to protect against glucose induced oxidation of human low density lipoproteins (LDL) *in vitro*. Equol was a more effective antioxidant than daidzein or genistin in this system and was of similar antioxidant potency to the dietary flavonols quercetin and kaempferol and to the endogenous antioxidant 17 β-oestradiol (Vedavanam et al., 1999).

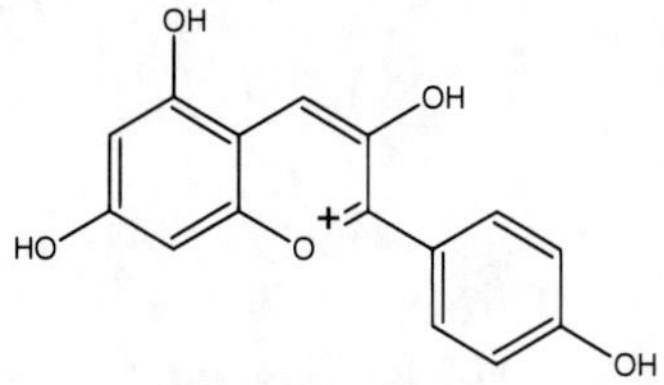

188
Pelargonidin

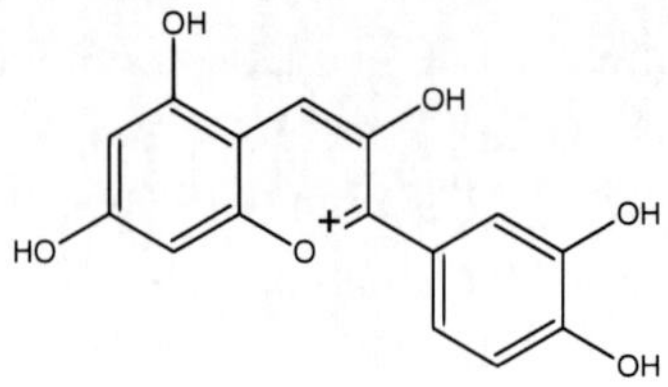

189
Cyanidin

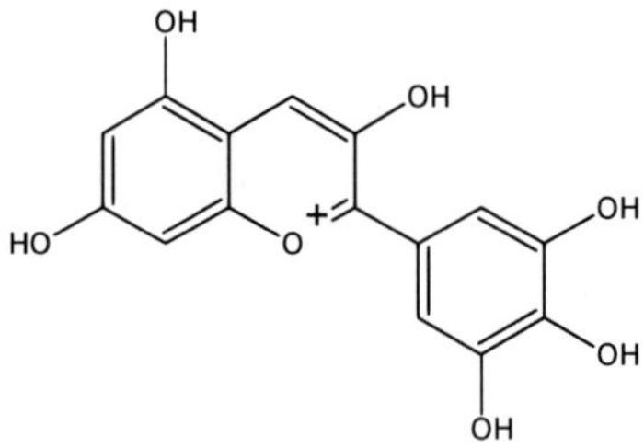

190
Delphinidin

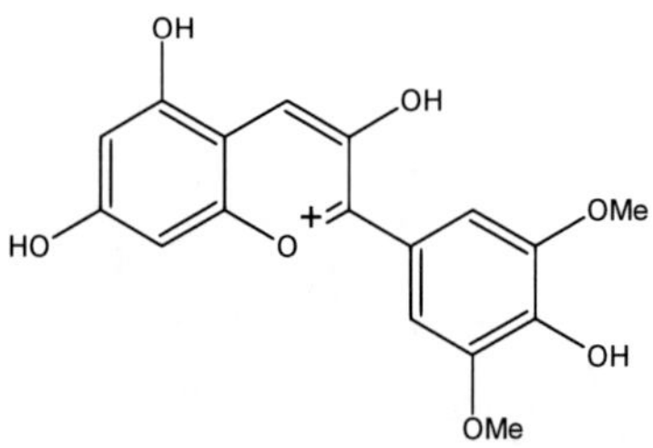

191
Malvidin

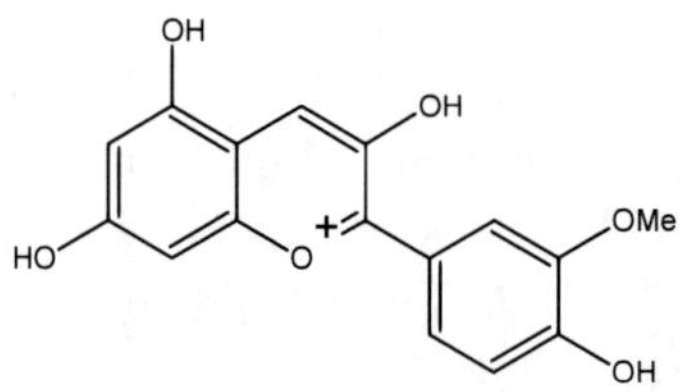

192
Peonidin

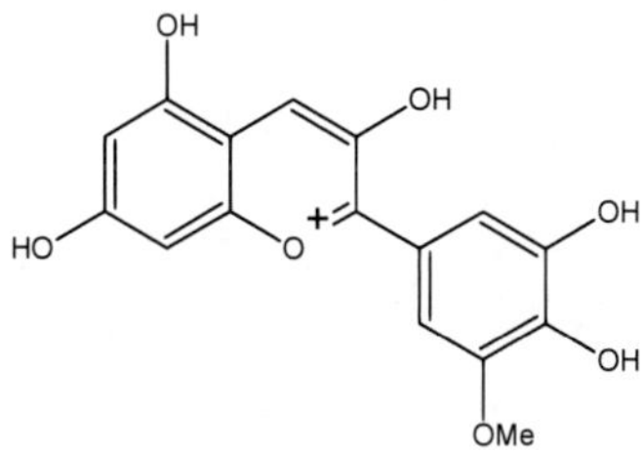

193
Petunidin

Anthocyanins represent one of the most widely distributed classes of flavonoids in plants. The activity of anthocyanidins in preventing the oxidation of human LDL at the 10 and 25 μM addition levels decreased in the order delphinidin > cyanidin > malvidin > peonidin > pelargonidin > petunidin. At the lowest level (2.5 μM), only malvidin, pelargonidin, and peonidin reduced the extent of hexanal formation. Petunidin showed slight pro-oxidant activity, that is enhanced the formation of hexanal at all tested concentrations (Kahkonen, and Heinonen 2003).

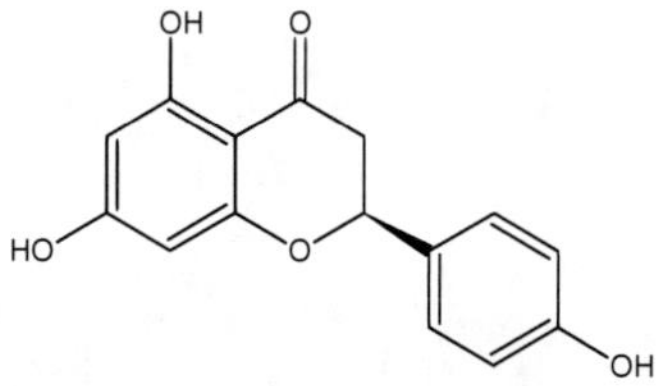

194
Naringenin

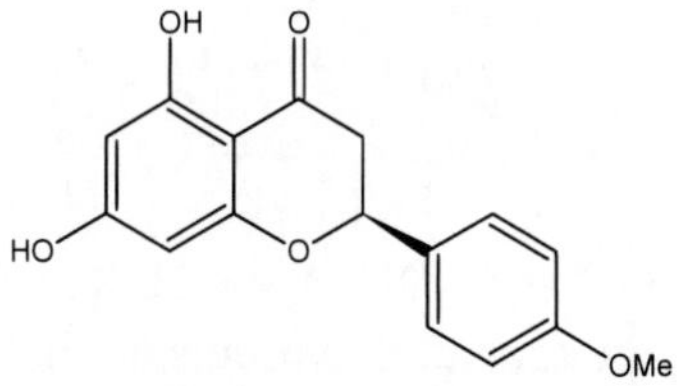

195
Isosakuranetin

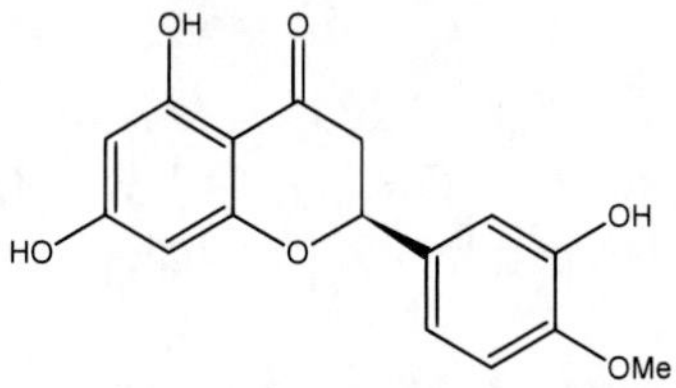

196
Hesperetin

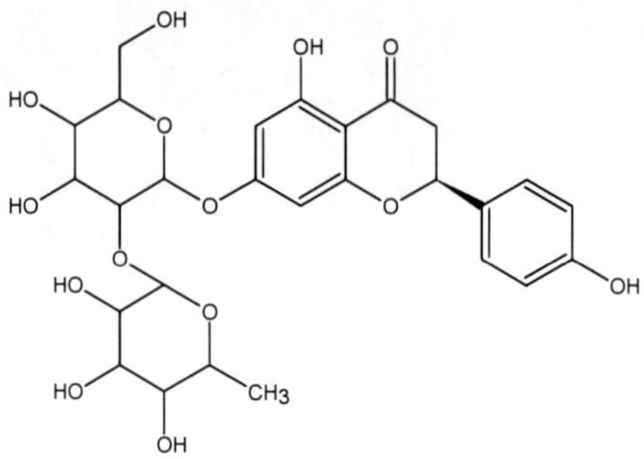

197
Naringin

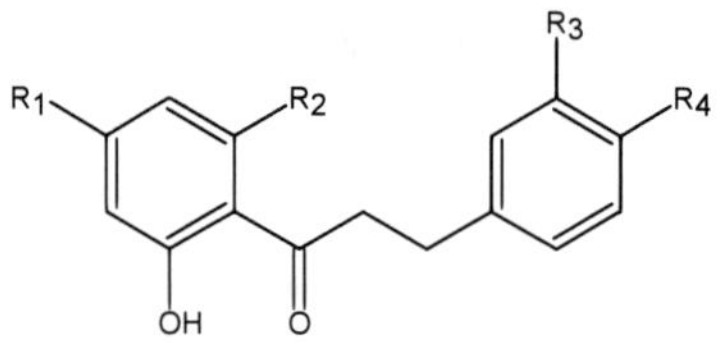

198
R_1 = OH, R_2 = OH, R_3= H, R_4 = OH
Phloretin

199
R_1 = OH, R_2 = OH, R_3= H, R_4 = OCH_3
Isosakuranetin dihydrochalcone

200
R_1 = OH, R_2 = OH, R_3= OH, R_4 = OCH_3
Hesperetin dihydrochalcone

201
R_1 = OH, R_2 = OH, R_3= OH, R_4 = OH
Eriodictyol dihydrochalcone

202
R_1 = ONeo, R_2 = OH, R_3= OH, R_4 = OCH_3
Neohesperidin dihydrochalcone

203
R_1 = ORut, R_2 = OH, R_3= OH, R_4 = OCH_3.
Hesperidin dihydrochalcone

204
R_1 = OH, R_2 = OCH_3, R_3= H, R_4 = OH
2'-O-Methyl-phloretin

205
$R_1 = OCH_3$, $R_2 = OH$, $R_3 = H$, $R_4 = OH$
4'-O-Methyl-phloretin

206
$R_1 = ONeo$, $R_2 = OH$, $R_3 = H$, $R_4 = OH$
Naringin dihydrochalcone

207
$R_1 = OH$, $R_2 = OGlu$, $R_3 = H$, $R_4 = OH$
Asebotin

208
$R_1 = OCH_3$, $R_2 = OGlu$, $R_3 = H$, $R_4 = OH$
Phloridzin

Dihydrochalcones are a family of bicyclic fiavonoids, are common in the genus *Citrus* and specific *Citrus* cultivar. All dihydrochalcones exhibited higher antioxidant antioxidant activities against the stable free radical (1, 1-diphenyl-2-picrylhydrazyl) and lipid peroxidation in the erythrocyte membrane. The active dihydrochalcone has a time-averaged conformation in which the aromatic A ring is orthogonal to the carbonyl group, while the inactive dihydrochalcone such as 2'-o-methyl-phloretin has a strongly hydrogen-bonded phenolic hydroxyl group, suggestive of a coplanar conformation. A hydroxyl group at the 2'-position of the dihydrochalcone A ring, newly formed by reduction of the flavanone C ring, is an essential pharmacophore for its radical scavenging potential (Yoshimasa et al., 2003).

Hesperidin dihydrochalcone: FAB-MS (matrix; glycerol) *m/z*: 613 ($[M + H]^+$; ^{1}HNMR (500 MHz, CD_3OD): δ 1.17 (3H, d, J = 6 Hz, rha-CH3), 2.76 (2H, t, J = 7 Hz, α), 3.24 (2H, t, J = 7 Hz, β), 3.30-3.96 (10H, d, sugar), 3.78 (3H, s, 4-OCH_3), 4.80, 5.04 (2H, d, J = 7 Hz, anomeric proton), 6.10 (2H, s, 3', 5'-H) 6.66 (1H, dd, J = 2, 8 Hz, 6-H), 6.72 (1H, d, J = 2 Hz, 2-H), 6.84 (1H, d, J = 8 Hz, 5-H); ^{13}CNMR (75 MHz, CD_3OD): δ 17.4 (Rha-CH_3), 30.8 (β), 46.4 (α), 56.9 (OCH_3), 66.8 (Glu-6), 69.6, 70.4, 71.1, 71.3, 73.2 (Glu-4, Rha-2, 3, 4, 5), 73.8, 76.0, 76.6 (Glu-2, 3, 5), 96.6 (3'), 96.6 (5'), 99.9 (Glu-1), 101.3 (Rha-1),107.1 (1'), 113.6, 116.5 (2), 121.2 (6), 135.7 (1), 144.5 (3), 146.6 (4), 163.8 (4'), 164.1 (2'), 164.1 (6'), 208.1 (C=O).

Naringin dihydrochalcone: FAR-MS (matrix; glycerol) *m/z*: 583 ($[M + H]^+$; ^{1}H NMR (500 MHz, $(CD_3)_2SO$): δ 1.21 (3H,d, J = 6 Hz. Rha-CH_3), 2.81 (2H, m, α), 3.24 (2H, t, J = 7 Hz, β), 3.25-5.30 (sugar proton), 5.13 (2H, d, J = 7 Hz, anomeric proton), 6.04 (2H, s, 3', 5'-H), 6.69 (1H, d, J = 8 Hz, 3, 5-H), 7.05 (2H, d, J = 8 Hz, 2, 6-H), 9.10 (1H, brs, 4-OH), 12.31 (2H, brs, 2', 6'-OH); ^{13}C NMR (75 MHz, $(CD_3)_2SO$): δ 18.1 (Rba-CH_3), 29.3 (β), 45.7 (α), 60.3 (Glu-6), 68.3, 69.5, 70.4, 70.5, 71.9 (Glu-4, Rha-2, 3, 4, 5),76.6, 76.8,77.1 (Glu-2, 3,5), 94.8 (3'), 94.8 (5'), 97.1 (Rha-l), 100.6 (Glu-l), 105.3 (1'), 115.1 (3), 115.1 (5), 129.1 (2), 129.1 (6), 131.5 (1), 155.4 (4), 162.9 (4'), 163.7 (2'), 163.7 (6'), 205.1 (C=O).

Asebotin: FAB-MS (matrix; glycerol) *m/z*: 451 ($[M + H]^+$; ^{1}H NMR (500 MHz, $(CD_3)_2SO$): δ 2.80 (2H, m, β), 3.30-3.70 (8H, α and sugar proton), 3.78 (3H, s, 4'-OCH_3), 4.62 (1H, t, J = 5 Hz, Glu-6-OH), 5.01 (1H, d, J = 7 Hz, anomeric proton), 5.05, 5.14, 5.30

(3H, d, J = 5 Hz, Glu-2,3.4-OH), 6.13 (1H, d, J = 2 Hz, 5'-H), 6.29 (2H, d, J = 2 Hz, 3'-H), 6.64 (2H, d, J = 8 Hz, 3, 5-H), 7.03 (2H, d, J = 8 Hz, 2, 6-H), 9.08 (1H, s, 4-OH), 13.34 (1H, s, 6'-OH); ^{13}CNMR (75 MHz, $(CD_3)_2SO$): δ 28,8 (β), 45.1 (α), 55.6 (OCH_3), 60.7 (Glu-6), 69.7 (Glu-4), 73.2 (Glu-2), 76.7 (Glu-5), 77.4 (Glu- 3), 93.5 (3'), 95.1 (5'), 100.8 (Glu-1), 106.2 (1'), 115.0 (3), 115.0 (5), 129.1 (2), 129.1 (6), 131.4 (1), 155.3 (4), 160.2 (2'), 164.9 (6'), 165.1 (4'), 205.1 (C=O).

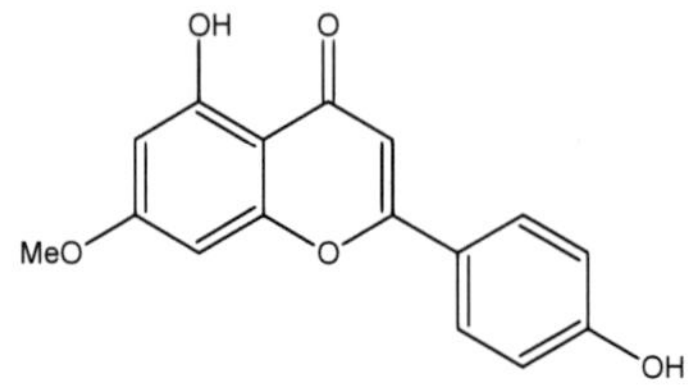

209
Genkwanin

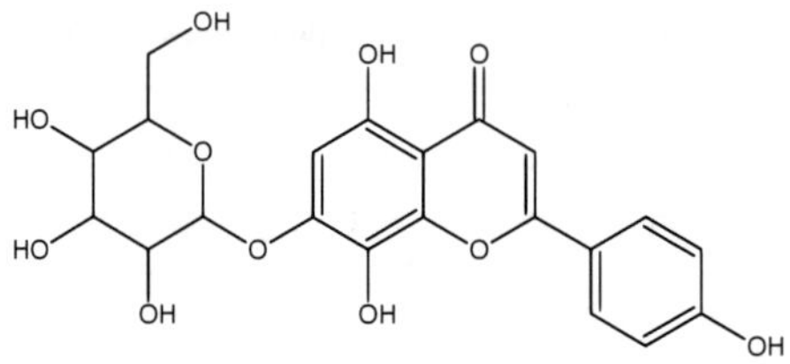

210
Isoscutellarein 7-O-glucoside

Found in *Rosmarinus officinalis* (Labiatae). These flavones are excellent antioxidants in both aqueous and lipid systems. (Dorman et al., 2003). The antioxidant activity mechanism of rosemary flavones is similar to that of other polyphenols and flavonoids.

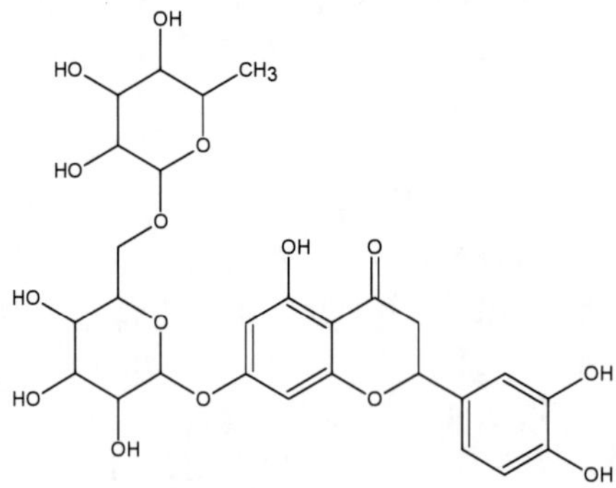

211
Eriocitrin

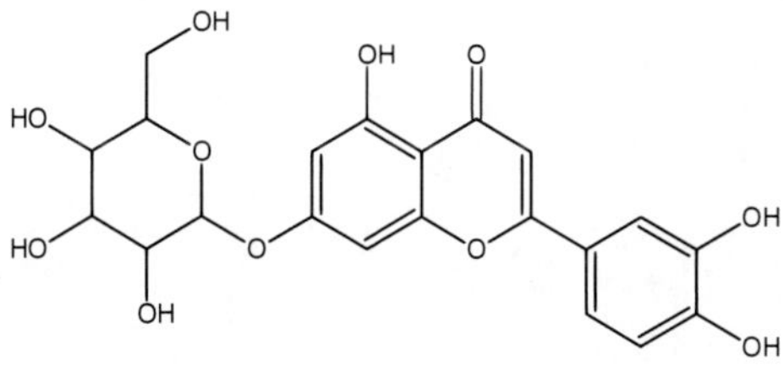

212
Luteolin-7-O-glucoside

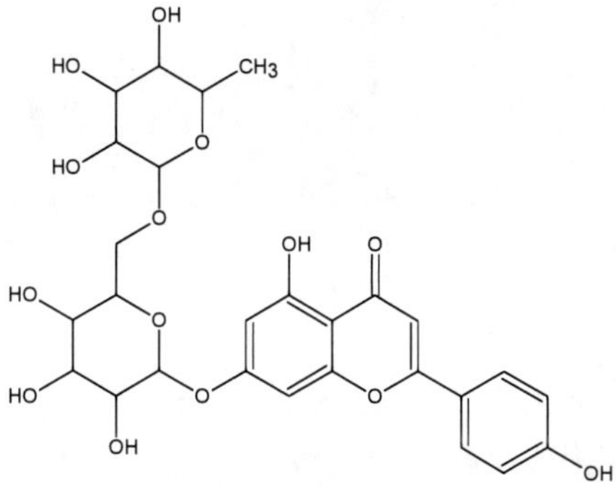

213
Isorhoifolin

Found in *Mentha aquatica* and *M. arvensis* (Lamiaceae). These compounds demonstrated varying degrees of efficacy within each antioxidant assay included iron (lll) reduction, iron (ll) chelation, 1, 1-diphenyl-2-picrylhydrazyl radical scavenging, and the ability to inhibit iron (III)-ascorbate-catalyzed hydroxyl radical-mediated brain phospholipid peroxidation (Dorman et al., 2003).

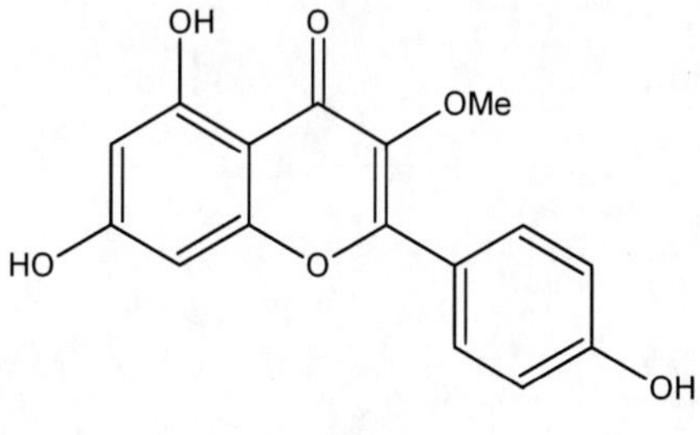

214
Isokaempferide

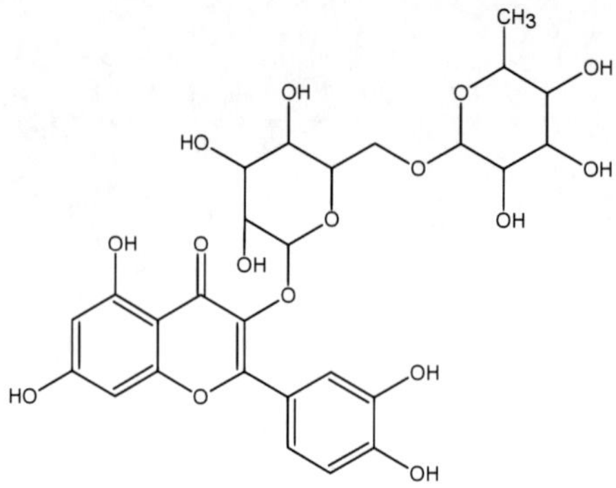

215
Rutin

These compounds were isolated from the n-butanol soluble fraction of the leaves of *Peucedanum japonicum* Thunb. Showed antioxidant activity on the test of radical scavenging on 1, 1-diphenyl-2-picrylhydrazyl radical and for inhibition of oxidation of liposome induced by 2,2'-azobis{2-amidinopropane)dihydrochloride (Hisamoto et al., 2003).

Rutin: UV λmax (MeOH): 258, 266sh, 296sh, 360; (+NaOMe) 271, 322sh, 410, (+$AlCl_3$) 273, 303sh, 430; (+$AlCl_3$/HCl): 269, 304sh, 360sh, 400; (+NaOAc) 270, 320, 398; (+NaOAc/H_3BO_3) 263, 304, 383; ^{1}HNMR (400 MHz, CD_3OD): δ 6.25 (1 H, d J = 2 Hz, H-6), 6.40 (1H, d, J=2Hz, H-8), 6.95 (1H, d, J= 8Hz, H-5'), 7.50 (1H, dd, J_1 = 2 Hz, J_2 = 8 Hz, H-6'), 7.75 (1H, d, J = 2 Hz, H-2').

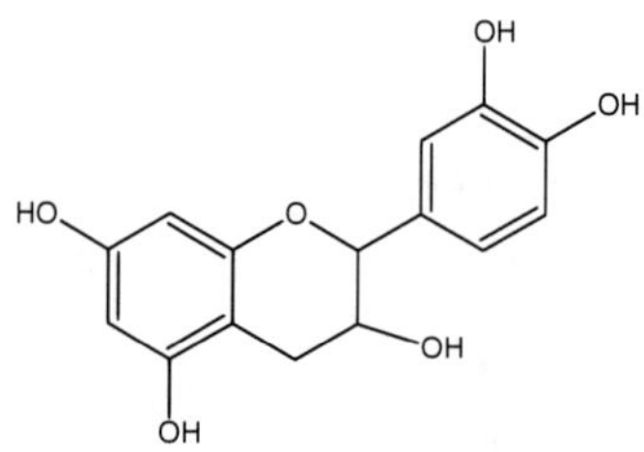

216
Epicatechin

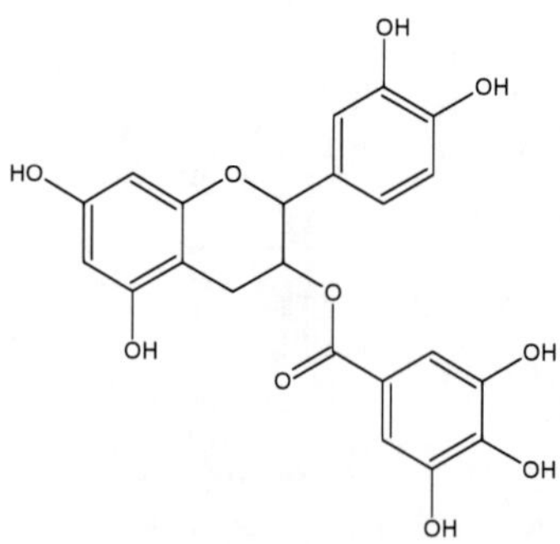

217
Epicatechin gallate

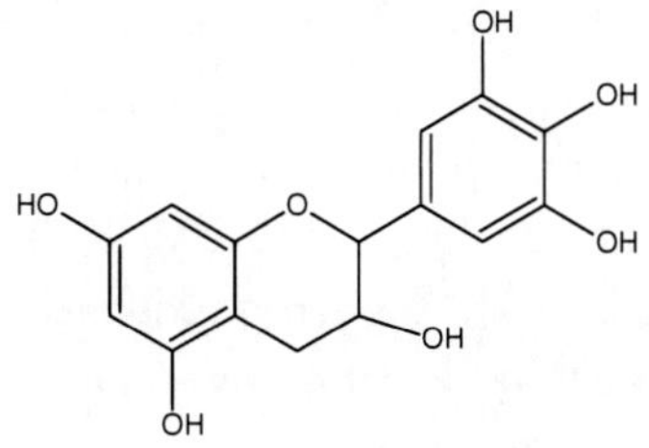

218
Epigallocatechin

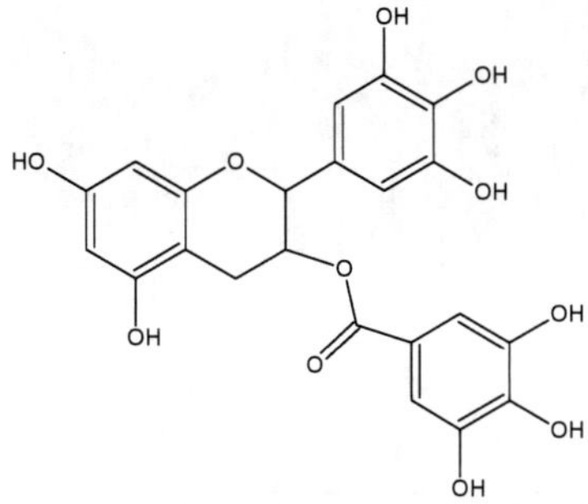

219
Epigallocatechin gallate

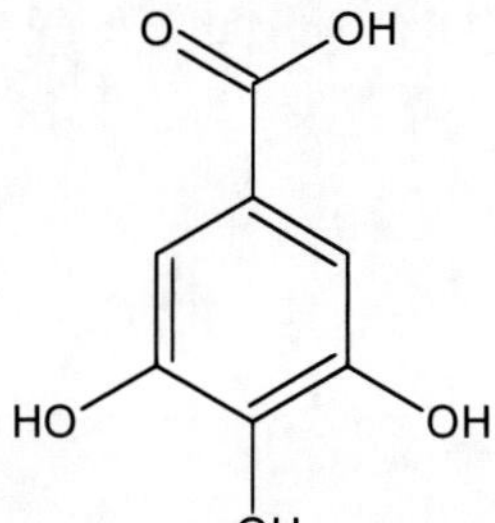

220
Gallic acid

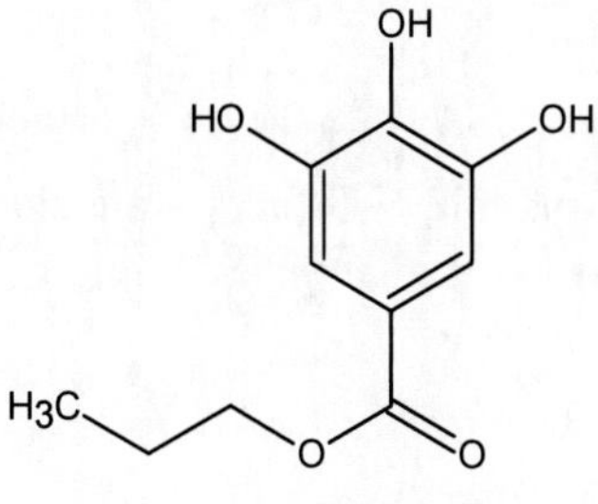

221
Propyl gallate

Found in green tea (*Camellia sinensis* L. Kuntze, Theaceae) catechins showed different trends in relative antioxidant activity in different lipid systems. In corn oil triglycerides oxidized epigallocatechin, epigallocatechin gallate, and epicatechin gallate were better antioxidants than epicatechin and catechin at 140 μM. However, in the corresponding corn oil-in-water emulsions, all tea catechins, were pro oxidants at 5 and 20 μM by accelerating hydroperoxide and hexanal formation. The improved antioxidant activity observed for tea catechins in liposomes compared to emulsions can be explained by the greater affinity of the polar catechins toward the polar surface of the lecithin bilayers, thus affording better protection (Huang and Frankel, 1997, He and Shahidi, 1997).

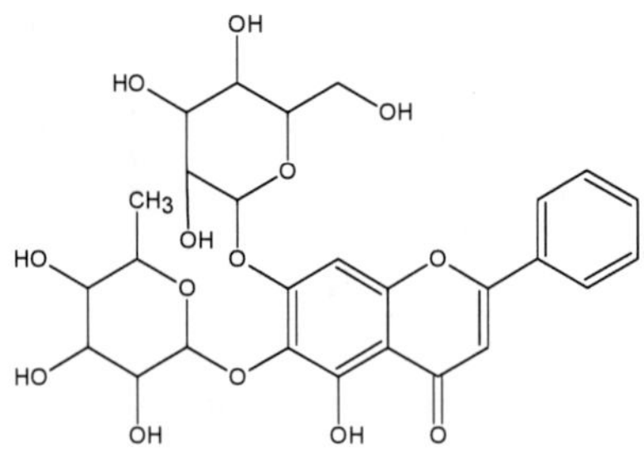

222
Chrysin 7-O-β-D-glucopyranosyl-6-C- fucopyranoside

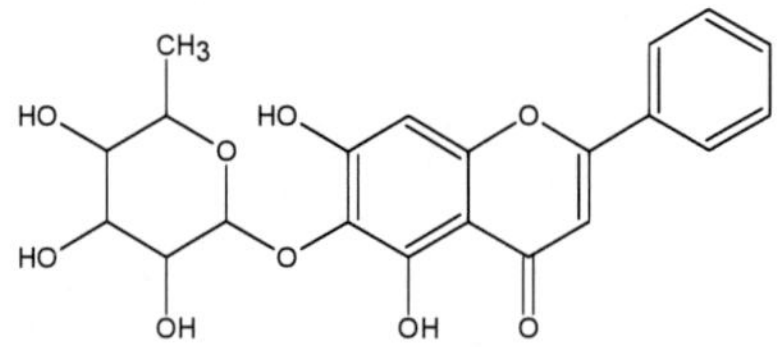

223
Chrysin-6-C-fucopyranoside

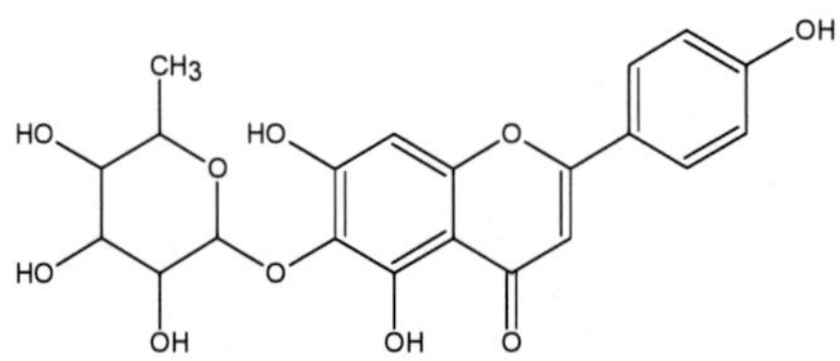

224
Apigenin-6-C-fucopyranoside

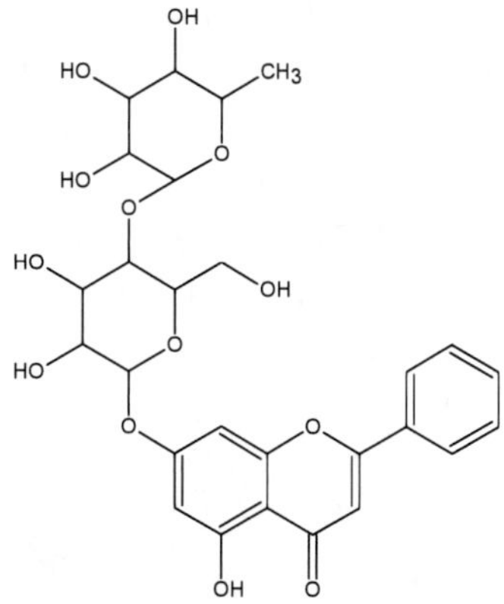

225
Chrysin 7-O-β-D-gluopyranosyl-(1-4)-α-L-rhamnopyranoside

All of the compounds isolated from *Cyclanthera pedata* (Cucurbitaceae) fruit had radical scavenging activity similar to the activity of kaempferol but lower than that of quercetin. The TEAC values indicate that the number of free phenolic OH groups (excluding OH groups on the sugar moiety) makes a statistically significant contribution to TEAC, (Montoro et al., 2001).

Chrysin 7-O-β-D-glucopyranosyl-6-C-fucopyranoside: ($C_{27}H_{30}O_{13}$) was obtained as a yellow amorphous powder; $[\alpha]_D25$ -51.3° (MeOH, c 0.l); UV (MeOH): λmax 330, 275, and 210 nm; HRESIMS, *m/z* 585.1596 $[M + Na]^+$; ESIMS: *m/z* 563 $[M + H]^+$, 585 $[M + Na]^+$, and 401 $[M + H - 162]^+$; ^{1}HNMR: δ 6.87 (s, H-3), 7.11 (s, H-8), 8.06 (br, d, J = 8.0 Hz, H-2′), 7.61 (m, H-3′), 7.59 (m, H-4′), 7.61 (m, H-5′), 8.06 (br, d, J = 8.0 Hz, H-6′); Fuc: 4.84 (d, J = 7.0 Hz, H-1), 4.64 (br, t, J = 9.0 Hz, H-2), 3.58 (dd, J = 9.0, 3.5 Hz, H-3), 3.76 (m, H-4), 3.79 (m, H-5), 1.29 (d, J = 6.5 Hz, H-6); Glc: 4.94 (d, J = 7.5 Hz, H-1), 3.66 (br,t, J = 9.0 Hz, H-2), 3.54 (br,t, J = 9.0 Hz, H-3), 3.40 (t, J = 9.0 Hz, H-4), 3.62 (m, H-5), 3.75 (dd, J = 12, 5.0 Hz, H-6a), 4.04(dd, J = 12.0, 2.5 Hz, H-6b); ^{13}C NMR: δ 166.2 (C-2), 107.0 (C-3), 184.6 (C-4), 162.0 (C-5), 112.2 (C-6), 164.8 (C-7), 96.0 (C-8), 158.9 (C-9), 106.8 (C-10), 131.7 (C-1′), 128.0 (C-2′), 133.3 (C-3′), 130.5 (C-4′), 133.3 (C-5′), 128.0 (C-6′); Fuc: 74.9 (C-1), 70.0 (C-2), 77.1 (C-3), 76.2 (C-4), 73.8 (C-5), 17.5 (C-6); Glc: 103.9 (C-1), 74.6 (C-2), 77.0 (C-3), 71.6 (C-4), 78.4 (C-5), 62.7 (C-6).

Chrysin-6-C-fucopyranoside: ($C_{21}H_{20}O_8$) was obtained as a yellow amorphous powder; $[\alpha]_D25$ -26.6° (MeOH, c 0.1); UV (MeOH): λmax 330, 275, and 210 nm; HRESIMS, *m/z* 401.1236 $[M + H]^-$; ESIMS: *m/z* 401 $[M + H]^+$ and 423 $[M + Na]^+$; ^{1}HNMR: δ 6.85 (s, H-3), 7.09 (s, H-8), 8.06 (br, d, J = 8.0 Hz, H-2′), 7.61 (m, H-3′), 7.59 (m, H-4′), 7.61 (m, H-5′), 8.06 (br d, J = 8.0 Hz, H-6′); Fuc: 4.88 (d, J = 7.0 Hz, H-1), 4.65 (br,t, J = 9.0 Hz, H-2), 3.57 (dd, J = 9.0, 3.5 Hz, H-3), 3.75 (m, H-4), 3.81 (m, H-5), 1.28 (d, J = 6.5 Hz, H-6); ^{13}C NMR: δ 166.2 (C-2), 107.0 (C-3), 185.0 (C-4), 162.1 (C-5), 111.9 (C-6), 164.5 (C-7), 96.5 (C-8), 158.7 (C-9), 107.0 (C-10), 132.0 (C-1′), 128.2 (C-2′), 133.4 (C-3′), 131.0 (C-4′), 133.4 (C-5′), 128.2 (C-6′); Fuc: 74.1 (C-1), 71.1 (C-2), 77.0 (C-3), 74.1 (C-4), 76.7 (C-5), 17.6 (C-6).

Apigenin-6-C-fucopyranoside: ($C_{21}H_{20}O_9$) was obtained as a yellow amorphous powder; $[\alpha]_D25$ -30.3° (MeOH, c 0.1); UV (MeOH); λmax 330, 275, and 210 nm; HRESIMS: *m/z* 417.1189 $[M + H]^+$; ESIMS, *m/z* 417 $[M+ H]^+$ and 439 $[M + Na]^+$; ^{1}HNMR: δ 6.87 (s, H-3),

7.12 (s, H-8), 7.87 (d, J = 8.5 Hz, H-2′), 6.97 (d, J = 8.5 Hz, H-3′), 6.97 (d, J= 8.5 Hz, H-5′), 7.87 (d, J = 8.5 Hz, H-6′); Fuc: 4.90 (d, J = 7.0 Hz, H-1), 4.66 (br,t, J = 9.0 Hz, H-2), 3.60 (dd, J = 9.0, 3.5 Hz, H-3), 3.76 (m, H-4), 3.81 (m, H-5), 1.30 (d, J = 6.5 Hz, H-6); ^{13}C NMR: δ 165.5 (C-2), 108.0 (C-3), 184.0 (C-4), 162.0 (C-5), 112.0 (C-6), 164.5 (C-7), 96.2 (C-8), 158.7 (C-9), 104.6 (C-10), 132.4 (C-1′), 128.4 (C-2′), 116.2 (C-3′), 155.0 (C-4′), 116.2 (C-5′), 128.4 (C-6′). Fuc: 76.3 (C-1), 71.0 (C-2), 77.0 (C-3), 74.1 (C-4), 76.8 (C-5), 17.5 (C-6).

Chrysin7-O-β-D-gluopyranosyl-(1-4)-α-L-rhamnopyranoside: ($C_{27}H_{30}O_{13}$) was obtained as a yellow amorphous powder; $[\alpha]_D$25 - 1.2° (MeOH, c 0.1); α:d -41.2; UV (MeOH) λmax 350, 275, and 210 nm; HRESIMS: *m/z* 585.1596 $[M + Na]^+$; ESIMS, m/z 563 $[M + H]^+$, 585 $[M + Na]^+$, 417 $[M + H -146]^+$, and 255 $[(M + H^+) -(162 + 146)]$; ^{1}HNMR: δ 6.84 (s, H-3), 6.45 (s, H-6), 6.60 (s, H-8), 8.06 (br d, J = 8.0 Hz, H-2′), 7.61 (m, H-3′), 7.59 (m, H-4′), 7.61 (m, H-5′), 8.06 (br, d, J = 8.0 Hz, H-6); Glc: 5.02 (H-1) 3.61 (br, t, J = 9.0 Hz, H-2), 3.50 (t, J = 9.0 Hz, H-3), 3.42 (,t, J = 9.0 Hz, H-4), 3.59 (m, H-5), 3.72 dd, J = 12.0, 5.0 Hz, H-6a), 4.02 (dd, J = 12.0, 3.0 Hz H-6b), Rha: 4.85 (d, J = 1.5 Hz, H-1), 3.78 (dd, J = 3.0, 1.5 Hz, H-2), 3.74 (dd, J = 9.5, 3.0 Hz, H-3), 3.50 (t, J = 9.5 Hz, H-4), 3.90 (m, H-5), 1.27 (d, J = 6.5 Hz, H-6); ^{13}C NMR: δ 166.2 (C-2), 107.0 (C-3), 184.6 (C-4), 162.0 (C-5), 99.0 (C-6), 165.3 (C-7), 96.8 (C-8), 158.9 (C-9), 106.8 (C-10), 131.6 (C-1′), 128.0 (C-2′), 133.2 (C-3′), 130.5 (C-4′), 133.3 (C-5′), 128.0 (C-6′); Glc: 103.5 (C-1), 73.9 (C-2), 78.0 (C-3), 79.0 (C-4), 77.0 (C-5), 62.2 (C-6); Rha: 101.2 (C-1), 72.1 (C-2), 73.0 (C-3), 72.5 (C-4), 70.0 (C-5), 18.2 (C-6).

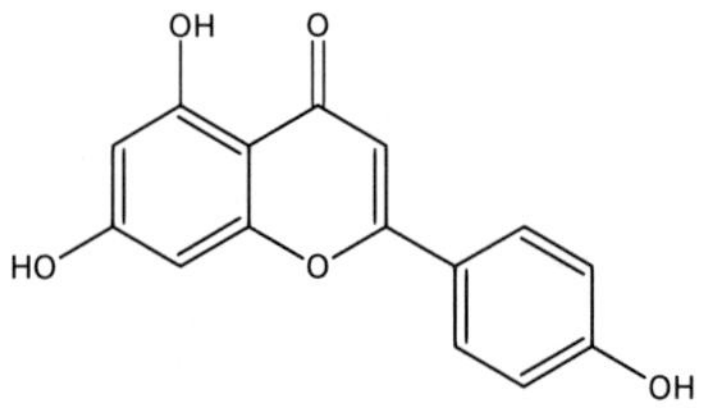

226
Apigenin

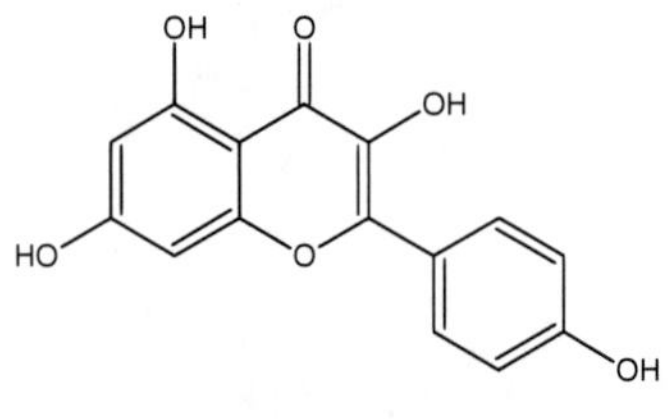

227
Kaempferol

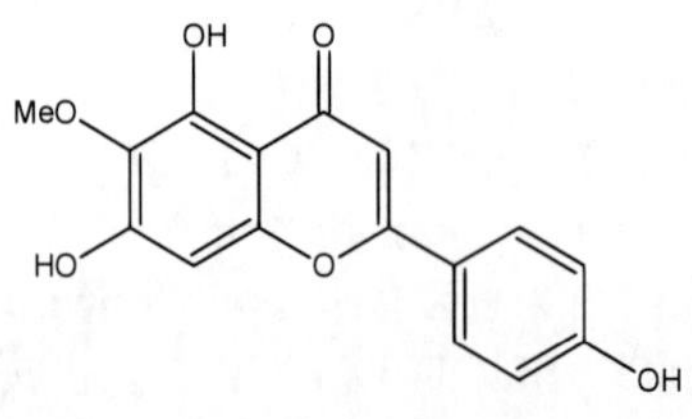

228
Hispidulin

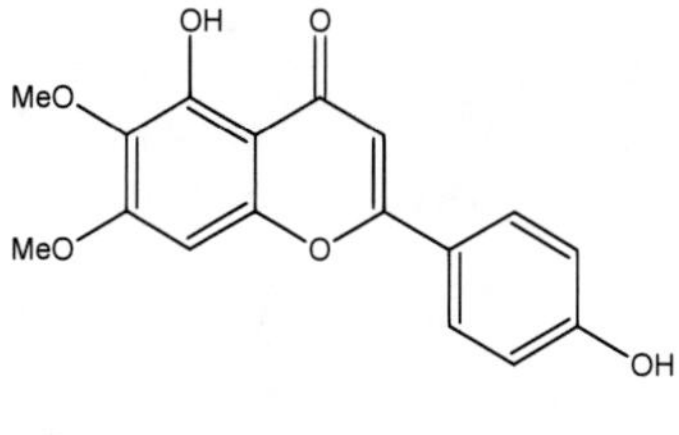

229
Cirsimaritin

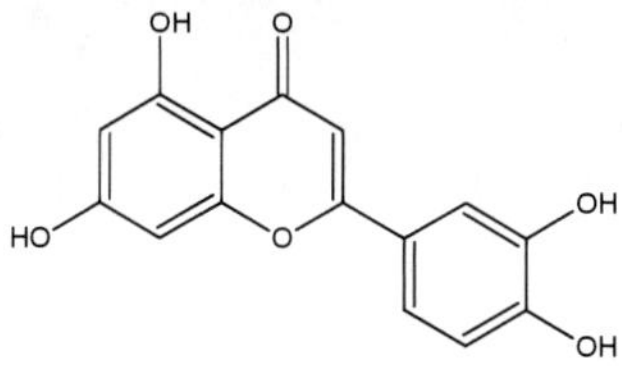

230
Luteolin

Found in *Salvia officinalis* (Labiatae). Sage is not only widely used as a natural source of food flavoring but also has medicinal properties for the treatment of various diseases. These compounds showed antioxidant activity (Zheng and Wang 2001).

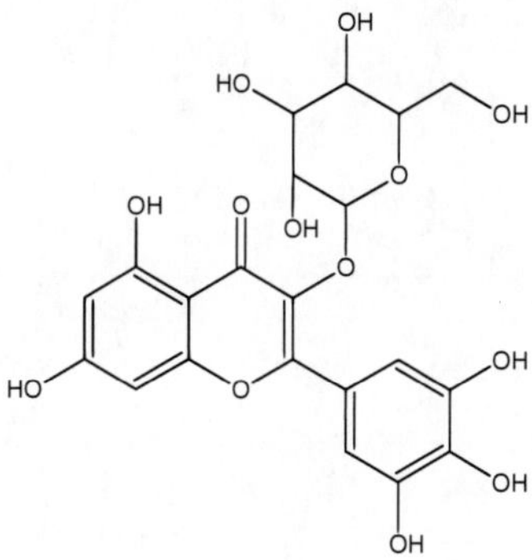

231
Myricetin 3-β-galactoside

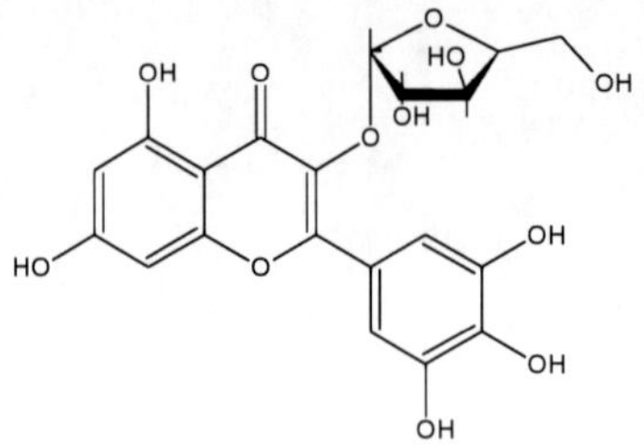

232
Myricetin 3-α-arabinofuranoside

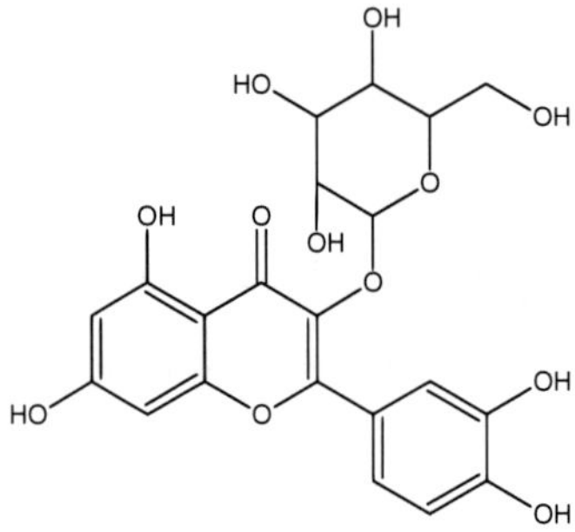

233
Quercetin 3-β-galactoside

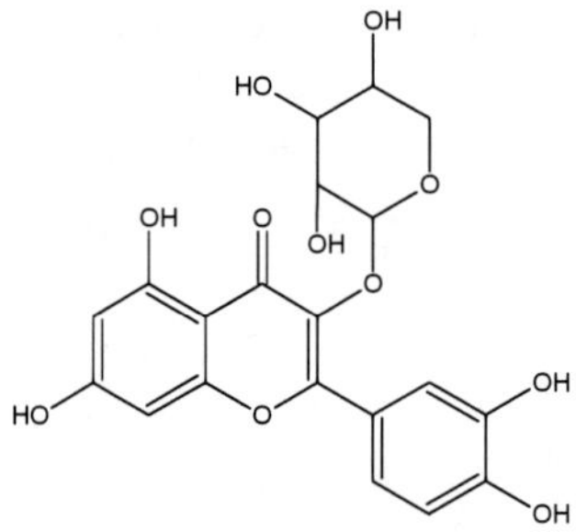

234
Quercetin 3-xyloside

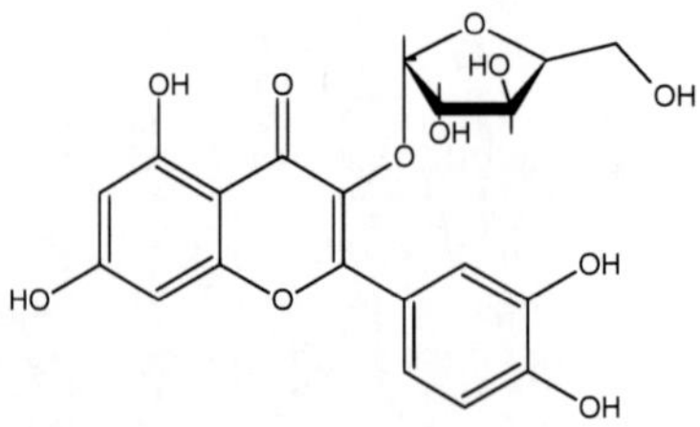

235
Quercetin 3-α-arabinofuranoside

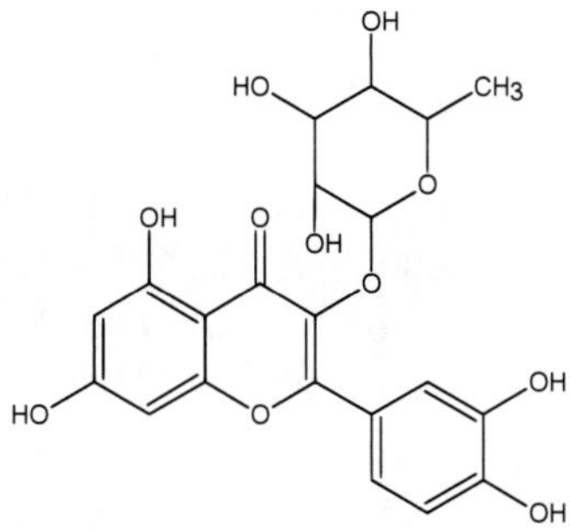

236
Quercetin 3-α-rhamnopyranoside

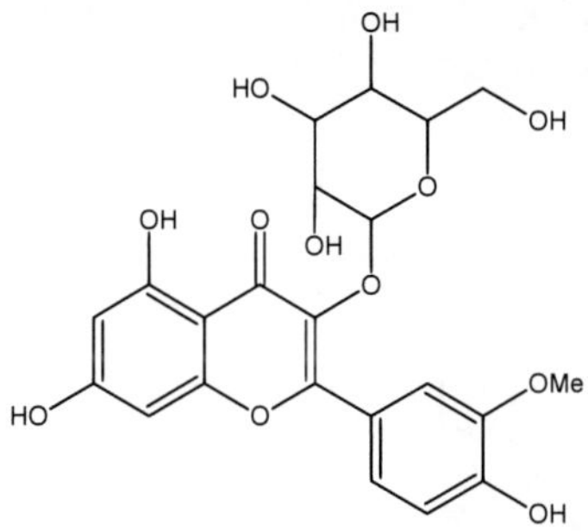

237
3′-Methoxyquercetin 3-β-galactoside

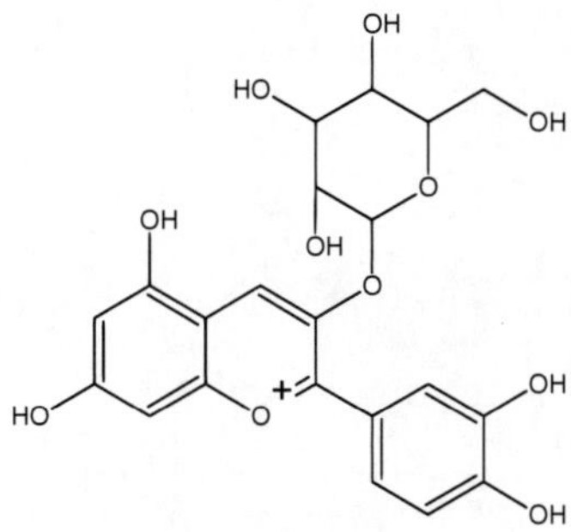

238
Cyanidin-3-galactoside

Found in *Vaccinium macrocarpon* Ait. (Ericaceae). Radical-scavenging activity was greatest when was evaluated for 1, 1-diphenyl-2-picrylhydrazyl and ability to inhibit low-density lipoprotein oxidation *in vitro*. Most of the flavonol glycosides showed antioxidant activity comparable or superior to that of vitamin E; cyanidin 3-galactoside showed activity superior to that of the flavonoids as well as vitamin E or Trolox in both antioxidant assays (Yan et al., 2002).

Quercetin 3-α-arabinofuranoside: FD-MS, *m/z* 435 (M + H); ^{1}HNMR: δ 12.65 (lH, s, 5-OH), 7.66 (lH, dd, J = 2.3, 8.0 Hz, H6'), 7.51 (lH, d, J = 2.3 Hz, H2'), 6.84 (lH, d, J = 8.6 Hz, H5'), 6.40 (lH, d, J = 2.3 Hz, H8), 6.20 (lH, d, J = 2.0 Hz, H6), 5.28 (lH, d, J = 5.1 Hz, G1), 3.75 (lH, dd, J = 5.1, 7.0 Hz, G6), 3.66 (lH, m, G4), 3.60 (lH, d, J = 5.5, 11.3 Hz, G5), 3.51 (lH, dd, J = 3.1, 7.0 Hz, G3), 3.21 (2H, dd, J = 2.3, 11.3 Hz, G5); ^{13}C NMR 156.2 (C-2), 133.7 (C-3), 177.5 (C-4), 103.9 (C-4a), 161.2 (C-5), 98.6 (6), 164.1 (C-7), 93.5 (C-8), 156.3 (C-8a), 121.1 (C-1´), 115.7 (C-2´), 144.9 (C-3´), 148.5 (C-4´), 115.3 (C-5´), 122.0 (C-6´), 101.4 (C-1´´), 70.7 (C-2´´), 71.6 (C-3´´), 66.0 (C-4´´), 64.2 (C-5´´).

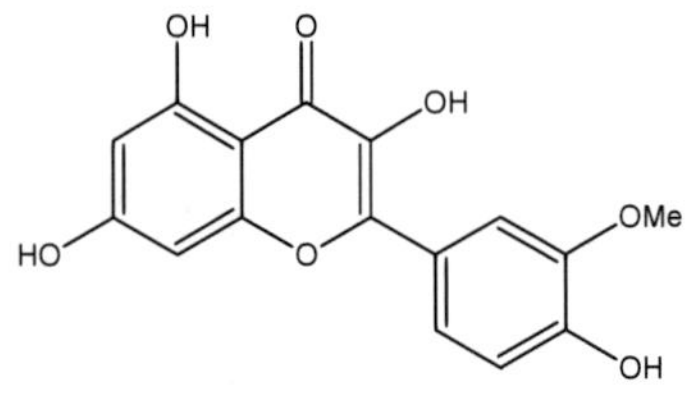

239
Isorhamnetin

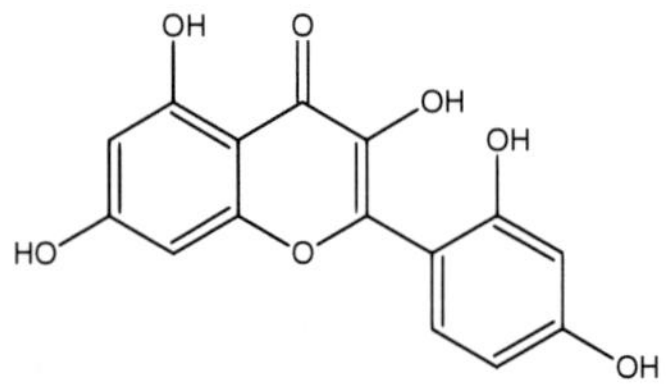

240
Morin

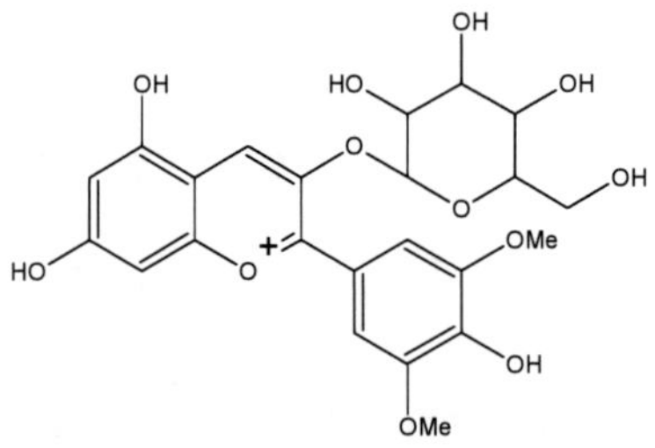

241
Malvidin-3-glucoside

These compound were isolated from red wine. Showed antioxidant and vasodilation activities. The results demonstrate that the different phenolic profiles of wines can produce varying antioxidant and vasodilatant activities, which opens up the possibility that some red wines may provide enhanced health benefits for the consumer (Burns et al, 2000).

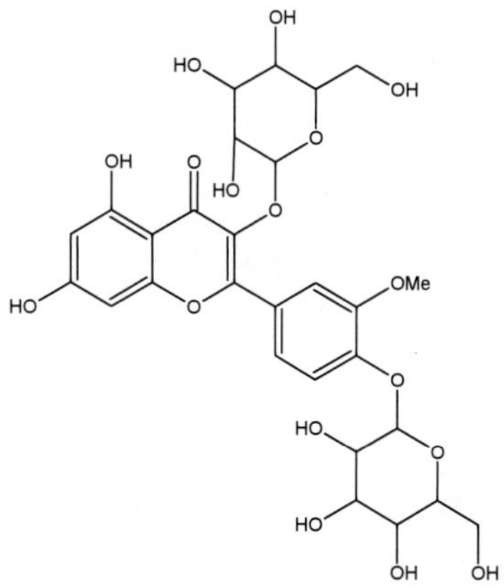

242
Isorhamnetin 3,4'-O-β-D-diglucopyranoside

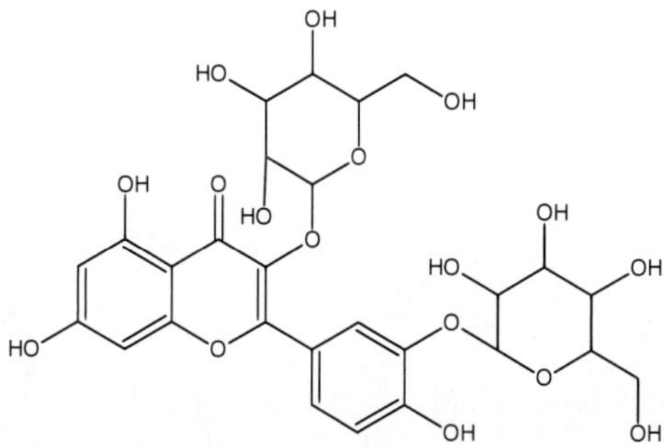

243
Quercetin 3,3'-O-β-D-diglucopyranoside

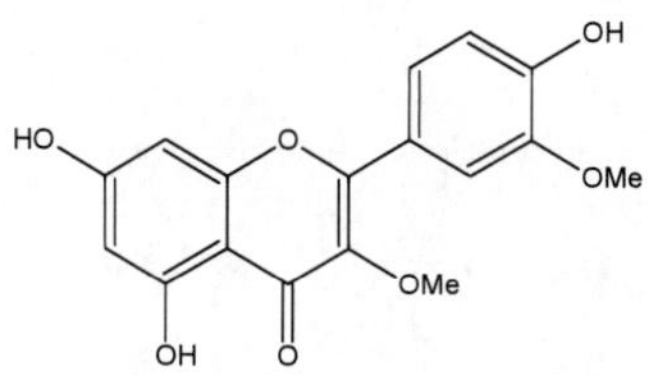

244
Quercetin 3,5'-dimethyl ether

Found in *Anoectochilus formosanus* Hayata (Orchidaceae). A comparative analysis of antioxidant activity on flavonoid glycosides and their flavonoid aglycones has demonstrated that the aglycone form can confer a higher antioxidant activity than their glycoside-conjugated counterparts. The glucose moiety of the flavonoid glycosides can apparently contribute to the decrease of free-radical scavenging activity of the compounds and, in combination with the methylation of hydroxyl group at the C-3 position, it can result in a much reduced effect on antioxidant activity. The new diarylpentonoid, kinsenone isolated from *A. formosanus,* contains two coumaroyl groups (E and Z forms) but was not found to have a comparable radical-scavenging activity relative to that of quercetin or *p*-coumaric acid. DPPH radicals were only inhibited by 20% using 30 μg/mL of kinsenone (Wang et al., 2002).

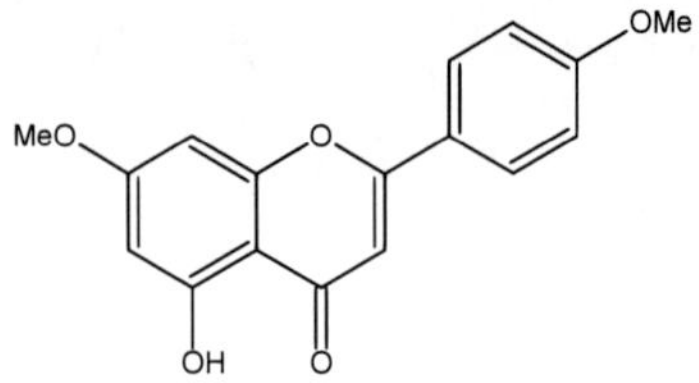

245
5-Hydroxy- 7 ,4'-dimethoxy flavone

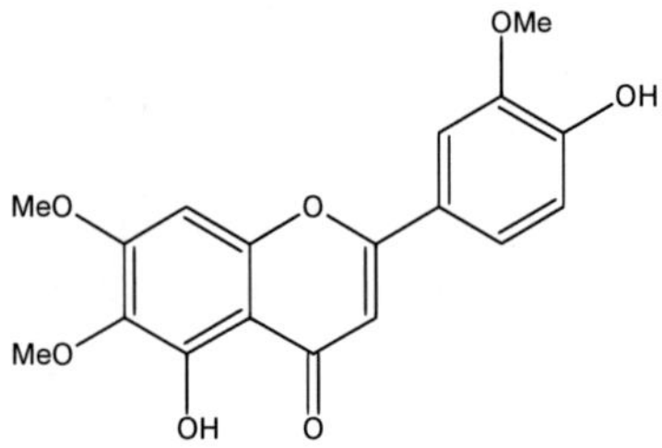

246
5, 4´-Dihydroxy-6,7,3´-trimethoxyflavone

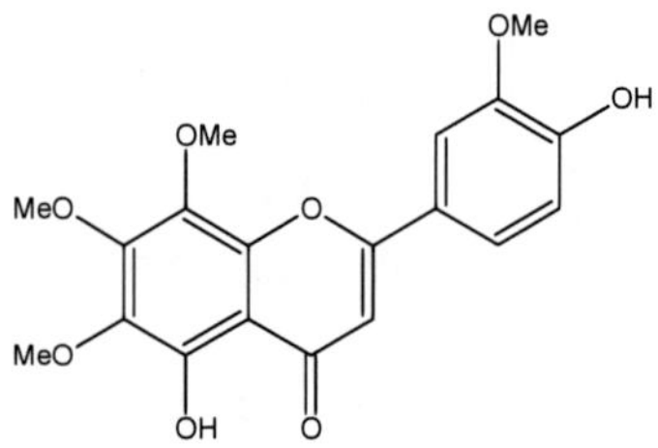

247
5,4´-Dihidroxy-6,7,8,3´- tetramethoxyflavone

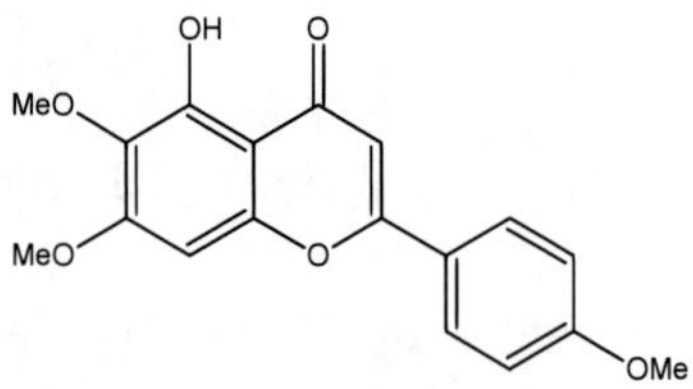

248
Salvigenin

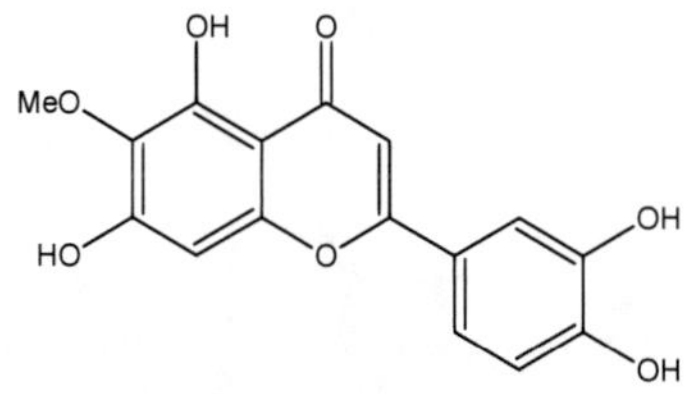

249
Eupafolin

Antioxidant activity of these compounds isolated from *Thymus vulgaris* L. (Labiatae); and *Salvia officinalis* L. (Labiatae) were evaluated by the oil stability index method using a model substrate oil including methyl linoleate in silicone oil. The activity of luteolin, 7-O-methyl luteolin, and eupafolin was comparable to that of butylated hydroxy toluene. The activity of these compounds was mainly due to the presence of ortho-dihydroxy groups. The 1,1-diphenyl-2-picrylhydrazyl radical scavenging activity of these compounds showed the similar result (Miura et al., 2002).

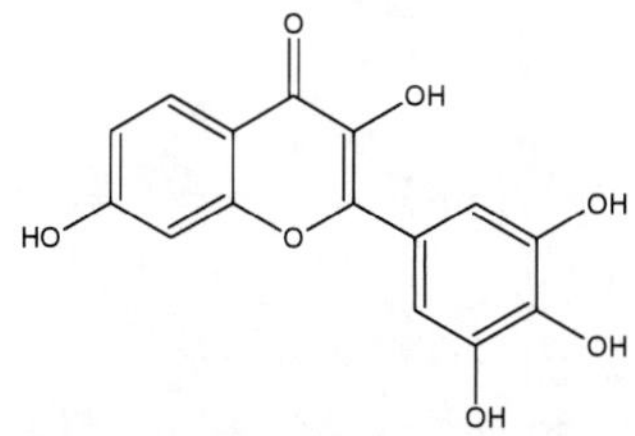

250
Robinetin

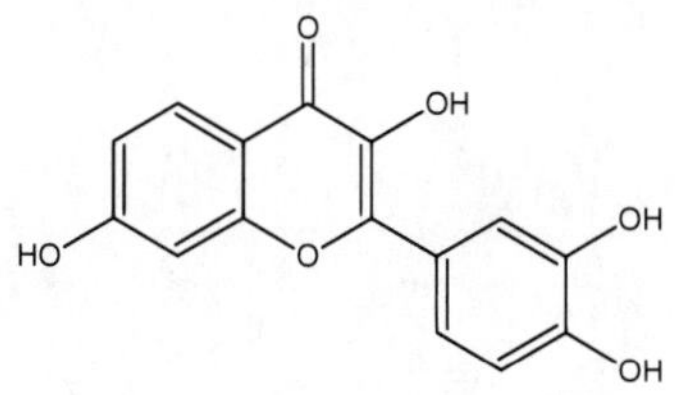

251
Fisetin

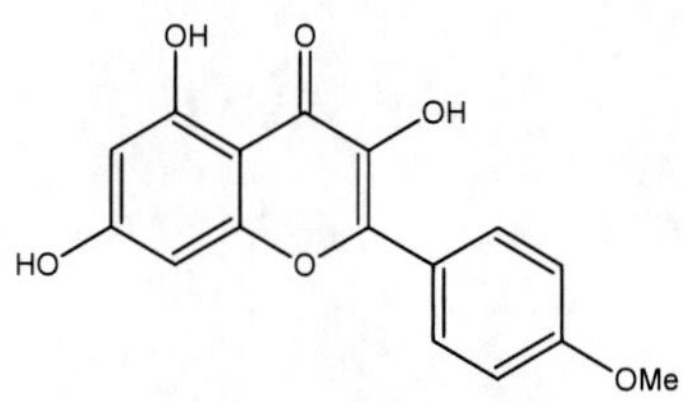

252
Kaempferide

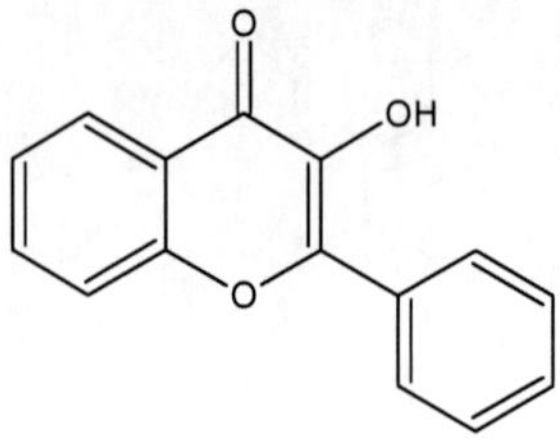

253
3-Hydroxyflavone

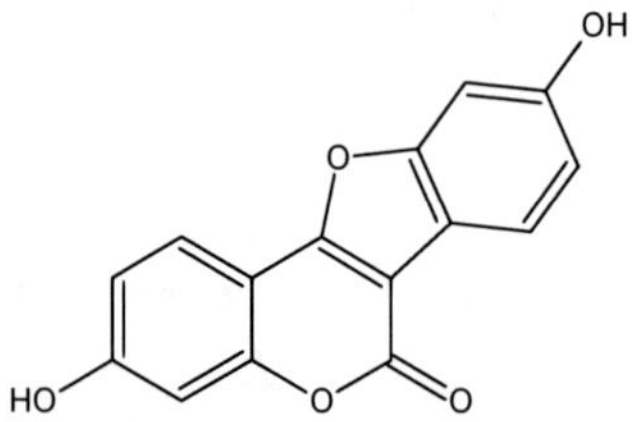

254
Coumestrol

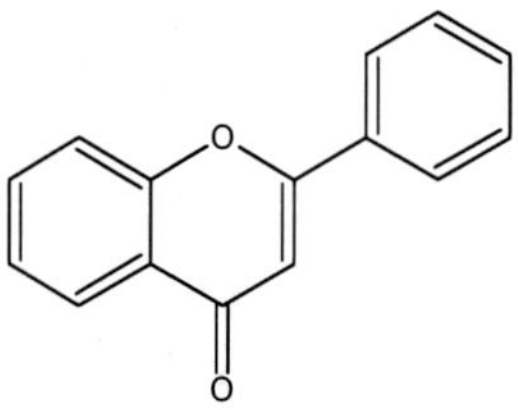

255
Flavone

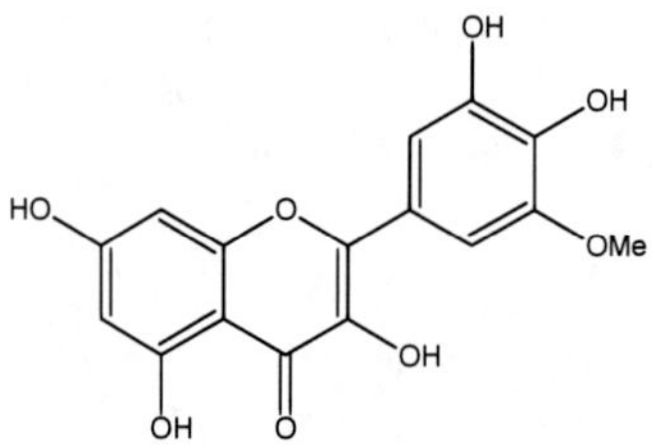

256
Laricytrin

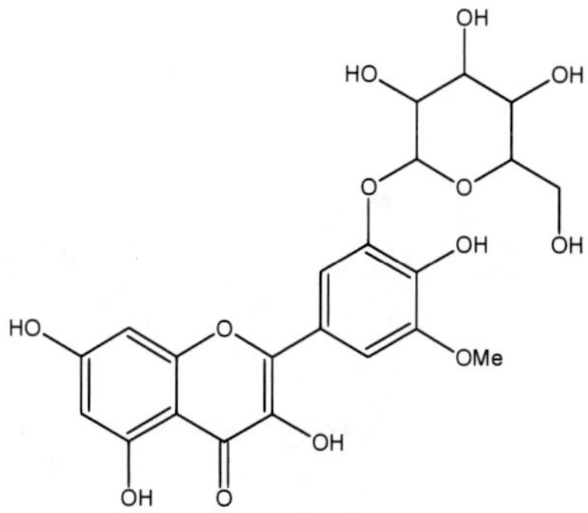

257
Laricytrin-3'-O-glucoside

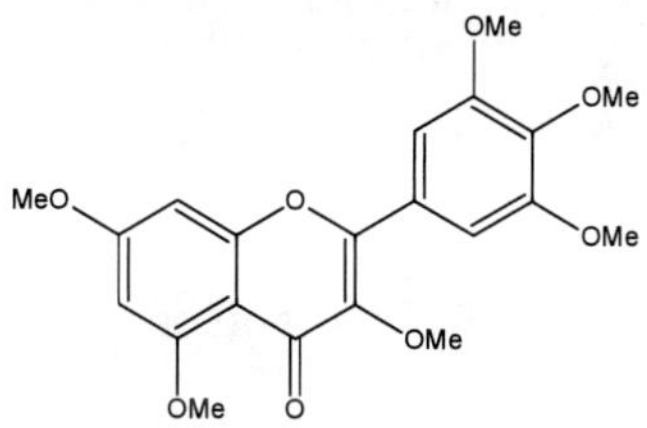

258
3,5,7,3',4',5'-Hexamethoxyfl avone

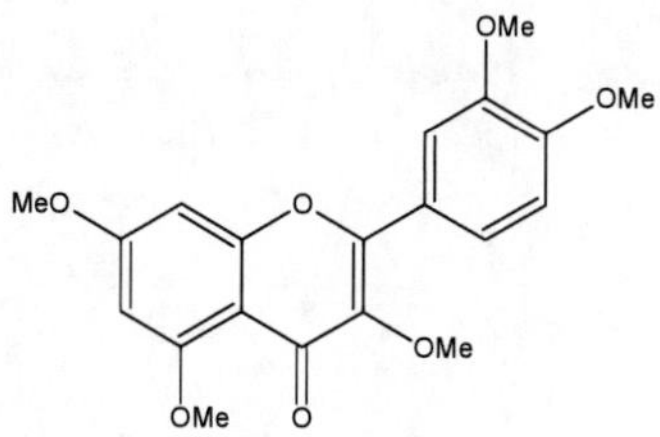

259
3,5,7,3',4'-Pentamethoxyfl avone

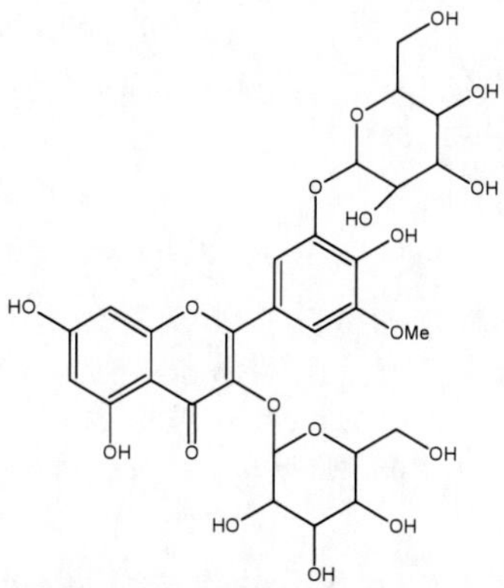

260
Laricytrin 3,3'-O-diglucoside

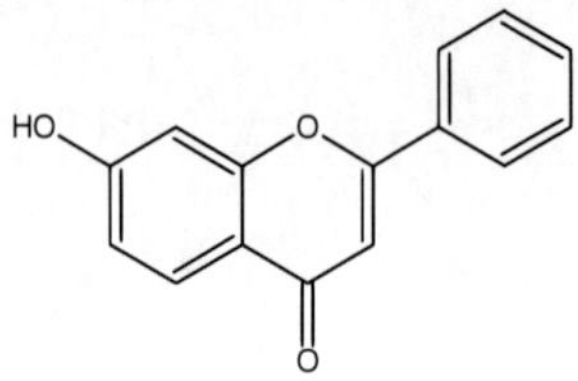

261
7-Hydroxyflavone

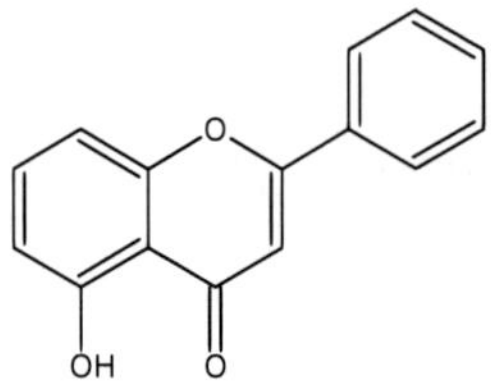

262
5-Hydroxyflavone

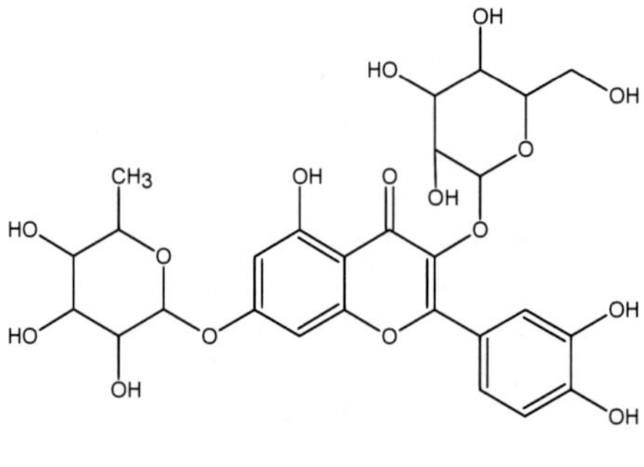

263
Quercetin 3-O-glucoside-7-O-rhamnoside

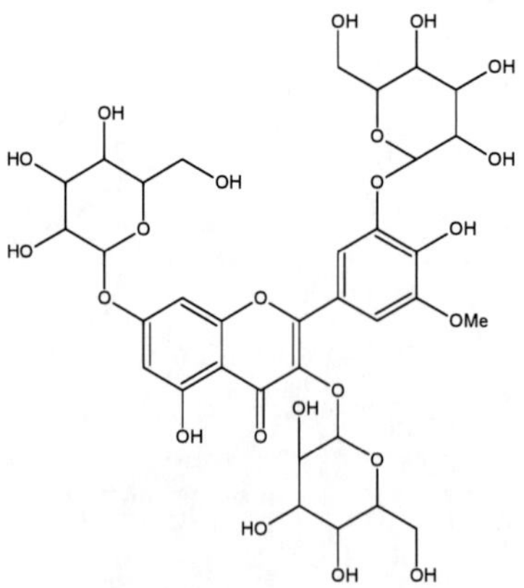

264
Laricytrin 3,7,3'-O-triglucoside

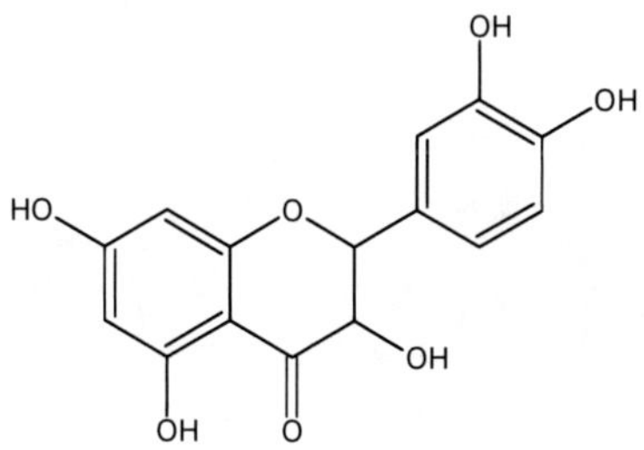

265
Taxifolin

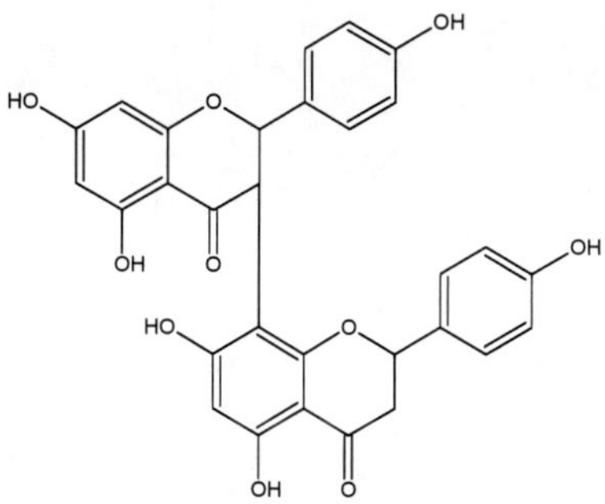

266
GB-1a

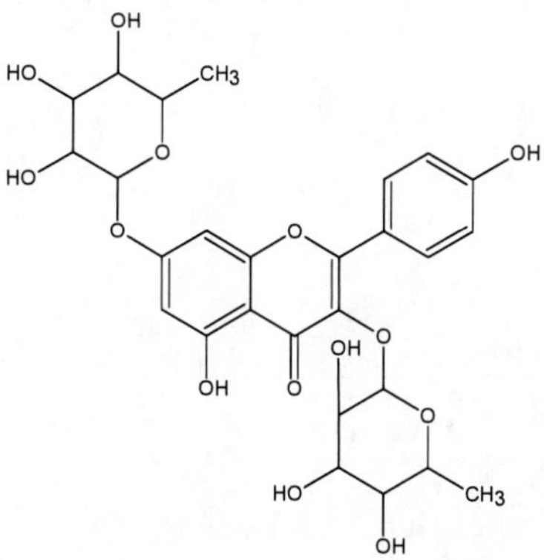

267
Kaempferol 3,7-O-dirhamnoside

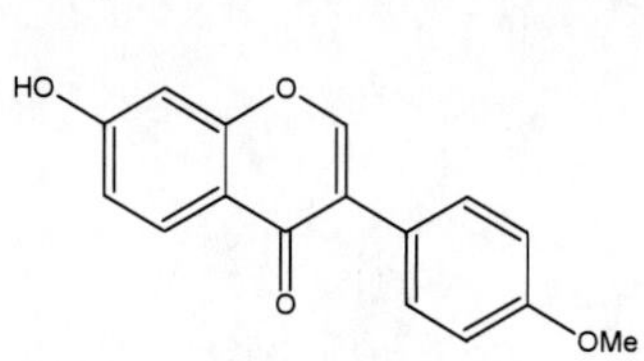

268
Formononetin

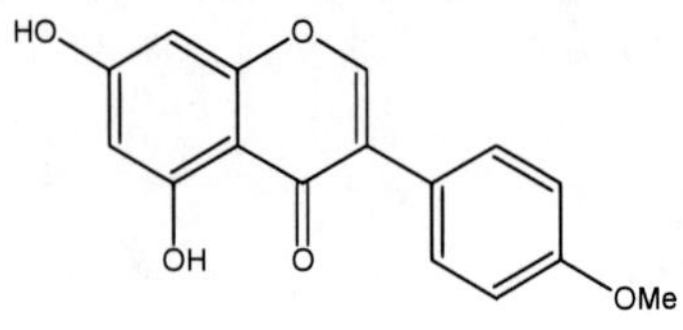

269
Biochanin A

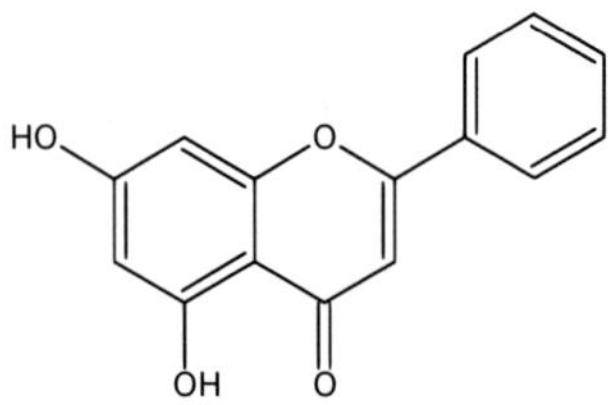

270
Chrysin

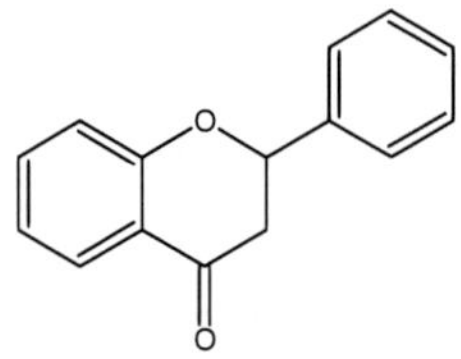

271
Flavanone

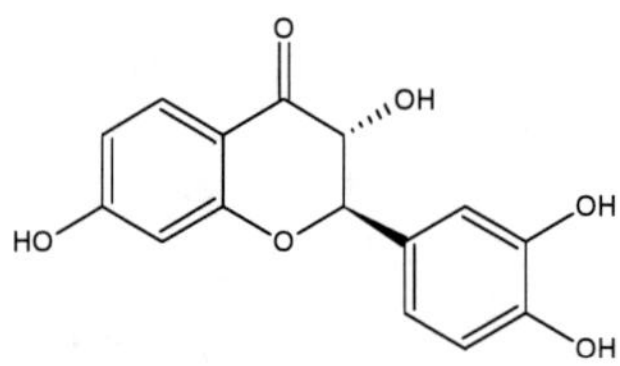

272
Fustin (dihydrofisetin)

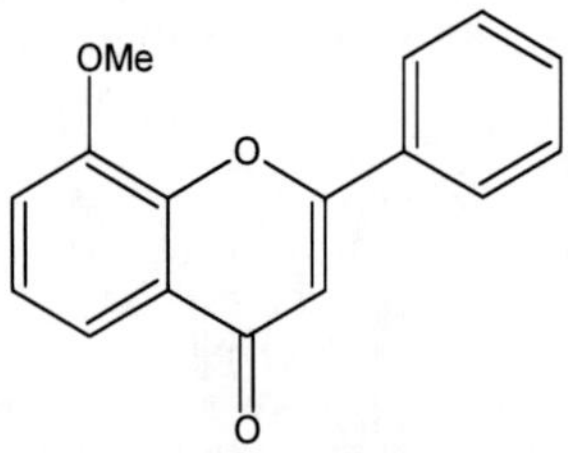

273
8-Methoxyfl avone

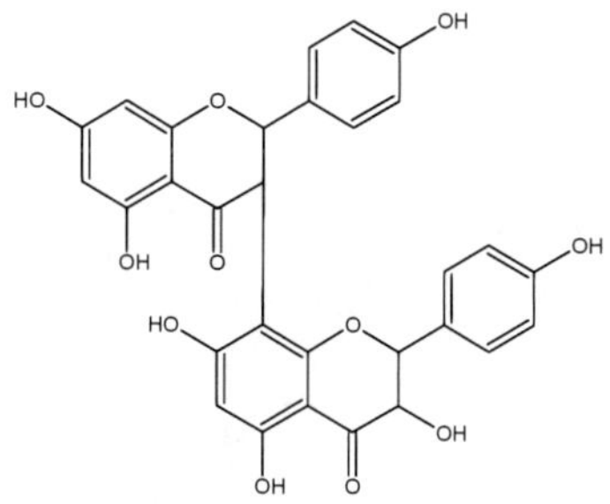

274
GB-1

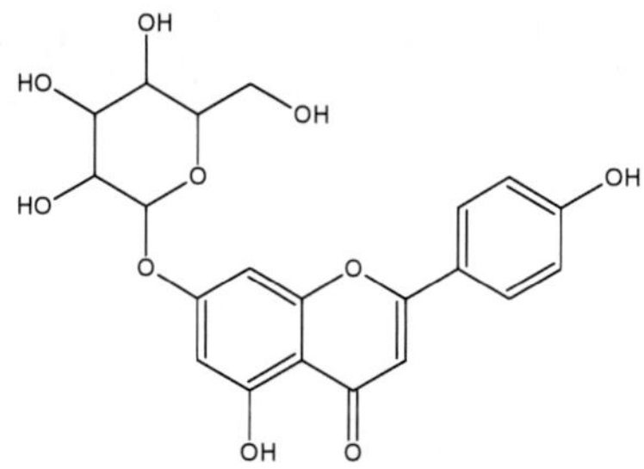

275
Apigenin 7-O-glucoside

Some compounds are flavonols with a free hydroxyl group at the C-3 position. This suggested that the flavonol C-3 hydroxyl group is responsible for the high inhibition of β-carotene oxidation in the heterogeneous system. Comparison of the antioxidant activity of flavonol aglycons with the activity of their glycosides or methyl derivatives showed that the blockage of the C-3 hydroxyl group resulted in a total loss of antioxidant activity. Glycosylation or methylation of other flavonol hydroxyls did not produce such an effect. The flavonol with a free 3-OH group had a higher antioxidant potential than its substituted derivatives. The highest antili-poperoxidant activity for free flavonols; taxifolin (dihy-droflavonol), rutin, and flavones had distintly lower activity. The ability of flavonoids to inhibit oxidation processes is controlled by the presence of the double bond between C-2 and C-3 and a free hydroxyl in C-3 position were found. The antioxidant activities obtained for laricytrin, laricytrin 3'-O-glucoside, and myricetin were low. The lower than expected antioxidant activity values obtained for these compounds can be explained by their high sensitivity to oxidation, which causes their rapid oxidation and partial decomposition during measurement oxidation (Burda and Oleszek, 2001).

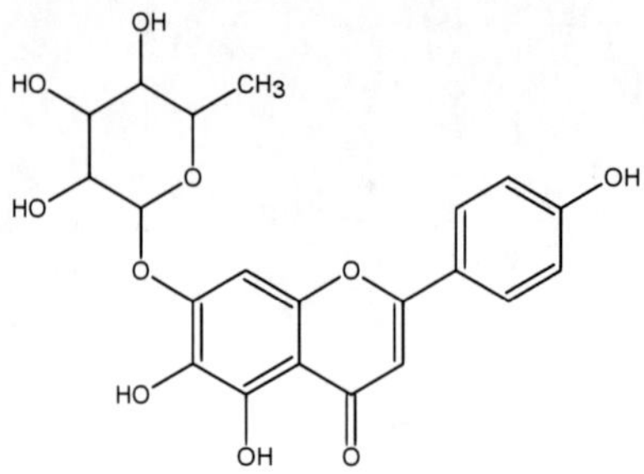

276
Sorbarin

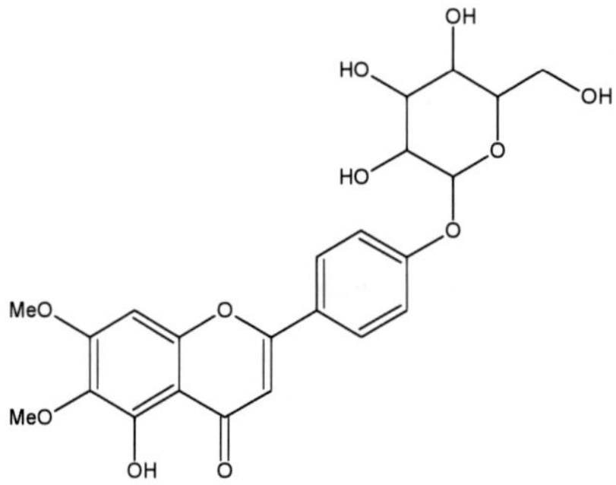

277
Cirsimarin

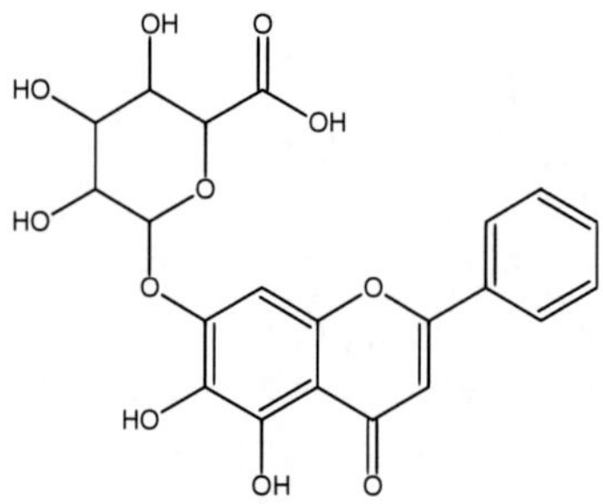

278
Baicalin

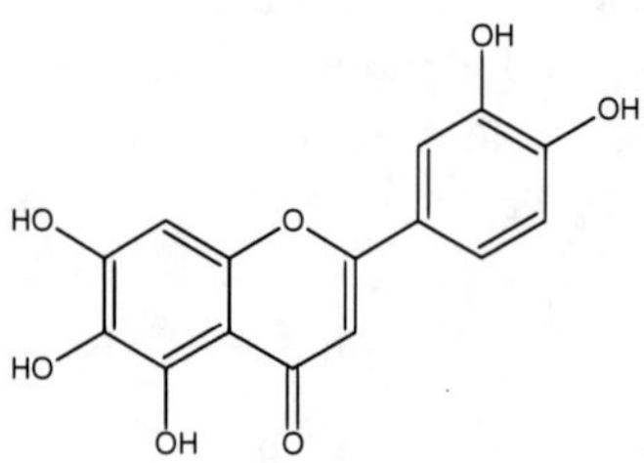

279
6-Hydroxyluteolin

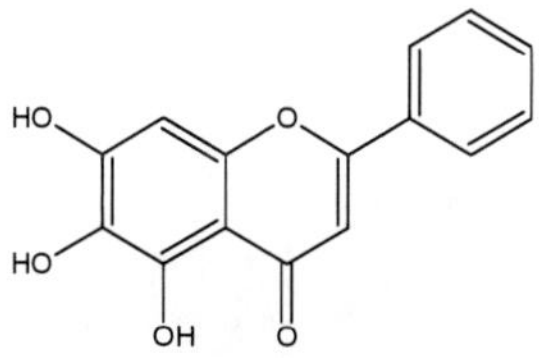

280
Baicalein

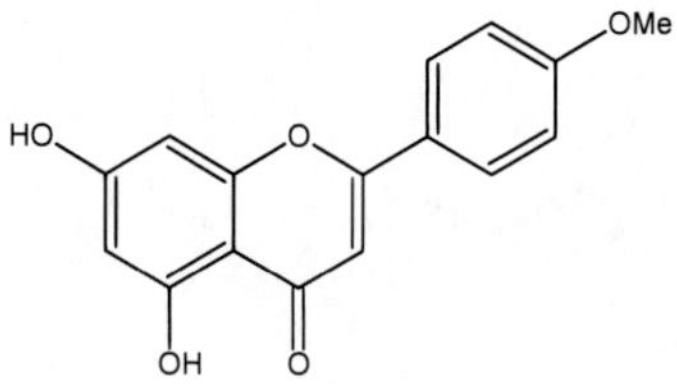

281
Acacetin

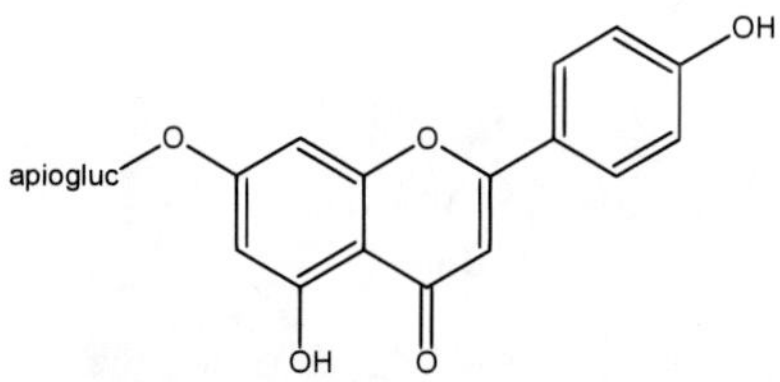

282
Apiin

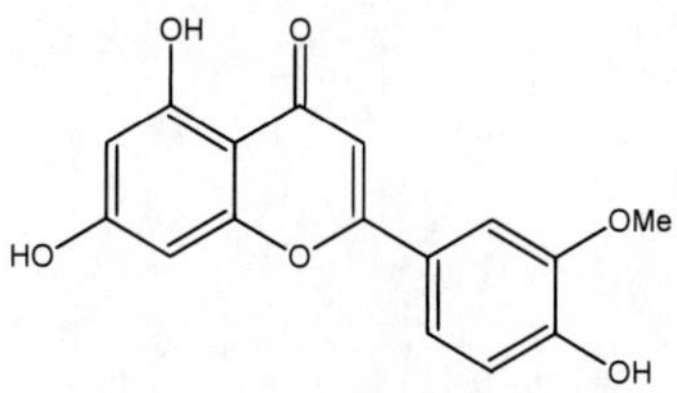

283
Chrysoeriol

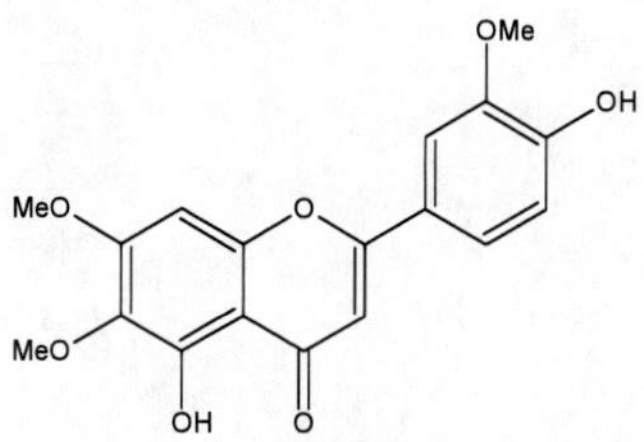

284
Cirsilineol

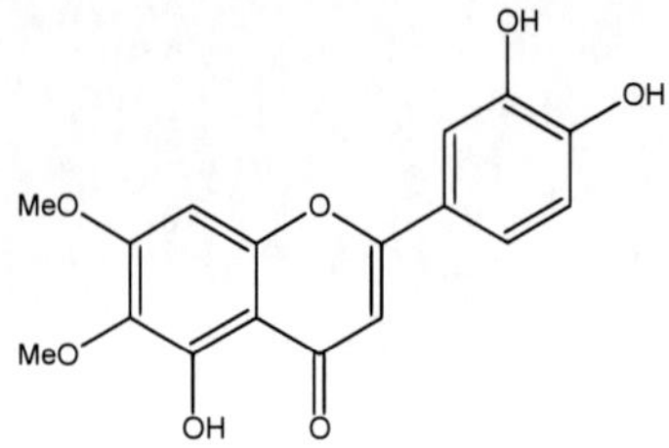

285
Cirsiliol

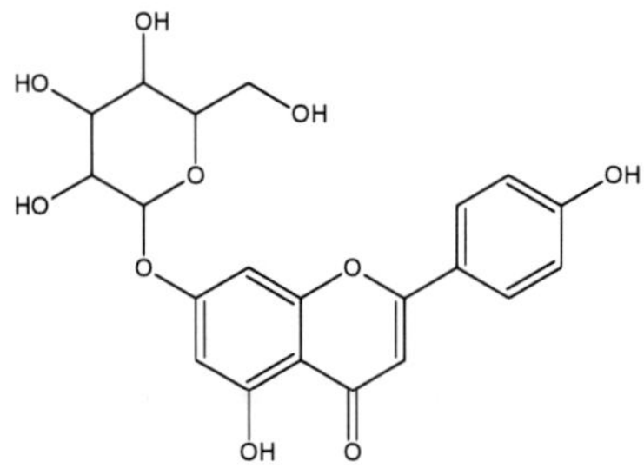

286
Cosmosiin

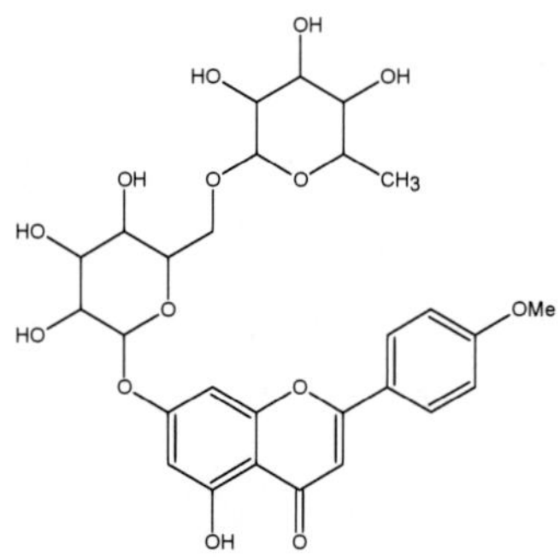

287
Linarin

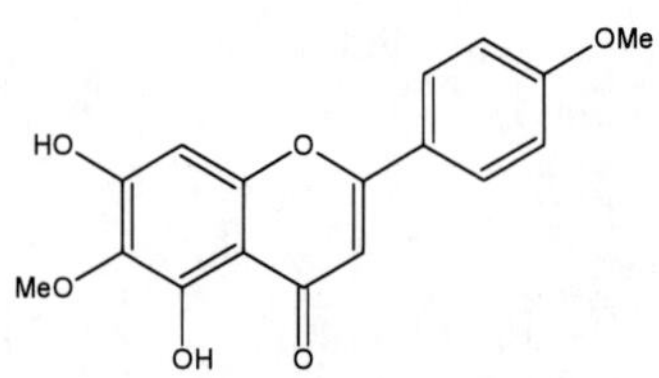

288
Pectolinarigenin

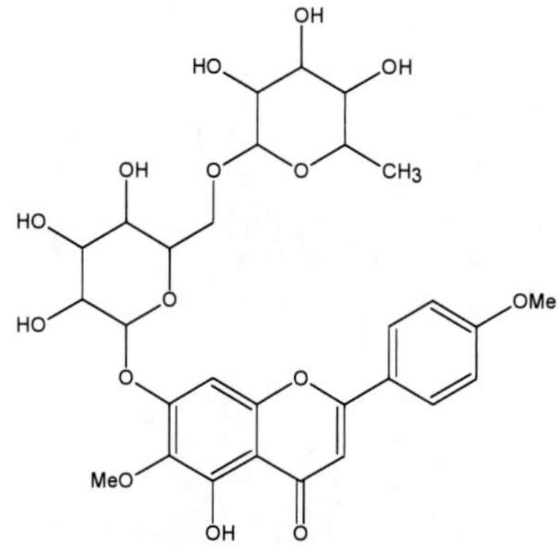

289
Pectolinarin

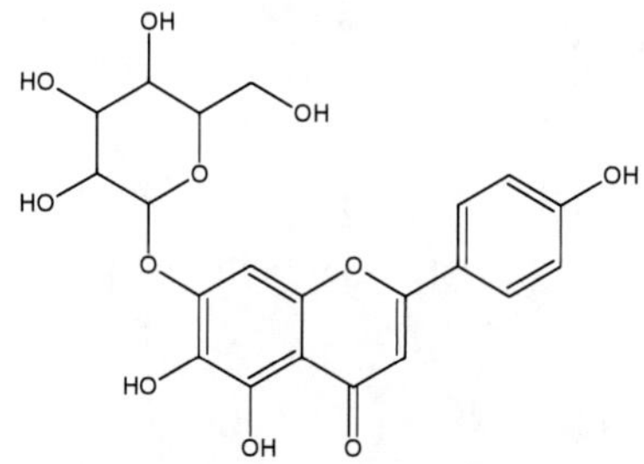

290
Plantaginin

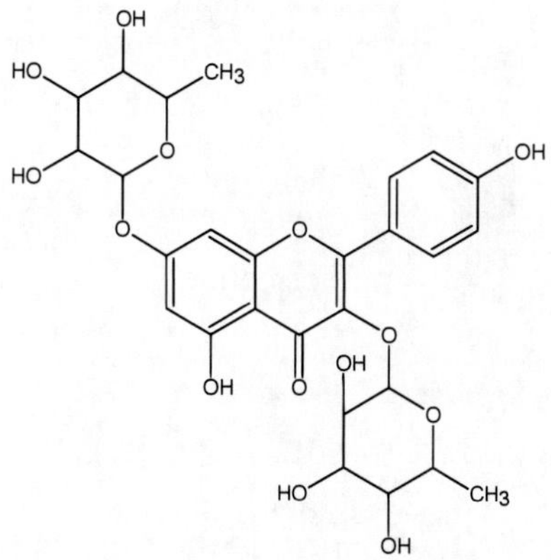

291
Kaempferitrin

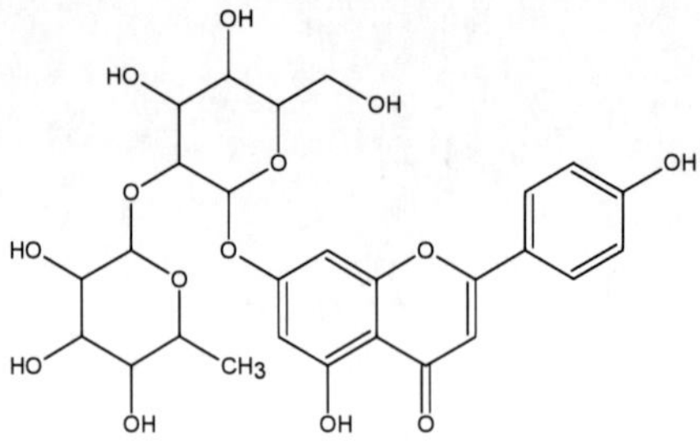

292
Rhoifolin

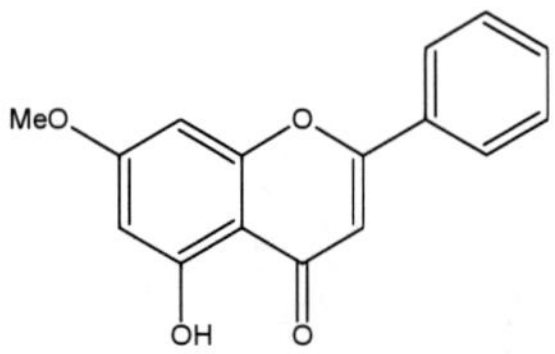

293
Tectochrysin

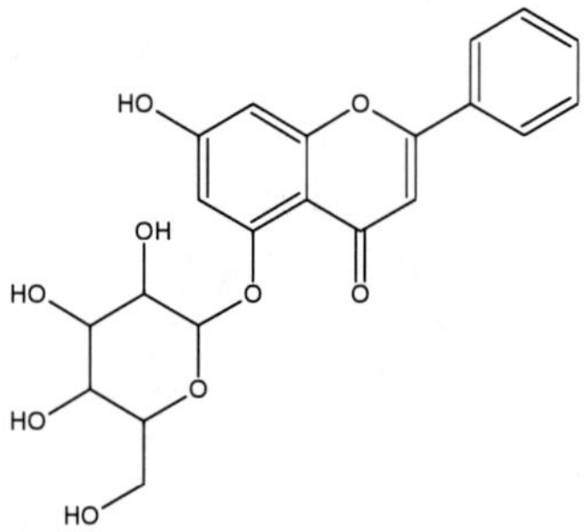

294
Toringin

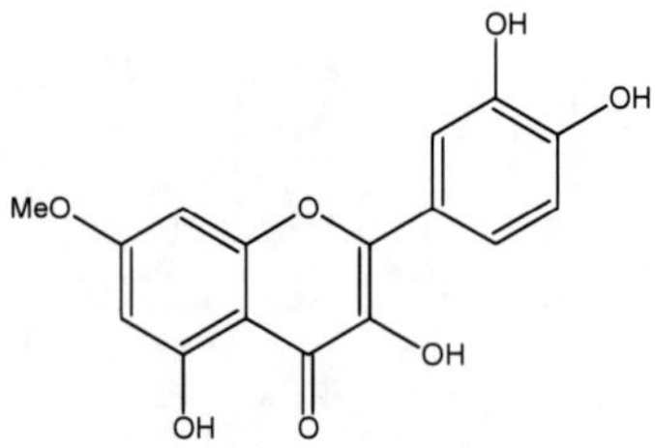

295
Rhamnetin

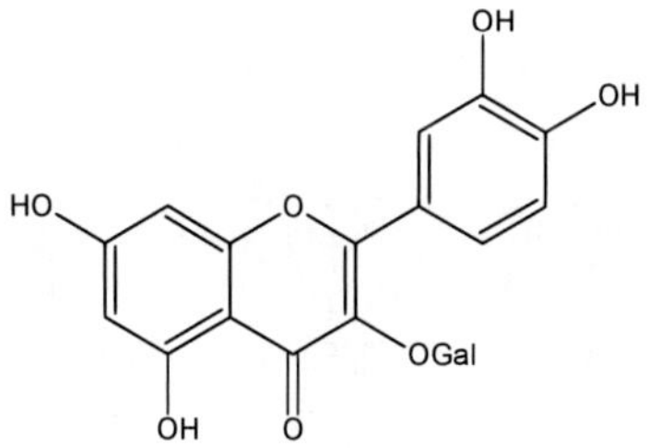

296
Hyperin

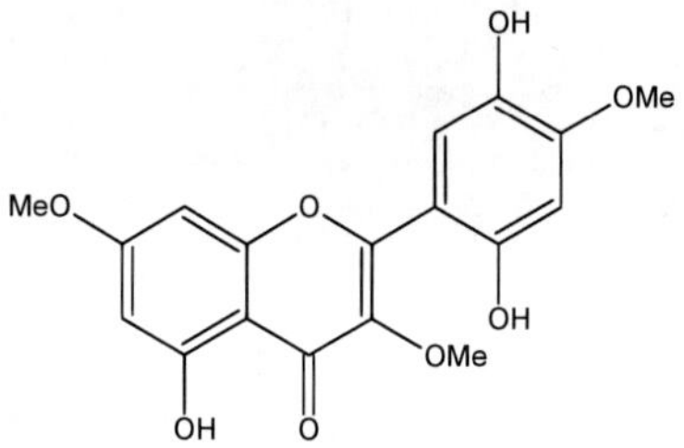

297
Oxyayanin A

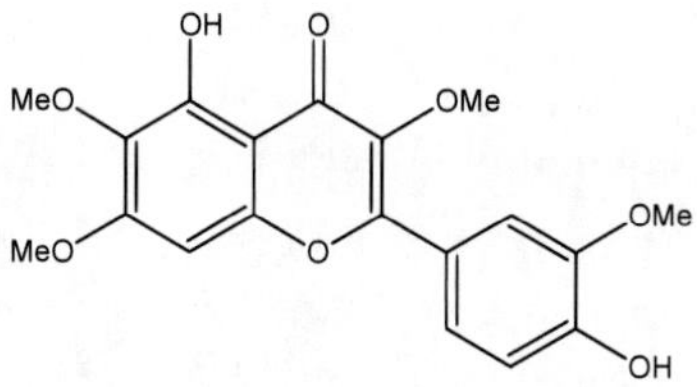

298
Chrysosplenol B

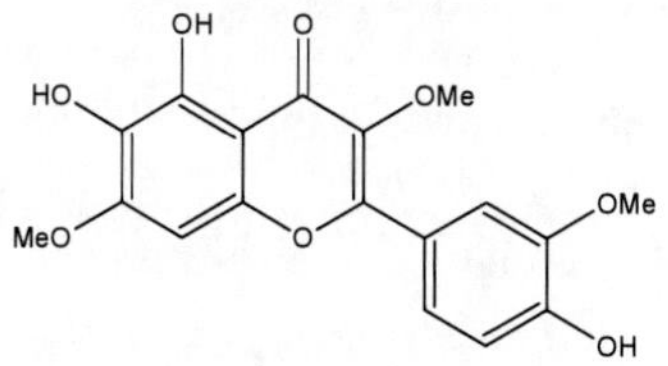

299
Chrysosplenol C

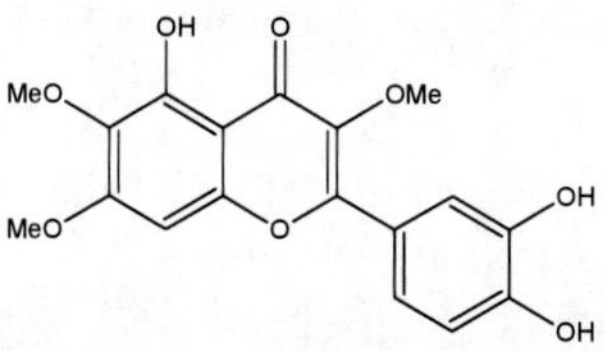

300
Chrysosplenol D

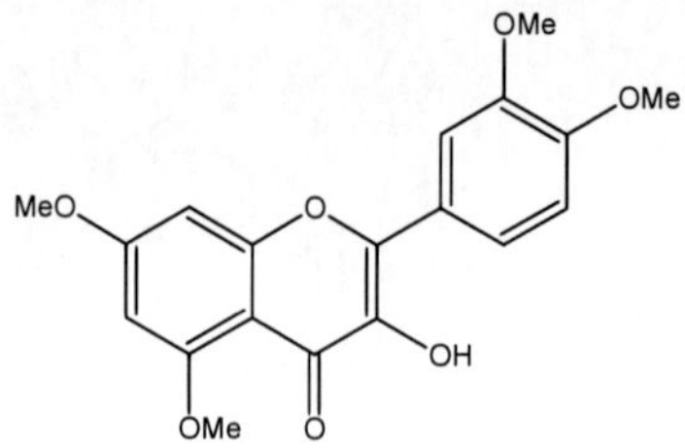

301
Tetramethylquercitin

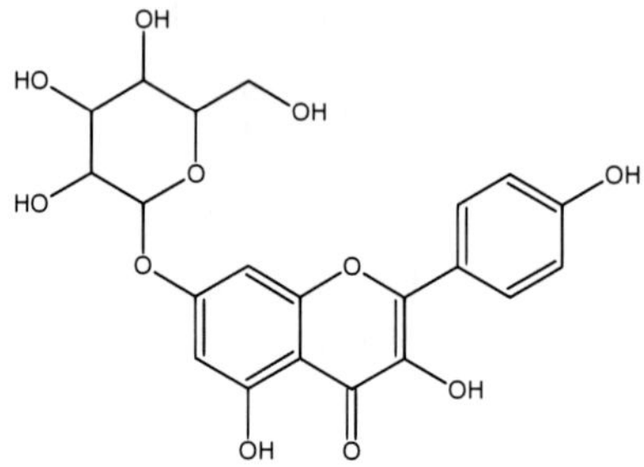

302
Kaempferol-7-glucoside

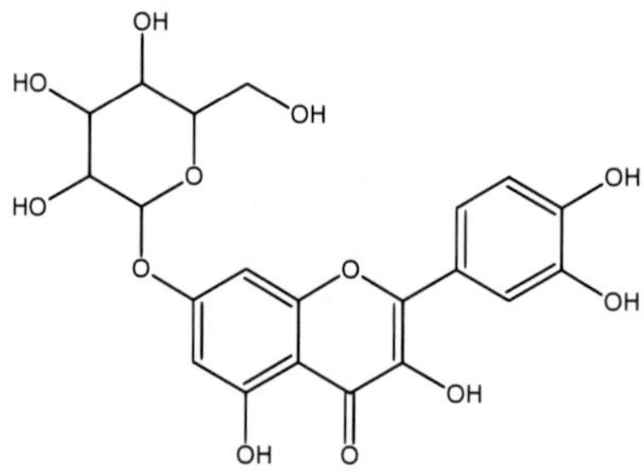

303
Quercimeritrin

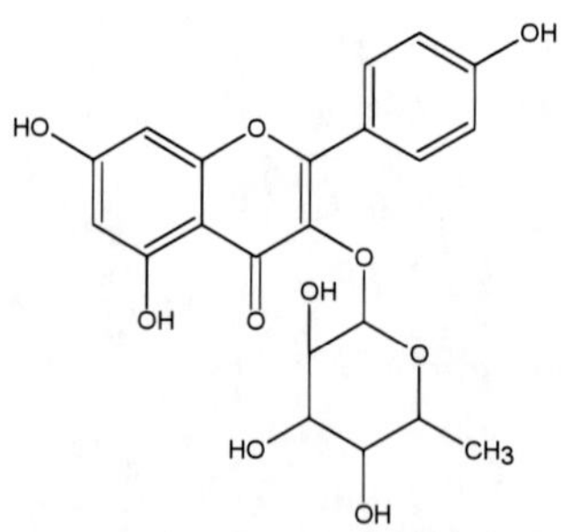

304
Afzelin

These compounds are widely distribuited. The antioxidant activity of these compounds against lipid peroxidation induced by H_2O_2 and Fe_{+2} (the so-called Fenton-type reaction). The activity, ranging from the strong to weak and occurred in the order: sorbarin > luteolin >cirsimarin > baicalin >6-hydroxyluteolin > baicalein> luteolin 7-glucoside and the IC_{50} values ranged from 0.82 to 125.07 μg/mL. The flavonol-type compounds had variable effects on the reaction with different IC_{50}; rhamnetin, hyperin, oxyayanin A, quercetin and chrysosplenol B were the strongest with IC_{50} values below 1.40 μg/mL, kaempferol and isoquercitrin were slightly weaker at 4.23 and 5.03 μg/mL respectively, and the rest showed an order of activity of chrysosplenol C > quercitrin > tetramethylquercitin > quercimeritrin, with IC_{50} values of 10.34, 20.56, 22.23 and 78.25 μg/mL, respectively. Generally, it is assumed that the radical scavenging and/or antioxidative potential for inhibiting TBARS formation caused by oxygen free radical species mediated by the Fenton-type reaction is decided mainly by the following three important structural groups: (a) the O-dihydroxy (catechol) structure in the B ring, which is the obvious radical target site for all flavonoids with a saturated 2,3-bond (flavan-3-Ols, flavanones); (b) the 2,3-double bond in conjugation with a 4-oxo function, which are responsible for electron delocalization from the B ring; and (c) the additional presence of both 3- and 5-hydroxyl groups for maximal radical-scavenging potential and strongest radical absorption; in fact, in tenns of their hydrogen bonds, the 3- and 5-hydroxyl groups are equivalent to the keto group (Yokozawa et al., 1997).

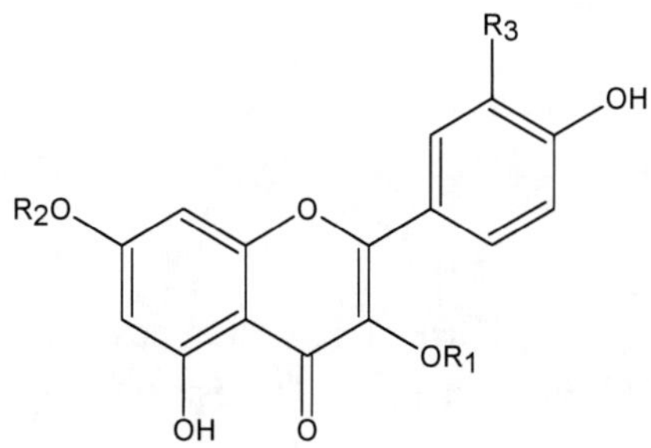

305

R_1= glc(l → 2)glc(1→ 6)caffeoyl, R_2 = rhamnose, R_3 = OH

Quercetin 3-O-(6-*trans*-caffeoyl)- β-gl ucopyranosyl-(1 → 2)- β-gl ucopyranoside-7-O-α-rhamnopyranoside

306

R_1= glc(l → 2)glc(1 → 6)caffeoyl, R_2 = rhamnose, R_3 = H

Kaempferol 3-O-(6-*trans*-caffeoyl)-β-glucopyranosyl-(1 → 2)- β-gl ucopyranoside- 7- O-α-rhamnopyranoside

307

R_1= glc(l → 2)glc(1 → 6)t-*p*-coumaroyl, R_2 = rhamnose, R_3 = OH

Quercetin 3-O-(6-*trans-p*-coumaroyl)- β-glucopyranosyl-(l → 2)- β-gl ucopyranoside- 7- O-α-rhamnopyranoside

308

R_1= glc(l → 2)glc(1→ 6)t-*p*-coumaroyl, R_2 = rhamnose, R_3 = H

Kaempferol 3-O-(6-*trans*-*p*-coumaroyl)-β-glucopyranosyl-(1 → 2)- β-g1ucopyranoside-7-O-α-rhamnopyranoside

309

R_1= glucose, R_2 = rha(l → 2)glc(1 → 6)caffeoyl, R_3 = OH

Quercetin 7-O-(6-*trans*-caffeoyl)- β-glucopyranosyl-(1 → 3)- α-rhamnopyranoside-3-O-β-glucopyranoside

310

R_1= glucose, R_2 = rha(l → 2)glc(1 → 6)caffeoyl, R_3 = H

Kaempferol 7-O-(6-*trans*-caffeoyl)- β-g1ucopyranosyl-(1 → 3)-α-rhamnopyranoside-3-O-β-glucopyranoside

311

R_1= glucose, R_2 = rha(l → 2)glc(1 → 6)t-*p*-coumaroyl, R_3 = H

Kaempferol 7-0-(6-*trans*-*p*-coumaroyl)-β-g1ucopyranosyl-(1→3)-α-rhamnopyranoside-3-O-β-glucopyranoside

312

R_1= ga1(l → 2)acetyl, R_2 = H, R_3 = H

Kaempferol 3-O-β-(2"-acetyl)galactopyranoside

313

R_1= gal(l→ 2)acetyl, R_2 = arabinose, R_3 = H

Kaempferol 3-O-β-(2"-acetyl)galactopyranoside-7-O-α-arabinopyranoside

314

R_1= ga1(l→ 2)acetyl, R_2 = arabinose, R_3 = OH

Quercetin 3-O-β-(2"-acetyl)galactopyranoside-7-O-α-arabinopyranoside

315

R_1= rhamnose, R_2 = rhamnose, R_3 = OH

Quercetin 3, 7-di-O-α-rhamnopyranoside

316

R_1= rhamnose, R_2 = rhamnose, R_3 = H

Kaempferol 3,7-di-O-α-rhamnopyranoside

317

R_1= glucose, R_2 = rhamnose, R_3 = OH

Quercetin 3-O-β-glucopyranoside- 7-O-α-rhamnopyranoside

Found in *Aconitum napellus* subsp. tauricum, *A. napellus* subsp. neomontanum, *A. paniculatum*, and *A. vulparia* (Ranunculaceae).Their antioxidant activity was determined by measuring free radical scavenging activity by DPPH test and the coupled oxidation of β-carotene and linoleic acid assay. The results showed that quercetin 7-O-(6-*trans*-caffeoyl)-β-glucopyranosyl-(1→3)-α-rhamnopyranoside-3-O-β-glucopyranoside is the most active compound in the DPPH free-radical scavenging test (IC_{50} 1.9 μ.M) while in the coupled oxidation of β-carotene and linoleic acid assay compound quercetin 3-O-(6-trans-caffeoyl)-β-glucopyranosyl-(1→2)-β-glucopyranoside-7-O-α-rhamnopyranoside has the highest inhibitory ratio after 1 h (58.9%), (Braca, et al., 2003).

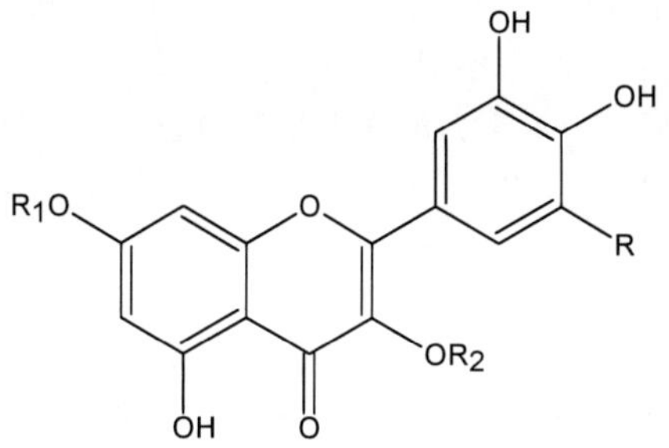

318

R = OH R_1=β-D-glc-(1→6)-β-D-glc R_2 = H

Myricetin 7-O-β-D-glucopyranosyl-(1→6)- β-D-glucopyranoside

319

R = OH R_1=α-L-rha-(1→6)-β-D-glc R_2 = H

Myricetin 7-O-α-L-rhamnopyranosyl-(1→6)- β-D-glucopyranoside

320

R = H R_1=H R_2 = β-D-glc

Quercetin 3-O-β-D-glucopyranoside

321

R = H R_1=H R_2 = β-D-glc

Quercetin 3-O-β-D-galactopyranoside

These compounds were isolated from the leaves of *Tachigalia paniculata* (Leguminosae).7-O-β-D-glucopyranosyl-(1-6)-β-D-glucopyranoside and myricetin 7-O-α-L-rhamnopyranosyl-(1-6)- β-D-glucopyranoside, showed activity in the assays, trolox equivalent antioxidant capacity (TEAC), the coupled oxidation of β-carotene and linoleic acid (autoxidation assay), and the inhibition of xanthine oxidase activity. Both compounds were the most potent of the isolates in the xanthine oxidase test (Cioffi et al., 2002).

7-O-β-D-Glucopyranosyl-(1→6)-β-D-glucopyranoside: Yellow amorphous powder; $[\alpha]_D$ -72° (c 0.1, MeOH); UV (MeOH): λmax (log ε) 270 (4.31),360 (3.75) nm; IR (KBr): νmax 3400, 3020, 1650, 1610, 1570 cm^{-1}; ESIMS: *m/z* 643 $[M + 1]^+$; anal. C 50.39%, H 4.73%, O

44.88%, calcd for $C_{27}H_{30}O_{18}$, C 50.47%, H 4.71%, O 44.82%; ^{1}HNMR (600 MHz, CD_3OD): δ 3.26 (1H, t, J = 9.0 Hz, H-3'''), 3.35 (lH, t, J = 9.0 Hz, H-4'''), 3.38 (lH, br, t, J = 9.0 Hz, H-2'''), 3.42 (lH, m, H-5'''), 3.48 (lH, m, H-5''), 3.58 (lH, t, J = 9.0 Hz, H-3''), 3.68 (lH, dd, J = 12.2, 2.0 Hz, H-6'''a), 3.76 (lH, dd, J = 12.0, 2.0 Hz, H-6''a), 3.83 (lH, br t, J = 9.0 Hz, H-2''), 3.86 (lH, t, J = 9.0 Hz, H-4''), 3.92 (lH, dd, J = 12.2, 5.0 Hz, H-6'''b), 3.98 (lH, dd, J = 12.0, 5.0 Hz, H-6''b), 4.72 (lH, d, J = 8.0 Hz, H-1'''), 5.25 (lH, d, J= 7.8 Hz, H-1''), 6.49 (lH, d, J = 1.8 Hz, H-6), 6.75 (lH, d, J = 1.8 Hz, H-8), 7.00 (2H, s, H-2', H-6'); ^{13}C NMR (600 MHz, CD_3OD): δ 158.1 (C-2), 136.5 (C-3), 177.0 (C-4), 164.5 (C-5), 100 (C-6), 165.0 (C-7), 94.5 (C-8), 159 (C-9), 103.5 (C-10), 123 (C-1´), 109.6 (C-2´), 147 (C-3´), 137.5 (C-4´), 147 (C-5´), 109.6 (C-6´), 99.7 (7-O-glc-1´´), 73.3 (C-2´´), 76.2 (C-3´´), 69.8 (C-4´´), 77.8 (C-5´´), 67 (C-6´´), 100.3 (glc-1´´´), 74 (C-2´´´), 78.1 (C-3´´´), 70.1 (C-4´´´), 78.2 (C-5´´´), 61.5 (C-6´´´).

Myricetin 7-O-α-L-rhamnopyranosyl-(1→6)-β-D-glucopyranoside: Yellow amorphous powder; $[\alpha]_D$ -121° (c 0.1, MeOH); UV (MeOH): λmax (log ε) 272 (4.03), 365 (3.67) nm; IR (KBr): νmax, 3450, 3000, 1660, 1605, 1555 cm^{-1}; ESIMS: *m/z* 627 $[M + 1]^+$; anal. C 51.71 %, H 4.84%, O 43.45%, calcd for $C_{27}H_{30}O_{17}$, C 51.76%, H 4.83%, O 43.41 %; ^{1}HNMR (600 MHz, CD_3OD): δ 1.10 (3H, d, J = 6.0 Hz, H.6'''), 3.45 (lH, t, J = 9.0 Hz, H-3''), 3.46 (lH, t, J = 9.0 Hz, H-4''), 3.50 (lH, br t, J = 9.0 Hz, H-2''), 3.52 (lH, m, H-5''), 3.55 (lH, t, J = 8.5 Hz, H-4'''), 3.70 (lH, dd, J = 12.0, 2.0 Hz, H-6''a), 3.84 (lH, m, H-2'''), 3.88 (lH, dd, J = 8.5, 3.5 Hz, H-3'''), 4.00 (lH, dd, J = 12.0, 5.0 Hz, H-6''b), 4.15 (lH, m, H-5'''), 4.80 (lH, d, J = 1.5 Hz, H-1'''), 5.10 (lH, d, J = 7.8 Hz, H-1''), 6.51 (lH, d, J = 1.8 Hz, H-6), 6.73 (lH, d, J = 1.8 Hz, H-8), 7.00 (2H, s, H-2', H-6'); ^{13}C NMR (600 MHz, CD_3OD): δ 158.8 (C-2), 136.3 (C-3), 178.0 (C-4), 164.3 (C-5), 100.1 (C-6), 165.2 (C-7), 95 (C-8), 159 (C-9), 103.8 (C-10), 123 (C-1´), 109.8 (C-2´), 147.2 (C-3´), 137.9 (C-4´), 147.2 (C-5´), 109.8 (C-6´), 101.2 (7-O-glc-1´´), 73.9 (C-2´´), 78.2 (C-3´´), 71.4 (C-4´´), 77.2 (C-5´´), 67.4 (C-6´´), 102.4 (rha-1´´´), 72.1 (C-2´´´), 72.4 (C-3´´´), 74 (C-4´´´), 69.7 (C-5´´´), 17.8 (C-6´´´).

Quercetin 3-O-β-D-glucopyranoside: FD-MS: *m/z* 465 (M + H); ^{1}HNMR: δ 12.65 (lH, s, 5 OH), 7.58 (lH; dd, H6'), 7.57 (1H, d, H2'), 6.84 (lH, ,d, J = 9.0 Hz, H5´), 6.40 (lH, d, J = 2.0 Hz, H8), 6.20 , (lH, d, J = 2.0 Hz, H6), 5.46 (lH, d, J = 7.4 Hz, G1),3.58 (lH, m, G6), 3.32 (lH, m, G6), 3.23 (2H, m, G2, G3), 3.09 (2H, m, G4, G5); ^{13}C NMR: δ 156.1 (C-2), 138.3 (C-3), 177.4 (C-4), 104.0 (C-4a), 161.2 (C-5), 98.6 (C-6), 164.1 (C-7), 93.5 (C-8), 156.3 (C-8a), 121.1 (C-1´), 116.2 (C-2´), 144.8 (C-3´), 148.4 (C-4´), 115.2 (C-5´), 121.6 (C-6´), 100.8 (C-1´´), 74.1 (C-2´´), 76.5 (C-3´´), 69.9 (C-4´´), 77.5 (C-5´´), 60.9 (C-6´´).

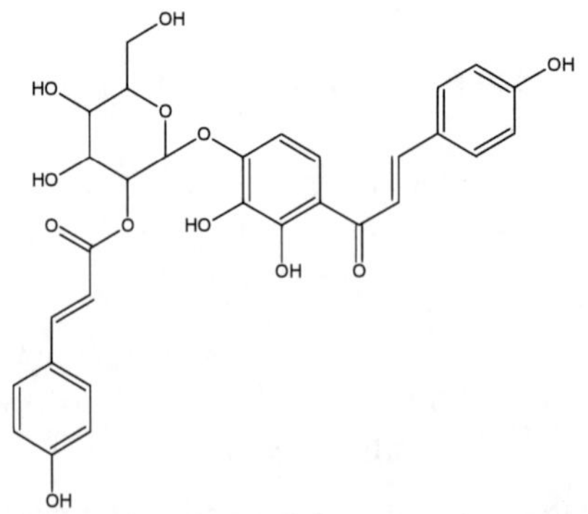

322
4'-O-β-D-(2''-*p*-Coumaroyl)glucopyranosyl-4,2',3'-trihydroxychalcone

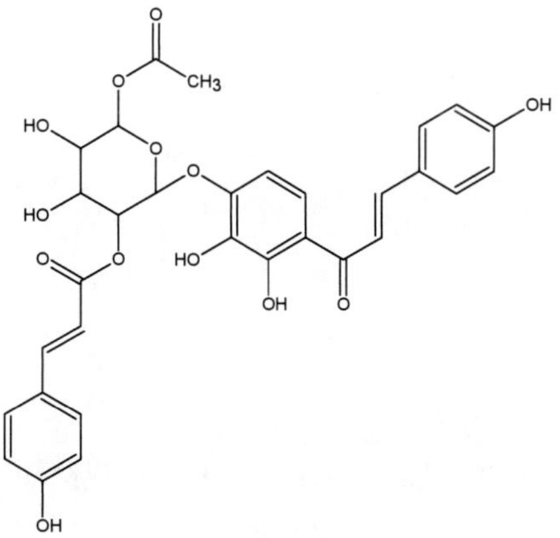

323

4'-O-β-D-(2"-*p*-Coumaroyl-6"- acetyl) glucopyranosyl-4,2',3'-trihydroxychalcone

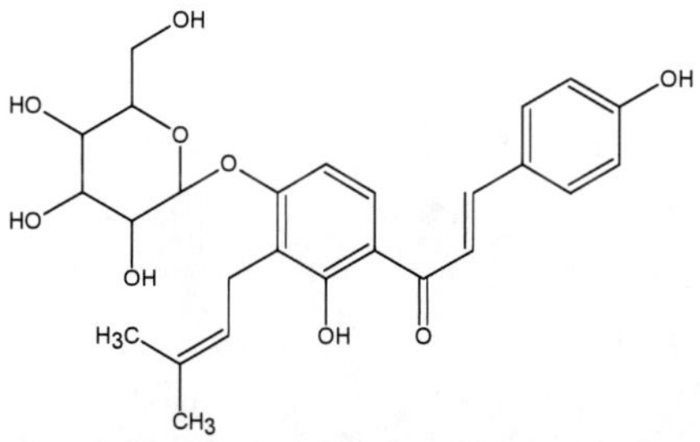

324

3'-(3-Methyl-2-butenyl)-4'-O-β-D- glucopyranosyl-4,2'-dihydroxychalcone

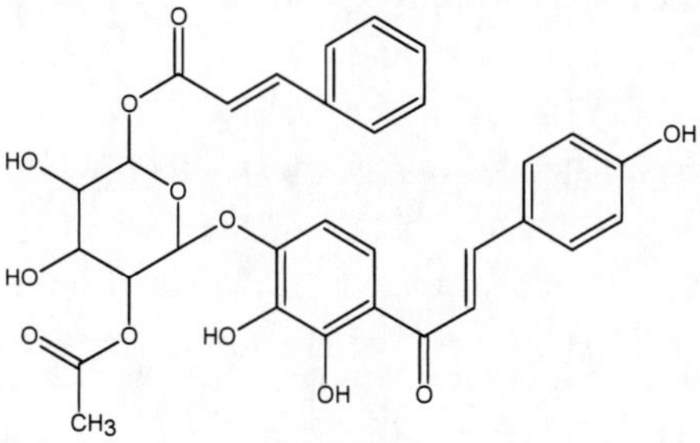

325

4'-O-β-D-(2"-Acetyl-6"-cinnamoyl) glucopyranosyl- 4,2',3'-trihydroxychalcone

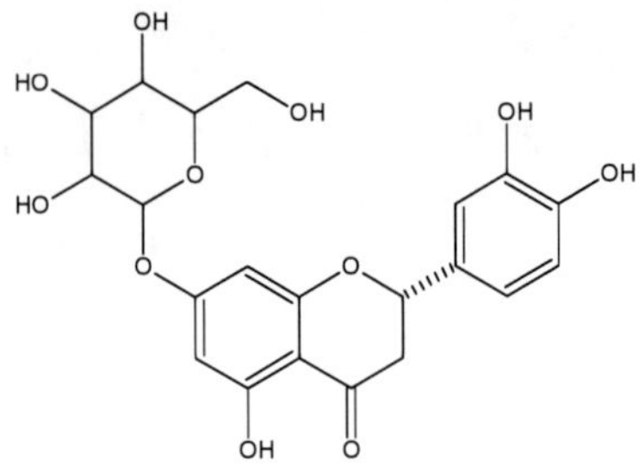

326
Eriodictyol 7-O-β-D-glucopyranoside

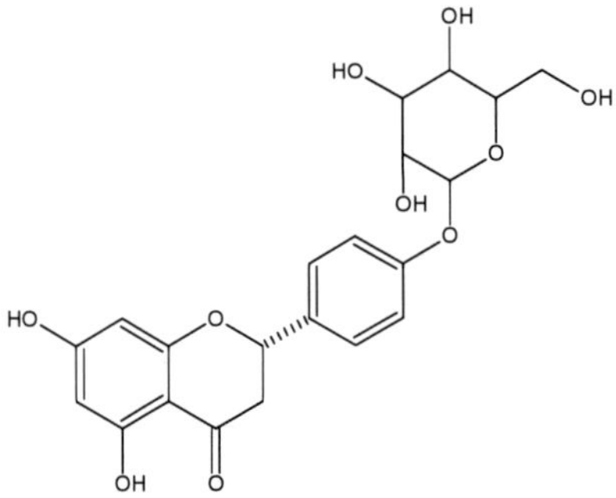

327
Naringenin 4'-O-β-D-glucopyranoside

These chalcone glycosides, and flavanones were isolated from the methanol extract of stem bark of *Maclura tinctoria* (Moraceae). The antioxidant activity of all the isolated compounds was determined by measuring free-radical-scavenging effects using two different assays, the trolox equivalent antioxidant capacity (TEAC) and the coupled oxidation of β-carotene and linoleic acid (autoxidation assay). 3'-(3-methyl-2-butenyl)-4'-O-β-D-glucopyranosyl-4,2'-dihydroxychalcone was the most active in both antioxidant assays (Cioffi et al., 2003).

4'-O-β-D-(2"-*p*-Coumaroyl)glucopyranosyl-4,2',3'-trihydroxy: Yellow amorphous powder; $[\alpha]_D$ +62° (c 0.1, MeOH); UV (MeOH): λmax (log ε) 293 (5.71),316 (4.75), 380 (3.42) nm; ESIMS: *m/z* 581 $[M + H]^+$, 435 $[(M + H) -146]^+$, 273 $[(M + H) -146 -162]^+$; anal. C 62.02%, H 4.89%, calcd for $C_{30}H_{28}O_{12}$, C 62.07%, H 4.86%; ^{1}HNMR (600 MHz, CD_3OD): δ 3.42 (lH, t, J = 9.0 Hz, H-4"), 3.48 (lH, t, J = 9.0 Hz, H-3´´), 3.50 (lH, m, H-5´´), 3.72 (lH, dd, J = 12.0, 3.5 Hz, H-6"a), 3.93 (lH, dd, J = 12.0, 5.5 Hz, H-6"b). 4.45 (lH, br, t, J = 8.6 Hz, H-2´´), 5.15 (lH, d, J = 7.5 Hz, H-1´´), 6.38 (lH, d, J = 16.0 Hz, H-8'"), 6.73 (2H, d, J = 8.0 Hz. H-3'", H-5'"), 6.80 (lH, d, J = 8.0 Hz, H-5'), 6.95 (2H, d, J = 8.0 Hz, H-2, H-6), 7.25 (2H, d, J = 8.0 Hz, H-3, H-5), 7.41 (lH, d, J = 16.0 Hz, H-7'"), 7.45 (lH, d, J = 15.0 Hz, H-α), 7.46 (2H, d, J = 8.0 Hz, H-2'", H-6'"), 7.55 (lH, d, J= 8.0 Hz. H-6'), 7.97 (lH, d, J = 15.0 Hz, H-β); ^{13}C NMR (600 MHz, CD_3OD): δ 127 (C-1), 130.8 (C-2), 118 (C-3), 162 (C-4), 118 (C-5), 130.8 (C-6), 118 (α), 147.3 (β), 198 (CO), 118 (C-1´), 155 (C-2´), 137 (C-3´), 151 (C-4´), 110 (C-5´), 121.1 (C-6´), 99.4 (C-1´´), 75.8 (C-2´´), 76.8 (C-3´´), 71.5 (C-4´´), 77.8 (C-5´´),

61.5 (C-6´´), 124 (C-1´´´), 130 (C-2´´´), 116.7 (C-3´´´), 161.2 (C-4´´´), 116.7 (C-5´´´), 130 (C-6´´´), 146.8 (C-7´´´), 118 (C-8´´´), 169 (C-9´´´).

4'-O-β-D-(2"-*p*-Coumaroyl-6"-acetyl)glucopyranosyl-4,2',3'-trihydroxychalcone: yellow amorphous powder; $[\alpha]_D$ +70° (c 0.1. MeOH): UV (MeOH) λmax (log ε) 295 (4.76), 320 (4.12), 384 (3.51) nm; ESIMS: *m/z* 623 $[M + H]^+$, 477 $[(M+ H) -146]^+$,435 $[(M + H) -146- 42]^+$, 273 $[(M + H) -146- 42- 162]^+$; anal. C 61.70%, H 4,87%, calcd for $C_{32}H_{30}O_{13}$, C6l,74%, H4.86%; ^{1}HNMR (600 MHz, CD_3OD): δ 1.98 (3H, s, CH_3CO), 3.48 (lH, t, J = 9.0 Hz, H-3´´), 3.49 (lH, t, J = 9.0 Hz, H-4´´), 3.62 (lH, m, H-5´´), 4.05 (lH, dd, J = 12.0, 3.5 Hz, H-6"a), 4.48 (lH, br t, J = 8.5 Hz, H-2´´), 4.68 (lH, dd, J =12.0, 5.5 Hz. H-6"b), 5.12 (lH, d, J = 7.6 Hz, H-1´´), 6.20 (lH, d, J = 15.5 Hz, H-8'''), 6.70 (2H, d, J = 8.0 Hz, H-3''', H-5'''), 6.81 (lH. d, J = 8.0 Hz, H-5'), 6.89 (2H, d, J = 8.0 Hz, H-2, H-6), 7.13 (lH, d, J = 8.0 Hz, H-6'), 7.30 (lH, d, J = 16.0 Hz, Hβ), 7.33 (2H, d, J = 8.0 Hz, H-3, H-5), 7.40 (lH, d. J = 16.0 Hz, Hα), 7.46 (2H. d, J = 8.0 Hz, H-2''', H-6'''), 7 .50 (lH, d, J = 15.5 Hz, H-7'''); ^{13}C NMR (600 MHz, CD_3OD): δ127 (C-1), 130.3 (C-2), 118.2 (C-3), 160.9 (C-4), 118.2 (C-5), 130.3 (C-6), 118.2 (α), 144.5 (β), 198.3 (CO), 117.2 (C-1´), 154 (C-2´), 137.9 (C-3´), 151 (C-4´), 109 (C-5´), 121.2 (C-6´), 99.8 (C-1´´), 75.6 (C-2´´), 76.9 (C-3´´), 71.0 (C-4´´), 76.4 (C-5´´), 64.8 (C-6´´), 127.3 (C-1´´´), 130.9 (C-2´´´), 116.4 (C-3´´´), 161.0 (C-4´´´), 116.4 (C-5´´´), 130.9 (C-6´´´), 147.3 (C-7´´´), 115 (C-8´´´), 170 (C-9´´´), 22.4 ($COCH_3$), 179 ($COCH_3$).

3' - (3 – Methyl - 2 - butenyl)-4' - O - β- D- glucopyranosyl - 4,2' - dihydroxychalcone: Yellow amorphous powder; $[\alpha]_D$ +80.5° (c 0.1, MeOH); UV (MeOH): λmax (log ε) 256 (4.25), 296 (3.68), 385 (3.04) nm; ESIMS: *m/z* 487 $[M + H]^+$, 444 $[(M + H) - 43]^+$,432 $[(M + H) -55]^+$,325 $[(M + H) -162]^+$: anal. C 64.15%, H 6.21%, calcd for $C_{26}H_{30}O_9$, C 64.19%, H 6.22%; ^{1}HNMR (600 MHz, CD_3OD): δ 1.68 (3H, d, J = 1.5 Hz, H-5´´´), 1.75 (3H, d, J = 1.5 Hz, H-4´´´), 3.38 (lH, m, H-5"), 3.41 (lH, t, J = 9.0 Hz, H-2"), 3.47 (lH, t, J = 9.0 Hz, H-3"), 3.50 (lH, t, J = 9.0 Hz, H-4"), 3.51 (lH, dd, J = 12.0, 6.5 Hz, H-1'''a), 3.58 (lH, dd, J = 12.0, 7.0 Hz, H-1'''b), 3.62 (lH, dd, J = 12.0, 2.5 Hz, H-6"a), 3.95 (lH, dd, J = 12.0, 5.5 Hz, H-6"b), 5.09 (lH, br t, J = 6.8 Hz, H-2´´´), 5.18 (lH, d, J = 7.5 Hz, H-1"), 6.78 (lH, d, J = 8.0 Hz, H-5'), 6.92 (2H, d, J = 8.0 Hz, H-2, H-6), 7.10 (lH, d, J = 8.0 Hz, H-6'), 7.28 (2H, d, J = 8.0 Hz, H-3, H-5), 7.40 (lH, d, J = 16.0 Hz, Hα), 7.82 (lH, d, J = 16.0 Hz, Hβ); ^{13}C NMR (600 MHz, CD_3OD): δ 127.8 (C-1), 130.0 (C-2), 119.0 (C-3), 162 (C-4), 119 (C-5), 130 (C-6), 118 (α), 147.5 (β), 198 (CO), 120 (C-1´), 163 (C-2´), 115 (C-3´), 159 (C-4´), 108 (C-5´), 120 (C-6´), 99.4 (C-1´´), 74 (C-2´´), 77.9 (C-3´´), 71.2 (C-4´´), 78 (C-5´´), 61.5 (C-6´´), 28.2 (C-1´´´), 127 (C-2´´´), 135.5 (C-3´´´), 17.5 (C-4´´´), 21.5 (C-5´´´).

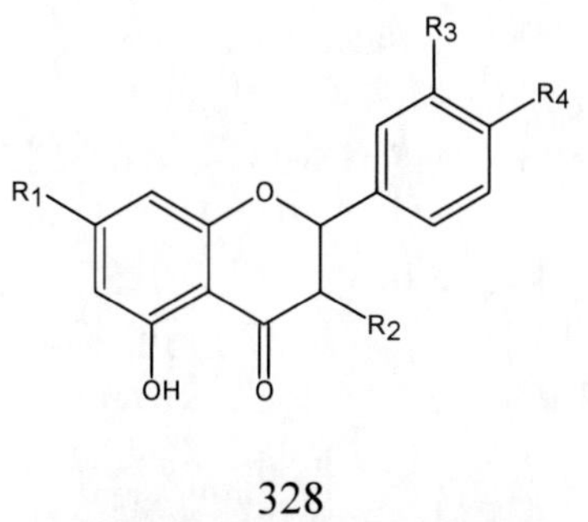

328
R_1=R_4=OH R_2=R_3=H
5,7,4'-Trihydroxyflavanone

329
R_1=OCH_3 R_2=H R_3=R_4=OH
5,3',4'-Trihydroxy-7-methoxyflavanone

330
R_1=OH R_2=OAc R_3=R_4=H
3-Acetoxy-5,7-dihydroxyflavanone

331
R_1=OH R_2=R_3=R_4=H
5,7-Dihydroxyflavanone

332
R_1= R_3=OH R_2=H R_4=OCH_3
5,7,3'-Trihydroxy-4'-methoxyflavanone

333
R_1=OH R_2= OCH_3 R_3=R_4=H
5,7-Dihydroxy-3-methoxyflavone

334
R_1= R_4=OH R_2= R_3=OCH_3
5,7,4'-Trihydroxy-3,3'-dimethoxyflavone

335
R_1=R_2=R_3=OCH_3 R_4=OH
5,4'-Dihydroxy-3,7,3'-trimethoxyflavone

These compounds were isolated from the resin of *Heliotropium sinuatum* (Boraginaceae). The antioxidant activities of the isolated phenols were evaluated by the consumption of the radical cations derived from ABTS and DPPH. Thus 5,3',4'-trihydroxy-7-methoxylavanone, with catechol substitution pattern, leading to a more stable quinone contributes most to the antioxidant activity of the resin. Flavones which exhibit total conjugation among the three aromatic ring come next (5,7-dihydroxy-3-methoxyflavone, 5,7 ,4'-trihydroxy-3,3'-dimethoxyflavone and 5,4'-dihydroxy-3, 7,3'-trimethoxylavone), followed by flavanones without catecholic groups (4'-trihydroxyflavanone, 3-acetoxy-5,7-dihydroxyflavanone and 5,7 ,3'-trihydroxy-4'-methoxyflavanone), (Modak et al., 2003).

5,7,4'-Trihydroxyflavanone: ^{1}H NMR (DMSO-d_6): δ 12.15 (lH; s, 5-OH), 10.80 (lH, s, 7-OH), 9.60 (lH, s, 4'-OH), 7.32 (2H, d, J = 8.5 Hz, H-2', H-6'), 6.81 (2H, d, J = 8.5 Hz, H-3', H-5'), 5.90 (2H, s, H-6, H-8), 5.43 (lH, dd, J = 12.7 Hz, 2.8 Hz, H-2), 3.26 (dd, J = 17.1 Hz, 12.7 Hz, H-3ax), 2.69 (dd, J = 17.1Hz, 2.8 Hz, H-3eq); ^{13}C NMR (DMSO-d_6): δ 196.6 (C-4), 166.8 (C-7), 163.4 (C-5), 162.1 (C-9), 157.8 (C-4'), 129.4 (C-1'), 128.8 (C-2', C-6'), 115.6 (C-3',5'), 102.2 (C-10), 96.1 (C-6), 95.4 (C-8), 78.9 (C-2), 42.4 (C-3).

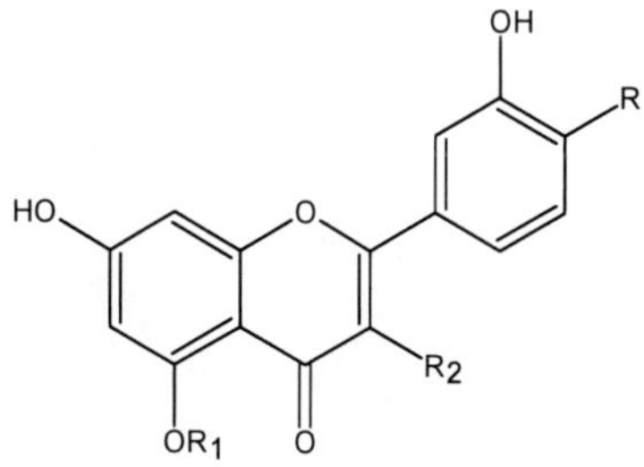

336
R= O-β-glc R_1=H R_2=H
Luteolin 4' β-glucoside

337
R=H R_1= CH_3 R_2= O-β-glc
Azaleatin 3β-glucoside

338
R= O-β-glc R_1=H R_2=OH
Quercetin 4'β-glucoside

These flavonoids glycosides bearing catechol structure in ring B were isolated from the butanol extract of *Spartium junceum* L. (Fabaceae). In case, the glycosylation to the C4, position of ring B as in quercetin 4'β-glucoside, the antioxidant activity is of 13.736 U/ml. However, the activity remarkably increased by the methylation of C5-OH group as in azaleatin 3β-glucoside (19.086 U/ml). Luteolin 4'β-glucoside showed a highest antioxidant activity (22.593 U/m1), (Yesilada et al., 2000).

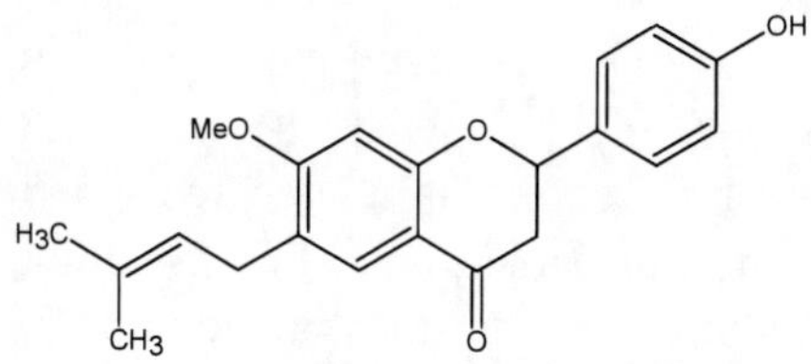

339
Bavachinin

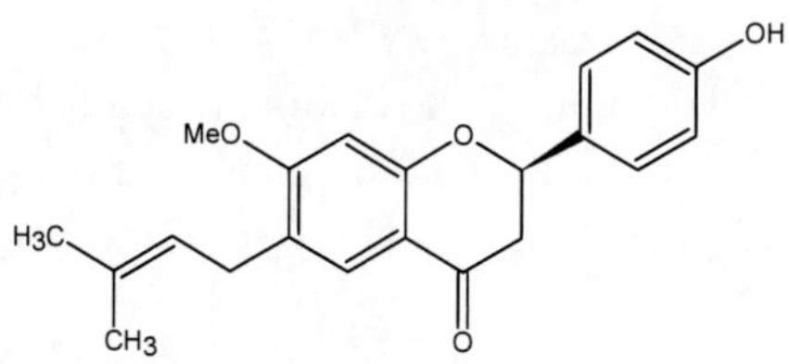

340
4'-Hydroxy-7-methoxy-6- prenylflavanone

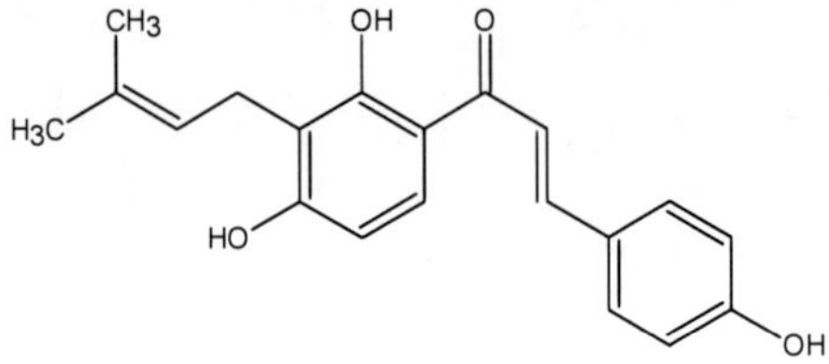

341
Isobavachalcone

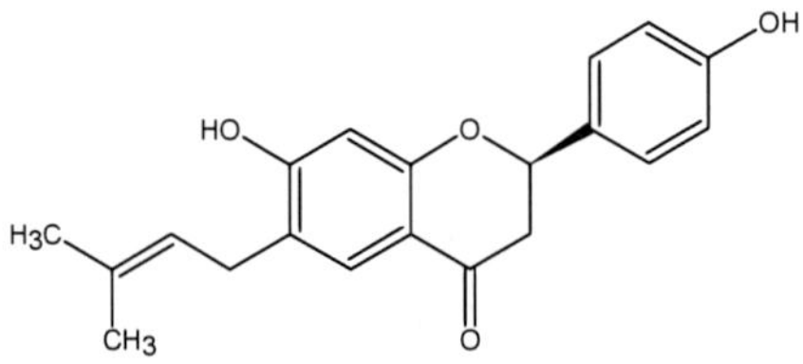

342
Bavachin

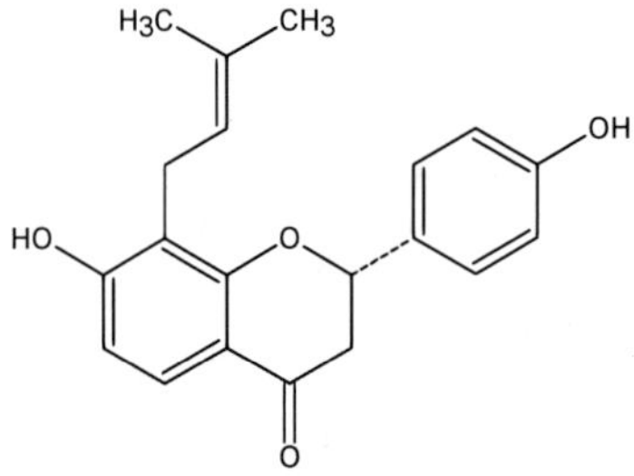

343
Isobavachin

The antioxidant activity of flavonoids is related to an inhibition of the formation of *trans,trans* hydroperoxide isomers of linoleic acid. This inhibition exhibited the great H-atom donating ability of flavonoids to peroxy radical, thus terminating the chain radical reaction (Bhalla et al., 1968, Torel et al., 1986). Four flavonoids were isolated from the seeds or *Psoralea corylifolia* (Leguminosae) as antioxidative components. In particular, flavonoids isobavachin and isobavachalcone showed broad antioxidative activities in rat liver microsomes and mitochondria. They inhibited NADPH-, ascorbate- t-BuOOH- and CCl_4-induced lipid peroxidation in microsomes. They also prevented NADH-dependent and ascorbate-induced mitochondrial lipid peroxidation. These phenolic compounds in *P. corylifolia* shown to be effective in protecting biological membranes against various oxidative stresses (Haraguchi et al., 2002).

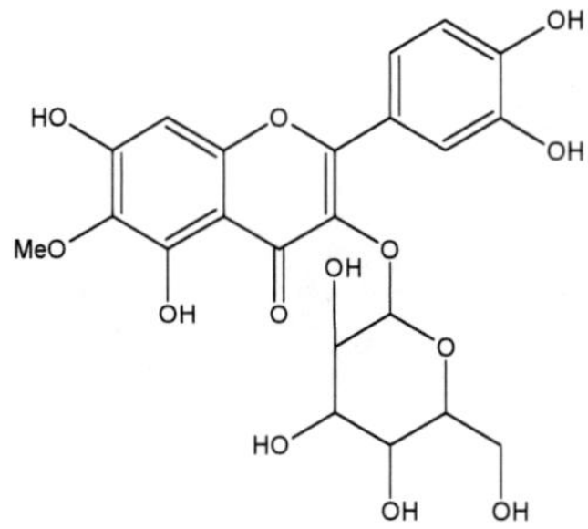

344
Patuletin 3-O-β-glucoside (P3-G)

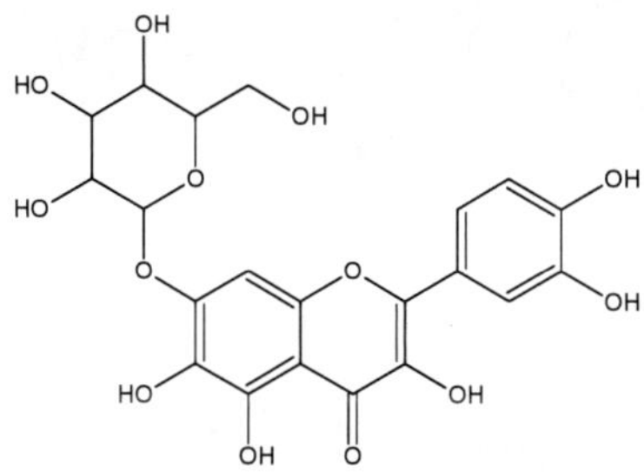

345
Quercetagetin 7-O-β-glucoside (Qu 7-G)

Found in *Arnica montana* (Compositae). An additional O-dihydroxy structure in the A-ring, or a 6-methoxy group, respectively, has no significant influence on antioxidant activity, the O-dihydroxy group of the B-ring as the most important structural feature for the radical scavenger activity the main effect of these tested flavonols is based on their inhibition of myeloper-oxidase, besides inhibition of enzymes involved in activating the NADPH-oxidase, and a direct reaction with oxygen radicals (Heilmann et al., 1995).

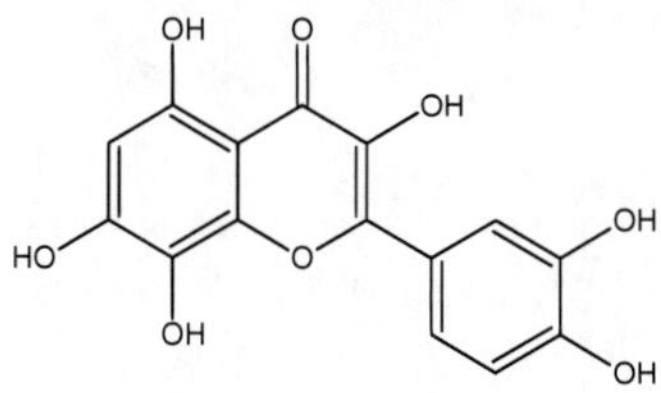

346
Gossypetin

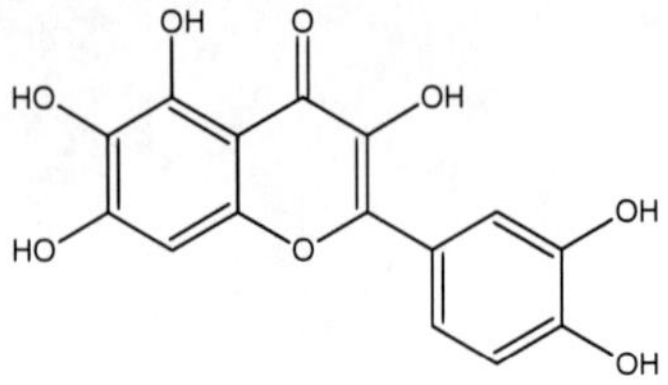

347
Quercetagetin

Found in *Malva sylvestris* (Malvaceae) showed antioxidant activity (Lea and Swoboda, 1956).

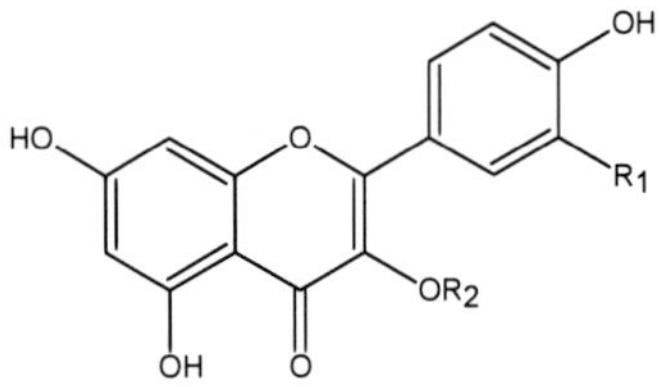

348
R_1=OH R_2=rhamnose
5,7,3',4'-Tetrahydroxy-flavonol-3-rhamnoside

349
R_1=H R_2=rutinose
5,7,4'-Trihydroxyflavone 3-rutinoside

350
R_1=OCH_3 R_2=rutinose
5,7,4'-Trihydroxy-3'-methoxyflavonol-3-rutinoside

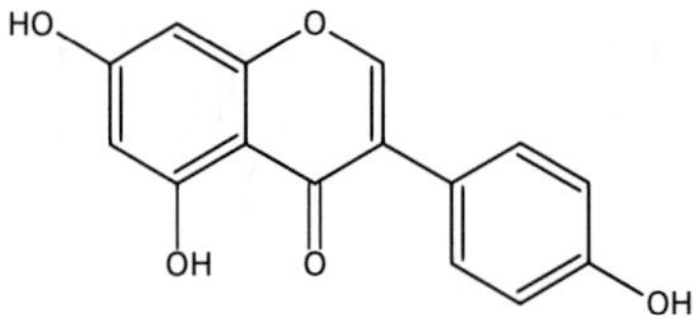

351
5,7,4'-Trihydroxyisoflavone

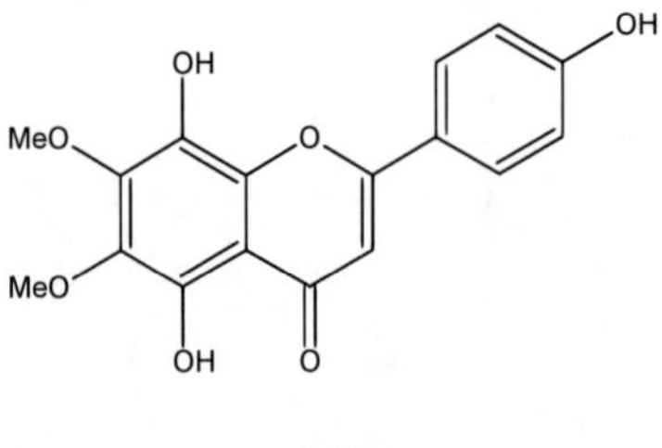

352
6,7-Dimethoxy-5,8,4'- trihydroxyflavone

Methanolic extracts of dried Balaton and Montmorency tart cherries *Prunus cerasus* (Rosaceae) inhibited lipid peroxidation induced by Fe^{2+} at 25 ppm concentrations. The antioxidant assays revealed that 6,7-dimethoxy-5,8,4'-trihydroxyflavone is the most active, followed by quercetin 3-rhamnoside, genistein, naringenin, and genistin, at 10 μM concentrations (Wang et al., 1999).

5,7,4'-Trihydroxyisoflavone: ^{1}H NMR (DMSO-d_6): δ 12.98 (lH, s, 5-OH), 10.92 (lH, s, 7-OH), 9.60 (lH, s, 4'-OH), 8.26 (lH, s, H-2), 7.38 (2H, d, J = 8.2 Hz, H-2', H-6'), 6.82 (2H, d, J = 8.2 Hz, H-3', H-5'), 6.39 (lH, d, J = 1.95 Hz, H-8), 6.21 (lH, d, J = 1.95 Hz, H-6); ^{13}C NMR (DMSO-d_6): δ 180.6 (C-4), 164.4 (C-7), 163.6 (C-5), 158.1 (C-9), 157.5 (C-4'), 154.4 (C-2), 130.6 (C-2!,6'), 122.8 (C-3), 121.7 (C-1'), 115.5 (C-3', C-5'), 104.9 (C-10), 99.3 (C-6), 94.2 (C-8).

5,7,3',4'-Tetrahydroxy- flavonol-3-rhamnoside: ^{1}H NMR (DMSO-d_6): δ 7.62 (lH, d, 2.2, H-2'), 7.58 (lH, dd, J = 8.6, 2.2, H-6'), 6.70 (lH, d, J = 8.6, H-5'), 6.10 (lH, d, J = 2.0, H-8), 5.96 (lH, d, J = 2.0, H-6), 4.96 (lH, s, H-l"), 3.82-3.22 (H-2'"H-5"), 1.15 (3H, d, 6.1, H-6").

5, 7 ,4'-Trihydroxyflavono 3-rutinoside: FABMS: m/z 594 $[M + H]^+$, 617 $[M + Na]^+$; ElMS *m/z* (% rel intensity) 286 (100); ^{1}H NMR (DMSO-d_6): δ 7.94 (2H, d, J = 8.8 Hz, H-2', H-6'), 6.85 (2H, d, J = 8.8 Hz, H-3', H-5'), 6.34 (lH, s, H-8), 6.12 (lH, s, H-6), 5.24 (lH, d, J = 7.3, H-l"), 4.35 (lH, s, H-1′′′), 3.51 (d, J = 10.5 Hz, H-6"), 3.37 (lH, m, H-4"), 3.28 (lH, m, H-5"), 3.27 (lH, m, H-3'"), 3.22 (lH, dt, J = 9.3, 6.0 Hz, H-5′′′), 3.22 (lH, m, H-3"), 3.13 (lH, m, H-4'"), 3.13 (lH, d, J = 5.1 Hz, H-2'"), 3.01 (lH, dd, J = 9.6, 7.3 Hz, H-2"), 0.94 (3H, d, J = 6.0 Hz, H-6'"); ^{13}C NMR (DMSO-d_6): δ 177.2 (C-4), 161.0 (C- 7), 159.9 (C-5), 156.9 (C-9), 156.8 (C-2), 156.8 (C-4'), 133.4 (C-3), 131.1 (C-2'), 131.1 (C-6'), 121.3 (C-l '), 115.2 (C-3'), 115.2 (C-5'), 103.6 (C-10), 101.7 (C-l"), 100.9 (C-l'"), 99.4 (C-6), 94.2 (C-8), 76.4 (C-3"), 75.8 (C-5"), 74.2 (C-2′′), 71.8 (C-4′′′), 70.6 (C-3′′′), 70.4 (C-2'"), 70.0 (C-4"), 68.4 (C-5'"), 67.1 (C-6"), 17.9 (C-6'").

5,7,4'-Trihydroxy-3' methoxyflavonol-3-rutinoside: FABMS: *m/z* 624 $[M + H]^-$, 647 $[M + Na]^-$; ElMS (% rel intensity) m/z 316 (100); ^{1}H NMR (DMSO-d_6): δ 7.81 (lH, d, J = 2.0 Hz, H-2'), 7.48 (lH, d, J = 8.4 Hz, 2.0, H-5'), 6.88 (lH, d, J = 8.4 Hz, H-6'), 6.34 (lH, s, H-8), 6.12 (lH, s, H-6), 5.39 (lH, d, J = 7.3 Hz, H-l′′), 4.38 (lH, s, H-l'"), 3.82 (OCH_3), 3.67 (lH,d, J = 10.5, H-6"), 3.37 (lH, m, H-4′′), 3.31 (lH, dd, J = 6.4, 5.1 Hz, H-3'"), 3.28 (lH, m, H-5"), 3.22 (lH, m, H-3"), 3.22 (lH, dt, J = 9.3, 6.0 Hz, H-5'"), 3.13 (d, 5.1, H-2'"), 3.13 (lH, m, H-4'"), 3.05 (lH, dd, J = 9.6, 7.3 Hz, H-2"), 0.95 (3H, d, J = 6.0 Hz, H-6'"); ^{13}C NMR (DMSO-d_6): δ 177.2 (C-4), 161.0 (C- 7), 159.9 (C-2), 159.9 (C-5), 156.7 (C-9), 133.2 (C-3), 121.2 (C-l '), 113.4 (C-2'), 99.3 (C-6), 94.2(C-8), 103.6 (C-10), 149.4 (C-3'), 147.0 (C-4'), 115.3 (C-5'), 122.4 (C-6'), 101.5 (C-l"), 101. (C-l '"), 76.4 (C-3"), 76.0 (C-5"), 74.3 (C-2"), 71.8 (C-4'"), 70.6 (C-3'"), 70.4 (C-2'"), 70.2 (C-4"), 68.5 (C-5'"), 67.1 (C-6"), 55.9 (OCH_3), 17.9 (C-6'").

5,7,4 '-Trihydroxyisoflavone- 7 –glucoside: ^{1}H NMR (DMSO-d_6): δ 8.41 (lH, s, H-2), 7.40 (2H, d, J = 8.7 Hz, H-2', H-6'), 6.82 (2H, d, J = 8.7 Hz, H-3', H-5'), 6.72 (lH, d, J = 2.1 Hz, H-8), 6.46 (lH, d, J = 2.1 Hz, H-6), 5.05 (lH, d, J = 7.5 Hz, H-l "),3.91-3.30 (5H, H-2"-6"); ^{13}C NMR (DMSO-d_6): δ 181.7 (C-4), 164.2 (C-7), 163.8 (C-5), 158.4 (C-9), 155.8 (C-2), 155.8 (C-4'), 131.3 (C-2', C-6'), 123.7 (C-3), 123.7 (C-l '), 116.3 (C-3', C-5'), 101.0 (C-10), 100.7 (C-l "), 100.1 (C-6), 95.9 (C-8), 78.4 (C-5"), 77.6 (C-3"), 74.2 (C-2"), 70.7 (C-4"), 61.8 (C-6").

6,7-Dimethoxy-5,8,4 '-trihydroxyflavone: UV λ.max (MeOH): 209, 222 (sh), 290 (sh), 303 nm; UV λ.max (MeOH + $AlCl_3$) 209, 235, 229, 321, 360 nm; UV λ.max (MeOH +

$AlCl_3$ + HCl) 209, 233, 289, 329, 355 nm; UV A.max (MeOH + NaOAc) 210, 305 nm; ^{1}H NMR (CD_3OD): δ 7.95 (2H, d, J = 8.5 Hz, H-2', H-6'), 6.92 (2H, d, J = 8.5 Hz, H-3', H-5'), 6.64 (lH, s, H-3), 4.02 (3H, s, OCH_3), 3.91 (3H, s, OCH_3); ^{13}C NMR (CD_3OD): δ 184.8 (C-4), 166.7 (C-2), 162.9 (C-4'), 149.2 (C- 7), 146.4 (C-9), 142.9 (C-5), 137.8 (C-6), 132.2 (C-8), 129.8 (C-2', C-6'), 123.3 (C-l'), 117.0 (C-3', C-5'), 107.9 (C-10), 103.5 (C-3), 62.0 (OCH_3), 61.3 (OCH_3).

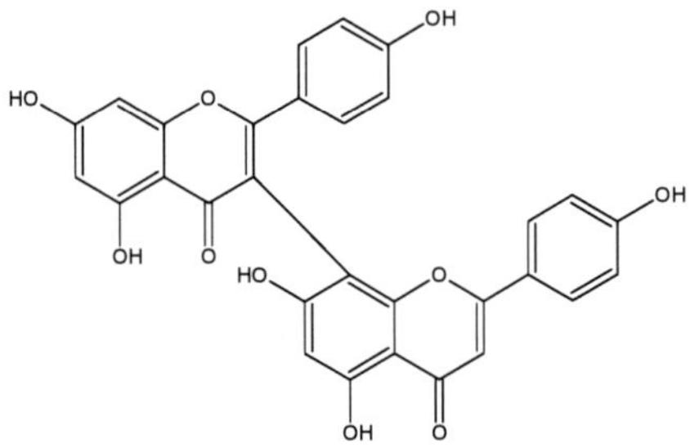

353
3-8´´-Biapigenin

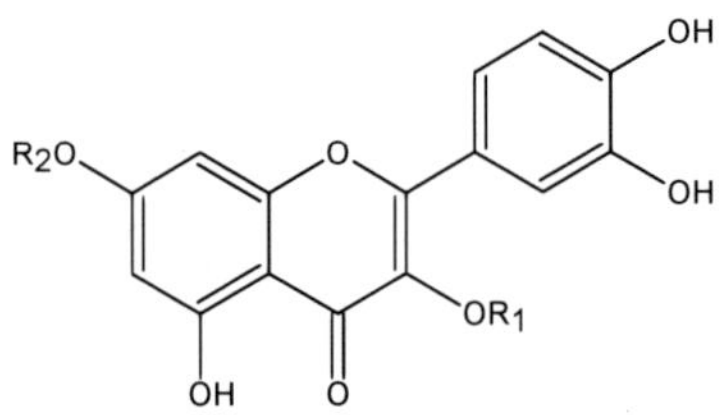

354
R_1=β-D-galactose, R_2=β-D-glucose Quercetin-3-O-β-D-galactopyranoside-7-O-β-D-glucopyranoside

3-8´´-Biapigenin together with the flavonoids quercetin, quercetin-3-O-α-L-arabinofuranoside, quercetin-3-O-β-D-galactopyranoside, and quercetin-3-O-β-D-galactopyranoside-7-O-β-D-glucopyranoside were isolated from *Hypericum hyssopifolium* (Guttiferae). All isolated compounds showed antioxidant and DPPH radical-scavenging activities. Although, 3-8´´-biapigenin was able to show highest antioxidant activity, they had the lowest DPPH radical-scavenging activity (Cakira et al., 2003).

3,8" Biapigenin. UV λmax (MeOH): 269, 327; (+NaOMe) 280, 323sh. 389; (+$AlCl_3$) 278sh. 346, 390, (+$AlCl_3$/HCl) 279, 297sh. 342, 390: (+NaOAc) 276, 375 (+NaOAc/H_3BO_3) 271, 336: ^{1}HNMR (400 MHz CD_3OD): δ 6.23 (1 H, d, J = 2 Hz, H-6), 6.25 (1 H, s, H-6"), 6.45 (1H, d, J = 2 Hz, H-8), 6.50 (1H, s, .H-3"), 6.60 (2H, dd, J = 2, 8 Hz, H-3', H-5'), 6.75 (2H, dd, J = 2, 8 Hz, H-3''', H-5'''), 7.25 (2H, dd, J = 2, 8 Hz, H-2', H-6'), 7.50 (2H, dd, J = 2, 8 Hz, H-2''', H-6''').

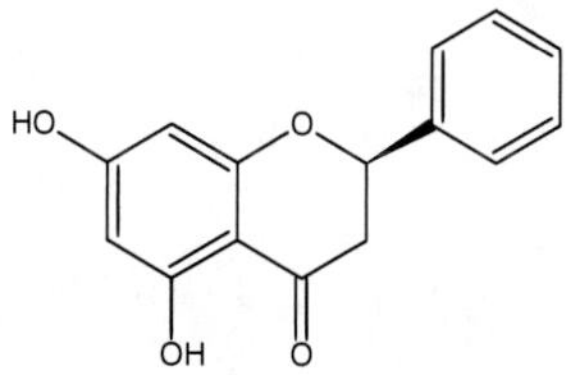

355
Pinocembrin

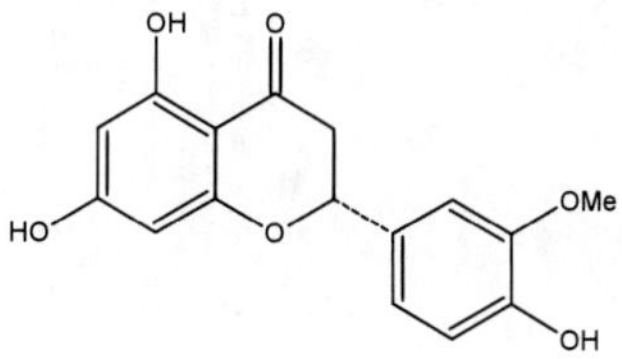

356
Homoeriodictyol

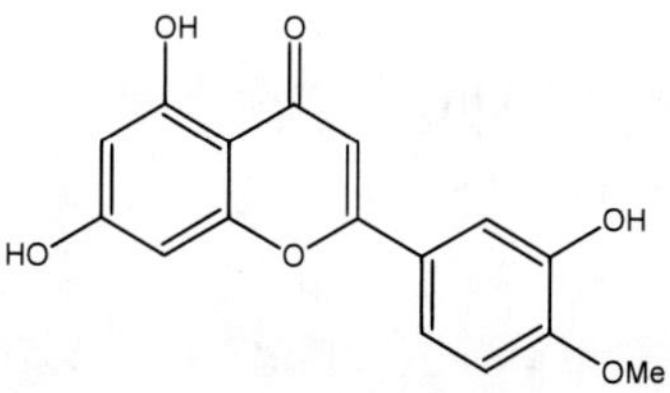

357
Diosmetin

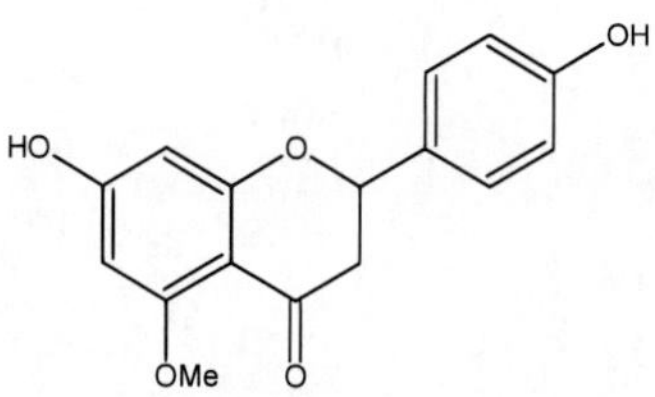

358
Thevetiaflavon

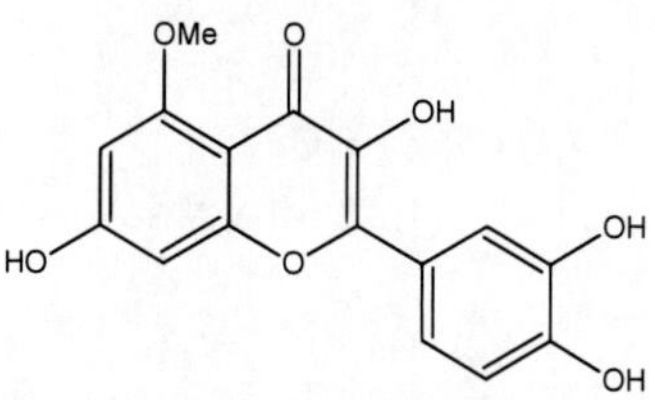

359
Azaleatin

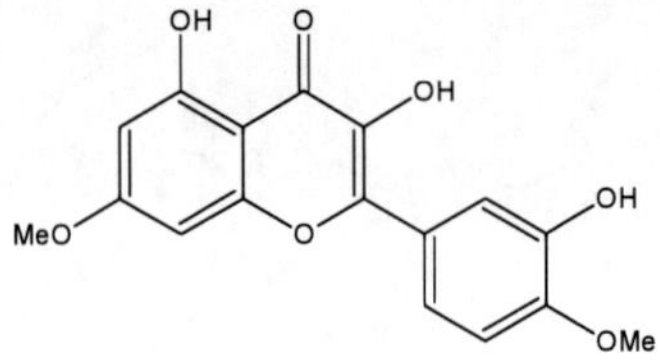

360
Ombuin

These compounds are widely distribuited, showed antioxidant activity (Wagner et al., 1976).

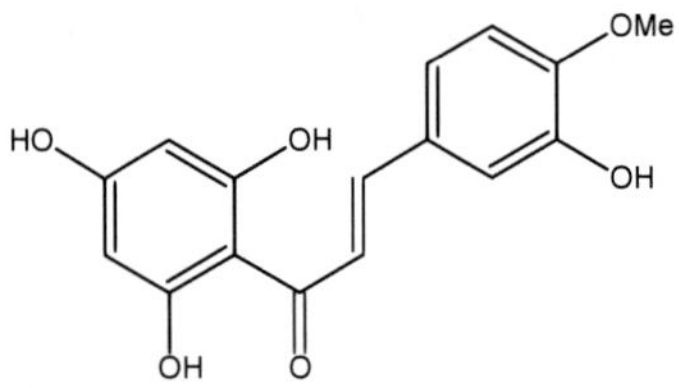

361
Dihydrohesperetin

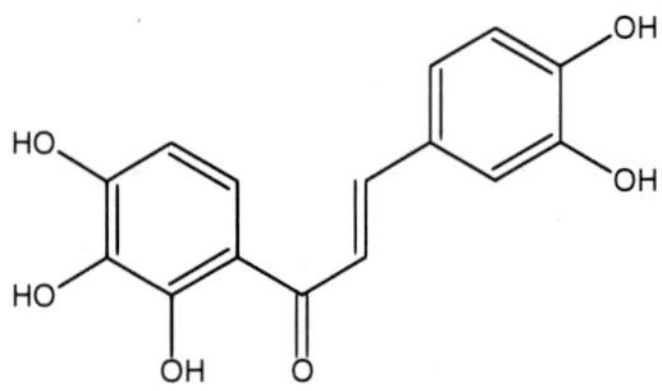

362
Okanin

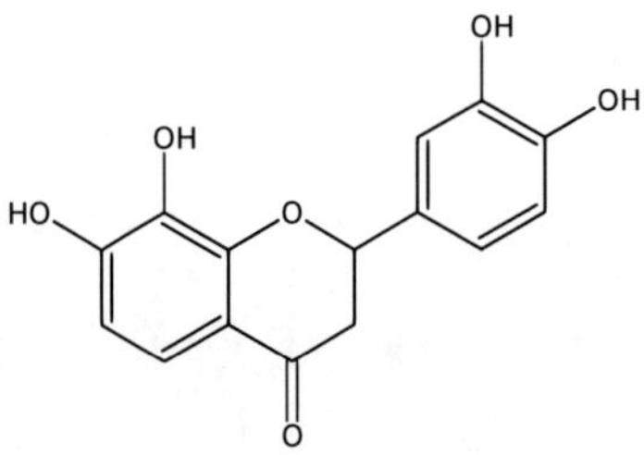

363
3´,4',7, 8-Tetrahydroxyflavanone

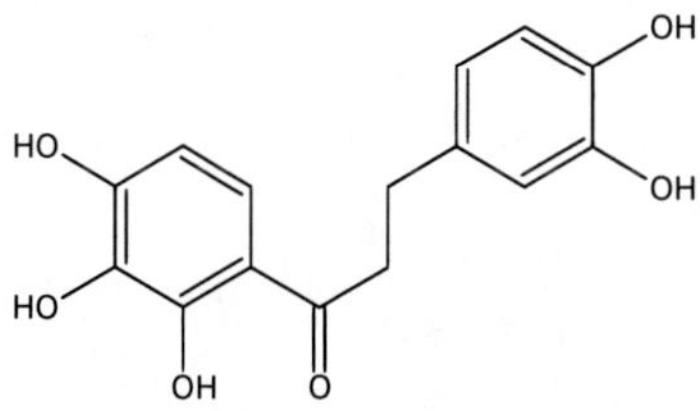

364
Dihydrookanin

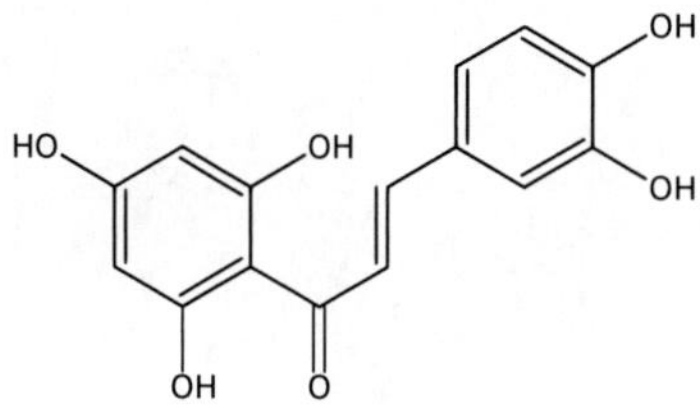

365
2´,4´,6',3,4-Pentahydroxychalcone

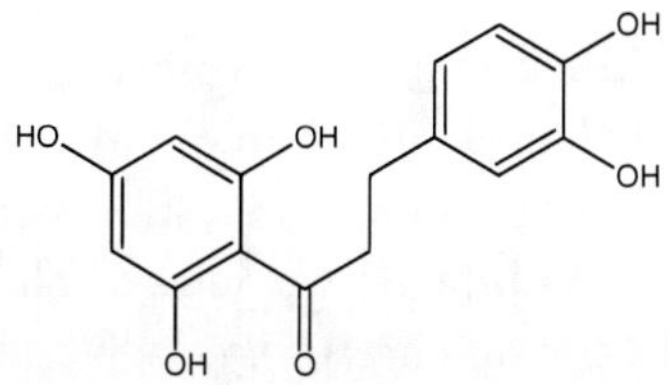

366
2´,4'6',3,4-Pentahydroxydihydrochalcone

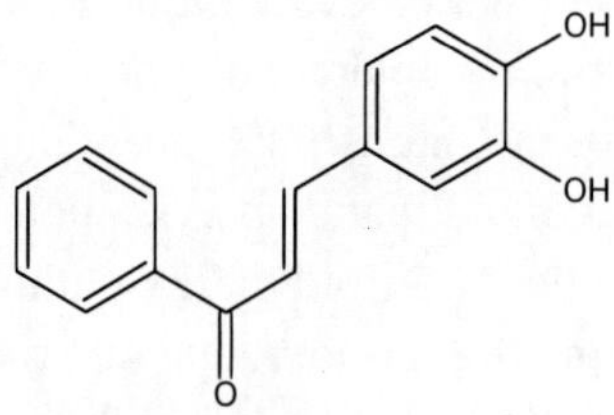

367
3,4-Dihydroxychalcone

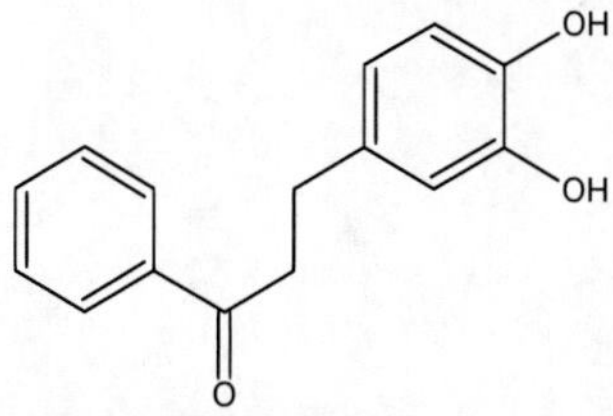

368
3,4-Dihydroxydihydrochalcone

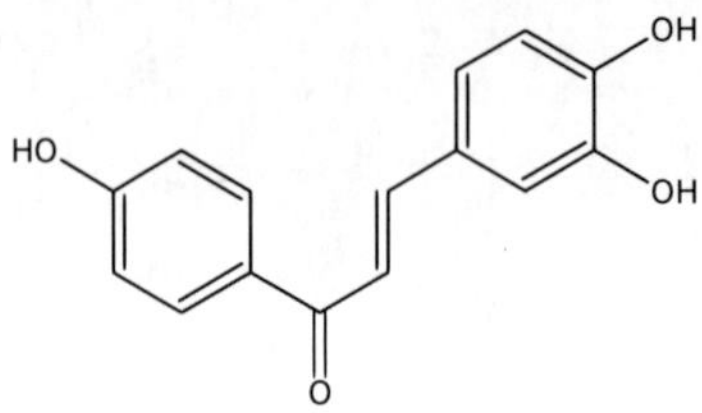

369
4',3,4-Trihydroxychalcone

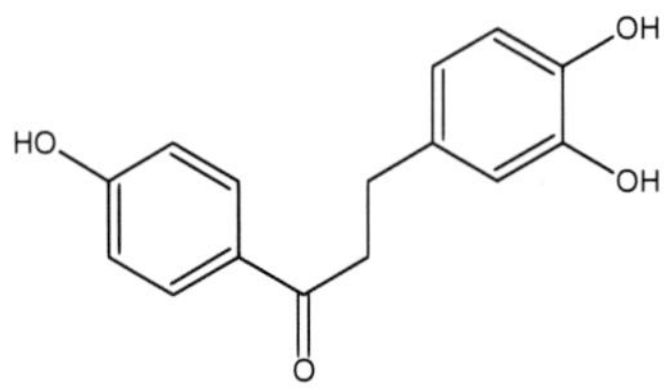

370
4',3,4-Trihydroxydihydrochalcone

These compounds are widely distribuited. The dihydrochalcones show more antioxidant activity than the corresponding chalcones, just as the polyhydroxydihydrocinriamic acids are more active than the corresponding cinnamic acids. Flavanones are significantly less active than the corresponding chalcones. With favorably located phenolic hydroxy groups, all these chemical classes exhibit marked antioxidant activity. It therefore became pertinent to enquire whether dihydrochalcones were also more active than the corresponding chalcones. A few dihydrochalcones have been reported to be present in plant material as natural products. The best known is the glucoside phloridzin, which occurs in the bark and leaves of the apple. They are largely confined to the Rosaceae and Ericaceae families and can be regarded as precursors of the polyhydroxyflavones. Certain dihydrochalcones also have interesting organoleptic properties. Neohesperidin dihydrochalcone is potentially important as an artificial sweetener. Phloretin is a dihydrochalcone flavonoid that displays a potent antioxidant activity in peroxy-nitrite scavenging and the inhibition of lipid peroxidation. Comparison with structurally related compounds revealed that the antioxidant pharmacophore of phloretin is 2,6-dihydroxyacetophenone. The potent activity of 2,6-dihydroxyacetophenone is due to stabilisation of its radical via tautomeri-sation (Rezk et al., 2002).

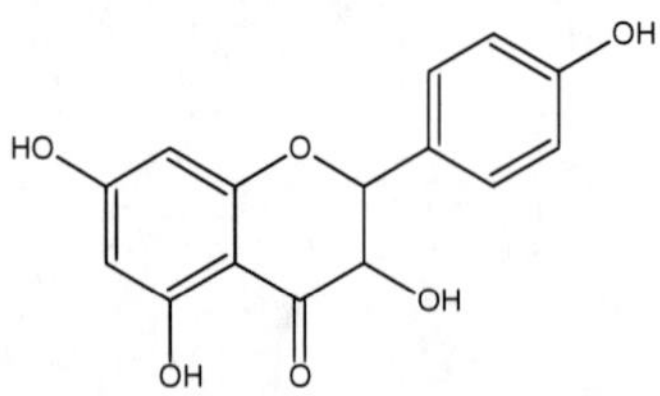

371
2R,3R-Dihydrokaempferol

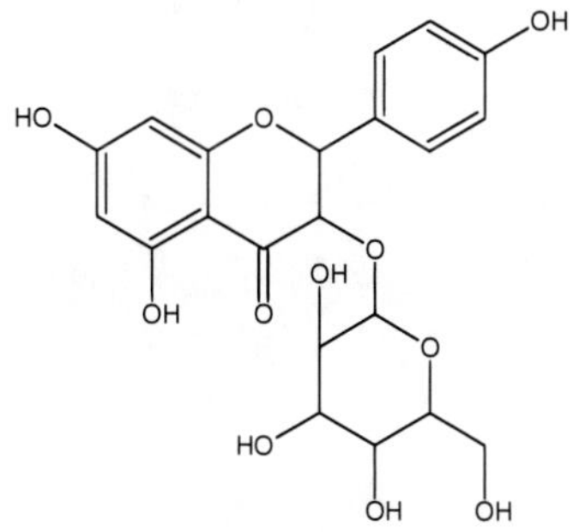

372
2R,3R-Dihydrokaempferol 3-O-β-D-glucoside

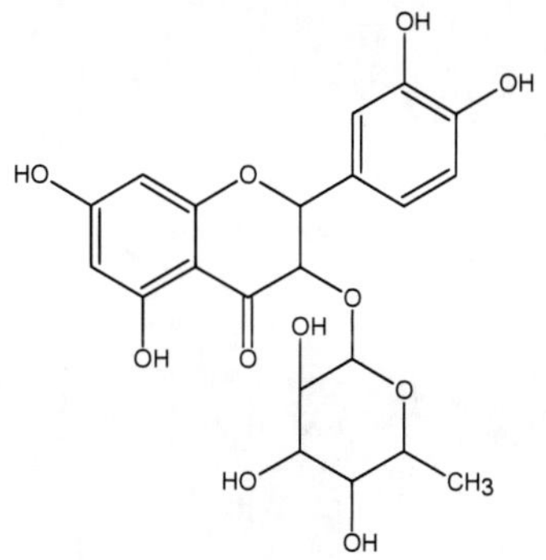

373
2R,3R-Dihydroquercetin 3-O-α-L-rhamnoside

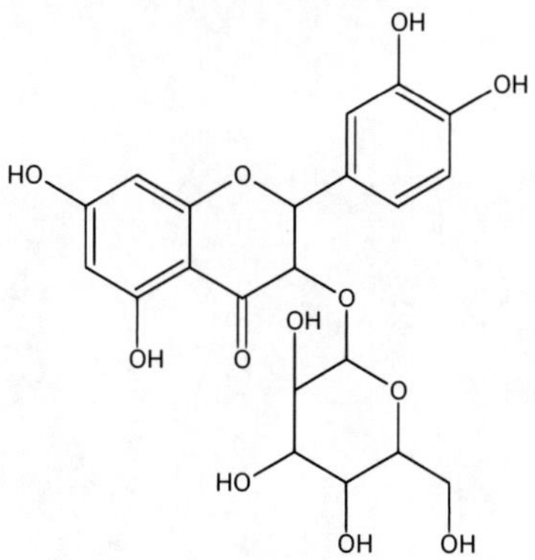

374
2R,3R-Dihydroquercetin 3-O-β-D-glucoside

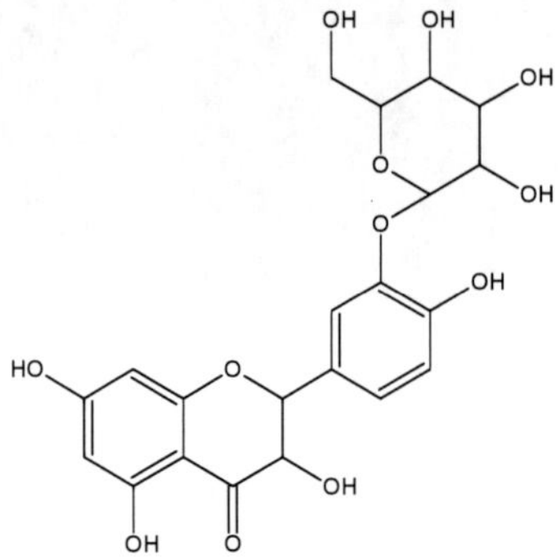

375
Dihydroquercetin 3'-O-β-D-glucoside

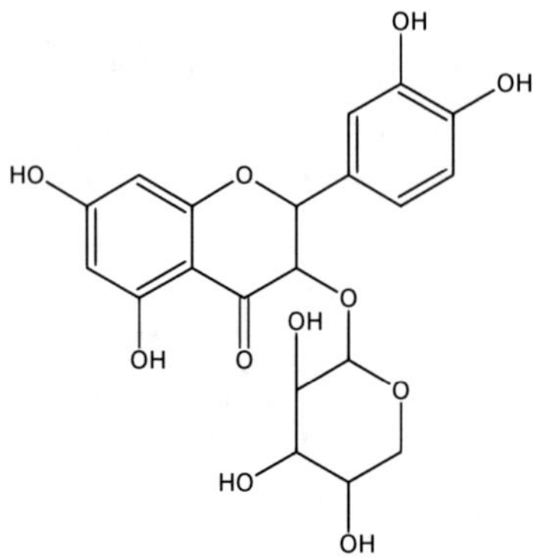

376
Dihydroquercetin-3-O-β-D- xyloside

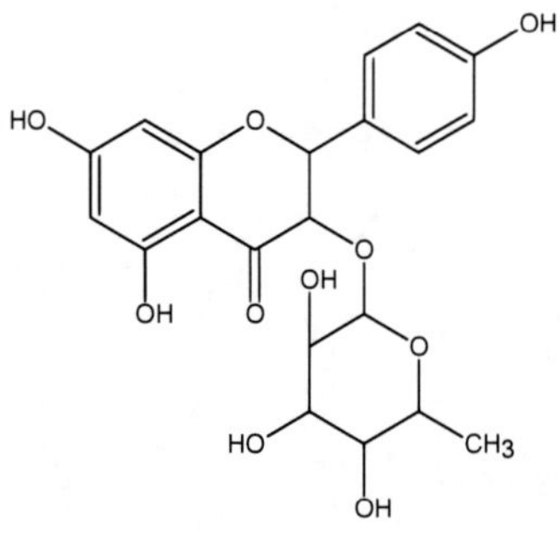

377
2R,3R-Dihydrokaempferol 3-O-α-L-rhamnoside

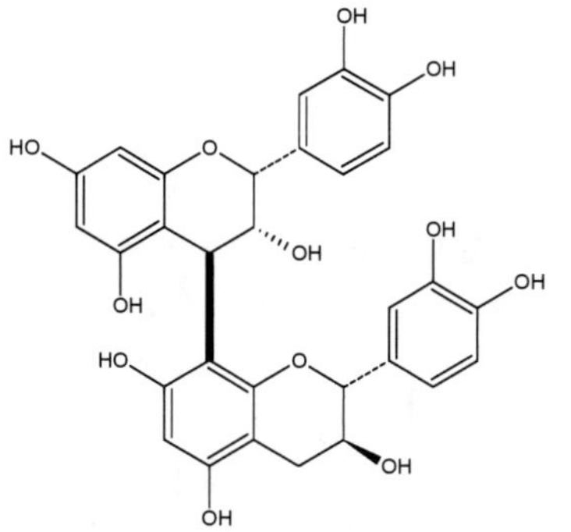

378
Procyanidin B_1

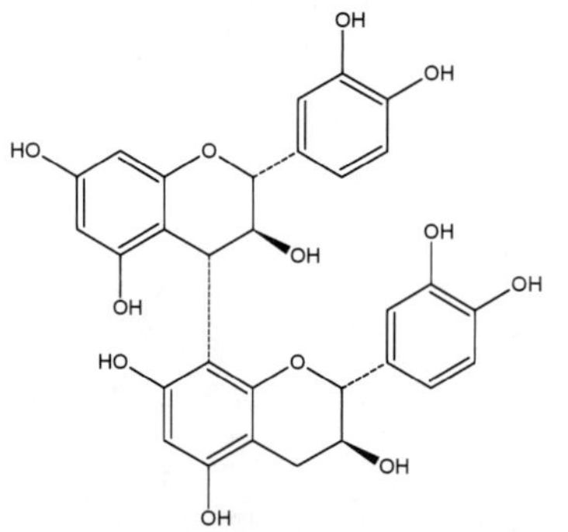

379
Procyanidin B_3

A German Riesling wine has been fractionated and the constituents is reported together with the determination of their antioxidant activity. The chemistry of flavonoids to be predictive for their free radical scavenging activity. Three criteria were found to enhance radical scavenging effectiveness: (i) an ortho-dihydroxy structure in the B ring; (ii) a 2,3-double bond in conjugation with a 4-oxo function in the C ring, and (iii) hydroxy groups in positions 3 and 5 in the A ring (Baderschneiderand Winterhalter, 2001).

2R,3R, Dihydrokaempferol: UV (MeOH): λmax 208, 292 nm; ESIMS (negative mode): pseudo molecular ion at *m/z* 287 $[M(288) -H^+]^-$; MS/MS of *m/z* 287 at *m/z* 259 $[M -H^+ -CO]^-$; ^{1}HNMR (CD_3OD): δ 4.54 (1H, d, J = 11 Hz, H3), 4.98 (1H, d, J = 11 Hz, H2); 5.88 (1H, d, J = 2 Hz, H8), 5.93 (1H, d, J = 2 Hz, H6), 6.83 (2 H, m, H3'/H5'), 7.35 (2 H, m, H2'/H6').

2R,3R-Dihydrokaempferol-3-O-β-D-glucoside: UV (MeOH): λmax 208,291 nm; circular dichroism (CD) (c = 0.0040% in MeOH): $[\theta]_{293}$ -4.1 x 10^4; $[\theta]_{337}$ + 1.0 x 10^4; DCIMS (reactant gas, NH_3); pseudo molecular ion at *m/z* 468 $[M(450) + NH_4]^+$, *m/z* 451 $[M + H]^+$. ^{1}HNMR (CD_3OD): δ 2.98 (1H, ddd, J = 9.5, 5.5, 2.5 Hz, H5"); 3.10 (1H, t, J = 9 Hz, H3"); 3.21 (1H, dd, J = 8, 9 Hz, H2"), 3.25 (1H, dd, J = 9, 9.5 Hz, H4"), 3.59 (1H, dd, J = 12.5, 5.5 Hz, H6"a), 3.75 (1 H, dd, J = 12.5, 2.5 Hz, H6"b), 3.82 (1H, d, J = 8 Hz, H1"), 4.95 (1H, d, J = 10 Hz, H3), 5.26 (1H, d, J = 10 Hz, H2), 5.89 (1 H, d, J = 2 Hz, H8), 5.91 (1H, d, J = 2 Hz, H6), 6.81 (2H, m, H3'/H5'), 7.36 (2H, m, H2'/H6'); ^{13}CNMR (90.6 MHz, CD_3OD): δ 62.6

(C6"), 71.2 (C4"), 74.6 (C2"), 77.2 (C3"), 77.6 (C5"), 78.2 (C3), 83.5 (C2), 96.9 (C6/C8), 102.2 (C10), 102.6 (C1"), 116.2 (C3'/C5'), 128.7 (C1 '), 130.4 (C2'/C6'), 159.3 (C4'), 164.2 (C9), 165.6 (C5), 170.9 (C7), 195.5 (C4).

2R,3R-Dihydroquercetin 3-O-β-D-glucoside: UV (MeOH): λmax 205, 292 nm. CD (c= 0.0056% in MeOH): $[\theta]_{297}$ -3.4 x 10^4; $[\theta]_{331}$ + 1.5 x 10^4. Thermospray-MS: pseudo molecular ion at *m/z* 484 $[M(466) + NH_4]^+$, *m/z* 467 $[M+H]^+$; ^{1}HNMR (360 MHz,CD_3OD): δ 3.01 (1H, ddd, J = 9.5, 5.5, 2.5 Hz, H5"), 3.13 (1H, t, J = 9 Hz, H3"), 3.22 (1H, dd, J = 7.5, 9 Hz, H2"); 3.24 (1H, dd, J = 9, 9.5 Hz, H4"), 3.61 (1H, dd, J = 12.5, 5.5 Hz, H6"a), 3.78 (1 H, dd, J = 12.5, 2.5 Hz, H6"b), 3.89 (1 H, d, J = 7.5 Hz, H1"); 4.92 (1 H, d, J = 9.5 Hz, H3), 5.23 (1H, d, J = 9.5 Hz, H2), 5.88 (1H, d, J = 2 Hz, H8), 5.90 (1 H, d, J = 2 Hz, H6), 6.78 (1 H, d, J = 8 Hz, H5'), 6.84 (1 H, dd, J = 8, 2 Hz, H6'), 6.94 (1H, d, J = 2 Hz, H2'); ^{13}CNMR (90.6 MHz, CD_3OD): δ 62.6 (C6"), 71.3 (C4"), 74.7 (C2"), 77.2 (C3), 77.7 (C5"), 78.2 (C3´´), 83.6 (C2), 96.5 (C6), 97.3 (C8), 102.5 (C10), 102.6 (C1 "), 115.9 (C2'), 116.2 (C5'), 121.1 (C6'), 129.1 (C1'), 146.4 (C3'), 147.3 (C4'), 164.1 (C9), 165.5 (C5), 169.8 (C7), 195.7 (C4).

Dihydroquercetin 3'-O-β-D-glucoside: Thermospray-MS: pseudo molecular ion at *m/z* 484 $[M(466) + NH_4]^+$, 467 $[M + H]^+$; ^{1}HNMR (360 MHz,CD_3OD): δ 3.37 (1H, t, J = 9.0 Hz, H4"), 3.44 (1H, ddd, J = 9.5, 5.5, 2.5 Hz, H5"), 3.49 (1H, t, J = 9 Hz, H3"), 3.52 (1 H, dd, J = 8.5, 9 Hz, H2"), 3.67 (1H, dd, J = 12.5, 5.5 Hz, H6´´a), 3.89 (1H, dd, J= 12.5, 2.5Hz, H6"b), 4:57 (1H, d, J = 11 Hz, H3), 4.83 (1H, d, J= 8.5 Hz, H1´´), 4.98 (1H, d, J= 11 Hz, H2), 5.88 (1H, d, J = 2 Hz, H8), 5.90 (1H, d, J = 2 Hz, H6), 6.89 (1 H, d, J = 8 Hz, H5'), 7.10 (1H, dd, J = 8, 2 Hz, H6'), 7.38 (lH, d, J= 2 Hz, H2'); ^{13}CNMR (90.6 MHz, CD_3OD): δ 62.6 (C6"), 71.5 (C4"), 73.4 (C3), 74.9 (C2"), 77.7 (C5"), 78.4 (C3"), 85.0 (C2), 96.3 (C8), 97.3 (C6), 102.5 (C10), 104.1 (C1"), 116.9 (C5'), 118.3 (C2'), 124.7 (C6'), 130.0 (C1'), 146.6 (C3'), 149.1 (C4'), 164.5 (C5), 165.3 (C9), 168.9 (C7), 198.5 (C4).

Dihydroquercetin-3-O-β-D-xyloside: Thermospray-MS: pseudo molecular ion at m/z 454 $[M(436) + NH_4]^+$, *m/z* 437 $[M + H]^+$. ^{1}HNMR (360 MHz,CD_3OD): δ 3.06 (1H, dd, J = 12, 9 Hz, H5"a), 3.23 (2H, m, H2"/H3"), 3.51 (1H, dt, J= 4.5, 9 Hz, H4"), 3.89 (1H, d, J = 7 Hz, H1"), 3.94 (1H, dd, J = 12, 4.5 Hz, H5"b), 4.74 (1H, d, J = 10 Hz, H3), 5.21 (1H, d, J = 10 Hz, H2), 5.92 (1H, d, J = 2 Hz, H8), 5.93 (1H, d, J = 2 Hz, H6), 6.78 (1H, d, J = 8 Hz, H5'), 6.82 (1H, dd, J =8, 2 Hz, H6'), 6.94 (1H, d, J = 2 Hz, H2'); ^{13}CNMR (90.6 MHz, CD_3OD): δ 65.9 (C5"), 70.8 (C4"), 73.5 (C2"), 75.7 (C3"), 77.5 (C-3), 83.6 (C-2), 96.3 (C-8), 97.4 (C-6), 101.9 (C10), 103.1 (C1"), 115.7 (C2'), 116.2 (C5'), 120.8 (C6'), 129.0 (C1 '), 146.1 (C3'), 147.4 (C4'), 164.1 (C5), 165.4 (C9), 168.8 (C7), 195.6 (C4).

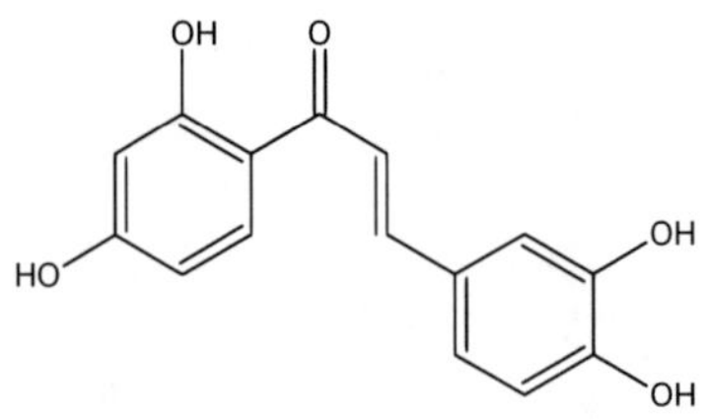

380
Butein

This compound is widely distribuited. Polyhydroxylated chalcones such as butein, which are biosynthetic intermediates between cinnamic acids and flavonoids, also show considerable antioxidant activity for lard. In this system, at 100°, butein at about 10^{-3} M was approximately twice as active as the flavonol quercetin or β-tocopherol. It prolonged the induction period for lard auto oxidation from 1.3 to 50 hr. Interestingly, chalcones with only two adjacent hydroxyl groups were almost fully effective, introduction of additional hydroxyl groups leading to only slight increases in inhibitory activity. Hydrogenation of the chalcone double bond increased their antioxidant activity to some extent; for example, the pentahydroxydihydrochalcone was ca 2-3 times as active as the corresponding unsaturated chalcone (Larson, 1988).

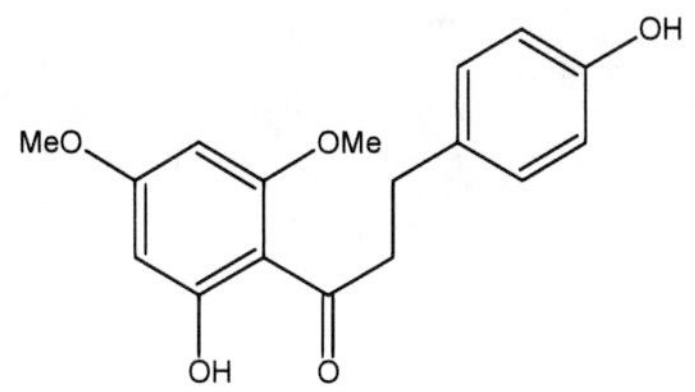

381
2',4-Dihydroxy-4',6'-dimethoxy- dihydrochalcone

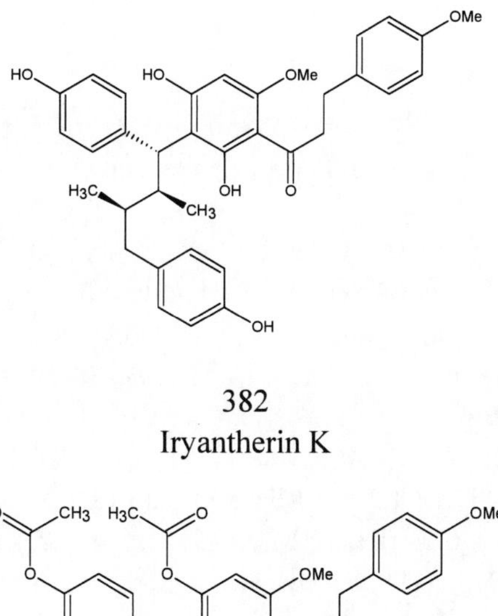

382
Iryantherin K

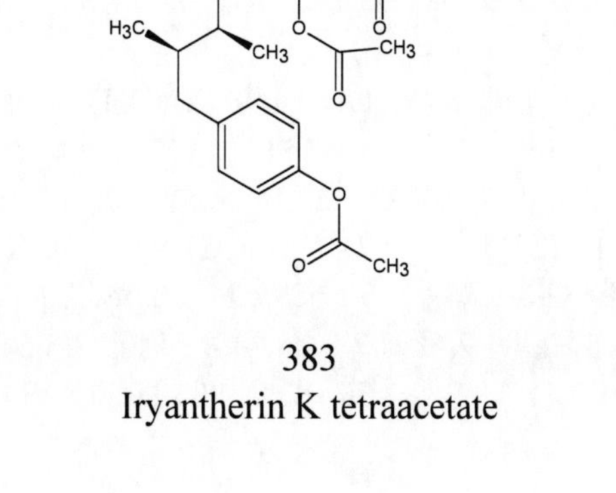

383
Iryantherin K tetraacetate

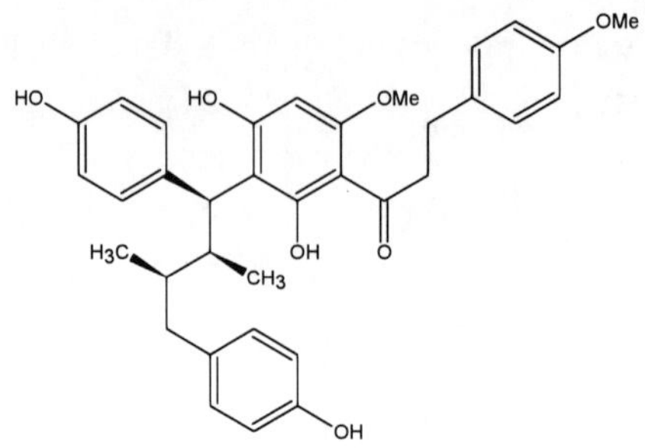

384
Iryantherin L

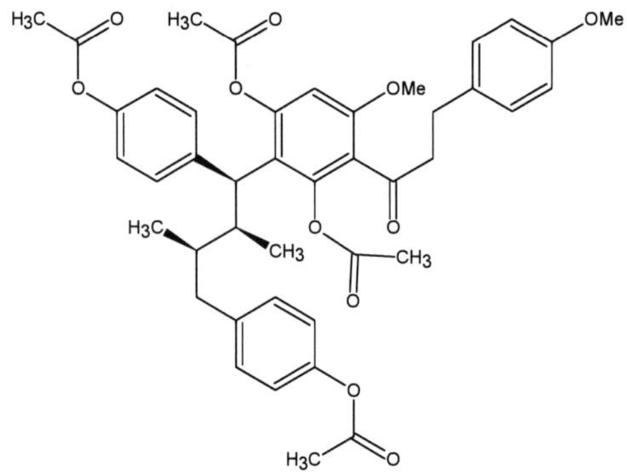

385
Iryantherin L tetraacetate

An extract from the pericarps of *Iryanthera lancifolia* (Asteraceae) afforded two dihydrochalcones and two flavonolignans. Showed antioxidant activity (Silva et a., 1999).

2',4-Dihydroxy-4',6'-dimethoxydihydrochalcone: Obtained as white needles (hexane); mp 182-184 ˚C; UV (MeOH) λmax (log ε) 232 (4.45), 285 (4.39) nm; IR (KBr): νmax 3257, 1648, 1633, 1567, 1511, 1297, 1196, 1164, 1110, 1030, 800 cm^{-1}; ElMS *m/z* 302 $[M]^+$ (9), 181 (14), 149 (5), 120 (46), 107 (100); anal. C 67.32%. H 6.15%. calcd for $C_{17}H_{18}O_5$, C 67.55%, H 5.96%; ^{1}HNMR ($CDCl_3$, 200 MHz): δ 7.15 (2H, d, J = 8.6 Hz, H-2. H-6), 6.82 (2H, d, J = 8.6 Hz, H-3, H-5), 6.01 (lH, s, H-3'). 5.97 (lH, s, H-5'), 3.84 (3H, s, CH_3O-6'), 3.77 (3H, s, CH_3O -4'), 3.27 (2H, t, J = 7.5 Hz, H-α), 2.91 (2H, t, J = 7.5 Hz, H-β); ^{13}CNMR ($CDCl_3$, 50 MHz): δ 204.7 (s, C=O), 167.2 (s, C-4'), 163.4 (s, C-6'), 163.2 (s, C-2'), 157.8 (s, C-4), 96.5 (d, C-3'), 90.9 (d, C-5'), 55.6, 55.2 (q, MeO-4'. MeO-6'), 45.9 (t, C-α), 29.0 (t, C-β).

Iryantherin K: $[\alpha]_D$ -36.0c (c 1.6, MeOH); UV (MeOH): λmax (log ε) 245 (4.81), 288 (4.31) nm; IR (KBr) νmax 3354. 1701. 1611, 1511. 1433. 1244, 1209, 1100, 1030, 825, 807 cm^{-1}; ElMS: *m/z* 570 $[M]^+$ (1), 407 (100), 302 (14), 287 (1), 268 (6), 245 (3), 167 (56), 163 (4), 162 (20), 161 (18), 147 (18), 134 (57), 121 (92), 107 (50), 91 (14), 77 (16); anal. C 73.43%, H 6.84%, calcd for $C_{35}H_{38}O_7$, C 73.68%, H 6.67%; ^{1}HNMR ($CDCl_3$, + DMSO-d_6, 200 MHz): δ 7.09 (d, J = 8.5 Hz, H-2,6), 6.77 (d, J = 8.5 Hz, H-3,5), 2.86 (t, J = 7.5 Hz, H-7), 3.16 (t, J = 7.5 Hz, H-8), 5.81 (s, H-14), 7.30 (d, J = 8.3 Hz, H-2′,6′), 6.61 (d, J = 8.3 Hz, H-

3′,5′), 4.15 (d, J = 11.6 Hz, H-7′), 2.98 (ddq, J =2.8, 6.8, 11.7 Hz, H-8′), 0.68 (d, J = 6.9 Hz, H-9′), 6.88 (d, J = 8.1 Hz, H-2′′, 6′′), 6.67 (d, J = 8.1 Hz, H-3′′, 5′′), α: 2.35 (dd, J = 8.2, 13.4 Hz, H-7′′), β: 2.50 (dd, J = 6.6, 13.4 Hz, H-7′′), 1.7-1.9 (m, H-8′′), 0.63 (d, J = 6.9 Hz, H-9′′), 3.68 (s, MeO-4), 3.73 (s, MeO-15), 14.34 (s, HO-11); ^{13}CNMR ($CDCl_3$, + DMSO-d_6, 50 MHz): δ137.4 (C-1), 129.9 (C-2,6), 114.4 (C-3,5), 158.4 (C-4), 31.0 (C-5), 46.8 (C-8), 205.3 (C-9), 105.0 (C-10),161.7 (C-11), 111.5 (C-12), 161.7 (C-13), 91.2 (C-14), 161.7 (C-15), 134.1 (C-1′), 130.5 (C-2′,6′), 115.0 (C-3′,5′),154.9 (C-4′), 45.4 (C-7′), 37.1 (C-8′), 12.6 (C-9′), 134.6 (C-1′′),130.6 (C-2′′,6′′), 115.3 (C-3′′,5′′), 154.9 (C-4′′), 42.7 (C-7′′), 37.1 (C-8′′),12.65 (C-9′′), 55.6 (MeO-4), 55.7 (MeO-15).

Iryantherin K Tetraacetate : ^{1}HNMR (CDCI3. 200 MHz): δ 7.07 (d, J = 7.5 Hz, H-2,6) 6.88 (d, J = 7.7 Hz, H-3,5), 2.86 (t, J =7.6 Hz, H-7), 3.24 (t, J = 7.6 Hz, H-8), 6.03 (s, H-14), 7.29 (d, J = 8.5 Hz, H-2′,6′), 6.77 (d, J = 8.5 Hz, H-3′, 5′), 3.96 (d, J = 11.4 Hz, H-7′), 2.9-3.0 (m, H-8′), 0.83 (d, J = 7.4Hz, H-9′), 7.02 (d, J = 7.7 Hz, H-2′′, 6′′), 6.83 (d, J = 7.5 Hz, H-3′′, 5′′), 2.2-2.4 (m, H-7′′), 1.6-1.8 (m, H-8′′), 0.64 (d, J = 6.5 Hz, H-9′′), 3.67 (s, MeO-4), 3.71 (s, MeO-15); ^{13}CNMR ($CDCl_3$, 50 MHz): δ 133.5 (C-1), 129.3 (C-2,6), 120.9 (C-3,5), 157.9 (C-4), 29.1 (C-7), 46.6 (C-8), 201.0 (C-9), 121.3 (C-10), 113.7 (C-12), 108.8 (C-14),159.9 (C-15), 139.0 (C-1′),129.5 (C-2′,6′), 121.1 (C-3′, 5′), 148.7 (C-4′), 42.1 (C-7′), 36.3 (C-8′), 12.3 (C-9′), 140.8 (C-1′′), 130.1 (C-2′′,6′′), 121.1 (C-3′′,5′′),148.8 (C-4′′), 41.8 (C-7′′), 36.3 (C-8′′), 12.3 (C-9′′), 55.2 (MeO-4), 55.6 (MeO-15), 21.0 (CH_3CO), 21.4 (CH_3CO), 169.2 (CH_3CO).

Iiryantherin L: $[\alpha]_D$ + 45.6c (c 1.9. MeOH); UV (MeOH): λmax (log ε) 248 (4.22), 285 (4.17); IR (KBr) νmax 3360, 1699, 1612, 1511, 1454, 1441, 1242, 1210, 1096, 1030, 825, 806, 668 cm^{-1}; EIMS *m/z* 570 $[M]^+$ (1), 407 (100), 302 (12), 287 (1), 268 (4), 245 (4), 167 (25), 163 (2), 162 (10), 161 (10), 147 (10), 134 (22), 121 (58), 107 (23), 91 (5), 77 (7); anal. C 73.31%, H 6.92%, calcd for $C_{35}H_{38}O_7$, C 73.68%, H 6.67%; ^{1}HNMR ($CDCl_3$, + DMSO-d_6, 200 MHz): δ 7.08 (d, J = 8.1Hz, H-2,6), 6.77 (d, J = 8.1Hz, H-3, 5), 2.83 (t, J = 6.6 Hz, H-7), 3.15 (t, J = 6.6Hz, H-8), 5.64 (s, H-14), 7.08 (d, J = 8.5 Hz, H-2′,6′), 6.59 (d, J = 8.1 Hz, H-3′,5′), 4.39 (d, J = 11.0 Hz, H-7′), 2.8-2.9 (m, H-8′), 0.65 (d, J = 6.6 Hz, H-9′), 6.88 (d, J = 8.1, Hz, H-2′′, 6′′), 6.71 (d, J = 8.8 Hz, H-3′′, 5′′), α:2.41 (d, J = 6.6 Hz, H-7′′), β:2.42 (d, J= 6.6 Hz, H-7′′), 1.7-1.9 (m, H-8′′), 0.74 (d, J = 5.9 Hz, H-9′′), 3.58 (s, MeO-4), 3.73 (s, MeO-15), 14.48 (s, HO-11); ^{13}CNMR ($CDCl_3$, + DMSO-d_6, 50 MHz): δ 135.3 (C-1), 130.1 (C-2,6), 113.6 (C-3,5), 157.5 (C-4), 30.1 (C-7), 45.9 (C-8), 204.3 (C-9), 104.6 (C-10), 160.8 (C-11), 111.5 (C-12), 162.3 (C-13), 90.8 (C-14), 164.8 (C-15), 133.0 (C-1′), 130.0 (C-2′,6′), 114.4 (C-3′,5′), 153.9 (C-4′), 41.8 (C-7′), 35.0 (C-8′), 11.0 (C-9′), 133.8 (C-1′′), 130.1 (C-2′′,6′′), 114.9 (C-3′′,5′′), 154.2 (C-4′′), 41.3 (C-7′′), 32.6 (C-8′′), 13.1 (C-9′′), 55.0 (MeO-4), 55.1 (MeO-15).

Iryantherin L tetraacetate ^{1}HNMR ($CDCl_3$, 200 MHz): δ 7.16 (d, J= 8.0 Hz, H-2,6), 7.03 (d, J = 8.4 Hz, H-3,5), 2.89 (t, J =7.0 Hz, H-7), 3.09 (t, J = 7.0 Hz, H-8), 6.43 (s, H-14), 7.16 (d, J = 8.0 Hz, H-2′,6′), 6.83 (bs, H-3′,5′), 4.08 (d, J = 10.0 Hz, H-7′), 2.9-3.0 (m, H-8′), 0.64 (d, J = 6.8 Hz, H-9′), 7.16 (d, J = 8.0 Hz, H-2′′, 6′′), 7.03 (d, J = 8.4 Hz, H-3′′, 5′′), 2.2-2.6 (m, H-7′′α), 1.9-2.1 (m, H-8′′), 0.85 (d, J= 6.5 Hz, H-9′′), 3.77 (s, MeO-4), 3.86 (s, MeO-15); ^{13}CNMR ($CDCl_3$, 50 MHz): δ 132.9 (C-1), 129.3 (C-2,6), 120.8 (C-3,5), 157.0 (C-4), 29.4 (C-7), 45.3 (C-8), 206.2 (C-9), 121.4 (C-10), 113.9 (C-12), 105.0 (C-14), 158.0 (C-15), 138.8 (C-1′), 130.2 (C-2′,6′), 121.1 (C-3′,5′), 148.8 (C-4′), 42.0 (C-7′), 35.3 (C-8′), 11.0 (C-9′), 138.8 (C-1′′), 130.2 (C-2′′,6′′), 121.1 (C-3′′,5′′), 148.2 (C-4′′), 41.0 (C-7′′), 35.3 (C-

8´´), 14.0 (C-9´´), 55.2 (MeO-4), 55.2 (MeO-15), 20.6 (CH_3CO), 21.2 (CH_3CO), 169.5 (CH_3CO).

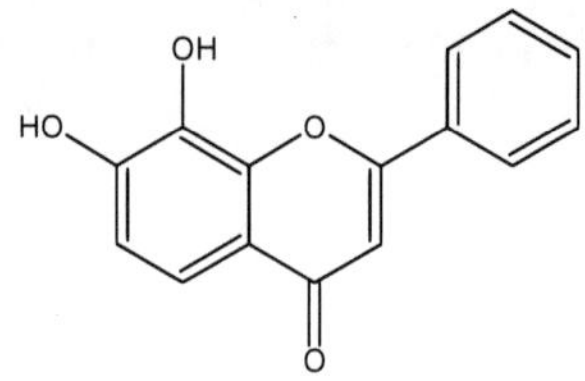

386

7,8-Dihydroxy-flavone

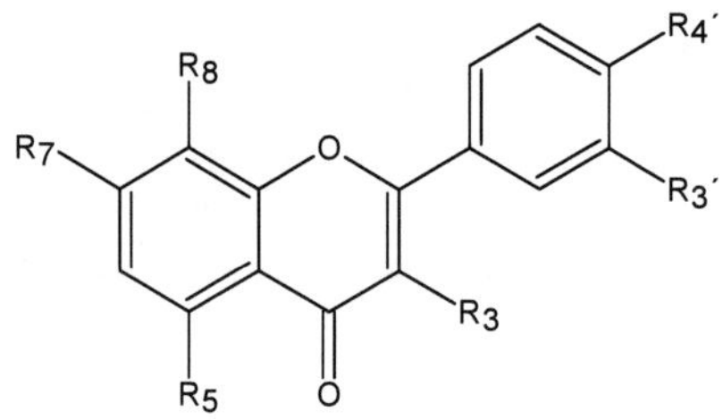

387

R_3=H, R_5= OH, R_7= OAAG, R_8= OH, R_3'= H, R_4´= OMe

AAG=6'''- acetylallopyranosyl (1-2) glucopyranose

4'-O-Methylisoscutellarein 7-O-[6´´´, -acetylallopyranosyl-(1→2)glucopyranoside]

388

R_3=H, R_5= OH, R_7= OAG, R_8= OH, R_3´= H, R_4´= OMe

AG=allopyranosyl (1-2) glucopyranose

4'-O-Methylisoscutellarein 7-O-allopyranosyl(1→2)glucopyranoside

389

R_3=H, R_5= OH, R_7= OAAG, R_8= OH, R_3´= OH, R_4´= OMe

AAG=6'''- acetylallopyranosyl (1-2) glucopyranose

3'-Hydroxy-4'-O-methylisoscutellarein 7-O-[6'''-acetylallopyranosyl(1→2) glucopyranoside]

390

R_3=H, R_5= OH, R_7= OAAG, R_8= OH, R_3´= H, R_4´= OH

AAG=6'''- acetylallopyranosyl (1-2) glucopyranose

Isoscutellarein 7-O-[6'''-acetylallopyranosyl(1→2)glucopyranoside]

391

R_3=H, R_5= OH, R_7= OH, R_8= OG, R_3´= OH, R_4´= OH

G=glucose

Hypolaetin-8-glucoside

Five flavonoid glycosides, have been isolated from *Sideritis javalambrensis* Pau (Labiatae) aerial parts. These glycosides have been tested for their antioxidant properties using $FeSO_4$/cysteine-induced microsomal lipid peroxidation. Superoxide scavenging activity was assessed in the nitroblue tetrazolium test. Among this series of flavonoids, hypolaetin-8-glucoside is the most potent inhibitor of non-enzymic lipid peroxidation. The antiperoxidative activity of these flavonoids may be related to their superoxide scavenging ability (Rios et al., 1992).

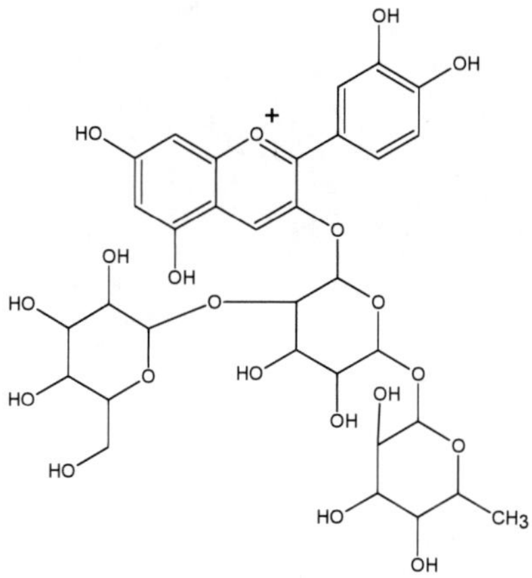

392

3-Glucuronopyranoside cyanidin

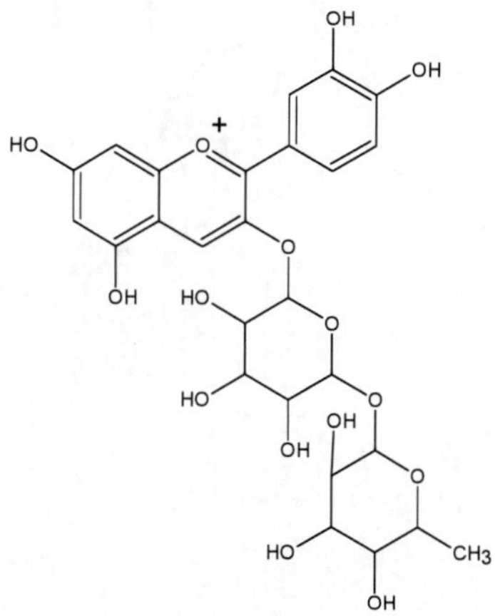

393

3-Rhamnose cyanidin

The anthocyanins were isolated from tart cherries (*Prunus cerasus,* Rosaceae) exhibited *in vitro* antioxidant and anti-inflammatory activities. The antioxidant activities of compounds were comparable to the antioxidant activities of tert-butylhydroquinone and butylated hydroxytoluene and superior to vitamin E at 2-mM concentrations. In the antiinflammatory assay, cyanidin gave IC_{50} values of 90 and 60 mM, respectively, for prostaglandin H endoperoxide synthase-l and prostaglandin H endoperoxide synthase-2 enzymes (Wang et al., 1999c).

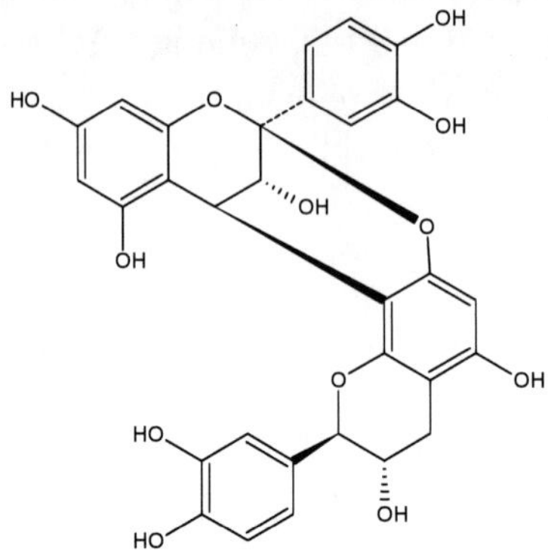

394
Proanthocyanidin A_1

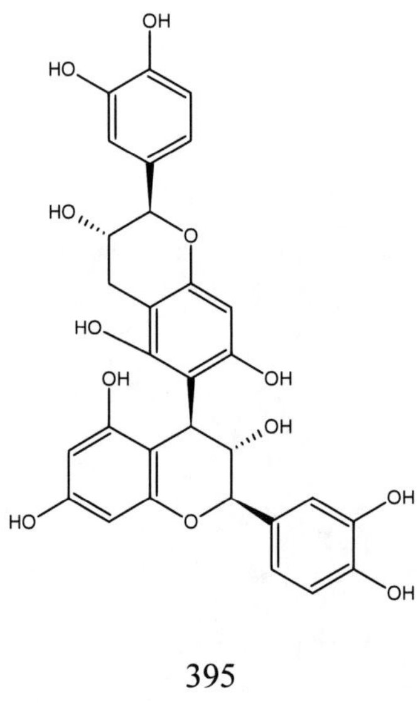

395
Procyanidin B_6

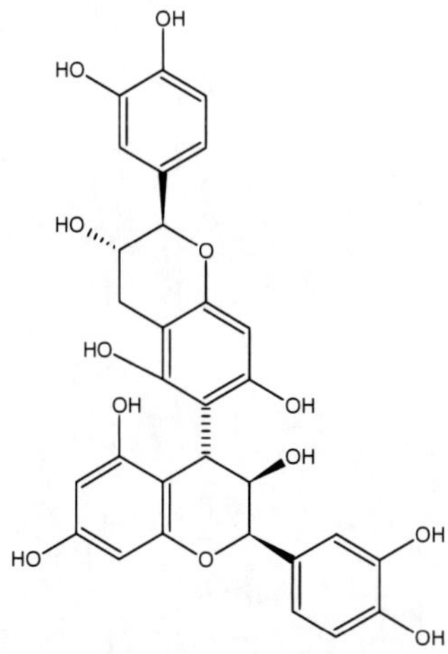

396
Procyanidin B_7

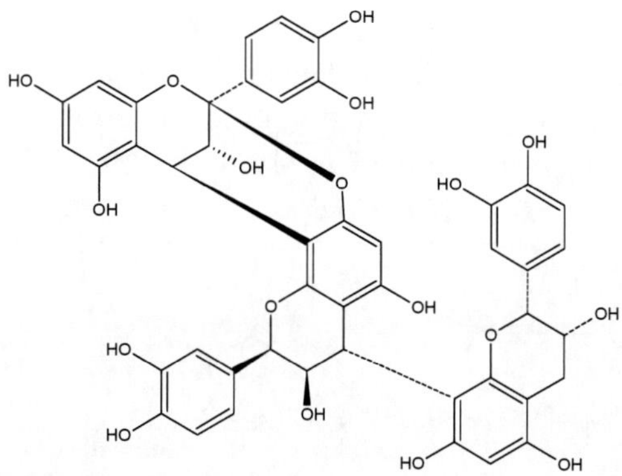

397
Cinnamtannin B_1

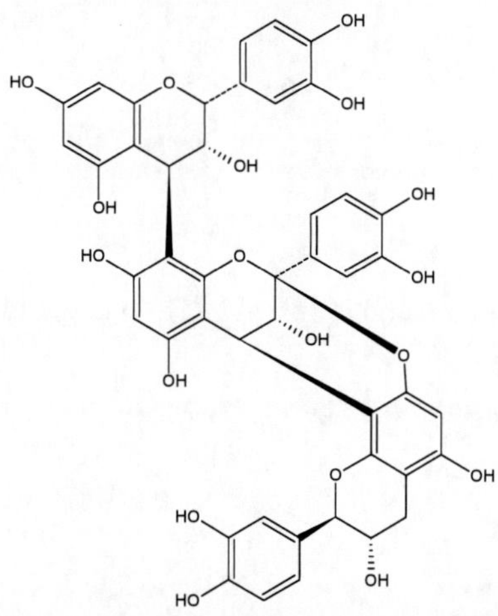

398
Epicatechin-(4β→ 8)-epicatechin-(4β→ 8, 2β→ O→ 7)-catechin

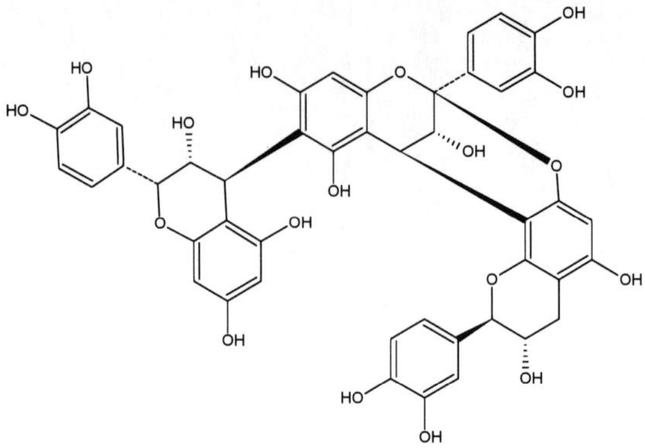

399
Epicatechin-(4β→6)-epicatechin-(4β→ 8, 2β→ O → 7)-catechin

Isolated from *Vaccinium vitis-idaea* L. (Vacciniaceae) exhibited multiple antioxidant activity. Reactive oxygen molecules have been implicated as important pathological mediators in many clinical disorders and periodontal disease. To provide possible alternative treatment of periodontal disease. Cinnamtannin B_1 displayed the strongest anti-lipid peroxidation activity, proanthocyanidin A_1 displayed the strongest superoxide scavenging activity, and epicatechin-(4β→6)-epicatechin-(4β→ 8, 2β→ O→ 7)-catechin had the strongest anti-superoxide formation effect Ho et al., 1999).

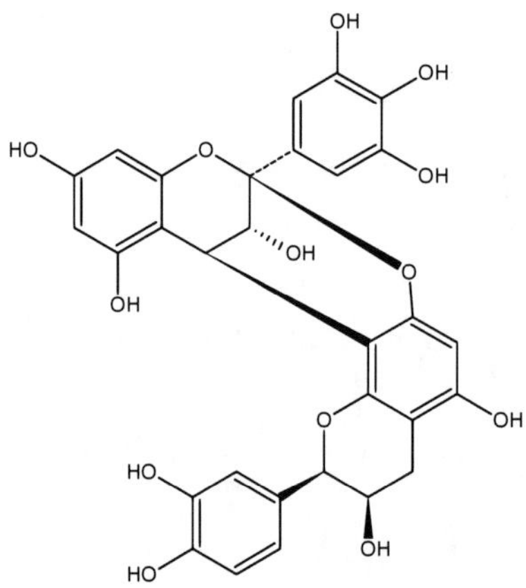

400
Epicatechin-(2β→ 7,4β)-epicatechin

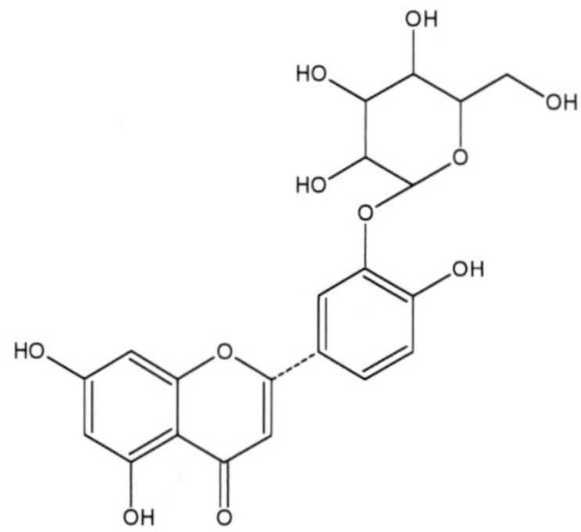

401
Luteolin 3'β-D-glucopyranoside

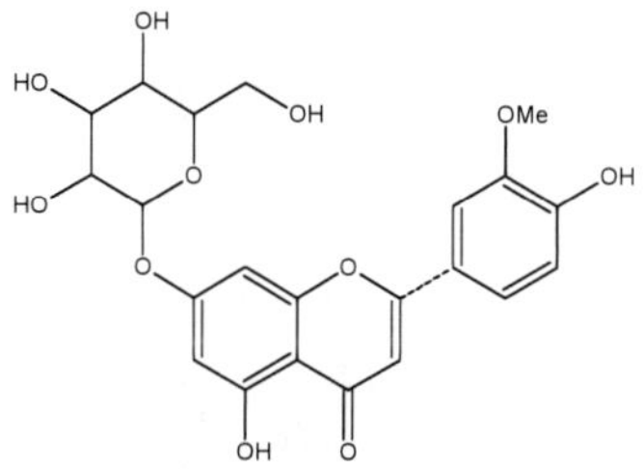

402
Chrysoeriol 7β-D-glucopyranoside

Found in *Dioclea lasiophylla* (Leguminosae). The proanthocyanidin epicatechin-(2β→7,4β)-epicatechin possesses considerable antioxidant activity (AA = 35) compared to propyl gallate (AA = 49) and α-tocopherol (AA = 27), (Barreiros et al., 2000).

Epicatechin-(2β→7,4β)-epicatechin: Brown crystals: mp 223-225°C; $C_{30}H_{24}O_{13}$ (Found: C, 54.1, H, 5.0%. $C_{30}H_{24}O_{13}$ .4H_2O requires: C. 54.2, H 4.8%). CD: [ɵ] 211-79573. $[ɵ]_{227}$ +119573. UV λmx (MeOH) (log ε): 210 (5.16), 225 (5.10), 277 (3.98); POS FABMS $[M+H]^+$, *m/z* (rel. int.): 593 (5), 441 (4), 303 (10), 287 (8), 277 (8), 207 (31), 149 (9), 115 (100); ^{1}HNMR: δ 4.06 (d, J = 3.4 Hz, H3μ), 4.41 (d, J = 3.4 Hz, H4μ), 6.01 (d, J = 2.4 Hz, H6μ), 6.08 (d, J = 2.4 Hz, H8μ), 6.74 (s, H2μ), 6.74 (s, H6μ), 4.93 (br s, H2ι), 4.24 (br m, H3ι), 2.95 (dd, J = 17.2, 4.9 Hz, H4aι), 2.76 (dd, J = 17.2, 1.9 Hz, H4bι), 6.11(s, H6ι), 7.16 (d, J = 2.0 Hz, H2ι), 6.82 (d, J = 8.2 Hz, , H5ι), 6.98 (dd, J = 8.2, 2.0 Hz, H6ι); ^{13}CNMR: δ 100.13 (C-2μ), 68.02 (C-3μ), 29.14 (C-4μ), 156.51 (C-5μ), 98.26 (C-6μ), 158.02 (C-7μ), 96.60 (C-8μ), 154.17 (C-9μ), 104.25 (C-10μ), 131.60 (C-1μ), 107.48 (C-2μ), 146.26 (C-3μ), 134.58 (C-4μ), 146.26 (C-5μ), 107.48 (C-6μ), 81.68(C-2ι), 66.90 (C-3ι), 29.85 (C-4aι),156.90 (C-5ι), 96.48 (C-6ι), 152.21(C-7ι), 107.14 (C-8ι), 152.06 (C-9ι), 102.39 (C-10ι), 131.14 (C-1ι), 115.90 (C-2ι), 145.91 (C-3ι), 146.21 (C-4ι), 116.03 (C-5ι), 120.36 (C-6ι).

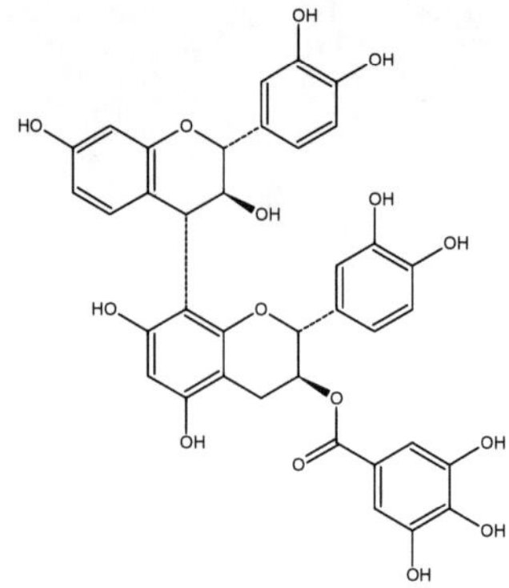

403
Proanthocyanidin 1 or Fisetinidol-(4α → 6 ,4α → 8)-catechin 3-gallate

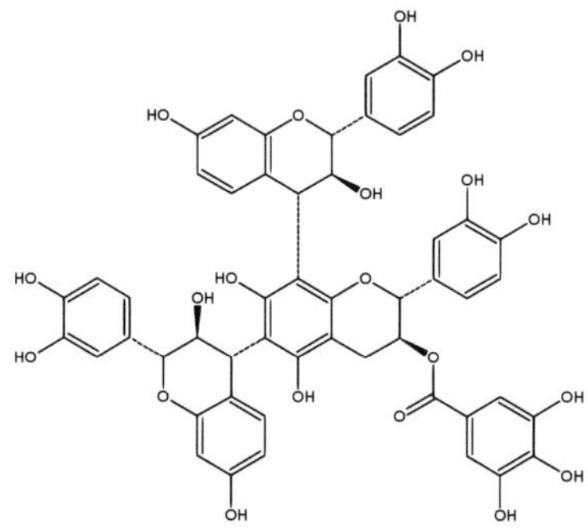

404
Proanthocyanidin 2 or bis-fisetinidol-(4α → 6 ,4α → 8)- catechin 3-gallate

Found in *Burkea Africana* Hook (Leguminosae). Proanthocyanidins have numerous reported biological activities, and some proanthocyanidin preparations, such as hawthorn extracts and pycnogenol, are widely used as pharmaceuticals. It may be noted that the radical scavenging activity of *Burkea* preparations compares well with that of pycnogenol, a highly efficient antioxidant and radical scavenger. Showed activities as radical scavengers and 15-lipoxygenase inhibitors, but substance bis-fisetinidol-(4α → 6,4α → 8)-catechin 3-gallate was a better inhibitor of iron-induced phospholipid peroxidation (Mathisen et al., 2002).

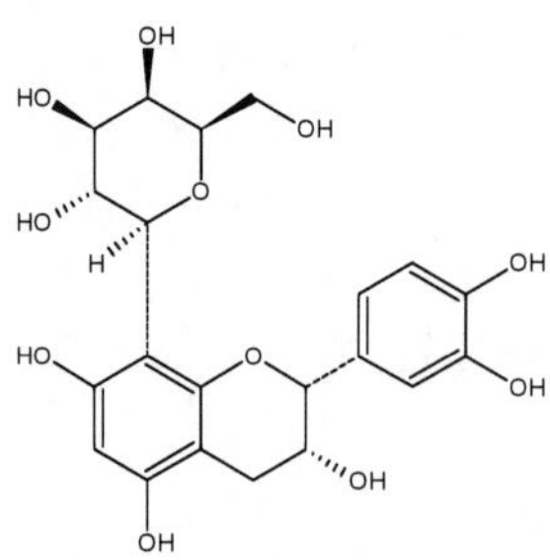

405
(-)-Epicatechin 8-C-galactopyranoside

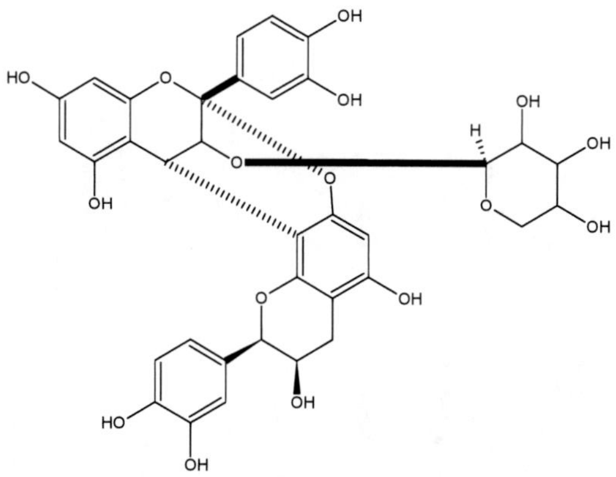

406

3-O-Arabinopyranosyl-*ent*-epicatechin-(2α→ 7, 4α→8)-catechin

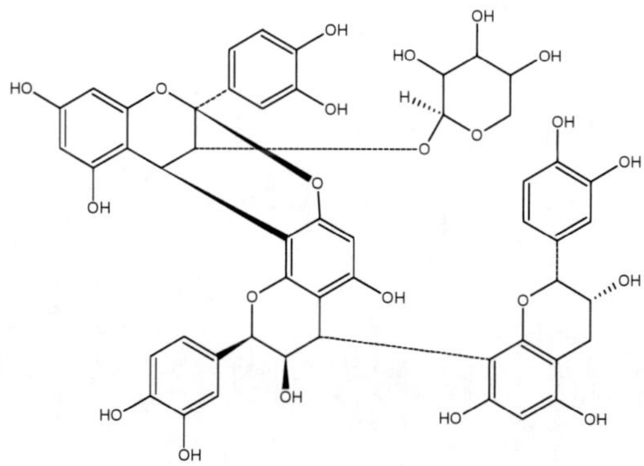

407

3-O-α-L-Arabinopyranosylcinnamtannin B_1

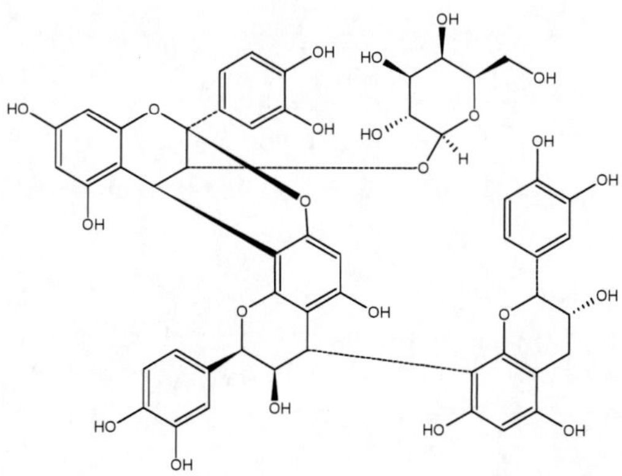

408

3-O-β-D-Galactopyranosylcinnamtannin B_1

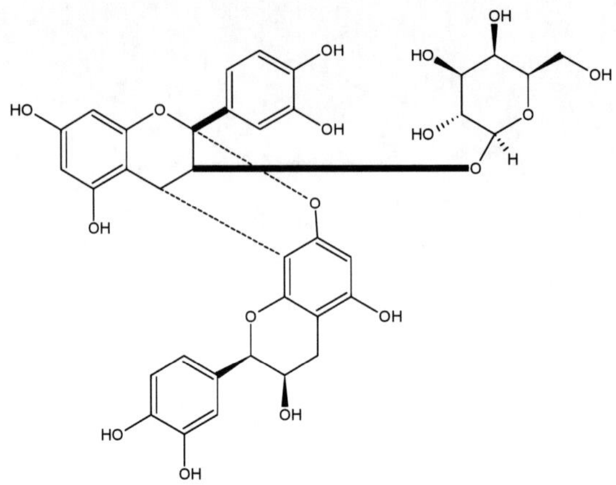

409

3-O-β-D-Galactopyranosyl-*ent*-epicatechin-(2α→ 7, 4α→8)-catechin

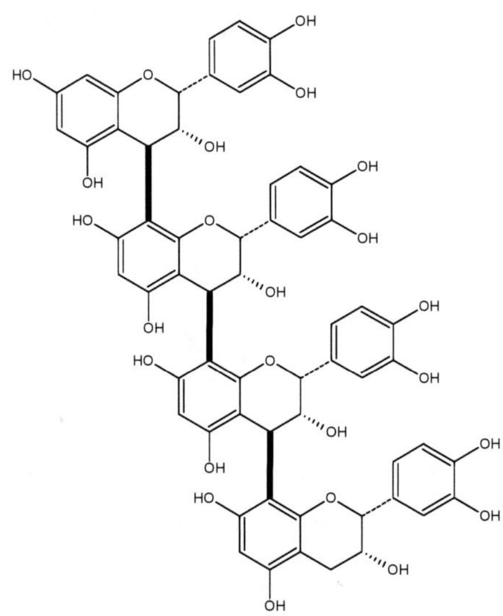

410

Cinnamtannin A_2

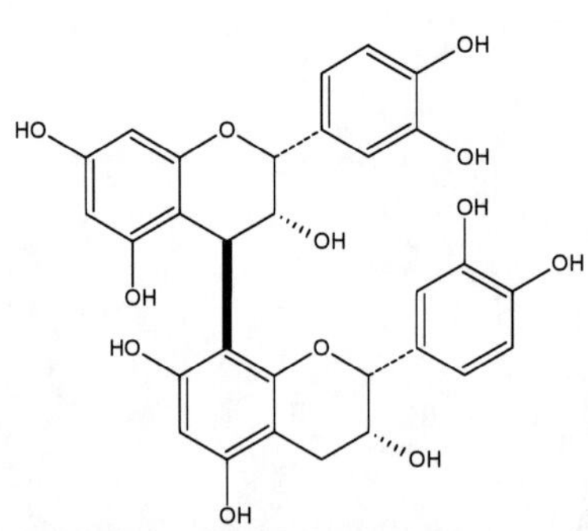

411

Procyanidin B_2

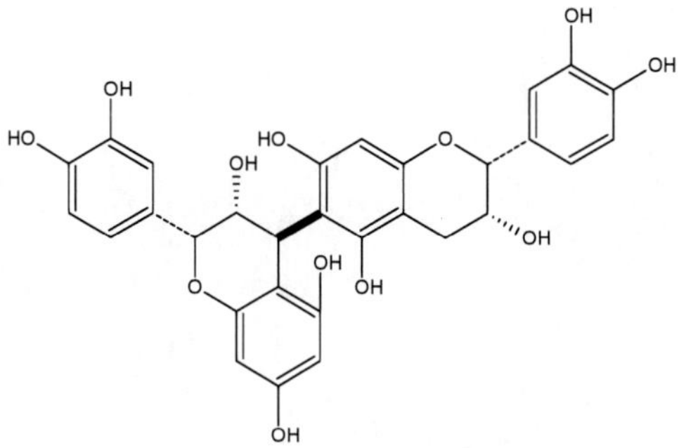

412
Procyanidin B_5

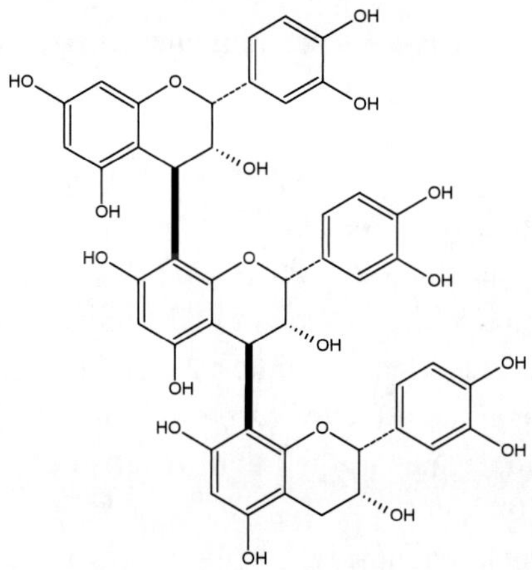

413
Procyanidin C_1

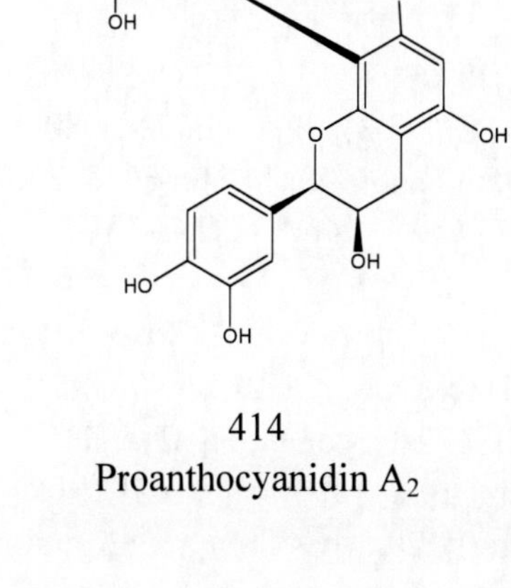

414
Proanthocyanidin A_2

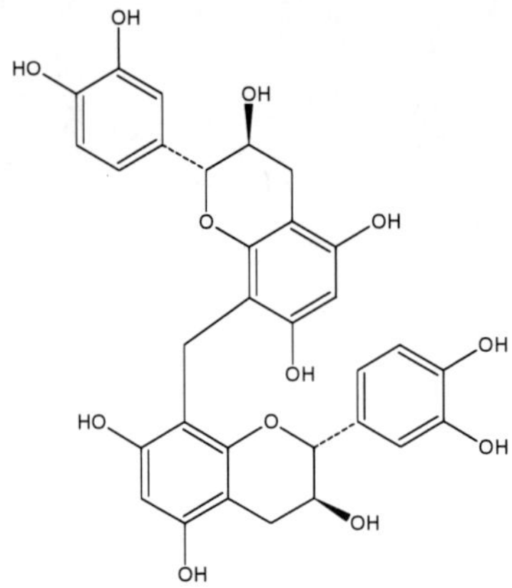

415
Bis-8,8'-catechinylmethane

Found in *Theobroma cacao* L. (Sterculiaceae). The polyphenolic constituents of cacao liquor have been reported to have antioxidant effects (Osakabe et al., 2000). The NADPH-dependent lipid peroxidation in rat liver microsomes was inhibited by almost all of the polyphenols tested. All polyphenols except for (-)-epicatechin 8-C-galactoside had inhibitory effects on the lipid peroxidation with IC_{50} values of 12-68 μM. Although these values are no better than that α-tocopherol. The effects of almost all of the poly-phenols on the oxidation of linoleic acid were stronger than the effect of α-tocopherol. The difference may be attributed to the presence of other biomolecules in the microsomal solution, which interacted with the polyphenol (Hatanoa et al., 2002).

(-)-Epicalechin 8-C-galaclopyranoside: A light-brown amorphous powder, $[\alpha]_D$, -25.8°(c=0.9, MeOH). ESI MS *m/z*: 453 ($[M+H]^+$, 470 ($[M + NH4]^+$); High-resolution ESI MS m/z: 470.1649 (($[M+NH4]^+$), Calc. for $C_{21}H_{24}O_{11}$ + NH4, 470.1662; UV λmx (log ε): 280 (3.32); CD (MeOH): $[\theta]_{207}$ -4.4 x 10^4, $[\theta]_{250}$ + 1.8 x 103, $[\theta]_{279}$-3.4 x 103; ^{1}HNMR (in Me_2CO-d_6+ D_2O, 24 °C): δ 2.80 (1H, dd, J=2.5, 17 Hz, epicatechin (EC) H-4), 2.86 (1H, dd, J=4, 17 Hz, EC H-4), 3.62 (1H, br dd, J = 5, 6.5 Hz, galactose (Gal) H-5), 3.66 (IH, dd, J = 3, 9.5 Hz, Gal H-3), 3.70 (1H, dd, J = 5, 11.5 Hz, Gal H-6), 3.73 (1H, dd, J = 6.5 Hz, 11.5 Hz, Gal H-6), 4.03 (1H, br d, J = 3 Hz, Gal H-4), 4.05 (1H, t, J = 9.5 Hz, Gal H-2), 4.10 (1H, m, EC H-3), 4.85 (1H, d, J = 9.5 Hz, Gal H-1), 4.92 (1H, br s, EC H-2), 5.98 (1H, s, EC H-6), 6.76 (1H, d, J = 8 Hz, EC H-5'), 6.82 (1H, dd, J = 2,8 Hz, EC H-6'), 7.16 (1H, d, J = 2 Hz, EC H-2'); ^{13}C NMR (in MeOH-d_4, 40 °C): δ 30.5 (EC C-4), 63.8 (Gal C-6), 68.2 (EC C-3), 71.4 (Gal C-4), 73.1 (Gal C-2), 77.0 (Gal C-3), 77.7 (Gal C-l), 80.9 (EC C-2), 81.1 (Gal C-5), 98.2 (EC C-6), 100.5 (EC C-4a), 105.8 (EC C-8), 116.0 (EC C-2'), 116.9 (EC C-5'), 119.8 (EC C-6'), 133.2 (EC C-1'), 146.5 (EC C-4'), 146.8 (EC C-3'), 155.4 (EC C-8a), 157.9 (EC C-7), 158.9 (EC C-5).

3 - O - Arabinopyranosyl - *ent* - epicatechin- (2α→ 7, 4α→8) - catechin: A light-brown amorphous powder, $[\alpha]_D$, -16.4) (c= 1, MeOH) ESI MS *m/z*: 709 ($[M + H]^+$). High-resolution ESI MS *m/z* 709.1713 ($[M+H]^+$). Calc. for $C_{35}H_{32}O_{16}$ +H, 709.1769. UV λmx (log ε): 280 (4.17); CD (MeOH): $[\theta]_{203}$ -2.3 x 10^{-3} , $[\theta]_{271}$,+ 4.9 x 10^{-3}, $[\theta]_{228}$,- 6.9 x 10^{-3}, $[\theta]_{211}$,+1.9 x 10^{-3}; ^{13}C NMR: δ 24.8 [C-4 (C-ring)], 29.3 [C-4 (E)], 63.2 [arabinose (Arb) C-5], 66.2 (Arb C-

4), 67.6 [C-3 (E)], 70.9 (Arb C-2), 71.0 [C-3 (C)], 72.5 (Arb C-3), 84.3 [C-2 (E)], 96.0 [C-8 (C)], 96.6 [C-6 (E)], 98.0 [C-6 (C)], 98.9 (Arb C-1), 99.2 [C-4a(E)], 103.2 [C-2(C), C-4a (A)], 105.2 [C-8 (D)], 115.4 (2C) [C-2' (B), C-5 '(B)], 115.9 (2C)[C-2' (E), C-5' (E)], 119.5 [C-6, (B)], 120.5 [C-6′ (E)], 130.5 [C-1' (E)], 131.6 [C-1′ (B)], 145.3 [C-4' (B)], 145.7 [C-4' (E)], 146.2 [C-3' (E)], 146.5 [C-3' (B)],152.1 [C-7 (D)], 152.4 [C-8a (D)], 153.7 [C-8a (A)], 155.5. 155.6 [C-5 (A/D)], 158.6 [C- 7 (A)].

3-O-α-L-Arabinopyranosylcinnamtannin B_1: A light-brown amorphous powder. $[\alpha]_D$, + 12.6 (c =1, MeOH). ESI MS *m/z* 997 ($[M + H]^+$). High-resolution ESI-MS *m/z* 1014.2593 ($[M + NH_4]^+$). Calc. for $C_{50}H_{44}O_{22}$ + NH_4. 1014.2668. UV λmx (log ε): 280(4.20). CD (MeOH): $[\theta]_{283}$ -5.2 x 10^{-3} , $[\theta]_{272}$ -3.1 x 10^{-3} , $[\theta]_{257}$ -3.7 x 10^{-3} , $[\theta]_{225}$ -4.5 x 10^{-3} , $[\theta]_{209}$ -1.8 x 10^{-4} , $[\theta]_{202}$ -2.3 x 10^{-4}; ^{13}C NMR: δ 24.9 [C-4(C)], 36.4 [C-4 (F)], 64.3 (Arb C-5), 66.1 [C-3 (F)], 67.1 (ArbC-4), 71.0 (Arb C-2), 71.2 [C-3(C)], 72.5 (Arb C-3), 72.6 [C-3 (F)], 77.8 [C-2 (F)], 79.1 [C-2 (1)], 95.9 [C-8 (A)], 96.2 [C-6(D)], 96.9 [C-6 (G)], 97.9 [C-6 (A)], 99.3 [C-2 (C)], 99.5 (Arb C-1), 100.1 [C-4a (G)], 102.6 [C-8 (D)], 104.1 [C-4a (D)], 105.1 [C-4a (A)], 106.7 [C-8 (G)], 115.0, 115.4, 115.6 (3C), 115.8 [C-2' (B/E/H), C-5' (B/E/H)],118.9, 119.2, 119.5 [C-6' (B/E/H)], 131.2, 131.3, 131.4 [C-1' (B/E/H)], 144.9-146.3 [C-3' (B/S, H), C-4' (B/E/H)], 152.2-158.3 [C-5 (A/D/G), C-7 (A/D/G), C-8a (A/D/G)].

3-O-β-D-Galactopyranosylcinnamtannin B_1: A light-brown amorphous powder, $[\alpha]_D$, + 17.1° (c= 1, MeOH) ESI MS *m/z*: 1044 ($[M + NH_4]^+$), 1027 ($[M + H]^+$), 865 ([M-galactose + H] $^+$). High-resolution ESI MS 1027.2531 ([M + H]+). Calc. for $C_{51}H_{46}O_{23}$ + H, 1027.2508; UV MeOH λmx (log ε): 281 (5.02).CD (MeOH): $[\theta]_{204}$ -9.1 x 10^{-3} , $[\theta]_{258}$ -1.3 x 10^{-3} , $[\theta]_{234}$ -5.5 x 10^{-3} , $[\theta]_{209}$ -3.4 x 10^{-4} , $[\theta]_{234}$ -5.6 x 10^{-4} ; ^{1}HNMR δ: 2.71 (1H, br, H-4), 2.84 [1H, br d, J = 15 Hz, H-4 (1)], 3.95 [1H, brs, H-3 (F)], 4.31 [1H. br, H-3 (1)], 4.48 [1H, d, J=3.5 Hz, H-3 (C)], 4.73 [1H, br.s,11-4 (F)], 4.74 [1 H, br. s, H-4 (C)], 4.94 [1 H, br, H-2 (1)], 5.35 [1H, br .s, H-2 (F)], 5.88 [1 H, d, J= 2 Hz, H-6 (A)], 5.96 [2H. H-6 (G) and H-8 (A)], 6.22 [1 H, s, H-6 (D)], 6.6- 7.3 (9H, B-ring protons); ^{13}C NMR: δ 25.7 [C-4 (C)], 36.3 [C-4 (F)], 61.6 (Gal C-6), 66.0 [C-3 (1)], 69.2 (Gal C-4), 71.6 (Gal C-2), 72.1 [C-3 (C)], 72.7 [C-3 (F)], 73.7 (Gal C-3), 76.0 (Gal C-5), 77. 7 [C-2 (F)], 79,0 [C-2 (1)], 95.9 [C-8 (A)], 96.3 [C-6 (D)], 96.9 [C-6 (G)], 97.8 [C-6 (A)], 99.4 [C-2 (C)], 100.1 [C-4a (G)], 101.0 (Gal C-1), 102.9 [C-8 (D)], 104.2 [C-4a (D)], 105.5 [C-4a (A)], 106.7 [C-8 (G)], 114.8, 115.0, 115.1, 115.6. 115.8, 116.2 [C-2' (B E H), C-5' (B E H)], 118.9, 119.0, 119.7 [C-6' (B E H)], 131.2, 131.4, 131.5 [C-1' (B/E/H)], 144.6-146.2 [C-3' (B/E/H), C-4' (B/E/H)], 152.3-158.1 [C-5(A/D/G), C- 7 (A/D/G), C-8a (A/Q/G)].

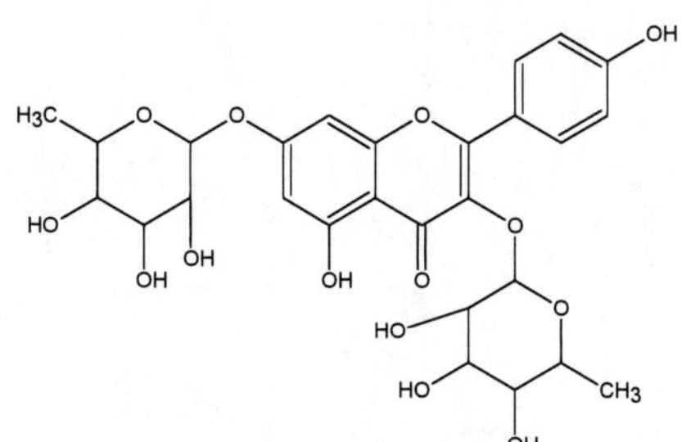

416

7-O-α-L-Rhamnopyranoside-kaempferol-3-O-α-L-rhamnopyranoside

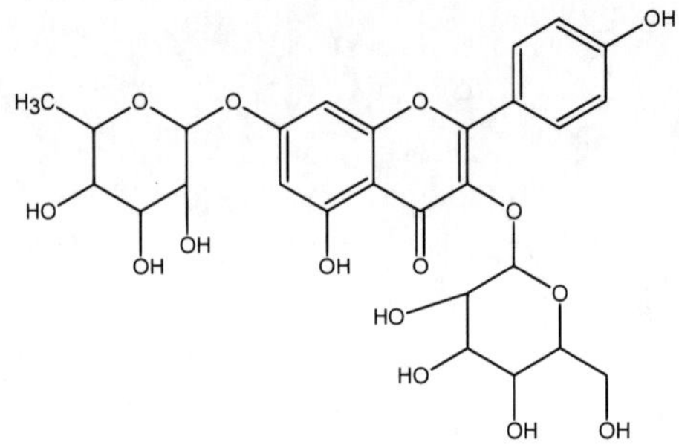

417

7-O-α-L-Rhamnopyranoside-kaempferol-3-O-β-D-glucopyranaside

Found in *Ligustrum sinense* (Oleaceae). Showed antioxidant activity (Ouyang et al., 2003).

7-O-α-L-Rhamnopyranoside-kaempferol-3-O-α-L-rhamnopyranoside: ^{1}H NMR: δ 7 .78 (H-2', H-6' , 2H, d, J = 8.4 Hz), 6.90 (H-3', H-5' , 2H, d, J = 8.5 Hz), 6.78 (H-8), 6.45 (H-6), 5.54 (1H. br,s, Rha), 5.29 (1H, br,s, Rha), 1.01 (3H, d, J=6.4 Hz), 0.79 (3H, d, J=6.0 Hz); ^{13}C NMR: δ 177.8 (C-4), 161.6 (C-7), 160.9 (C-5), 160.0 (C-4'), 157.6 (C-2), 156.0 (C-9), 134.5 (C-3), 130.5 (C-2', C-6'), 120.3 (C-1'). 115.3 (C-3', C-5'), 105.7 (C-10), 98.4 (C-6), 94.5 (C-8), 101.8 (C_R-1, 3-O-Rha), 71.5 (C_R-4), 70.5 (C_R-3), 70.2 (C_R-2), 69.9 (C_R-5), 17.8 (C_R-6), 99.3 (C_R-1, 7-O-Rha), 71.1 (C_R-4), 70.3 ($C_{R,}$-3), 70.2 ($C_{R,}$-2), 69.7 ($C_{R,}$-5), 17.3 ($C_{R,}$-6).

7-O-α-L-rhamnopyranoside-kaempferol-3-O-β-D-glucopyranasido: ^{1}H NMR δ 8.08 (H-2', H-6' ,2H. d, J= 8.8 Hz), 6.90 (H-3', H-5', 2H, d, J= 8.7 Hz), 6.81 (H-8). 6.44 (H-6), 5.54 (1H, br.s, Rha), 5.47 (1H, d, J=7.2 Hz), 1.10 (3H, d, J=6.4 Hz); ^{13}C NMR: δ 177.6 (C-4), 161.6 (C-7), 160.8 (C-5), 160.0 (C-4'), 156.7 (C-2), 156.0 (C-9), 133.5 (C-3), 130.9 (C-2', C-6'), 120.7 (C-1'), 115.1 (C-3', C-5'), 105.6 (C-10), 98.4 (C-6), 94.2 (C-8), 100.8 (C_G-1, 3-O-Glc), 77.4 (C_G-5), 76.4 (C_G-3), 74.2 (G_G-2), 70.2 (C_G-4), 60.8 (C_G-6), 99.3 (C_R-1, 7-O-Rha), 71.6 (C_R-4), 70.1 (C_R-3), 70.2 (C_R-2), 69.7 (C_R-5), 17.8 (C_R-6).

Chapter 5

MONOTERPENOIDS, DITERPENOIDS AND SESQUITERPENOIDS

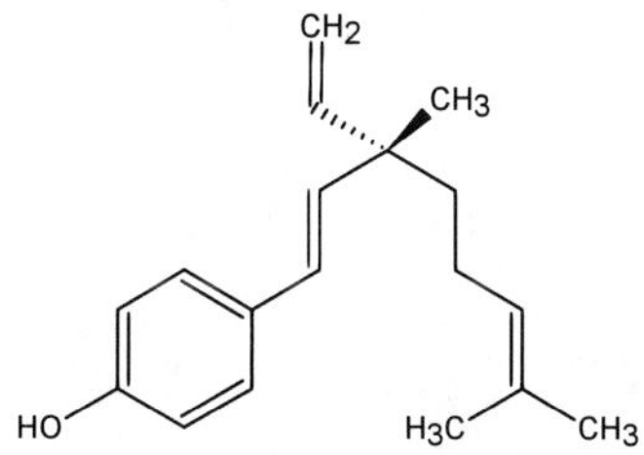

418

Bakuchiol

A meroterpene was isolated from the seeds or *Psoralea corylifolia* (Leguminosae) as antioxidative components. Showed broad antioxidative activities in rat liver microsomes and mitochondria. It inhibited NADPH, ascorbate, BuOOH and CCl_4 induced lipid peroxidation in microsomes. Also prevented NADH-dependent and ascorbate-induced mitochondrial lipid peroxidation. Bakuchiol is a potent antioxidant in microsomes and the inhibition or oxygen consumption induced by lipid peroxidation was time-dependent. Furthermore, also protected human red blood cells against oxidative haemolysis. This phenolic compound in *P. corylifolia* was shown to be effective in protecting biological membranes against various oxidative stresses (Haraguchi et al., 2002).

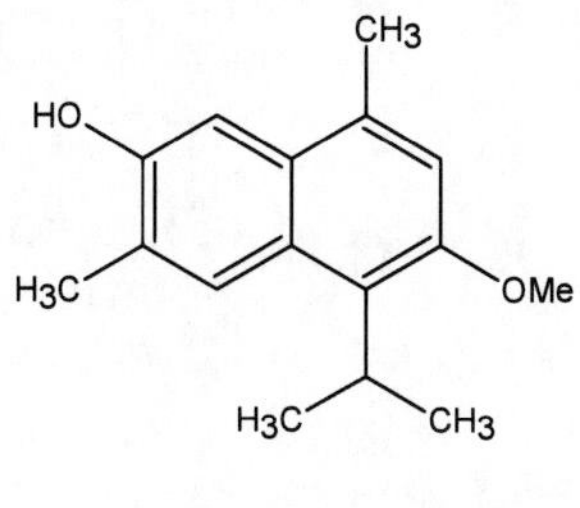

419

7-Hydroxy-3-methoxy-cadalene

7-Hydroxy-3-methoxy-cadalene extracted from *Zelkova serrata* exerts antioxidative effect on NNK-induced lung tumorigenesis in A/J mice (Kim et al., 2004).

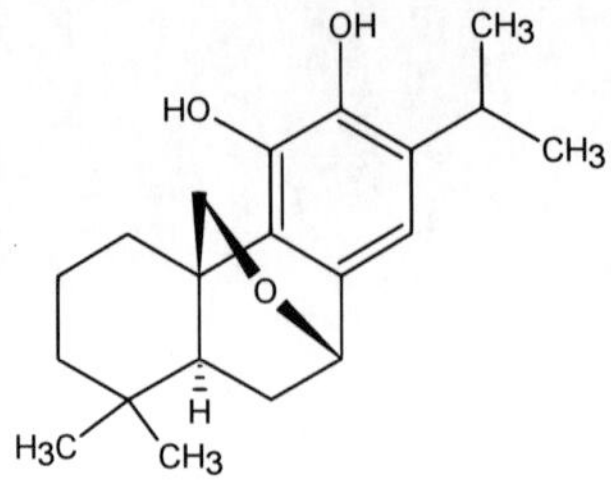

420
20-Deoxocarnosol
Found in *Sphacele salviae* (Lamiaceae). It is more efficient scavenger than BHT and that carnosic acid is even better than vitamin E (Escuder et al., 2002).

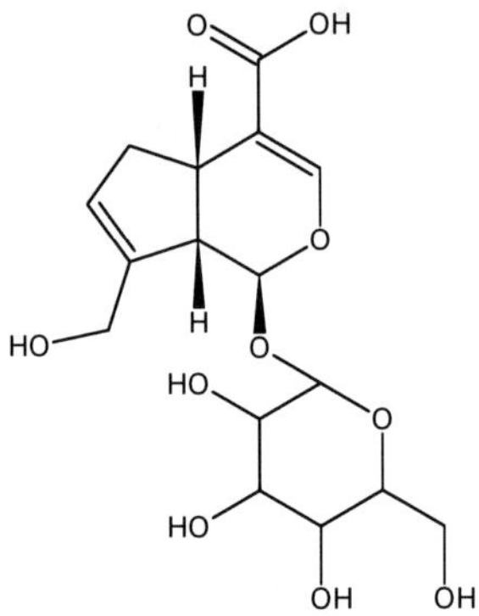

421
Geniposidic acid

A monoterpene glycoside, geniposidic acid, isolated from *Plantago asiatica* (Plantaginaceae), was as nearly as effective as BHA or BHT at inhibiting the air oxidation of linoleic acid. A concentration ofabout 4 x 10^{-4} M inhibited 50% of the oxidation as measured by two different tests. Structure-activity experiments with several related glycosides suggested that both the presence of the carboxyl group and the substituent pattern on the five-membered ring were important determinants of the antioxidant activity (Larson, 1988).

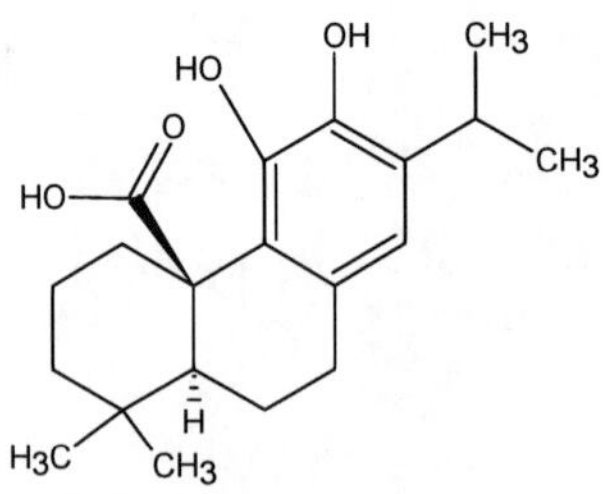

422
Carnosic acid

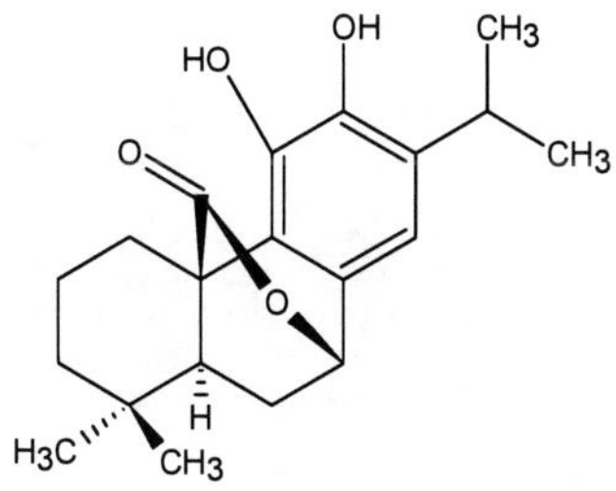

423
Carnosol

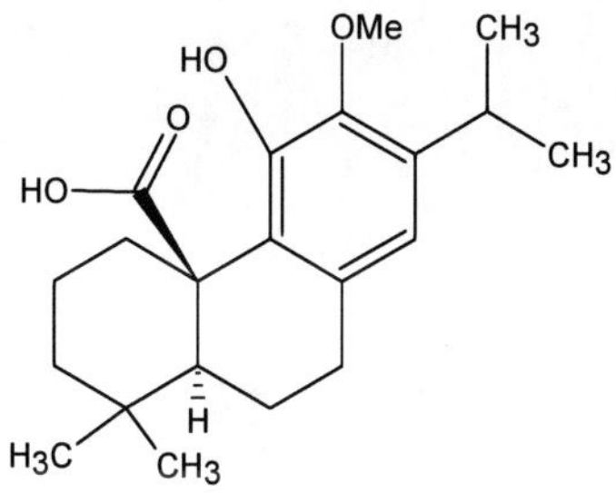

424
12-O-Methylcarnosic acid

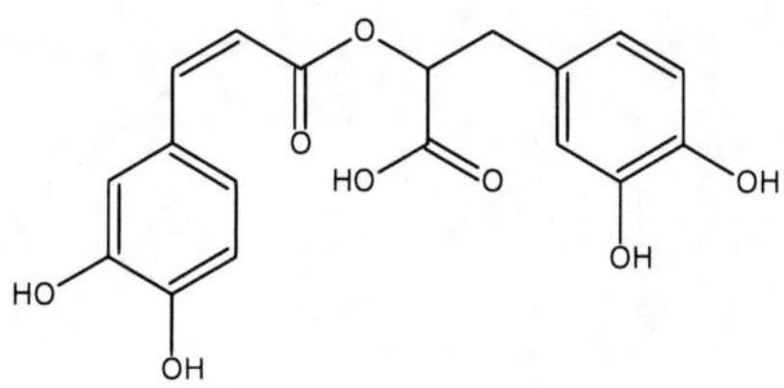

425
Rosmarinic acid

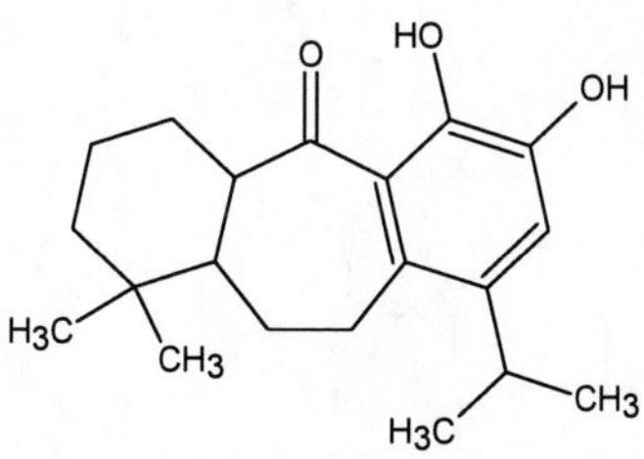

426
Rosmaridiphenol

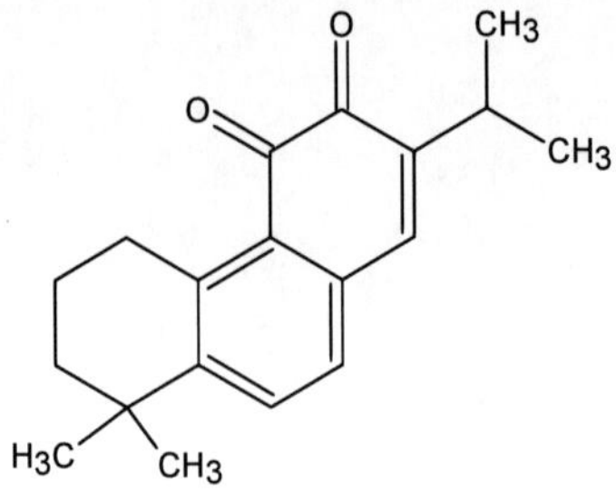

427
Rosmariquinone

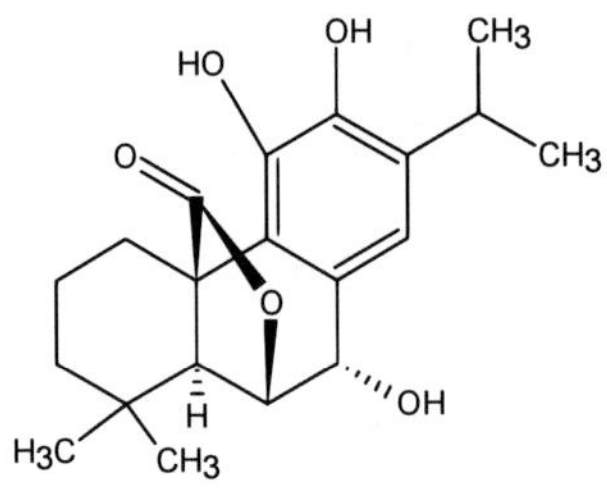

428
Rosmanol

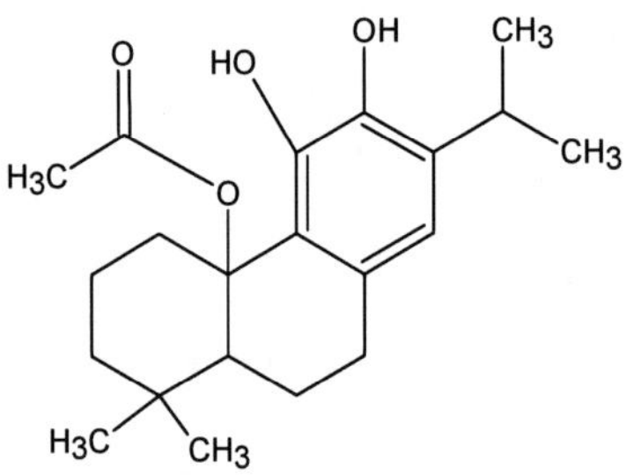

429
Methyl carnosote

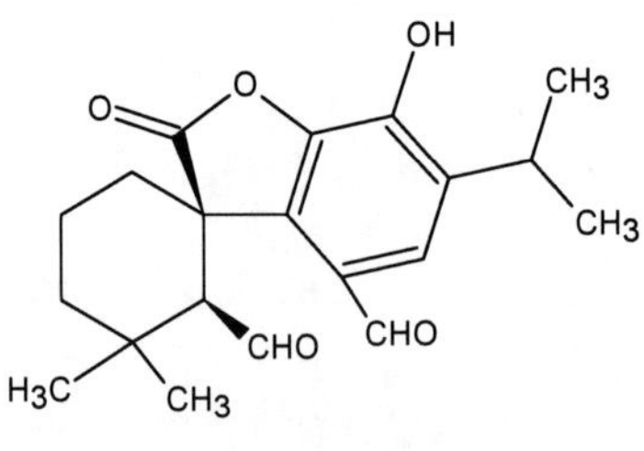

430
Rosmadial

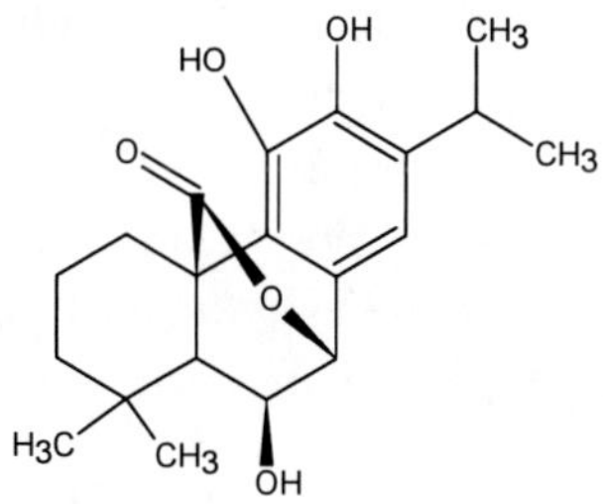

431
Epiisorosmanol

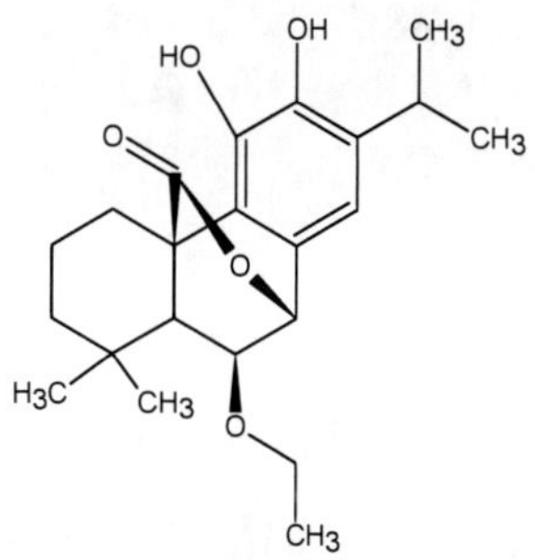

432
Epiisorosmanol ethyl ether

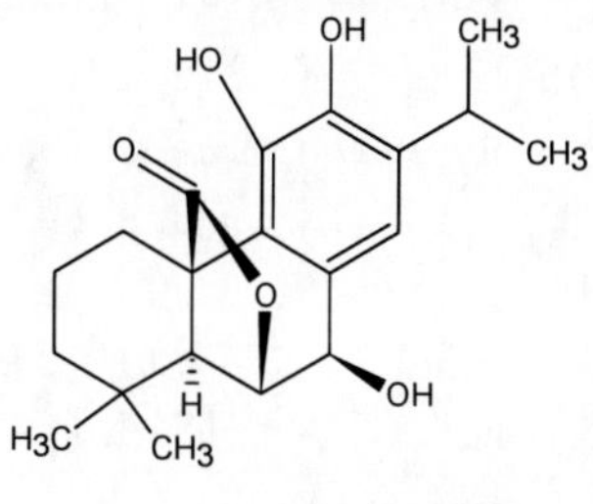

433
Epirosmanol

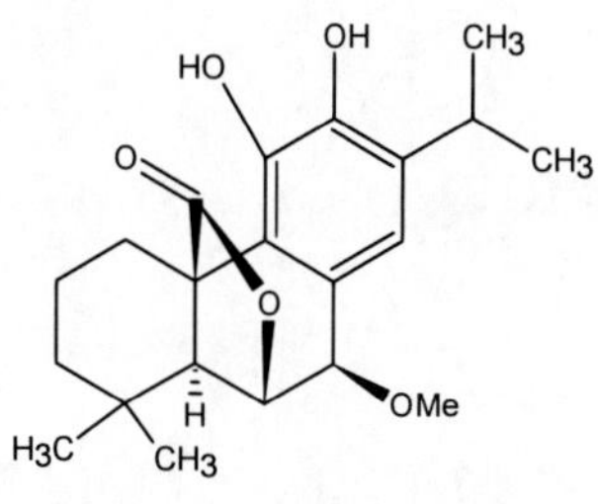

434
Epirosmanol methyl ether

Found in *Rosmarinus officinalis* (Labiatae). Especially carnosic acid and carnosol, are responsible for the high antioxidant activity of rosemary extracts, as was reported previously (Aruoma et al., 1992). These compounds are therefore excellent antioxidants in both aqueous and lipid systems . The presence of a catechol group in the aromatic ring (C11-C12) of the rosemary phenolic diterpene skeleton is probably the most important structural element in the antioxidant activity of these compounds. In rosmarinic acid, the presence of two catechol structures conjugated with a carboxylic acid group increases the antioxidant activity in aqueous or aqueous-alcoholic media (Frankel et al., 1996). The presence of flavones in the rosemary extract suggests a possible synergistic behavior in its radical scavenging capacity (Dorman et al., 2003, Ibañez et al., 2003). Rosmaridiphenol, a diterpene derivative with adjacent OH groups, its antioxidant activity in heated lard exceeded that of BHA and approached that of BHT (Larson, 1988). The antioxidant activity of rosmariquinone was superior to BHA, but it was slightly less than BHT (Houlihan et al., 1985).

Rosmarinic acid: Amorphous powder; DCI-MS m/z: 378[M + $NH_4]_+$; ^{1}HNMR (in CD_3COCD_3, 300 MHz): δ 3.03 (2H, m, H-7'), 5.20 (lH, m, H-8'), 6.29 (lH, d, J = 16.0 Hz, H-8), 6.61-6.87 (4H, m, H-5, H-2', H-5', H-6'), 7.03 (lH, dd, J = 8.0, 2.0 Hz, H-6), 7.16 (lH, d, J = 2.0 Hz, H-2), 7.55 (lH, d, J = 16.0 Hz, H-7); ^{13}C NMR (in CD_3COCD_3, 75 MHz): δ 171.2 (s, C-9'), 166.8 (s, C-9), 149.0 (s, C-4), 146.5 (s, C-3), 146.3 (d, C-7), 145.7 (s, C-3'), 144.7 (s, C-4'), 129.1 (S, C-1'), 127.3 (s, C-1), 122.7 (d, C-6), 121.6 (d, C-6'), 117.3 (d, C-2'), 116.3 (d, C-5), 115.5 (d, C-5'), 115.2 (d, C-2), 114.8 (d, C-8), 73.7 (d, C-8'), 37.4 (t, C-7'), (Wang et al., 1998).

Rosmariquinone: $C_{19}H_{22}O_2$, m.p. 94-95 C. Elemental composition was calculated as 80.82% carbon, 7.85% hydrogen; IR: ν max cm^{-1}: 3090, 1649 , 2960, 2940, 1570, 1560, 1450, 1410, 1380, 1300, 1250, 1140, 920; (ElMS): *m/z* (%) 282 ([M]$^+$, *m/z* (relative intensity) 267 (5%), 254 (52%), 239 (100%), 224 (17%), 219 (6%), 209 (9%), 197 (8%), 165 (10%); ^{1}HNMR: δ 3.17 (2H, t, J = 6.3 Hz, H-1), 1.79 (2H, m, H-2), 1.64 (2H, m, H-3), 7.60 (lH,d, J = 8.0 Hz, H-6), 7.12 (lH,d, .J = 8.0 Hz, H-7), 7.09 (lH, s, H-14), 3.02 (lH, sept, J = 7.1 Hz, H-15), 1.17 (6H, d, J = 6.8 Hz, H-16 and Me-17), 1.30 (6H, s, H-18 and Me-19); ^{13}CNMR: δ 29.9 (C-1), 19.1 (C-2), 37.8 (C-3), 34.5 (C-4), 144.5 (C-5), 133.8 (C-6), 139.9 (C-7), 128.2 (C-8), 134.4 (C-9), 145.0 (C-10), 182.4 (C-11), 181.5 (C-12), 149.7 (C-13), 127.9 (C-14), 26.9 (C-15), 21.5 (C-16,17), 31.8 (C-18,19).

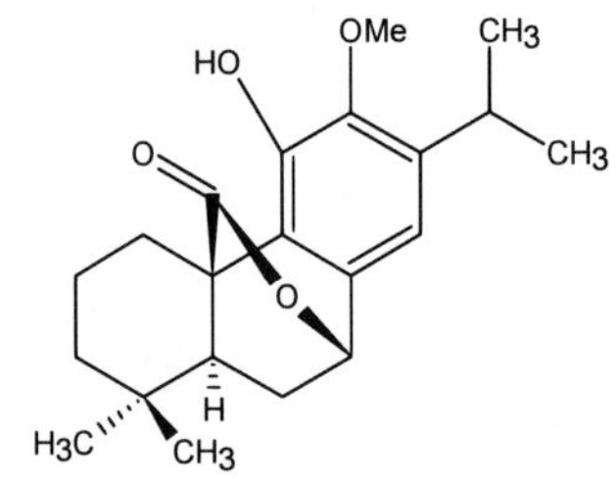

435
12-O-Methyl carnosol

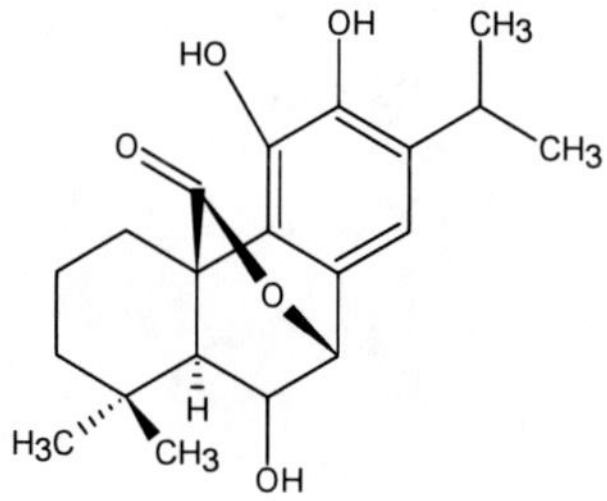

436
Isorosmanol

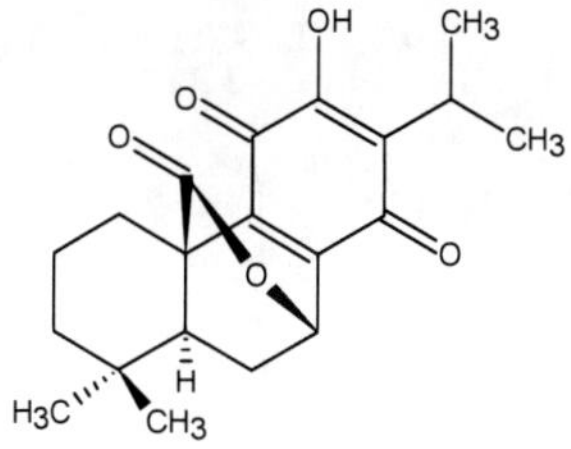

437
Columbaridione

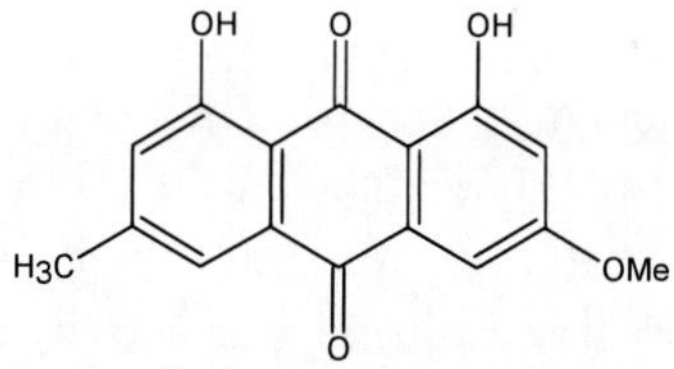

438
Physcion

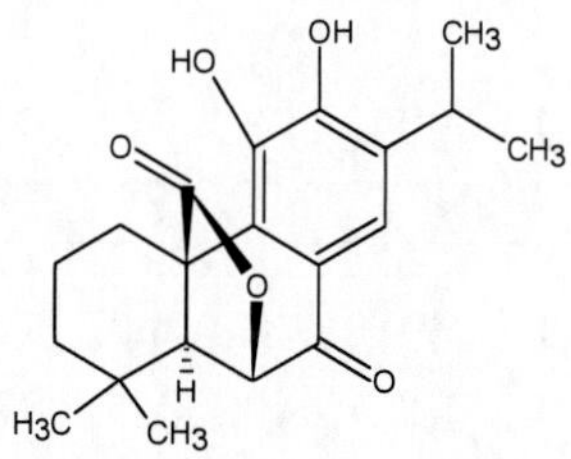

439
Galdosol

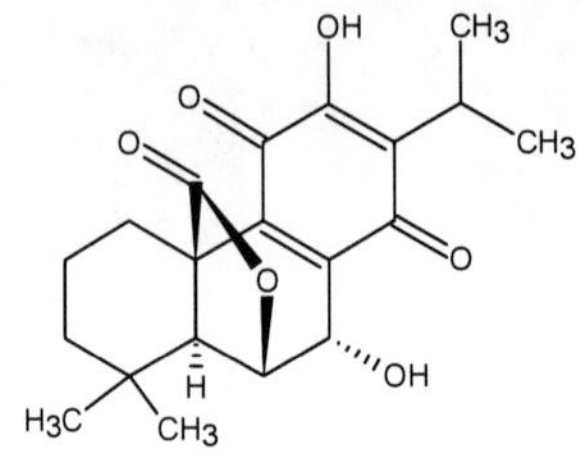

440
Atuntzensin A

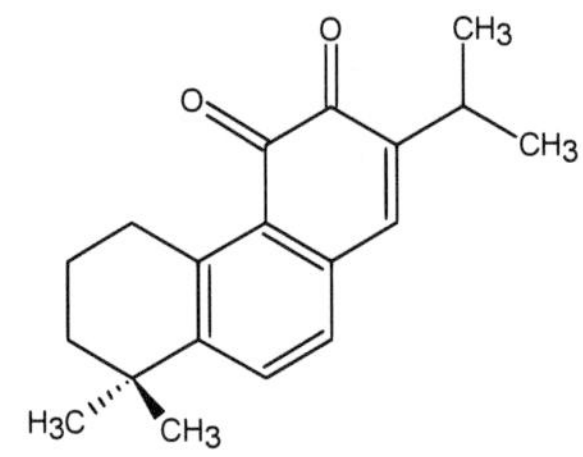

441
Miltirone

Antioxidant activity of these compounds isolated from *Thymus vulgaris* L. (Labiatae); and *Salvia officinalis* L. (Labiatae) were evaluated by the oil stability index method using a model substrate oil including methyl linoleate in silicone oil at 90 °C. Carnosol, rosmanol, epirosmanol, isorosmanol, galdosol, and carnosic acid exhibited remarkably strong activity, which was comparable to that of α-tocopherol. The activity of miltirone, atuntzensin A, was comparable to that of butylated hydroxy toluene. The activity of these compounds was mainly due to the presence of ortho-dihydroxy groups. The 1,1-diphenyl-2-picrylhydrazyl radical scavenging activity of these compounds showed the similar result (Miura et al., 2002). These abietane diterpenoids showed remarkably strong activity, which was comparable to that of α–tocopherol. The mechanism of antioxidant activity depends on the type and polarity of the lipid system used.

12-O-Methyl Carnosol: Colorless solid, mp 69.0 °C. Anal. Calcd for $C_{21}H_{28}O_4$: C, 73.22; H, 8.19. Found: C. 73.6; H, 8.12. Electron impact mass spectrometry (ElMS): *m/z* (%) 344 ($[M]^+$ (22), 300 (100), 298 (17), 285 (21), 283 (16), 231 (14), 229 (20), 218 (18), 215 (18); UV λmax, ETOH nm (log ε): 207 (4.53), 284 (3.44). IR vmax Nujol (cm^{-1}): 3320, 1728, 1615, 1580, 1245, 1232, 1220, 1160, 1120, 1040, 1025, 997; ^{1}HNMR (500Mz. $(CD_3)_2CO$): δ 2.46 (ddd, J = 4.4, 13.9, 14.2 Hz, H-1α), 2.88 (dddd, J = 1.7, 3.2, 4.9, 14.2 Hz, 1β), 1.65 (ddddd, J = 3.6, 4.4, 4.9, 4.9, 13.9 Hz, H-2α), 1.98 (ddddd, J = 3.2, 3.3, 13.6, 13.9, 13.9 Hz, 2β), 1.30 (ddd, J = 3.6, 13.2, 13.6 Hz, 3α), 1.54 (dddd, J = 1.7, 3.3, 4.9, 13.2 Hz, 3β), 1.74 (dd, J = 5.7, 10.6 Hz, H-5), 2.22 (ddd, J = 4.0, 5.7, 13.7 Hz, H-6α), 1.88 (ddd, J = 1.7, 10.6, 13.7 Hz, H-6β), 5.38 (dd, J = 1.7, 4.0 Hz, H-7), 5.985 (s, OH), 6.66 (s, H-14), 3.22 (sept, J = 6.8 Hz, H-15), 1.21 (d, J = 6.8 Hz, H-16), 1.21 (d, J = 6.8 Hz, H-17), 0.865 (s, H-18), 0.91 (s, H-19), 3.75 (s, H-21); ^{13}C NMR (125Mz, $(CD_3)_2CO$): δ 28.8 (C-1α), 18.9 (C-2α), 41.1(C-3α), 34.5 (C-4), 45.4 (C-5), 29.5 (C-6α), 77.7 (C-7), 136.4 (C-8), 121.5 (C-9), 48.4 (C-

10), 146.0 (C-11), 144.8 (C-12), 140.1 (C-13), 112.0 (C-14), 26.6 (C-15), 23.7 (C-16), 23.7 (C-17), 31.7 (C-18), 19.7 (C-19), 175.9 (C-20), 62.0 (C-21).

Columbaridione: Orange-yellow needles. mp 179 °C; ^{1}HNMR(400Mz.CDCl$_3$): δ 02.31 (1H,ddd, J = 4.4, 13.8, 14.1 Hz, H-1α), 2.71 (1H, dddd, J = 1.5, 3.3, 3.3, 14.1 Hz, H-1β), 1.65 (1H, ddddd, J = 3.2, 3.3, 3.3, 4.4, 14.1 Hz, H-2α), 1.92 (1H, ddddd, J = 3.2, 3.3, 13.8, 13.8, 14.1, Hz, H-2β), 1.25 (1H, ddd, J = 3.3, 13.2, 13.8 Hz, H-3α), 1.55 (1H, dddd, J = 1.5, 3.2, 3.2, 13.2 Hz, H-3β), 1.69 (1H, dd, J= 5.9, 10.6, Hz, H-5), 2.24 (1H, ddd, J= 4.2, 5.9, 14.1, Hz, H-6α), 1.84 (1H, ddd, J= 1.5, 10.6, 14.1, Hz, H-6β), 5.80 (1H, dd, J= 1.5, 4.2, Hz, H-7), 3.18 (1H. sept, J = 7.1, Hz, H-15), 1.22 (3H, d, J = 7.1 Hz, H-16), 1.22 (3H, d, J = 7.1 Hz, H-17), 0.90 (3H, s, H-18), 0.87 (3H, s, H-19); ^{13}C NMR (100Mz. (CD$_3$)$_2$CO): δ 027.4 (C-1), 18.5 (C-2), 40.7 (C-3), 34.8 (C-4), 45.3 (C-5), 28.0 (C-6), 70.3 (C-7), 138.8 (C-8), 146.9 (C-9), 48.9 (C-10), 179.6 (C-11), 150.7 (C-12),123.7 (C-13),181.6 (C-14), 24.4 (C-15), 20.1 (C-16), 20.1 (C-17), 32.4 (C-18), 19.7 (C-19), 172.9 (C-20).

Atuntzensin A: Yellow solid;mp 200°C. ^{1}HNMR (500Mz, CDCl$_3$): δ 1.61 (1H, m, H-1α), 3.21 (lH, br d, J = 14.1 Hz, H-1β), 1.64 (1H, m, H-2α), 1.58 (1H, m, H-2β), 1.17 (1H, ddd, J = 3.2, 13.4, 14.6 Hz, H-3α), 1.46 (1H, m, H-3β), 2.20 (1H, s, H-5), 4.60 (1H, dd, J = 0.7, 3.2 Hz, H-6), 4.74 (1H, d, J = 3.2 Hz, H-7), 3.38 (1H, br s, OH-7), 7.22 (1H, br s, OH-12), 3.16 (1H, sept, J = 7.1 Hz, H-15), 1.21 (3H, d, J = 7.1 Hz, H-16), 1.22 (3H, d, J = 7.1 Hz. H-17), 1.04 (3H,. s, H-18), 0.90 (3H, s, H-19); ^{13}C NMR (125 Mz CDCl$_3$): δ 25.5 (C-1), 18.5 (C-2), 37.9 (C-3), 31.4 (C-4), 50.6 (C-5), 75.9 (C-6), 65.5 (C- 7), 144.1 (C-8), 140.5 (C-9), 46.2 (C-10), 181.5 (C-11), 151.2 (C-12), 125.1 (C-13), 187.4 (C-14), 24.0 (C-15), 19.6 (C-16 or 17), 19.8 (C-16 or 17), 31.3 (C-18), 21.9 (C-19), 175.5 (C-20).

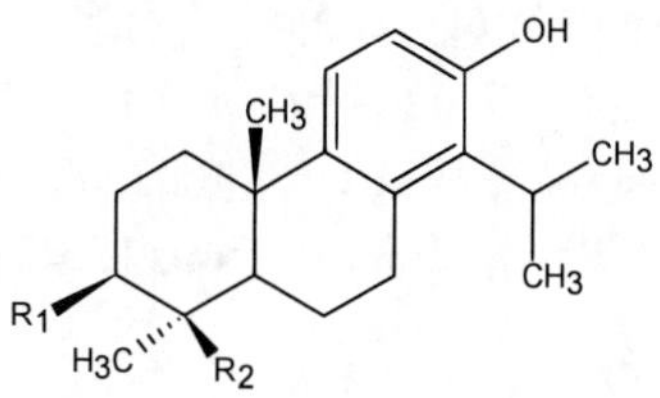

442
R_1= H R_2= Me
Totarol

443
R_1= OH R_2= Me
Totaradiol

444
R_1= H R_2= CH_2OH
19-Hydroxytotarol

445
R_1= H R_2= CHO
Totaral

446

R_1= H R_2= COOH

4β-Carboxy-19-nortotarol

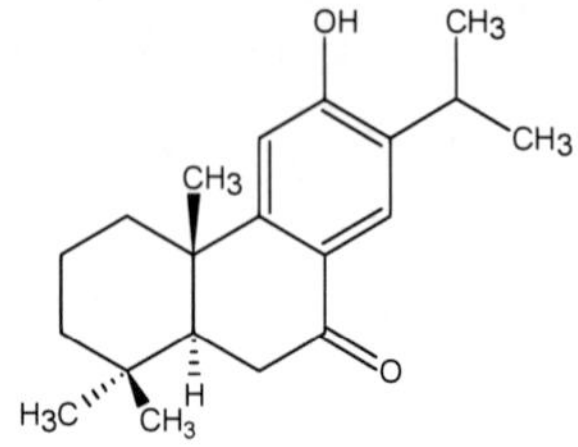

447

Sugiol

These diterpenoids were isotated from *Podocarpus nagi* (Podocarpaceae). They inhibited linoleic acid autoxidation but not generation of superoxide anion. Totarol protected mitochondrial respiratory enzyme activities against NADPH induced oxidative injury. Totarane diterpenes from *P. nagi* were shown to be effective to protect biological systems and function against various oxidative stresses (Haraguchi et al., 1997, Ying and Kubo, 1991).

Totarol: Mp 126-128°C; MS *m/z* (rel. int.): 286 $[M]^+$ (46), 271 (100), 256 (14), 243 (9), 236 (19), 229 (12), 215 (9), 201 (34); ^{1}HNMR: δ1.32 (ddd, J = 12.8, 12.4, 3.0 Hz, H-1α), 2.25 (br, d, J = 12.4 Hz, H-1β), 1.49 (ddd, J = 13.7, 3.8, 3.0 Hz, H-2α), 1.65 (H-2β), 1.14 (ddd, J = 13.7, 13.3, 3.8 Hz, H-3α), 1.40 (br, d, J = 13.3 Hz, H-3β), 1.26 (d, J = 12.0 Hz, H-5), 1.89 (dd, J = 12.0, 8.1 Hz, H-6α), 1.67 (H-6β), 2.81 (ddd. J = 16.7, 8.1, 6.4 Hz, H-7α), 3.04 (dd, J = 16.7, 6.4 Hz, H-7β), 7.10 (d, J = 8.5 Hz, H-11), 7.01 (d, J = 8.5 Hz, H-12), 3.46 (br, H-15), 1.64 (d, J = 6.8 Hz, H-16), 1.67 (d, J = 6.8 Hz, H-17),0.93 (s, H-18), 0.89 (s, H-19), 1.21 (s, H-20), 10.65 (s, OH); ^{13}C NMR: δ 40.03 (C-1α)19.88 (C-2α), 41.90 (C-3α), 33.35 (C-4), 50.11 (C-5), 19.82 (C-6α), 29.22 (C-7α), 133.77 (C-8), 141.91(C-9), 37.93 (C-10), 123.07 (C-11), 115.11 (C-12), 154.99 (C-13),131.72 (C-14), 28.08 (C-15), 20.72 (C-16), 20.76 (C-17), 33.35 (C-18), 21.68 (C-19),25.53 (C-20).

19-Hydroxytotarol: Mp 225-22 °C; MS *m/z* (rel. int.): 302 $[M]^+$ (95), 287 (99), 269 (32), 241 (7), 227 (26), 215 (10), 201 (30), 189 (24), 175 (100); ^{1}HNMR: δ1.50 (ddd, J = 12.4, 12.4, 3.4 Hz, H-1), 2.37(d, J = 12.4 Hz, H-1β), 1.64 (H-2α), 1.86 (H-2β), 1.11 (ddd, J = 10.3, 10.3, 3.4 Hz, H-3α), 2.22 (H-3β), 1.57 (d, J = 12.8 Hz, H-5), 2.20 (H-6α), 1.86 (H-6β), 2.90 (ddd, J = 17.1, 8.1, 6.0 Hz, H-7α), 3.12 (dd, J= 17.1, 6.0 Hz, H-7β), 7.11 (C-11), 6.90 (C-12), 3.48 (C-15), 1.62 (C-16), 1.64 (C-17), 1.28 (C-18), 3.73 (C-19a), 4.08 (C-19b), 1.33 (C-20), 4.46 (OH), 10.31 (OH); ^{13}C NMR: δ 40.45 (C-1), 19.87 (C-2α), 35.96 (C-3α), 38.12 (C4), 51.22 (C-5), 20.20 (C-6α), 29.96 (C-7α), 133.80 (C-8), 141.97 (C-9), 39.32 (C-10), 123.16 (C-11), 115.12 (C-12), 154.96 (C-13),131.84 (C-14), 28.15 (C-15), 20.85 (C-16), 20.88 (C-17), 27.80 (C-18), 64.30 (C-19a),26.69 (C-20).

Totaradiol: Mp 170-173°C; MS m/z (rel. int.): 302 $[M]^+$ (69), 287 (100),273 (47), 269 (75), 245 (99), 227 (64), 199 (40), 189 (29), 175 (48); ^{1}HNMR: δ1.58 (ddd, J = 13.3, 3.9, 3.9 Hz, H-1α), 2.34 (d, J = 13.3 Hz, H-1β), 2.00 (m, H-2), 3.50 (dd, J = 10.7, 5.1 Hz, H-3), 1.31 (H-5), 2.02 (m, H-6α), 1.83 (ddd, J = 24.8, 12.4, 6.4 Hz, H-6β), 2.89 (ddd, J = 17.1, 12.4, 7.5 Hz, H-7α), 3.13 (dd, J = 17.1, 6.0 Hz, H-7α), 7.12 (H-11), 6.98 (H-12), 3.50 (H-15), 1.67 (H-

16), 1.68(H-17), 1.32 (H-18), 1.14 (H-19), 1.31 (H-20), 5.67 (OH), 10.55 (OH); ^{13}C NMR: δ 38.60 (C-1α), 29.17 (C-2), 78.23 (C-3), 37.95 (C-4), 49.87 (C-5), 19.91 (C-6α), 29.72 (C-7α), 133.84 (C-8), 141.66 (C-9), 39.53 (C-10), 123.12 (C-11), 115.16 (C-12), 155.04 (C-13), 131.77 (C-14), 28.21 (C-15), 20.81 (C-16), 20.85 (C-17), 28.90 (C-18), 16.27 (C-19), 25.77 (C-820).

Totaral: Mp 176-178°C; MS *m/z* (rel. int.): 300 $[M]^+$ (l00), 285 (43), 271 (6),257 (83),215 (10), 201 (42), 189 (9), 175 (97); ^{1}HNMR: δ1.30 (ddd, J = 13.3, 13.3, 3.4 Hz, H-1α), 2.20 (H-1β), 1.50 (H-2α), 1.80 (H-2β), 0.98 (ddd, J = 9.8, 9.8, 3.8 Hz, H-3α),2.25 (H-3β), 1.55 (d, J = 13.7 Hz, H-5), 2.23 (H-6α), 1.98 (H-6β), 2.78 (dddd, J = 18.8, 13.7, 12.8, 6.0 Hz, H-6β), 3.07 (dd, J = 17.0, 5.3 Hz, H-7α), 7.08 (d, J = 8.5 Hz, H-11), 7.02 (d, J = 8.5 Hz, H-12), 3.46 (br, H-15), 1 .64 (d, J = 6.8 Hz, H-16), 1.69 (d, J = 6.8 Hz, H-17), 1.03 (s, H-18), 9.91 (s, H-19), 1.08 (s, H-20), 10.77 (OH); ^{13}C NMR: δ 39.54(C-1α), 19.81 (C-2α), 34.05 (C-3α), 48.51 (C-4), 51.51 (C-5), 19.57 (C-6α), 29.75 (C-7α), 133.50 (C-8), 139.44 (C-9), 38.14 (C-10), 123.98 (C-11), 115.33 (C-12), 155.23 (C-13), 131.79 (C-14), 28.15 (C-15), 20.64 (C-16), 20.76 (C-17), 24.04 (C-18), 205.59 (C-19), 24.45 (C-20).

4β-Carboxy-19-nortotarol: Gum; MS *m/z* (rel. int.): 316 $[M]^+$ (l00), 310, 287, 271, 255, 241, 227, 213, 199, 185, 175,157; ^{1}HNMR δ 1.43 (H-1α), 2.35 (br, d, J = 12.4 Hz, H-1β), 1.65 (H-2α), 2.42 (H-2β), 1.13 (ddd, J = 12.8, 12.4, 2.1 Hz, H-3α), 2.55 (br, d, J = 12.8 Hz, H-3β), 1.63 (H-5), 2.56 (dd, J = 12.8, 11.5 Hz, H-6α), 2.40 (dd, J = 16.7, 3.4 Hz, H-7β), 2.82 (ddd, J = 16.7, 11.5, 6.4 Hz, H-7α), 3.14 (dd, J = 16.7, 3.4 Hz, H-7β), 7.17 (d, J = 8.5 Hz, H-11), 7.04 (d, J = 8.5 Hz, H-12), 3.51 (br, H-15), 1.64 (d, J = 6.8 Hz, H-16), 1.70 (d, J = 6.8 Hz, H-17), 1.48 (s, H-18), 1.46 (s, H-20), 5.65 OH), 10.65 (OH); ^{13}C NMR δ 40.96 (C-1α), 20.97 (C-2α), 38.33 (C-3α), 44.07 (C-4), 52.56 (C-5), 22.23 (C-6α), 30.67 (C-7α), 134.38 (C-8), 140.45 (C-9), 38.96 (C-10), 124.54 (C-11), 115.35 (C-12), 155.01 (C-13), 131.60 (C-14), 28.20 (C-15), 20.71 (C-16), 20.8829.14 (C-18), 180.08 (C-19), 23.94 (C-820).

Sugiol: Recrystallized from MeOH, prisms, mp 273-275°C; IR νcm $^{-1}$: 3120,1680, 1600, 1585,1560,1310, 1265; ^{1}HNMR (pyridine-d_5): δ 0.81 (3H, s, Me-18), 0.85 (3H, s, Me-19), 1.09 (lH, dd, H-3α), 1.14 (3H, s, Me-20), 1.35 (lH, m, H-3β), 1.36 (3H, s, Me-16),1.38 (3H, s, Me-17), 1.38 (lH, m, H-1β), 1.40 (lH, m, H-2α), 1.58 (lH, dddd, H-2β), 1.78 (lH, dd, H-5α), 2.08 (lH, br d, H-lα), 2.68 (lH, dd, H-6β), 2.78 (lH, dd, H-6α), 3.61 (lH, qui, H-15), 7.10 (lH, s, H-11), 8.40 (lH, s, H-14); ^{13}C NMR (pyridine-d_5): δ 19.11 (C-2), 21.32 (C-19), 22.63 (C-16), 22.81 (C-17), 23.35 (C-20), 27.30 (C-15), 32.53 (C-18), 33.24 (C-4), 36.38 (C-6), 37.99 (C-l), 38.10 (C-10), 41.52 (C-3), 49.79 (C-5), 110.09 (C-11l), 122.97 (C-8), 126.51(C-14), 134.01 (C-13), 156.53 (C-9), 161.64 (C-12), 197.15 (C-7).

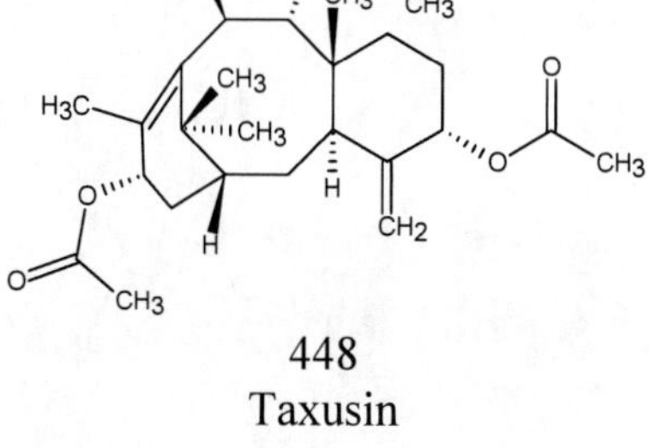

448

Taxusin

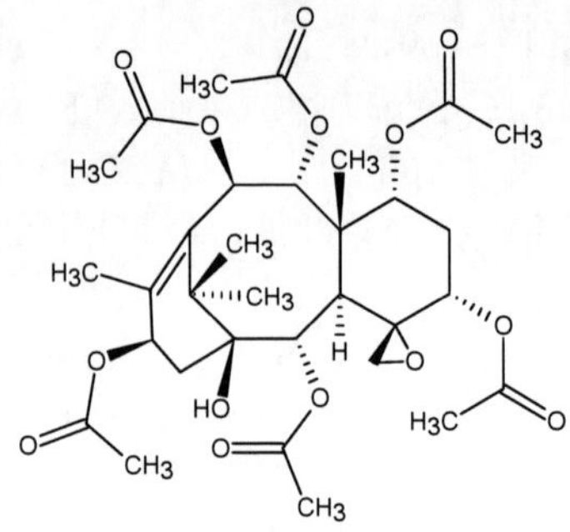

449
1β-Hydroxybaccatin I

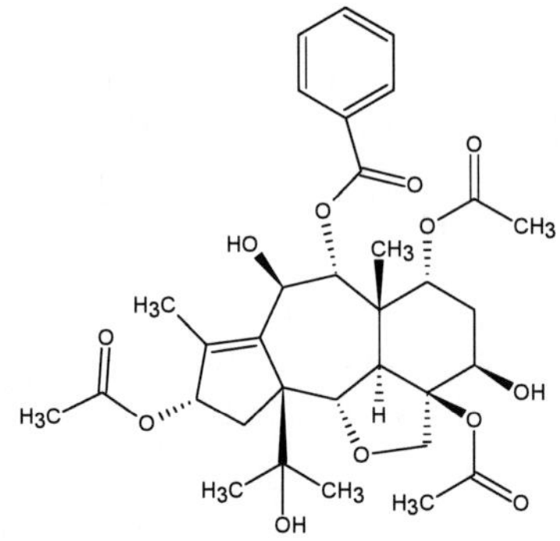

450
Taxuyunnanine E

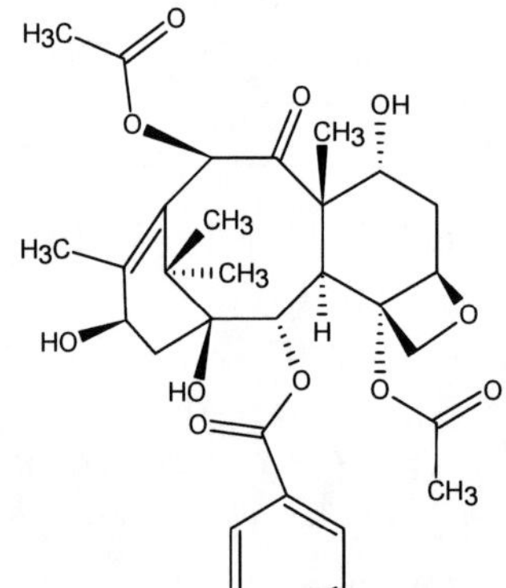

451
Baccatin III

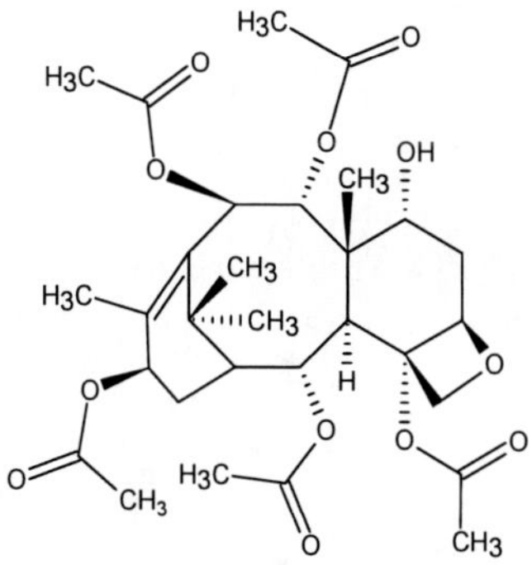

452
1-Dehydroxybaccatin IV

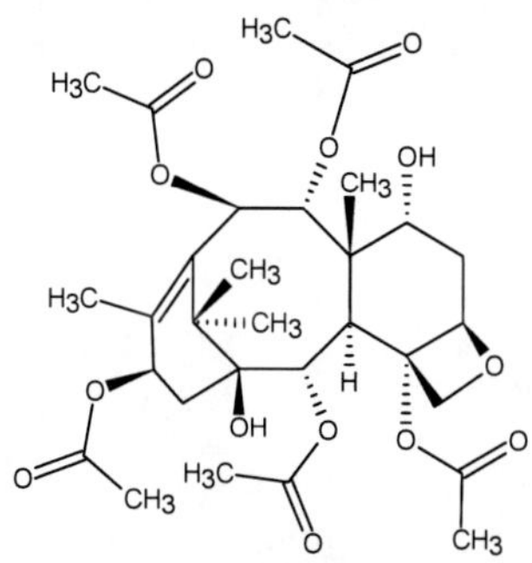

453
Baccatin IV

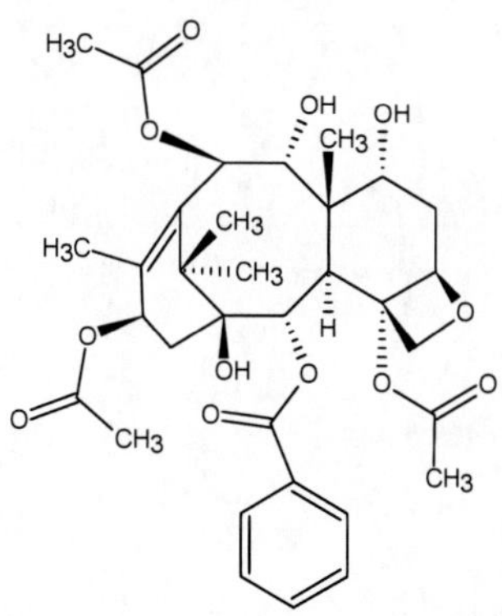

454
9-Dehydro-13-acetylbaccatin III

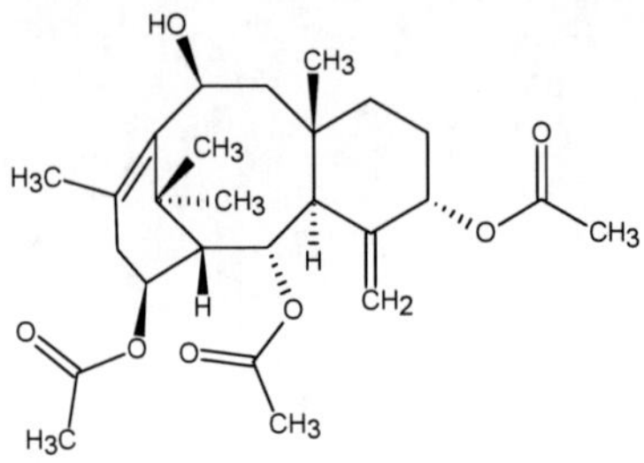

455
10-Deacetyltaxuyunnanine C

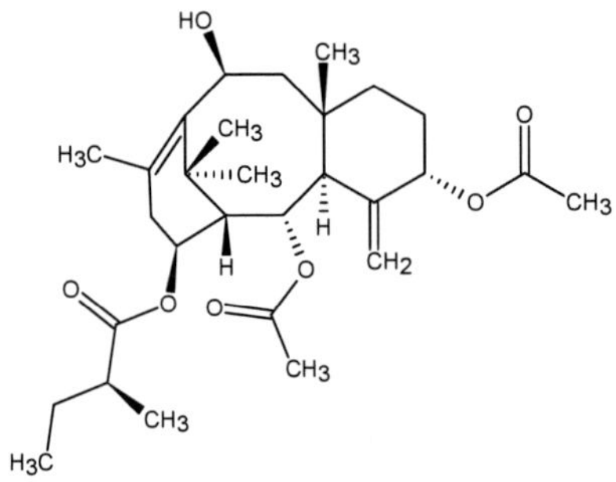

456
Hongdoushan A

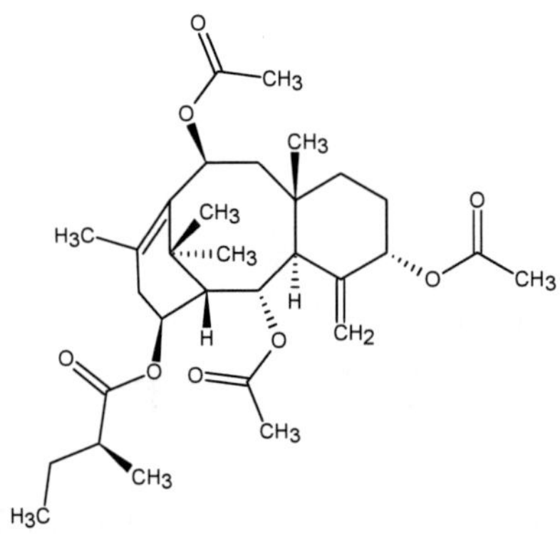

457
2α,5α,10β-Triacetoxy-14β-[(S)-2-methyl-butyryloxy]- 4(20),11-taxadiene

Found in *Taxus yunnanensis* (Taxaceae). The diterpenes possessed stronger NO inhibitory activities having IC_{50} values less than 100 μM, except for baccatin III (IC_{50} 120 μM). Taxusin, 10-deacetyltaxuyunnanine C, hong-taxane type diterpenes showed weak antiproliferative activities towards colon 26-L5 carcinoma cell line (Banskota et al., 2003).

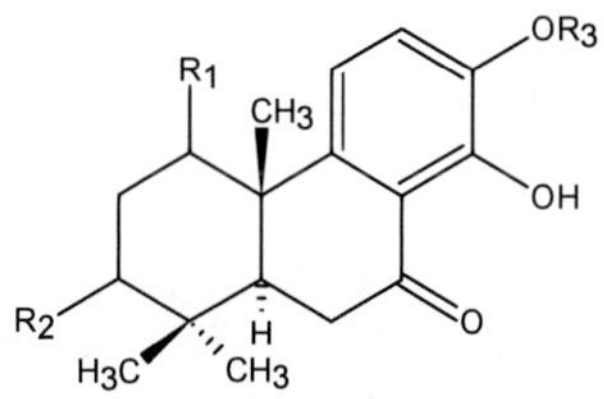

458
R₁= OH R₂= H R₃= H
1β,13,14-Trihydroxy-8,11,13-podocarpatrien-7-one

459
R_1= H R_2= O R_3= Me
14-Hydroxy-13-methoxy-8,11,13-podocarpatriene-3,7-dione

460
R_1= OH R_2= H R_3= Me
1β,14-Dihydroxy-13-methoxy-8,11,13-podocarpatrien-7-one

461
R_1=H R_2= O R_3= H
13,14-Dihydroxy-8,11,13-podocarpatriene-3,7-dione

462
R_1= H R_2= H/$^{\beta}$-OH R_3= H
3β,13,14-Trihydroxy-8,11,13-podocarpatrien-7-one

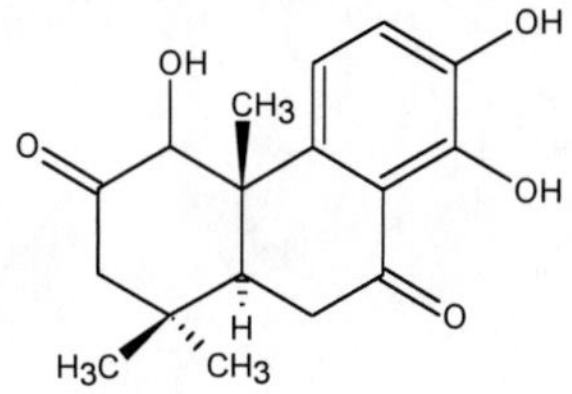

463
1β,13,14-Trihydroxy-8,11,13-podocarpatriene-2,7-dione

Five new l-oxopodocarpane-type trinorditerpenes were isolated from the bark of *Taiwania cryptomerioides*. These compounds exhibited strong antioxidative activity (Kuo et al., 2002).

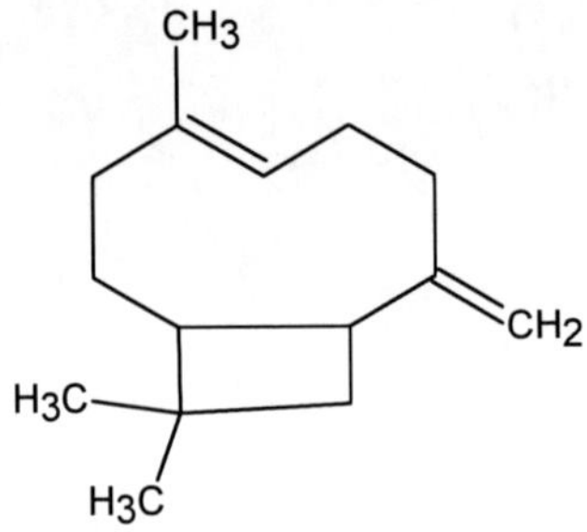

464
β-Caryophyllene

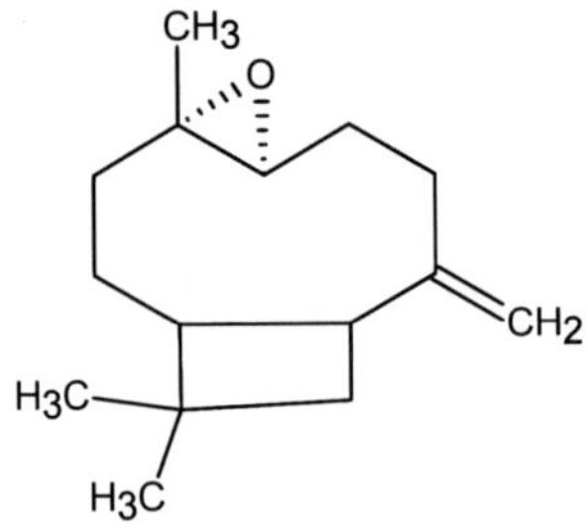

465
β-Caryophyllene 4,5 α-oxide

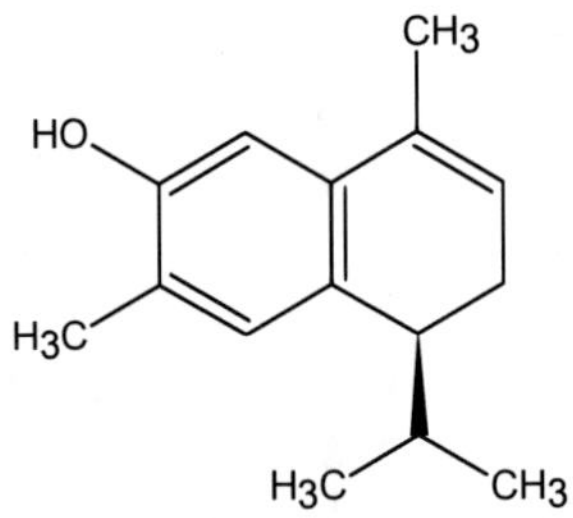

466
7-Hydroxy-3,4-dihydrocadalin

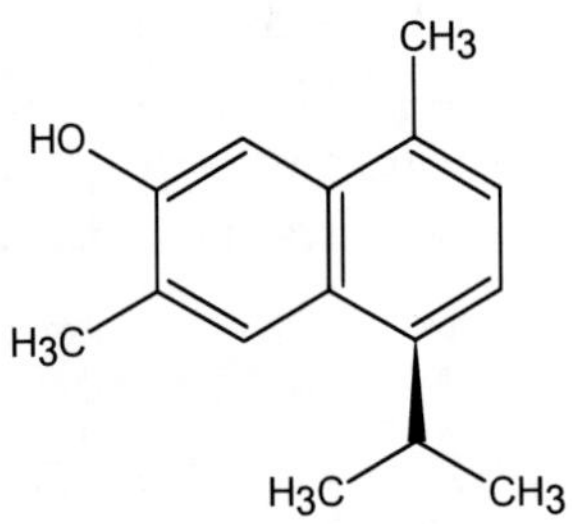

467
7-Hydroxycadalin

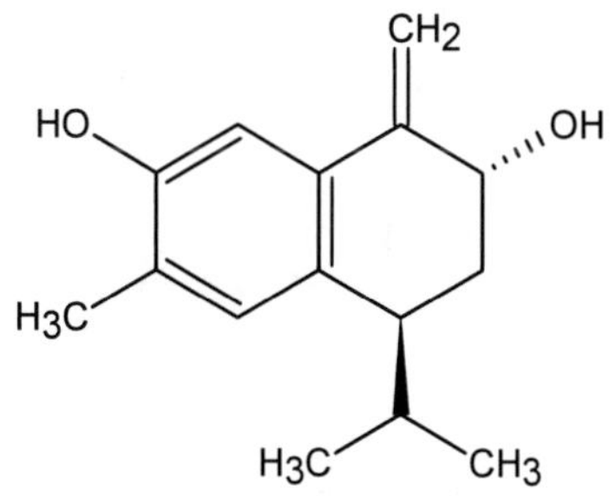

468
2,7-Dihydroxy-β-calacoren

Found in the dried flower *Heterotheca inuloides* (Asteraceae) locally known as "arnica" in Mexico. Hydroxy-3,4-dihydrocadalin was more effective at preventing the lipid peroxidation. Thus, complete inhibition was observed at 10 μg/ml and 50% inhibition was observed at 1 μg/ml also showed potent antioxidative activity: 70 % inhibition was observed at 10 μg/ml, have been found to exhibit cytotoxic activity (Kubo et al., 1996). β-Caryophyllene is a constituent of *Artemisia caryopteris* (Compositae). The oral administration of β-caryophyllene to rats significantly inhibited gastric mucosal injuries induced by necrotizing agents such as absolute ethanol and 0.6 N HCl, although it failed to prevent water immersion stress- and indomethacin –induced gastric lesions. In addition, this compound hardly affected the secretion of gastric acid and pepsin. Thus β-caryophyllene elicited anti-inflammatory effects without any indication of gastric mucosal damage typical of non-steroidal anti-inflammatory agents. Furthermore, this compound manifested cytoprotective effects, rendering the two-dimensional (Tambe et al., 1996).

Caryophy llene 4,5α-oxide: Needles, mp 60-61.5° (cold hexane); mp 63.5-64°; [α]-70.12° ($CHCl_3$ c 0.6460), [α]-79.4" ($CHCl_3$; c 2.32); UV MeOH: no maximum UV absorption above 220 nm; IR νmax cm^{-1}: 3075, 1630, 900; ElMS 70 eV, m/z (rel. int.): 220 $[M]^+$ (1), 205 (5), 187 (5), 177 (11), 161 {11), 138 (24), 121 (31), 109 (66), 93 (99), 79 {100), 69 (68), 55 (48); Found: C, 81.53; H, 11.27. Calc. for $C_{15}H_{24}O$: C, 81.76; H, 10.98%. ^{1}HNMR (400MHz, $CDCl_3$): δ 51.76 (1H, t, J = 10 Hz, H-1), 1.32 (1H, m, H-2a), 1.55-1.72 (1H, m, H-2b), 0.95 (1H, m, H-3a), 2.10 (1H, dt, J = 3.5, 12.5 Hz, H-3b), 2.87 (1H, dd, J = 10.6,4.1 Hz, H-5), 1.42 (1H, m, H-6a), 2.25 (1H, ddd, J = 16.4,7.8,4.1 Hz, H-6b), 2.11 (1H, t, J = 12.4 Hz, H-7a), 2.34 (1H, ddd, J = 12.4, 8.1, 4.4 Hz, H-7b), 2.61 (1H, br q, J = 10 Hz, H-9), 1.55-1.72 (2H, m, H-10), 0.98 (3H, s, H-12), 1.00 (3H, s, H-13), 1.20 (3H, s, H-14), 4.97 (1H, s, H-15a), 4.85 (1H, s, H-15b); ^{13}CNMR (100 MHz, $CDCl_3$): δ 50.8 (C-l), 27.2 (C-2), 39.1 (C-3), 59.8 (C-4), 63.7 (C-5), 30.2 (C-6), 29.8 (C-7), 151.8 (C-8), 48.7 (C-9), 39.8, (C-10), 34.0 (C-11), 29.9 (C-12), 21.6 (C-13), 17.0 (C-14), 112.7 (C-15).

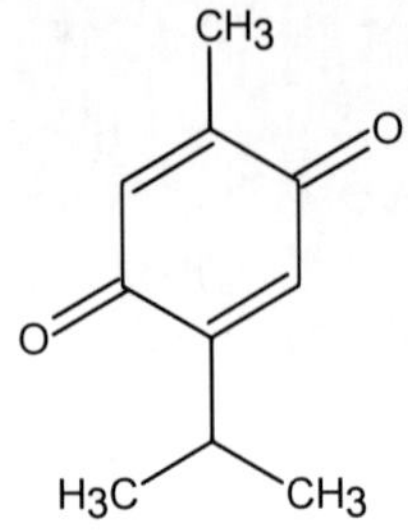

469
Thymoquinone

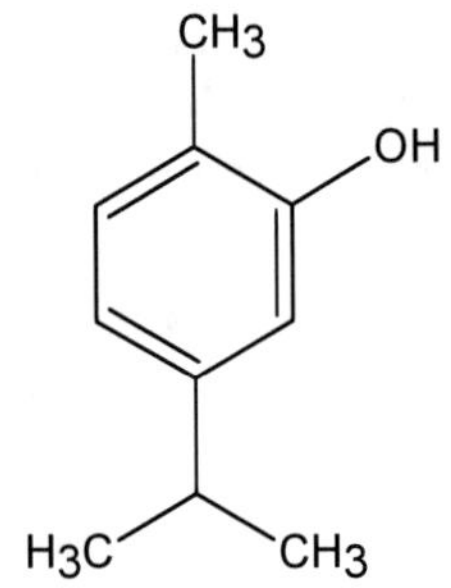

470
Carvacrol

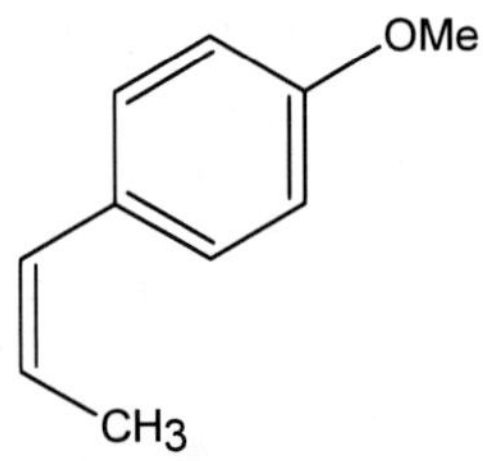

471
Anethole

These compounds isolated from the oil of *Nigella sativa* Linn (Ranunculaceae) have appreciable free radical scavenging properties. The different compounds in the oil were found to act in a synergistic manner more than the mere summation of the actions of the individual compounds. Although the antioxidant property of *N. sativa* is multifactorial, it does not seem to involve iron-complexing activity (Ali and Blunden, 2003).

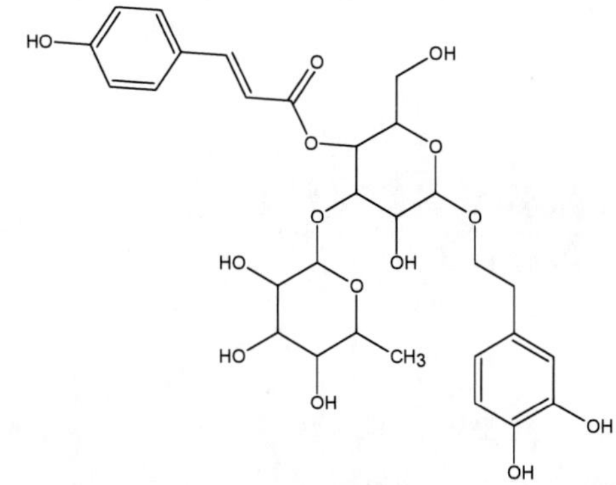

472
Lipedoside AI

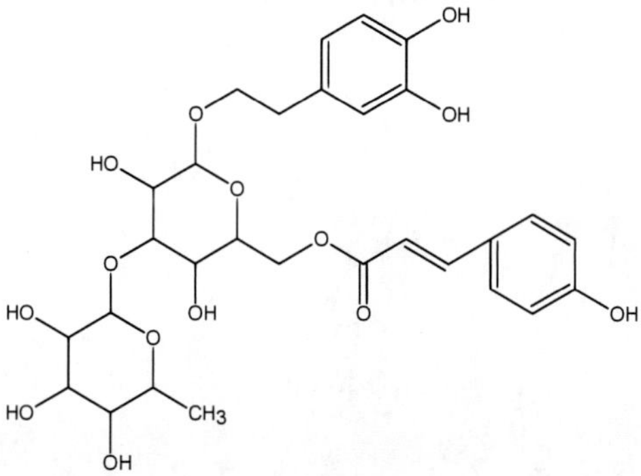

473
Lipedoside AII

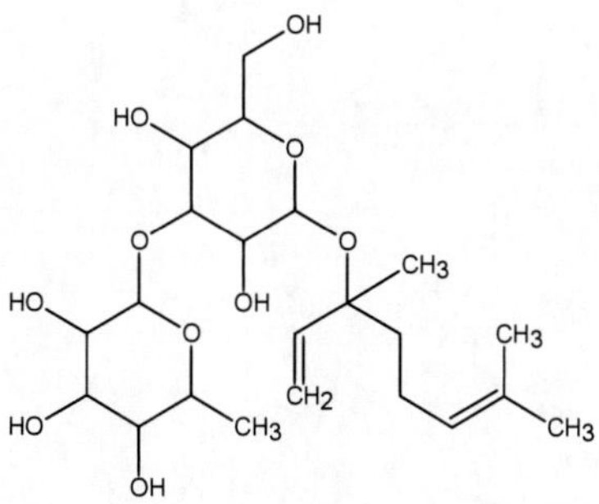

474
Lipedoside BI

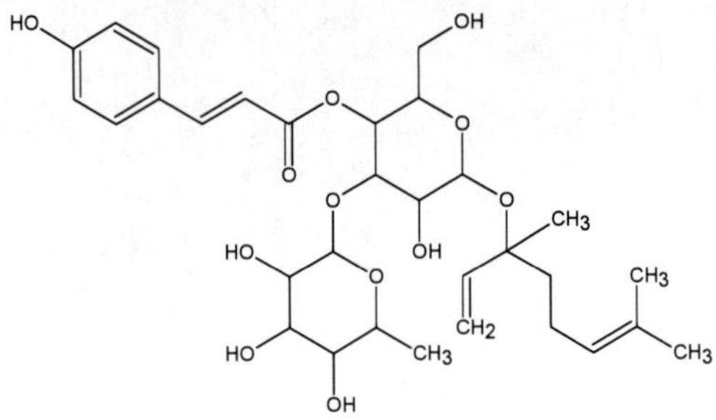

475
Lipedoside BIII

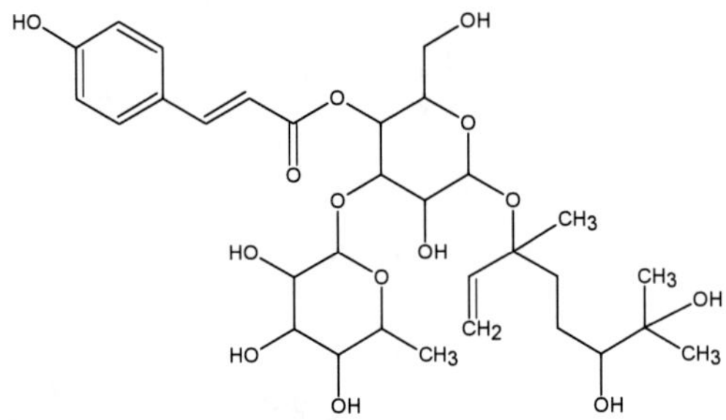

476
Lipedoside BV

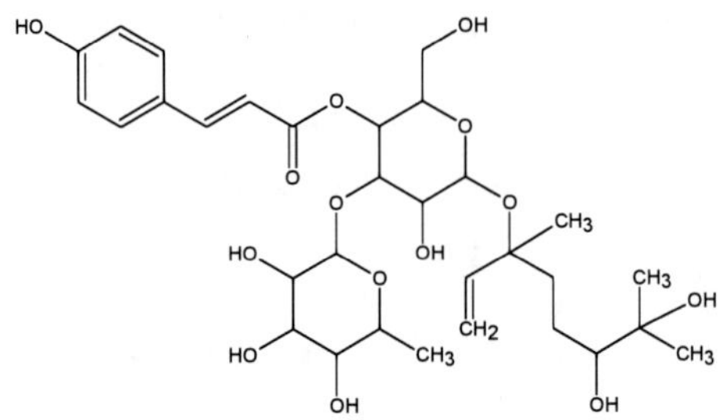

477
Lipedoside BVI

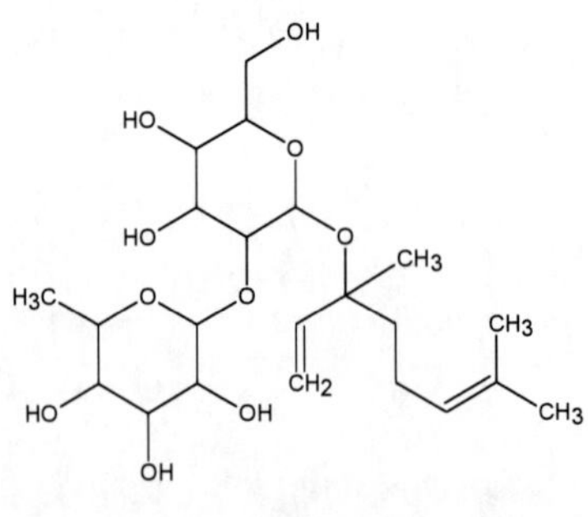

478
Anatolioside

The crude glycoside fraction from *Ligustrum pedunculare* strongly protected human low-density lipoprotein (LDL) from oxidation. The monoterpene glycosides: lipedoside A-I, lipedoside A-ll, lipedoside B-I, lipedoside B-Ill, lipedoside B-V, lipedoside B-VI, osmanthuside B, and anatolioside were isolated. It was found that lipedoside A-I, lipedoside A-ll, lipedoside B-V, and lipedoside B-VI were protective, whereas the other four compounds did not protect human LDL from Cu^{2+}-medicated oxidation. Upedoside A-I, lipedoside A-ll, lipedoside B-V, and lipedoside B-VI also had a scavenging effect on 2,2-diphenyl-1-picrylhydrazyl free radical (DPPH), comparable to that of α-tocopherol. The inhibitory effect of these four phenylethanoid or monoterpene glycosides on oxidation of human LDL and a-tocopherol was dose-dependent at concentrations of 5-40 μM (Che et al., 2002).

Chapter 6

PHENOLIC, LIGNANS, NEOLIGNANS AND TANNINS

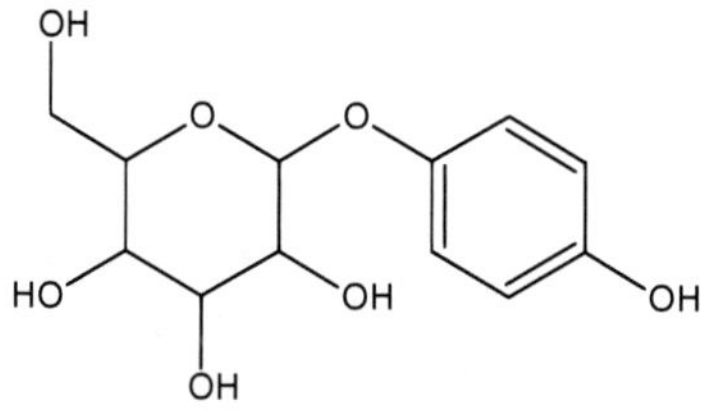

479
Arbutin

The ethyl acetate extract from the pericarp from the seed of *Xanthoxylum piperitum* (Rutaceae) showed strong antioxidative activity against linoleic acid by the ferric thiocyanate and thiobarbituric acid (TBA) methods. The contents of arbutin evaluated as the trifluoroacetate derivative by GC-MS were 35 and 3.0 mg/100 g of dw in the pericarp and seed respectively. Arbutin exhibited antioxidative activity against linoleic acid and radical-scavenging activity against the DPPH radical (Hisatomi et al., 2000).

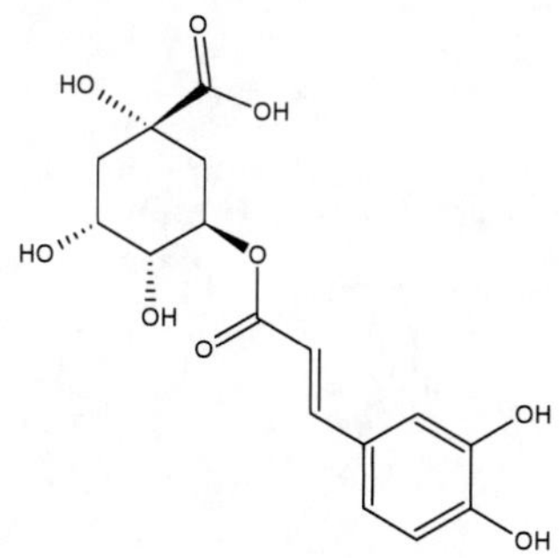

480
3-O-Caffeoylquinic acid

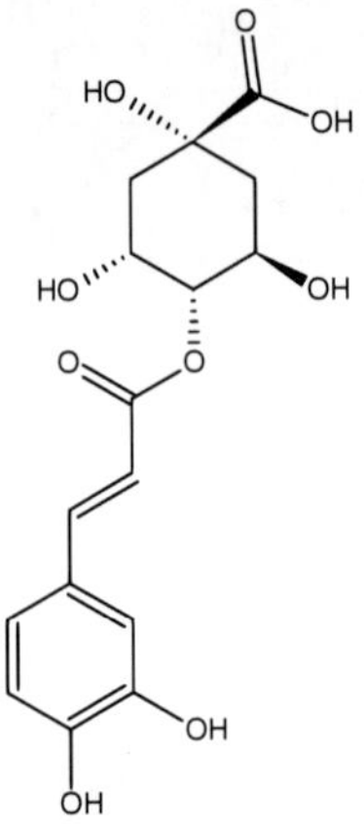

481
4-O-Caffeoylquinic acid

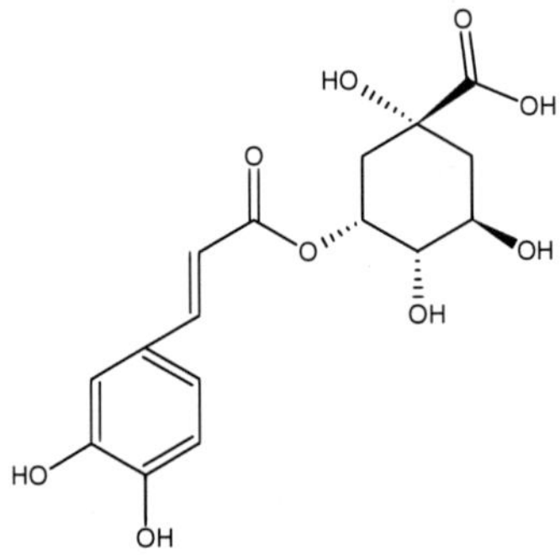

482
5-O-Caffeoylquinic acid

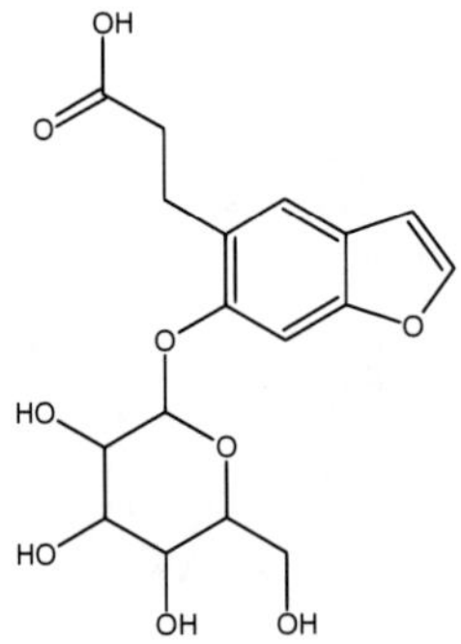

483
Cnidioside A

These compounds were isolated from the leaves of *Peucedanum japonicum* Thunb. (Umbelliferae). Showed antioxidant activity on the test of radical scavenging on 1, 1-diphenyl-2-picrylhydrazyl radical and for inhibition of oxidation of liposome induced by 2,2'-

azobis{2-amidinopropane)dihydrochloride (Hisamoto et al., 2003, Nakatani et al., 2000). 5-O-Caffeoylquinic acid was found to be the major potent constituent, which contribute to the antioxidant activity of *P. japonicum* leaves.

3-O-Caffeoylquinic acid: APCI MS. *m/z* 355 $[M+1]^+$; ^{1}HNMR (500 MHz. CD_3OD): δ 1.95 (lH, dd, J = 9, 14 Hz, H-6ax), 2.13 (2H, m, H-2eq and 6eq), 2.20 (lH, dd, J = 4, 15 Hz, H-2ax), 3.63 (lH, dd, J = 3, 9 Hz, H-4), 4.14 (lH, ddd, J = 3,9, 9 Hz, H-5), 5.34 (lH, ddd, J = 3, 3, 4 Hz, H-3), 6.30 (lH, d, J = 16 Hz, H-8'), 6.76 (lH, d, J = 8 Hz, H-5'), 6.93 (lH, dd, J = 2, 8 Hz, H-6'), 7.04 (lH, d, J = 2 Hz, H-2'), 7.58 (lH, d, J = 16 Hz, H-7'); ^{13}C NMR (125 MHz. CD_3OD): δ 36.7 (C-2), 41.5 (C-6), 68.3 (C-5), 73.0 (C-3), 74.8 (C.4), 75.4 (C-1), 115.1 (C-2'), 115.8 (C-8'), 116.4 (C-5'), 122.9 (C-6'), 127.9 (C-1'), 146.79 (C-3'), 146.80 (C-7'), 149.4 (C-4'), 169.0 (C-9'); 178.3 (C-7).

4-O-Caffeoylquinic acid: APCI MS, *m/z* 355 $[M+1]^+$; IH NMR (500 MHz, CD_3OD): δ 2.00 (lH, dd, J = 11,13 Hz, H-6ax), 2.06 (lH, ddd, J = 3, 4, 14 Hz, H-2eq), 2.17 (lH, dd, J = 4, 14 Hz, H-2ax), 2.20 (lH, ddd, J = 3, 5, 13 Hz, H-6eq), 4.27 (lH, ddd, J = 4, 9,11 Hz, H-5), 4.28 (lH. ddd, J = 3, 3, 4 Hz, H-3), 4.79 (lH, dd, J = 3, 9 Hz, H-4), 6.37 (lH, d, J = 16 Hz, H-8'), 6.77 (lH, d, J = 8 Hz, H-5'), 6.96 (lH, dd, J = 2,8 Hz, H-6'), 7.06 (lH, d, J = 2 Hz, H-2'), 7.65 (lH, d, J = 16 Hz, H-7'); ^{13}C NMR (125 MHz, CD_3OD): δ 38.4 (C-2), 42.7 (C-6), 65.5 (C-5), 69.6 (C-3), 76.6 (C-1), 79.3 (C-4), 115.1 (C-2′), 115.4 (C-8'), 116,5 (C-5'), 123.0 (C-6′), 127.8 (C-l'), 146:8 (C-3'), 147.1 (C-7'), 149.6 (C-4'), 169.0 (C-9'), 177.3 (C-7).

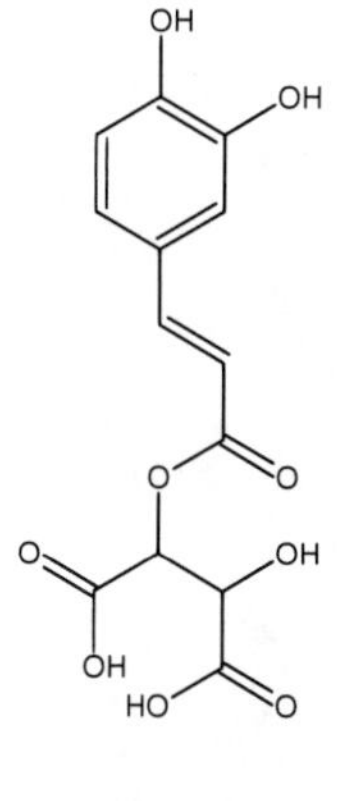

484
Caffeoyltartaric acid

Constituent of grapes. Also isolated from *Echinacea* spp. and *Orthosiphon aristatus.* Specific antioxidant activity (SAA) is the activity related to the molar or gallic acid equivalent amount of antioxidant. Benzoic acid series showed a higher SAA at low concentrations (Cartron et al., 2001).

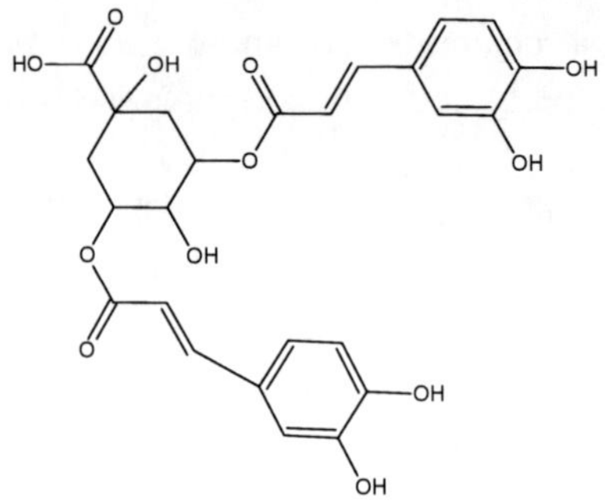

485
3,5-Dicaffeoylquinic acid

Found together with 5-cafeoylquinic acid in the leaves of *Corchorus olitorius* L.(Tiliaceae). The effects of this compound on the AAPH-induced peroxidation of linoleic acid no large differences in antioxidative activities among the isolated phenolic. 3,5-dicaffeoylquinic acid was more active than 5-cafeoylquinic acid. The antioxidative activities of these phenolic compounds were characterized by their extended stability. On the other hand, showed a high antioxidative activity at an initial stage of the peroxidation, but then it rapidly lost activity (Azuma et al., 1999).

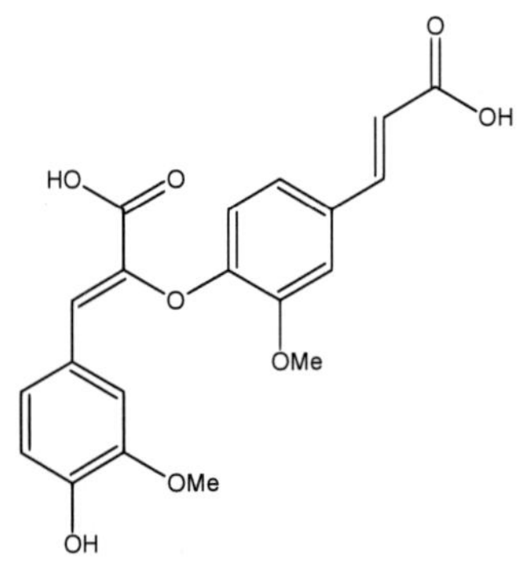

486
8-O-4-Diferulic acid

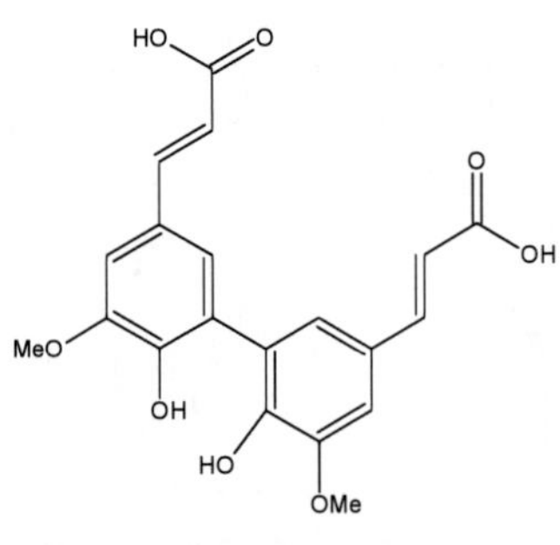

487
5,5- Diferulic acid

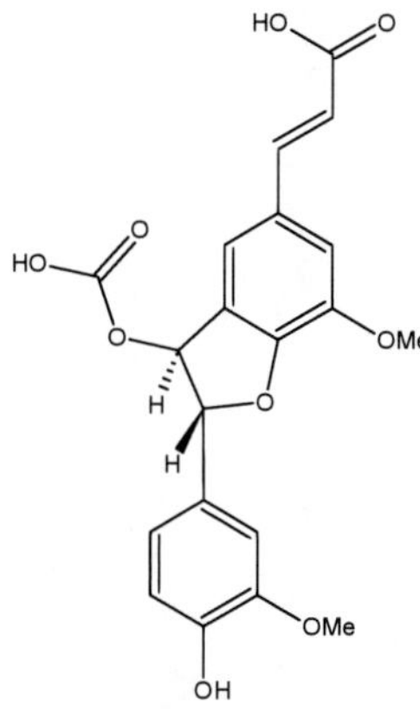

488
8,5- Diferulic acid-benzofuran

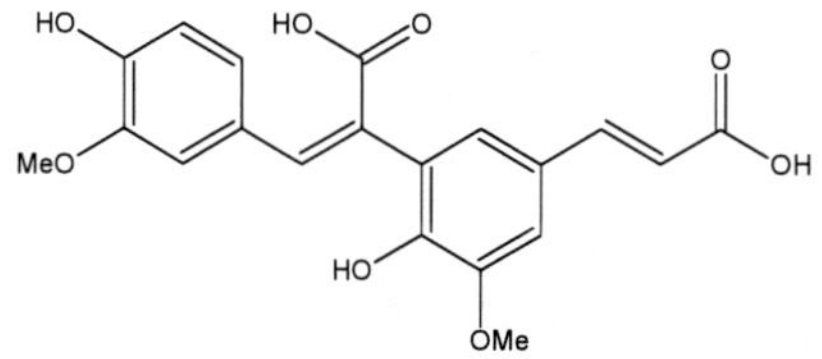

489
8,5- Diferulic acid

Dietary antioxidants that protect low-density lipoprotein (LDL) from oxidation may help to prevent artherosclerosis and coronary heart disease. These monomeric, dimeric and hydroxycinnamates were isolated from *Secale cereale* L. (Gramineae). The most abundant ferulic acid dehydrodimer (diFA) found in rye, 8-O-4-diFA, was a slightly better antioxidant than ferulic acid and p-coumaric acid. The antioxidant activity of the 8-5-diF A was comparable to that of ferulic acid, but neither 5-5-diFA nor 8-5-benzofuran-diFA inhibited LDL oxidation when added at 10-40 μM. The antioxidant activity of the monomeric hydroxycinnamates decreased in the following order: caffeic acid > sinapic acid > ferulic acid > p-coumaric acid (Andreasen et al., 2001).

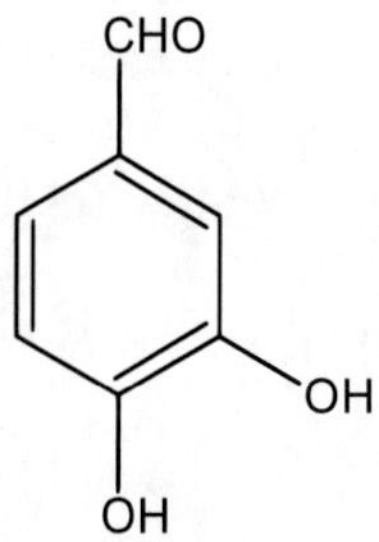

490
3,4-Dihydroxy benzaldehyde

Found in *Fagopyrum esculentum* showed antioxidant activity (Watanabe et al., 1997).

3,4-Dihydroxybenzaldehyde: The FAB-MS gave a molecular ion at *m/z* 139 $[M+H]^+$; The ^{1}HNMR: δ 9.68 (1H, s, H-1'), 7.30 (1H, d, J = 7.2 Hz, H-6), 7.29 (1H, s, H-2), 6.90 (1H, d, J = 7.2 Hz, H-5).

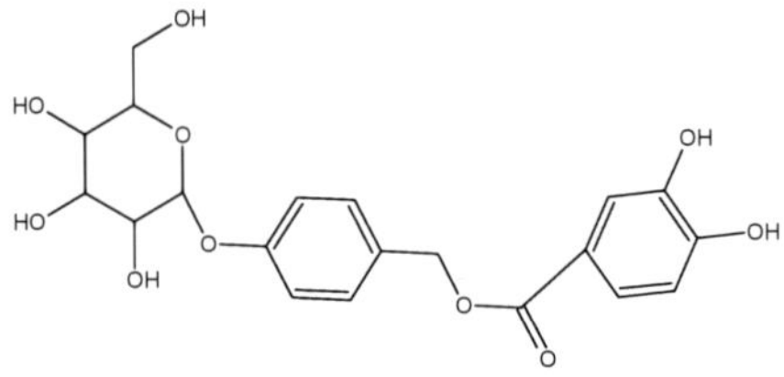

491

4-(3,4-Dihydroxybenzoyloxymethyl)phenyl-β-D-glucopyranoside

A antioxidative glucoside was isolated from the leaves of oregano (*Origanum vulgare* L.), one of the herbal spices belonging to the Family Labiatae. (Nobuji and Kikuzaki, 1987).

4-(3,4-Dihydroxybenzoyloxymethyl)phenyl-β-D-glucopyranoside: Recrystallized from water, mp 204-205 ˚C; Anal. Found: C, 56.48; H, 5.16. Calcd. for $C_{20}H_{22}O_{10}$: C 56.87; H, 5.25%. SIMS *m/z*: 607 $[M+ H+ 2G]^+$, 515 $[M+ H+ G]^+$, 422 (M^+), 269; UV λmax (MeOH): nm (log ε): 222 (4.49), 263 (3.13), 297 (2.86); $[\alpha]_D$ -112.1 (c =0.33, MeOH); IR νma, (Nujol) cm^{-1}: 3500-3150, 1680, 1610, 1515, 1235, 1045. ^{1}HNMR (CD_3OD): δ 3.35-3.50 (4H, m), 3.70 (1H, dd, J = 5.4, 12.1 Hz), 3.90 (1H, dd, J = 2.1, 12.1 Hz), 4.93 (1H, m), 5.22 (2H, s), 6.80 (1H, d, J = 8.1 Hz), 7.11 (2H, d, J = 8.8 Hz), 7.37 (2H, d, J = 8.8 Hz), 7.43 (2H, m); ^{13}CNMR(CD_3OD): δ 168.1 (s), 158.9(s),151.7(s),146.1 (s), 131.7 (s), 130.7 (d x 2), 123.7 (d), 122.6 (s), 117.7 (d x 2), 117.4 (d), 115.8 (d), 102.1 (d), 78.1 (d), 77.9 (d), 74.8 (d), 71.3 (d), 67.0 (t), 62.5 (t).

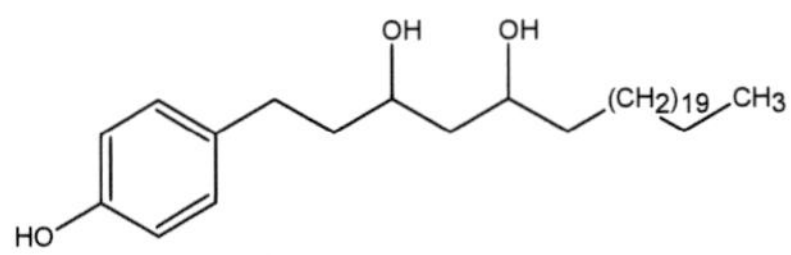

492

4-(3',5'-Dihydroxynonadecyl)phenol

This compound was isolated from the resin of *Heliotropium sinuatum* (Boraginaceae). The antioxidant activity is similar to that of other substituted monophenols, being reduced by the –I effect of the side-chain alcoholic groups (Modak et al., 2003).

4-(3',5'-Dihydroxynonadecyl)phenol: Yellow solid, mp 64-66°C $[\alpha]_D$ -5,7 I ° (MeOH; c 1.9). IR (KBr) cm^{-1}; 3375, 2915, 2849, 1616, 1449, 1175. MS 70eV *m/z* (rel.int): 390 ($[M+-2]^+$, 1), 226 (1), 133 (34), 107 (100), 57 (18), 43 (44); ^{1}HNMR ($CDCl_3$, 300MHz): δ 0.87 (3H, t, H-19'), 1.25 (24H, m, H-7' to H-18'), 1.45 (2H, m, H-6'), 1.59 (2H, m, H-4'), 1.75 (2H, m, H-2'), 2.64 (2H, m, H-1'), 3.84 (1H, m, H-5'), 3.87 (1H, m, H-3'), 6.75 (2H, d, J = 8.4 Hz, H-3 and H-5), 7.05 (2H, d, J = 8.4Hz, H-2 and H-6); ^{13}CNMR ($CDCl_3$ + MeOD, 75 MHz): δ 14.1 (C-19'), 22.7 (C-18'), 25.3 (C-17'), 29.3 (C-16'), 29.6 (C-15' to C-18'), 30.7 (C-1'), 31.9 (C-7'), 38.0 (C-6'), 39.9 (C-2'), 42.6 (C-4'), 72.2 (C-5'), 72.9 (C-3'), 115.3 (C-2 and C-6), 129.3 (C-3 and C-5), 133.1 (C-1), 154.6 (C-4).

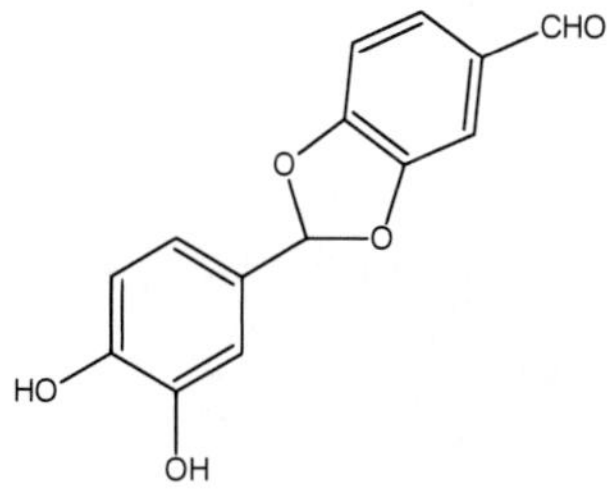

493

2-(3',4'-Dihydroxyphenyl)-1,3-benzodioxole-5-aldehyde

Found in the leaves of *Melissa officinalis* L. (lemon balm, Labiatae). The 1,1-diphenyl-2-picrylhydrazyl radical scavenging activity was measured. This compound showed a potent activity, about ten-fold that of ascorbic acid and of α-tocopherol. The compound had a characteristic 1,3-benzodioxole structure and was easily degraded into two molecules of protocatechualdehyde (Tagashira and Ohtake, 1998).

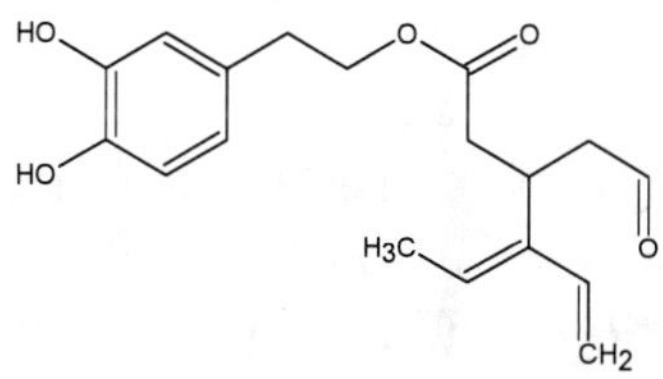

494

3,4-Dihydroxyphenylethyl-4-formyl-3-formylmethyl-4-hexenoate

From olive (*Olea europaea,* Oleaceae) leaves was isolated a nonglycosidic secoiridoid, 3,4-dihydroxyphenylethyl-4-formyl-3-formylmethyl-4-hexenoate (3,4-DHPEA-EDA). The antiradical activity on scavenging of 2,2-diphenyl-1-picrylhydrazyl radicals, was comparable to that of α-tocopherol (Paiva-Martins and Gordon, 2001).

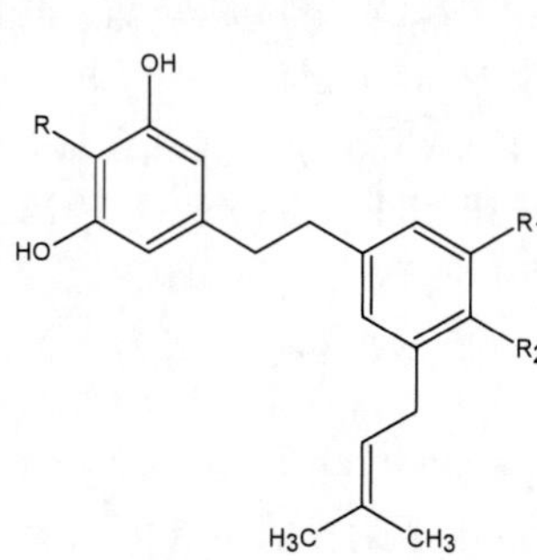

495

R= 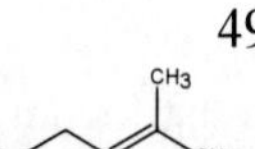, R_1=H, R_2=OH

α,α-Dihydro-3,5,4'-trihydroxy-4,5'-diisopentenylstilbene

496

R= 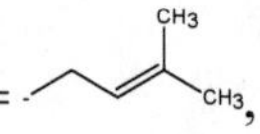

, R_1=OH, R_2=OH

α,α-Dihydro-3,5,3',4'-tetrahydroxy-4,5'-diisopentenylstilbene

497

R= H, R_1=H, R_2=OH

α,α-Dihydro-3,5,4'-trihydroxy-5'-isopentenylstilbene

498

R= H, R_1=OH, R_2=OMe

α,α-Dihydro-3,5,3'-trihydroxy-4'-methoxy-5'-isopentenylstilbene

499

R= H, R_1=OH, R_2=OH

α,α-Dihydro-3,5,3',4'-tetrahydroxy-5'-isopentenylstilbene

These compounds were isolated from a lipid extract of the leaves of Sicilian G*lycyrrhiza glabra* (liquorice, Leguminosae), showed antioxidant activity (Biondi et al., 2003).

α,α-Dihydro-3,5,4'-trihydroxy-4,5'-diisopentenylstilbene: Orange-yellow oil (yield 0.160%, fresh wt); HREIMS *m/z* 366.2190 (calcd for $C_{24}H_{30}O_3$ 366.2195); ElMS *m/z* (%) 366 [M^+] (8),192 (5),175 (100), 133 (9), 119 (5), 91 (6); UV (MeOH): λ max (log ε) 210 (4.68), 230 (4.36), 280 (3.64) nm; IR (CH_2Cl_2): ν max 3692, 3582, 3054, 2987, 1603, 1551, 1422, 896 cm^{-1}; ^{1}HNMR: δ 6.25 (bs, H-2), 6.25 (bs, H-6), 2.75 (m, H-α), 2.75 (m, H-α΄), 6.92 (dd, J = 2, 8 Hz, H-2΄), 6.73 (d, J = 8 Hz, H-3΄), 6.90 (d, J = 2 Hz, H-6΄), 3.40 (d, J = 7, Hz, H-7), 5.27 (t, J = 7 Hz, H-8), 1.83 (s, H-10), 1.77 (s, H-11), 3.34 (d, J = 7 Hz, H-7΄), 5.32 (t, J = 7 Hz, H-8΄), 1.78 (bs, H-10΄), 1.78 (bs, H-11΄); ^{13}C NMR: δ 142.1 (C-1), 108.0 (C-2), 155.2 (C-3), 111.5 (C-4), 155.2 (C-5), 108.0 (C-6), 37.1 (C-α), 38.2 (C-α΄), 134.6 (C-1΄), 127.6 (C-2΄), 116.1 (C-3΄), 152.7 (C-4΄), 127.2 (C-5΄), 130.4 (C-6΄), 22.8 (C-7), 122.4 (C-8), 135.0 (C-9), 18.3 (C-10), 26.2 (C-11), 30.2 (C-7΄), 122.3 (C-8΄), 135.5 (C-9΄), 18.3 (C-10΄), 26.2 (C-11΄).

α,α-Dihydro-3,5,3',4'-tetrahydroxy-4,5'–diisopentenylstilbene: Amorphous solid (yield 0.06%, fresh wt); HREIMS *m/z* 382.2141 (calcd for $C_{24}H_{30}O_4$ 382.2144); ElMS *m/z* (%) 382 [M^+] (10), 247 (3), 191 (100), 149 (7), 135 (6), 91 (1), 77 (2); UV (MeOH): λmax (log ε) 212 (4.44), 230 (4.16), 281 (3.33) nm; IR (CH_2Cl_2) νmax 3578, 3052, 2927, 2856, 1630. 1586, 1446, 1269, 1178,1044 cm^{-1}; ^{1}HNMR: δ 6.24 (bs, H-2), 6.24 (bs, H-6), 2.70 (m, H-α), 2.70 (m, H-α΄), 6.46 (d, J = 2 Hz, H-2΄), 6.59 (d, J = 2 Hz, H-6΄), 3.38 (d, J = 7 Hz, H-7), 5.25 (t, J = 7 Hz, H-8), 1.82 (s, H-10), 1.75 (s, H-11), 3.31 (d, J = 7 Hz, H-7΄), 5.28 (t, J = 7 Hz, H-8΄), 1.78 (bs, H-10΄), 1.78 (bs, H-11΄); ^{13}C NMR: δ 141.6 (C-1), 108.2 (C-2), 154.8 (C-3), 111.0 (C-4), 154.8 (C-5), 108.2 (C-6), 37.6 (C-α), 37.9 (C-α΄), 134.2 (C-1΄), 113.1 (C-2΄), 140.8 (C-3΄), 143.0 (C-4΄), 127.3 (C-5΄), 121.2 (C-6΄), 22.3 (C-7), 122.0 (C-8), 134.5 (C-9), 17.8 (C-10), 25.8 (C-11), 29.6 (C-7΄), 121.8 (C-8΄), 135.0 (C-9΄), 17.8 (C-10΄), 25.8 (C-11΄).

α,α-Dihydro-3,5,4'–trihydroxy-5'-isopentenylstilbene: Amorphous solid (yield 0.08%, fresh wt); HREIMS *m/z* 298.1563 (calcd for $C_{19}H_{22}O_3$ 298.1569); ElMS *m/z* (%) 298 [M^+] (13), 207 (6), 175 (100), 133 (11), 107 (10), 91 (7); UV (MeOH): λmax (log ε) 209 (4.62),

225 (4.32), 281 (3.74) nm: IR (CH_2Cl_2) νmax 3578, 2928, 2855, 1646, 1500, 1330, 1269, 1148, 833 cm^{-1}; ^{1}HNMR: δ 6.21 (d, J = 2 Hz, H-2), 6.18 (d, J = 2 Hz, H-4), 6.21 (d, J = 2 Hz, H-6), 2.70 (m, H-α), 2.70 (m, H-α′), 6.92 (dd, J = 2, 8 Hz, H-1′), 6.83 (d, J = 8 Hz, H-2′), 6.92 (d, J = 2 Hz, H-6′), 3.27 (d, J = 7 Hz H-7′), 5.31 (t, J = 7 Hz, H-8′), 1.69 (bs, H-10′), 1.69 (bs, H-11′); ^{13}C NMR: δ 144.3 (C-1), 107.3 (C-2), 158.4 (C-3), 100.6 (C-4), 158.4 (C-5), 107.3 (C-6), 37.2 (C-α), 38.7 (C-α′), 132.7 (C-1′), 126.8 (C-2′), 115.0 (C-3′), 152.9 (C-4′), 127.9 (C-5′), 130.0 (C-6′), 28.6 (C-7′), 123.5 (C-8′), 131.1 (C-9′), 17.4 (C-10′), 25.4 (C-11′).

α,α-Dihydro-3,5,3'-trihydroxy-4'-methoxy-5'–isopentenylstilbene: Colorless oil (yield 0.35%, fresh wt); HREIMS *m/z* 328.1670 (calcd for $C_{20}H_{24}0_4$ 328.L674); ElMS m/z (%) 328 [M+] (23), 281 (9), 205 (100), 173 (4), 44 (22); UV (MeOH): λmax (log ε) 206 (5.06), 229 (4.20), 280 (3.30) nm: IR (CH_2Cl_2): νmax 3579, 2930, 2959, 1602, 1496, 1452, 1333, 1148, 998 cm^{-1}; ^{1}HNMR: δ 6.24 (d, J = 2 Hz, H-2), 6.20 (d, J = 2 Hz, H-4), 6.24 (d, J = 2 Hz, H-6), 2.73 (m, H-α), 2.73 (m, H-α′), 6.61 (d, J = 2 Hz, H-2′), 6.48 (d, J = 2 Hz, H-6′), 3.31 (d, J = 7 Hz, H-7′), 5.24 (t, J = 7 Hz, H-8′), 1.74 (H-10′), 1.72 (s, H-11′), 3.76 (s, OMe); ^{13}C NMR: δ 145.4 (C-1), 108.7 (C-2), 156.8 (C-3), 101.1 (C-4), 156.8 (C-5), 108.7 (C-6), 37.1 (C-α) 37.7 (C-α′), 139.0 (C-1′), 114.1 (C-2′), 143.4 (C-3′), 148.6 (C-4′), 135.2 (C-5′), 122.0 (C-6′), 28.7 (C-7′), 123.1 (C-8′), 133.0 (C-9′), 18.2 (C-10′), 26.1 (C-11′), 61.6 (OMe).

α,α-Dihydro-3,5,3',4'–tetrahydroxy-5'– isopentenylstilbene: Yellow oil (yield 0.25 % fresh wt); HREIMS *m/z* 314.1512 (calcd for $C_{19}H_{22}0_4$ 314.1518); ElMS *m/z* (%) 314 [M^+] (22), 247 (2), 173 (5), 149 (14), 135 (11), 123 (8), 91 (6); UV (MeOH): λmax (log ε) 206 (4.67), 228 (4.10), 282 (3.54) nm; IR (CH_2Cl_2): νmax 3624, 3478, 3054, 2945, 2837, 1468, 1337, 1263, 1018 cm^{-1}; ^{1}HNMR: δ 6.19 (d, J = 2 Hz, H-2), 6.17 (d, J = 2 Hz, H-4), 6.19 (d, J = 2 Hz, H-6), 2.66 (m, H-α), 2.66 (m, H-α′), 6.56 (d, J= 2 Hz, H-2′), 6.46 (d, J = 2 Hz, H-6′), 3.27 (d, J = 7 Hz, 7′), 5.26 (t, J = 7 Hz, H-8′), 1.74 (s, H-10′), 1.73 (s, H-11′); ^{13}C NMR: δ 144.6 (C-1), 107.2 (C-2), 158.8 (C-3), 100.6 (C-4), 158.8 (C-5), 107.2 (C-6), 37.4 (C-α), 38.5 (C-α′), 133.2 (C-1′), 113.2 (C-2′), 144.8 (C-3′), 141.6 (C-4′), 128.2 (C-5′), 120.8 (C-6′), 28.6 (C-7′), 123.6 (C-8′), 131.2 (C-9′), 17.3 (C-10′), 25.4 (C-11′).

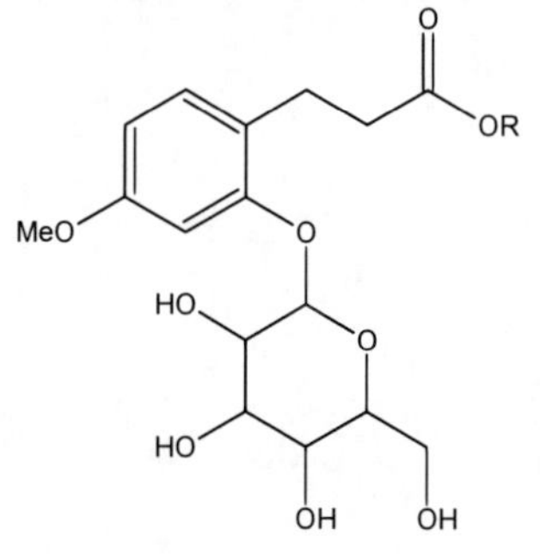

500
R=H
3-(2-O-β-D-Glucopyranosyl-4-methoxyphenyl)propanoic acid

501
R=CH_3
3-(2-O-β-D-Glucopyranosyl-4-methoxyphenyl) methylester

Analysis of a methanolic extract of *Tagetes lucida* Cav. (Asteraceae) leaves has resulted in the isolated of two phenolic acids, 3-(2-O-β-D-glucopyranosyl-4-methoxyphenyl) propanoic acid and its methylester. Using DPPH test showed a significant free-radical-scavenging effect in comparison to a-tocopherol and standard flavonols (Aquino et al., 2002).

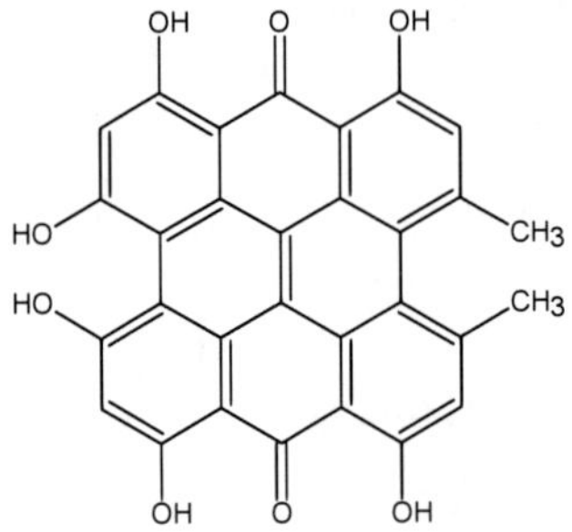

502
Hypericin

This compound was isolated from *Hypericum hyssopifolium* (Guttiferae). Showed antioxidant and DPPH radical-scavenging activitiy (Cakira et al., 2003).

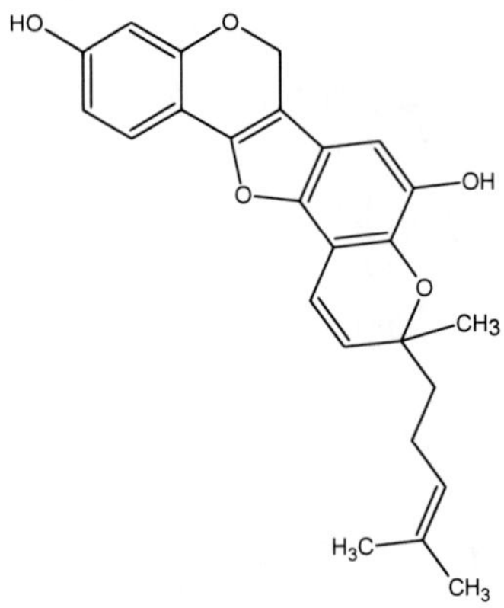

503
Lespedezol A_3

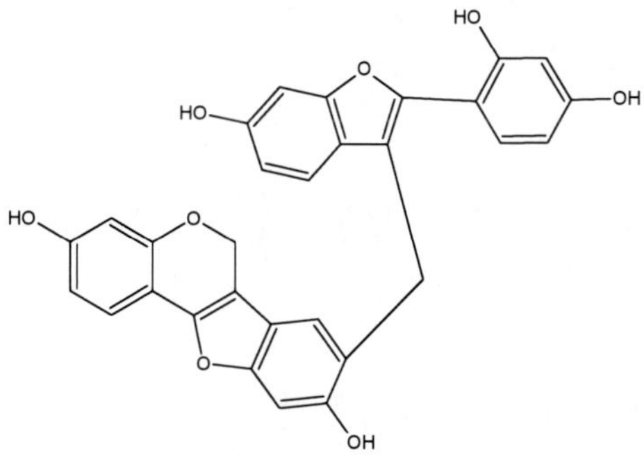

504
Lespedezol B_2

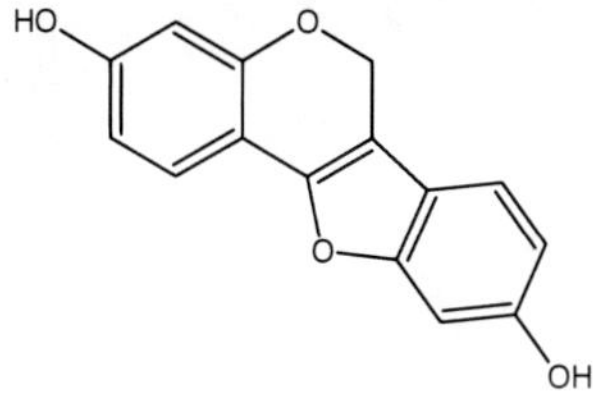

505
3,9-Dihydroxypterocarp-6^a-en

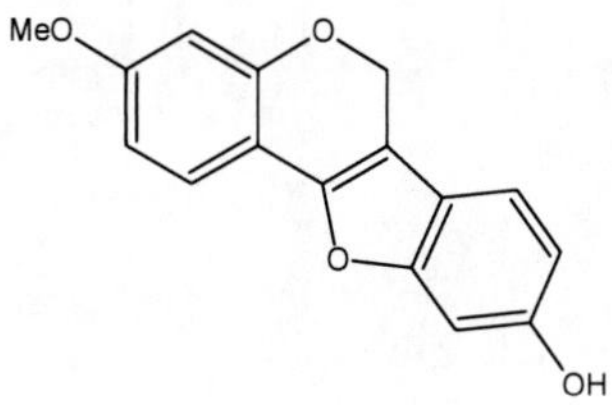

506
Lespedezol A_1

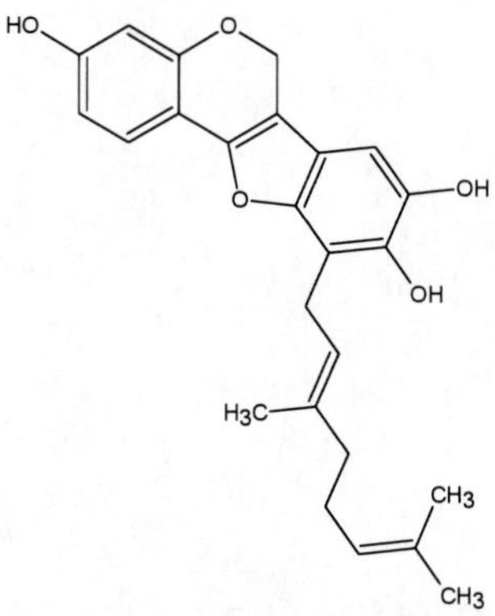

507
Lespedezol A_2

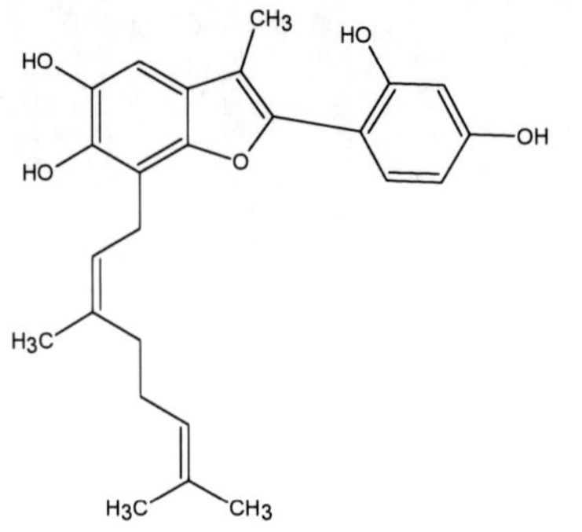

508
Lespedezol B_1

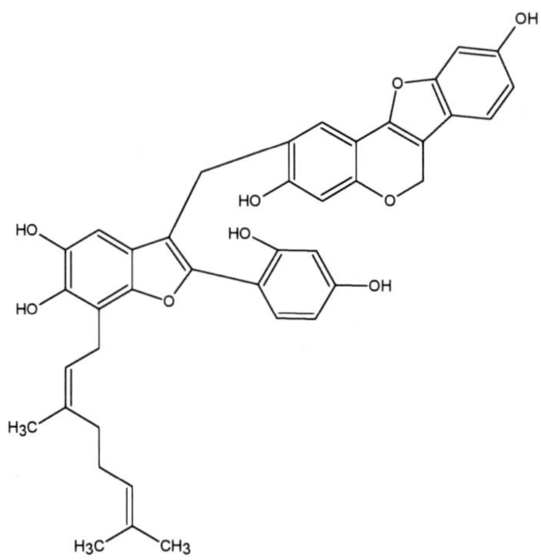

509
Lespedezol B_3

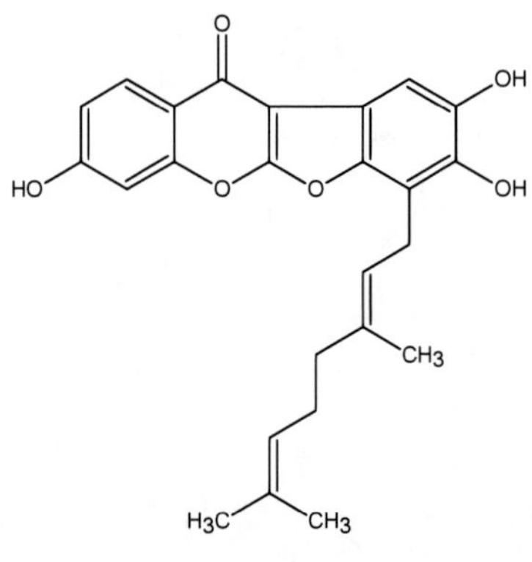

510
Lespedezol C_1

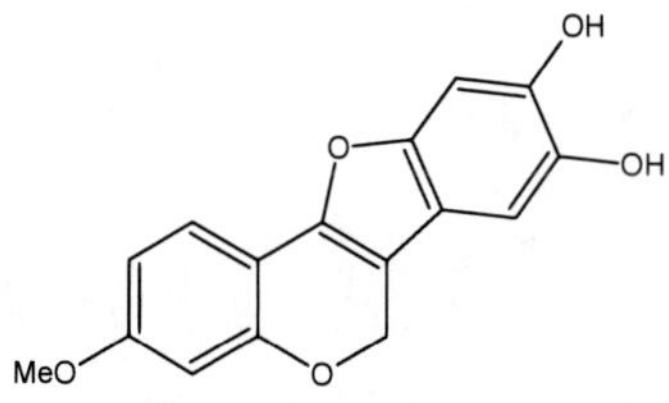

511
Lespedezol A_4

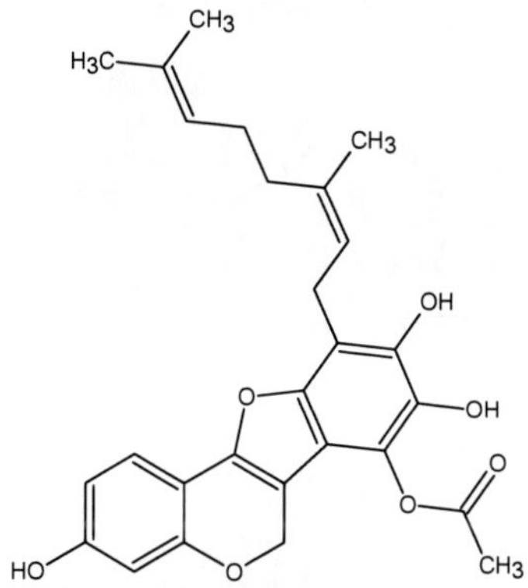

512
Lespedezol A_5

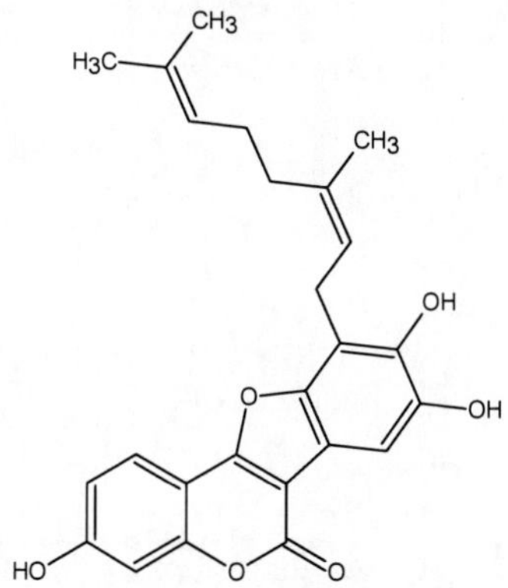

513
Lespedezol A_6

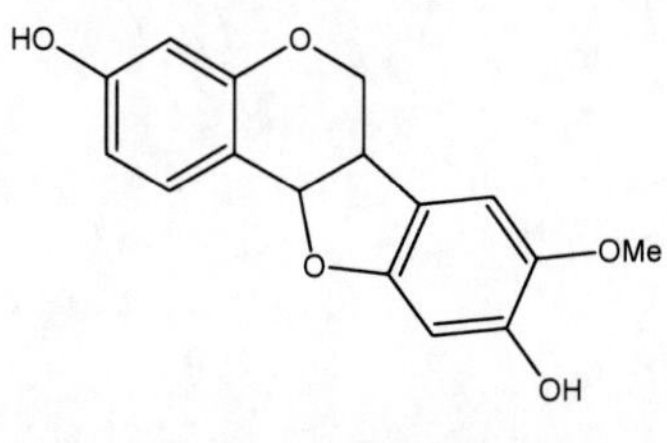

514
Lespedezol D_1

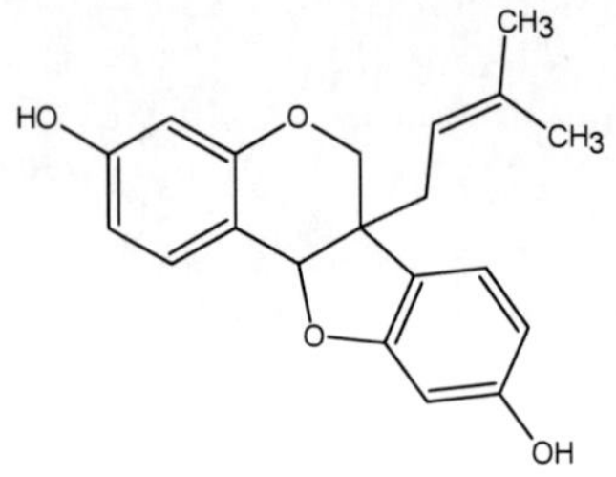

515
Lespedezol D_2

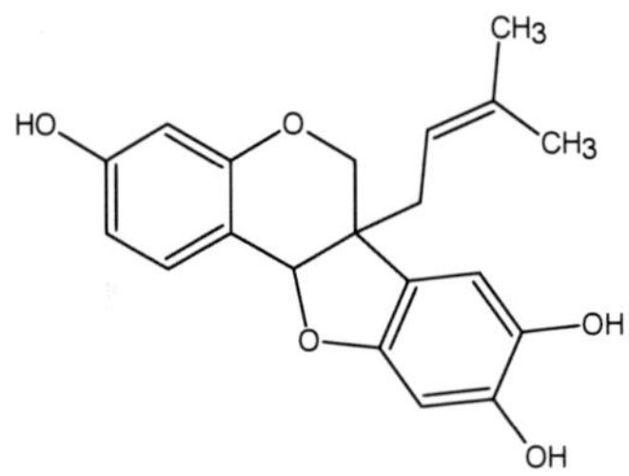

516
Lespedezol D_3

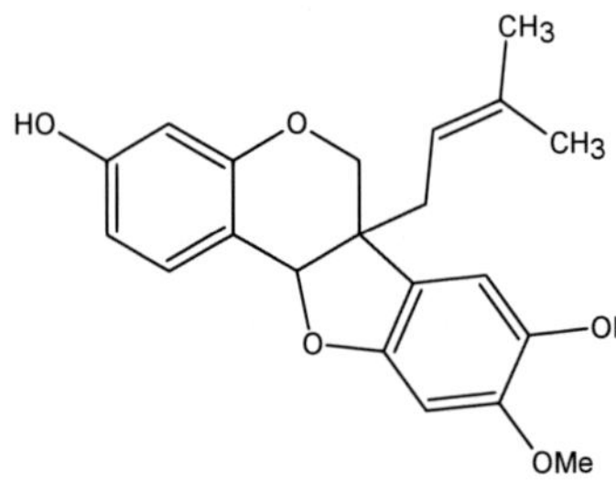

517
Lespedezol D_4

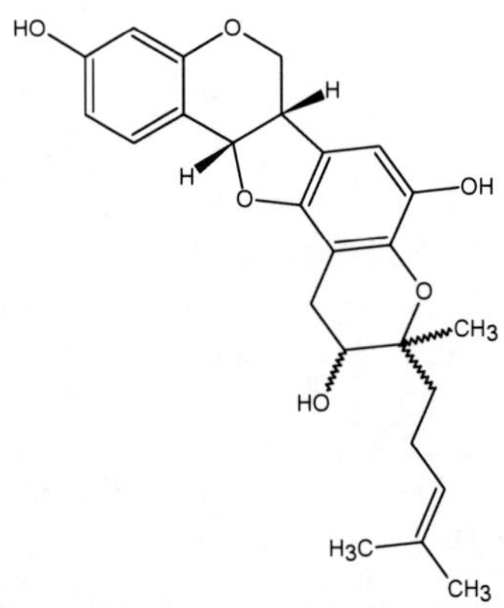

518
Lespedezol D_5

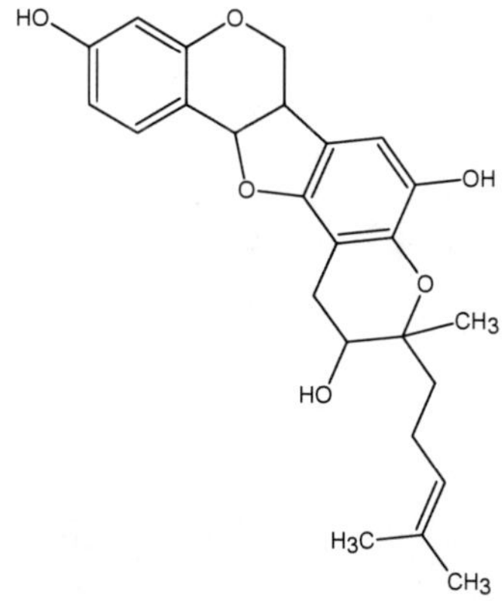

519
Lespedezol H_1

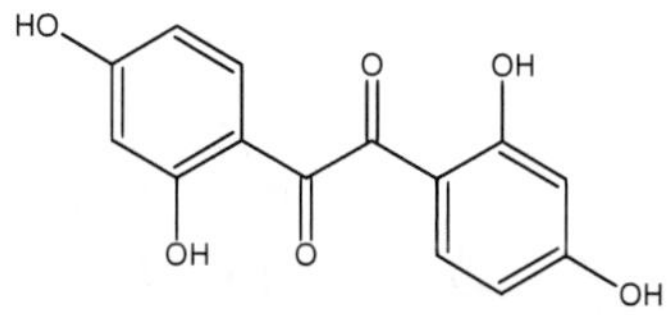

520
Lespedezol D_6

The stems of *Lespedeza homoloba* (Leguminosae) phenolic compounds.These compounds had strong antioxidative activity against lipid peroxidation in the rat brain homogenate test. 3,9-Dihydroxypterocarp-6a-en and lespedezol A_2 showed significant antiallergic activity in allergic (type I) mice (Miyasea et al., 1999, Miyase et al., 1999a). The *L. homoloba* is also very rich in isoflavonoids, a number of which have significant antioxidant activity. Isoflav-3-enes, lespedozol A_2, lespedozol A_3 and the 2-geranylbenzofuran, lespedezol B_2. The compounds containing a catechol group showed the strongest activity against lipid peroxidation in the rat brain homogenate together with superoxide anion scavenging activity. However, the effects of a geranyl and isoprenyl side chain, present in these isoflavonoid structures, on these activities was not clear (Harbone et al., 2000, Igarashi et al., 1990).

Lespedezol A_1: Amorphous powder, UV λmax nm (log ε): 229 (sh) (4.17), 289.5 sh (3.86), 3.26 (4.22). FAB-MS *m/z* 268 $[M]^+$; ^{1}HNMR (lower moiety): δ 5.57 (s, H-6), 7.35 (d, J = 8.5 Hz, H-1), 6.57 (dd, J = 8.5, 2.5 Hz, H-2), 6.49 (d, J = 2.5 Hz, H-4), 7.29 (d, J = 8.5 Hz, H-7), 6.84 (dd, J = 8.5, 2.5 Hz, H-8), 7.02 (d, J = 2.5 Hz, H-10), 3.80 (s, OMe); ^{13}C NMR: δ 66.1 (C-6), 107.1 (C-6a), 147.5 (C-11a), 110. 6 (C-1a), 121.5 (C-1), 107.9 (C-2), 162.0 (C-3), 103.3 (C-8), 156.0 (C-4a), 119.3 (C-7a), 119.9 (C-7), 113.2(C-8), 156.5 (C-9), 99.2 (C-10), 157.4 (C-10a).

Lespedezol A_2: Amorphous powder, UV λmax nm (log ε): 246.5 (4.17), 324.5 (4.24), 3.26 (4.22). FAB-MS *m/z* 406 $[M]^+$; ^{1}HNMR (lower moiety): δ 5.48 (s, H-6), 7.30 (d, J = 8.5 Hz, H-1), 6.50 (dd, J = 8.5, 2.0 Hz, H-2), 6.42 (d, J = 2.0 Hz, H-4), 6.77 (s, H-7); Side chain: 3.65 (br, d, J = 7.5 H-1), 5.44 (m, H-2), 1.99 (br, t, J = 7.5Hz, H-4), 2.08 (br, t, J = 7.5 Hz H-5), 5.06 (m, H-6), 1.57 (br, s, H-8), 1.52 (br, s, H-9), 1.90 (br, s, H-10); ^{13}C NMR (lower moiety): δ 66.1 (C-6), 106.9 (C-6a), 147.4 (C-11a), 110.2 (C-1a), 121.5 (C-1), 109.4 (C-2), 159.5 (C-3), 104.5 (C-4), 156.0 (C-4a), 117.9 (C-7a), 101.7 (C-7), 143.0 (C-8), 142.3 (C-9), 113.1 (C-10), 149.7 (C-10a); Side chain: δ 23.7 (C-1), 122.8 (C-2), 135.9 (C-3), 40.5 (C-4), 27.4 (C-5), 121.5 (C-6), 131.7 (C-7), 25.5 (C-8), 17.7 (C-9), 16.3 (C-10).

Lespedezol A_3: Amorphous powder, UV λmax nm (log ε): 288 (4.10), 302 (4.08), 357 (4.08). [α]D -10.7° (MeOH; c (0.79). FAB-MS *m/z* 404 $[M]^+$; ^{1}HNMR (lower moiety): δ 5.51(s, H-6), 7.32 (d, J = 8.5 Hz, H-1), 6.51 (dd, J = 8.5, 2.0 Hz, H-2), 6.43 (d, J = 2.0 Hz, H-4), 6.79 (s, H-7); Side chain: δ 6.87 (d, J = 10 H-1), 5.86 (d, J = 10 Hz, H-2), 1.78 (m, H-4), 2.17 (m, H-5), 5.12 (m, H-6), 1.63 (d, J = 1 Hz, H-8), 1.56 (br, s, H-9), 1.45 (s, H-10); ^{13}C NMR (lower moiety): δ 66.0 (C-6), 106.8 (C-6a), 148.1 (C-11a), 109.8 (C-1a), 121.7 (C-1a), 104.6(C-3), 156.2 (C-4), 119.3.0 (C-4a), 104.2 (C-7a), 138.0 (C-8), 143.7 (C-9), 107.4 (C-10), 145.7 (C-10a); Side chain: δ 23.7 (C-1), 122.8 (C-2), 135.9 (C-3), 40.5 (C-4), 27.4 (C-5), 121.5 (C-6), 131.7 (C-7), 25.5 (C-8), 17.7 (C-9), 16.3 (C-10).

Lespedezol B_1: Amorphous powder, UV λmax nm (log ε): 272 (3.93), 315 (4.02). FAB-MS *m/z* 408 $[M]^+$; ^{1}HNMR (upper moiety): δ 2.15 (s, H-4), 6.82 (s, H-5), 6.52 (d, J = 2.5, Hz, H-3′), 6.48 (dd, J = 8.5, 2.0 Hz, H-5′), 7.25 (d, J = 8.5 Hz, H-6′); Side chain: 3.61 (br, d, J = 7.5 H-1), 5.44 (m, H-2), 1.98 (m, H-4), 2.07 (m, H-5), 5.07 (m, H-6), 1.59 (d, J = 1 Hz, H-8), 1.53 (d, J = 1 Hz, H-9), 1.82 (d, J = 1 Hz, H-10); ^{13}C NMR (upper moiety): δ 148.9 (C-2), 122.4 (C-3), 9.3 (C-4), 112.4 (C-4a), 111.9 (C-5), 142.4 (C-6), 142.2 (C-7), 112.2 (C-8), 148.30 (C-8a), 111.5 (C-1′), 157.0 (C-2′), 104.1 (C-3′), 160.0 (C-4′), 108.3 (C-5′), 131.9 (C-6′); Side chain: δ 23.7 (C-1), 123.1 (C-2), 135.8 (C-3), 40.5 (C-4), 27.4 (C-5), 125.1 (C-6), 131.7 (C-7), 25.7 (C-8), 17.7 (C-9), 16.4 (C-10).

Lespedezol B_2: Amorphous powder, UV λmax nm (log ε): 209 sh (4.28), 320 (4.38). FAB-MS *m/z* 508 $[M]^+$; ^{1}HNMR (upper moiety): δ 4.10 (s, H-4), 7.11 (d, J = 8.5 Hz, H-5), 6.66 (dd, J = 8.5, 2.0 Hz, H-6), 6.56 (d, J = 2.0, Hz, H-3′), 6.47 (dd, J = 8.5, 2.0 Hz, H-5′), 7.27 (d, J = 8.5, Hz, H-6′), (lower moiety): 5.39 (s, H-6), 7.26 (d, J = 8.5, Hz, H-1), 6.48 (dd, J = 8.5, 2.0 Hz, H-2), 6.40 (d, J = 2.0, Hz, H-4), 7.12 (s, H-7), 7.06 (s, H-10); ^{13}C NMR: δ 150.1 (C-2), 115.3 (C-3), 25.1 (C-4), 123.4 (C-4a), 121.0 (C-5), 112.1 (C-6), 156.1 (C-7), 98.4 (C-8), 156.4 (C-8a), 110.7 (C-1′), 157.4 (C-2′), 104.0 (C-3′), 160.4 (C-4′), 108.3 (C-5′), 132.7 (C-6′), 66.0 (C-6), 106.5 (C-6a), 147.4 (C-11a), 109.8 (C-1a), 121.6 (C-1), 109.4 (C-2), 159.6 (C-3), 104.5 (C-4), 155.6 (C-4a), 124.1 (C-7a), 119.1 (C-7), 119.8 (C-8), 153.8 (C-9), 98.6 (C-10), 156.0 (C-10a).

Lespedezol B_3: Amorphous powder, UV λmax nm (log ε): 245 sh (4.39), 325 (4.43); FAB-MS *m/z* 660 $[M]^+$; ^{1}HNMR (upper moiety): δ 6.54 (d, J = 2.0 Hz, H-3′), 6.48 (dd, J = 8.5, 2.0 Hz, H-5′), 7.30 (d, J = 8.5 Hz, H-6′), lower moiety; δ 5.50 (s, H-6), 7.17 (s, H-1), 6.50 (s, H-2), 7.24 (d, J = 8.5 Hz, H-4), 6.78 (d, J = 8.5, Hz, H-7), 6.78 (dd, J = 8.5, 2.0 Hz, H-8), 6.90 (d, J = 2.0, Hz, H-10), 6.48 (dd, J = 8.5, 2.0 Hz, H-2), 6.40 (d, J = 2.0, Hz, H-4), 7.12 (s, H-7), 7.06 (s, H-10); Side chain: δ 3.64 (br, d, J = 7 Hz, H-1), 3.55 (m, H-2), 2.0 (m, H-4), 2.09 (m, H-5), 5.09 (m, H-6), 1.60 (d, J = 1.0 Hz H-8), 1.55 (br, s, H-9), 1.85 (d, J = 1.0 Hz, H-10); ^{13}C NMR (upper moiety): δ 150.0 (C-2), 115.1 (C-3), 24.7 (C-4), 121.9 (C-4a), 102.8 (C-5), 142.3 (C-6), 142.2 (C-7), 112.2 (C-8), 148.9 (C-8a), 111.1 (C-1′), 157.1 (C-2′), 104.1 (C-3′), 160.2 (C-4′), 108.4 (C-5′), 132.2 (C-6′); (lower moiety): 65.8 (C-6), 106.6 (C-6a), 147.7 (C-11a), 109.6 (C-1a), 122.0 (C-1), 120.6 (C-2), 157.0 (C-3), 104.3 (C-4), 154.2 (C-4a), 119.3 (C-7a), 119.6 (C-7), 113.1 (C-8), 156.2 (C-9), 99.1 (C-10), 157.1 (C-10a); Side chain: δ 23.8 (C-1), 123.1 (C-2), 135.8 (C-3), 40.5 (C-4), 27.4 (C-5), 125.2 (C-6), 131.7 (C-7), 25.8 (C-8), 17.7 (C-9), 16.4 (C-10).

Lespedezol C_1: Amorphous powder, UV λmax nm (log ε): 229 sh (4.37), 259 (4.35), 289 (4.31), 324 sh (3.92); FAB-MS *m/z* 421 $[M]^+$; ^{1}HNMR (upper moiety): δ 8.14 (d, J = 9.0 Hz, H-5), 7.06 (dd, J = 9.0, 2.0 Hz, H-6), 7.08 (d, J = 2.0 Hz, H-8), 7.45 (s, 6′); Side chain: δ 3.65 (br, d, J = 7.0 Hz, H-1), 5.44 (m, H-2), 2.01 (m, H-4), 2.09 (m, , H-5), 5.07 (m, H-6), 1.57 (d, J = 1.0 Hz, H-8), 1.53 (br, s H-9), 1.87 (d, J = 1.0 Hz, H-10); ^{13}C NMR (upper moiety): δ 166.9 (C-2), 99.9 (C-3), 173.5(C-4), 117.7 (C-4a), 128.3 (C-5), 115.5 (C-6), 163.0 (C-7), 104.0 (C-8), 155.8 (C-8a), 114.8 (C-1′), 144.0 (C-2′), 113.6 (C-3′), 143.0 (C-4′), 143.0 (C-5′), 104.7 (C-6′); Side chain: δ 23.6 (C-1), 122.3 (C-2), 136.5 (C-3), 40.4 (C-4), 27.3 (C-5), 125.0 (C-6), 131.7 (C-7), 25.7 (C-8), 17.7 (C-9), 16.4 (C-10).

Lespedezol A_4: λmax nm (log ε): 243 sh (4.09), 338 (4.36). FAB-MS *m/z:* 284 $[M]^+$; ^{1}HNMR: δ 7.33 (d, J = 8.5 Hz), 6.56 (dd, J = 8.5, 2.0 Hz, H-1), 6.48 (d, J = 2.0 Hz, H-4), 5.53 (s, H-6), 6.91 (s, H-7), 7.05, (s, H-10), 3.79 (OMe); ^{13}C NMR: δ 66.1 (C-6), 107.3 (C-6a), 147.2 (C-11a), 110.9 (C-1a), 121.4 (C-1), 107.9 (C-2), 161.8 (C-3), (C-4), 103.3 (C-4a),118.6 (C-7a), 104.4 (C-7), 143.6 (C-8), 144.7 (C-9), 99.2 (C-10), 150.6 (C-10a), 55.7 (OMe).

Lespedezol A_5: Amorphous powder λmax nm (log ε): 275 (4.14), 389 (4.16). FAB-MS *m/z*: 464 $[M]^+$. ^{1}HNMR: δ 5.60 (s, H-6), 7.30 (d, J = 8.5 Hz, H-1), 6.49 (dd, J = 8.5, 2.0 Hz, H-2), 6.38 (d, J = 2.0 Hz, H-4), 3.99 (s, COOMe); Side chain: 3.65 (br, d, J = 7.5 Hz, H-1), 5.38 (m, H-2), 1.99 (m, H-4), 2.06 (m, H-5), 5.03 (m, H-6), 1.55 (br, s, H-8),1.50 (br, s, H-9), (br, s, H-10); ^{13}C NMR: δ 68.5 (C-6), 107.1 (C-6a), 149.4 (C-11a), 109.2 (C-1a), 122.1 (C-1), 109.4 (C-2), 160.3 (C-3), 104.1 (C-4), 156.4 (C-4a), 116.8 (C-7a), 101.3 (C-7), 141.4 (C-8), 149.3 (C-9), 119.8 (C-10), 148.6 (C-10a), 171.5 (CO), 52.5 (COOMe); Side chain: 24.0 (C-1), 121.6 (C-2), 136.8 (C-3), 40.4 (C-4), 27.3 (C-5), 125.0 (C-6), 131.7 (C-7), 25.7 (C-8), 17.7 (C-9), 16.4 (C-10).

Lespedezol A_6: Amorphous powder, λmax nm (log ε): 254.5 (4.30), 355 (4.36). FAB-MS *m/z*: 421 $[M^+H]^+$. ^{1}HNMR: δ 7.88 (d, J = 8.5 Hz, H-1), 7.0 (dd, J = 8.5, 2.0 Hz, H-2), 6.93 (d, J = 2.0 Hz, H-4), 7.31 (s, H-7); Side chain: 3.73 (br, d, J = 7.0 Hz, H-1), 5.46 (m, H-2), 2.01 (m, H-4), 2.09 (m, H-5), 5.04 (m, H-6), 1.52 (d, J= 1.0 Hz, H-8), 1.50 (br, s, H-9), (d, J = 1.0 Hz, H-10); ^{13}C NMR: δ 158.5 (C-6), 104.1 (C-6a), 160.2 (C-11a), 106.4 (C-1a), 123.3 (C-1), 114.3 (C-2), 161.6 (C-3), 104.2 (C-4), 156.0 (C-4a), 115.4 (C-7a), 103.5 (C-7), 144.3 (C-8), 144.0 (C-9), 113.6 (C-10), 149.9 (C-10a); Side chain: 23.7 (C-1), 122.4 (C-2), 136.4 (C-3), 40.4 (C-4), 27.3 (C-5), 125.0 (C-6), 131.7 (C-7), 25.7 (C-8), 17.6 (C-9), 16.4 (C-10).

Lespedezol D_1: Amorphous powder, λmax nm (log ε):.227 sh (4.16), 287.5 (3.82), 2.99 (3.80), 329.5 sh (3.49). $[\alpha]_D$ -97.1° (MeOH; c 0.46). CD $[\theta]_{284}$ -24,280, $[\theta]_{292}$ + 6480 (MeOH; c 0.0106). FAB-MS *m/z*: 286 $[M]^+$; ^{1}HNMR: δ 3.60 (dd, J = 10,10 Hz, H-6), 3.54 (m, H-6a), 5.43 (br, d, J = 6.5, Hz, H-11a), 7.30 (d, J = 8.5 Hz, H-1), 6.55 (dd, J = 8.5, 2.5 Hz, H-2), 6.36 (d, J = 2.0, Hz, H-4), 6.33 (s, H-10), 3.80 (s, OMe); ^{13}C NMR: δ 67.2 (C-6), 41.2 (C-6a), 78.9 (C-11a), 113.1 (C-1a), 133.0 (C-1), 110.4 (C-2), 159.6 (C-3), 103.9 (C-4), 157.7 (C-4a), 117.9 (C-7a), 110.5 (C-7), 148.6 (C-8), 142.8 (C-9), 98.7 (C-10), 155.1 (C-10a), 57.6 (s, OMe).

Lespedezol D_2: Amorphous powder, λmax nm (log ε): 228 sh (4.17), 282 sh (3.92), 287 (3.96), 305 sh (3.37). $[\alpha]_D$ -136.9° (MeOH; c 0.16). CD $[\theta]_{235}$ -76,400, $[\theta]_{274}$ -7360, $[\theta]_{288}$ + 12,450 (MeOH; c 0.0229). FAB-MS *m/z:* 324 $[M]^+$, 325 $[M +H]^+$; ^{1}HNMR: δ 3.68 (d, J = 11

Hz, H-6), 5.12 (br, s, H-11a), 7.28 (d, J = 8.5, Hz, H-1), 6.54 (dd, J = 8.5, 2.5 Hz, H-2), 6.36 (d, J = 2.5 Hz, H-4), 7.07 (d, J = 8.0 Hz, H-7), 6.38 (dd, J = 8.0, 2.0 Hz, H-8), 6.26 (d, J = 2.0 Hz, H-10); Side chain: 2.46 (br, d, J = 7.5 Hz, H-1), 5.24 (m, H-2), 1.66 (d, J = 10 Hz, H-4), 1.49 (br, s, H-5); ^{13}C NMR: δ 70.8 (C-6), 47.2 (C-6a), 83.0 (C-11a), 112.4 (C-1a), 133.5 (C-1), 110.5 (C-2), 159.8 (C-3), 103.8 (C-4), 157.1 (C-4a), 122.3 (C-7a), 125.0 (C-7), 108.3 (C-8), 161.8 (C-9), 98.7 (C-10), 159.8 (C-10a); Side chain: 31.5 (C-1), 120.0 (C-2), 135.3 (C-3), 26.0 (C-4), 18.0 (C-5).

Lespedezol D_3: Amorphous powder, λmax nm (log ε): 225 sh (4.06),281 (3.69), 301 (3.65). $[\alpha]_D$ -113.2° (MeOH; c 0.70). CD $[\theta]_{233}$ -48,500, $[\theta]_{291}$ + 3630 (MeOH; c 0.0150). FAB-MS *m/z:* 340 $[M]^+$; ^{1}HNMR: δ 3.66 (d, J = 11 Hz, H-6), 5.03 (br, s, H-11a), 7.26 (d, J = 8.5, Hz, H-1), 6.54 (dd, J = 8.5, 2.5 Hz, H-2), 6.36 (d, J = 2.5 Hz, H-4), 6.78 (s, H-7), 6.32 (s, H-10); Side chain: 2.43 (br, d, J = 7.5 Hz, H-1), 5.25 (m, H-2), 1.67 (br,s, H-4), 1.50 (br, s, H-5); ^{13}C NMR: δ 70.7 (C-6), 47.6 (C-6a), 82.4 (C-11a), 112.5 (C-1a), 133.5 (C-1), 110.4 (C-2), 159.7 (C-3), 103.7 (C-4), 157.0 (C-4a), 121.4 (C-7a), 111.7 (C-7), 139.7 (C-8), 146.3 (C-9), 98.8 (C-10), 153.9 (C-10a); Side chain: 31.3 (C-1), 120.0 (C-2), 135.2 (C-3), 26.0 (C-4), 18.0 (C-5).

Lespedezol D_4: Amorphous powder, λmax nm (log ε): 231 sh (4.05), 302.5 (3.86), 330.5 sh (3.12). $[\alpha]_D$ -156.9; (MeOH: c 0.88). CD $[\theta]_{233}$ -89,000. $[\theta]_{277}$ -11,100.$[\theta]_{290}$ +6670 (MeOH; c 0.0191). FAB-MS *m/z*: 354 $[M]^+$; ^{1}HNMR: δ 3.70 (d, J = 11 Hz, H-6), 5.08 (br, s, H-11a), 7.27 (d, J = 8.0, Hz, H-1), 6.54 (dd, J = 8.0, 2.5 Hz, H-2), 6.35 (d, J = 2.5 Hz, H-4), 6.78 (s, H-7), 6.47 (s, H-10), 3.80 (s, OMe); Side chain: 2.45 (br, d, J = 7.0 Hz, H-1), 5.24 (m, H-2), 1.67 (d, J = 1.0, Hz, H-4), 1.50 (d, J = 1.0 Hz, H-5); ^{13}C NMR: δ 70.7 (C-6), 47.9 (C-6a), 82.5 (C-11a), 112.6 (C-1a), 133.5 (C-1), 110.5 (C-2), 159.8 (C-3), 103.8 (C-4), 157.1 (C-4a), 122.3 (C-7a), 111.1 (C-7), 141.6 (C-8), 148.6 (C-9), 96.1 (C-10), 153.3 (C-10a); Side chain: 31.5 (C-1), 120.0 (C-2), 135.3 (C-3), 26.0 (C-4), 18.0 (C-5).

Lespedezol D_5: Amorphous powder, λmax nm (log ε): 226.5 sh(4.24), 281.5 (3.87), 330.5 (3.92), 335 (3.79), 350.5 sh (3.66). $[\alpha]_D$ -49.9° (MeOH; c 1.03). CD [e]239 -41,370, $[e]_{290}$ + 12,180 (MeOH; c 0.0369). FAB-MS m/z: 424 $[M]^+$. ^{1}HNMR: δ 3.57 (dd, J = 10.5, 10.5 Hz, H-6), 5.40 (d, J = 7.0 Hz, H-11a), 7.31 (d, J = 8.5, Hz, H-1), 6.55 (dd, J = 8.5, 2.5 Hz, H-2), 6.36 (d, J = 2.5 Hz, H-4), 6.68 (s, H-7); Side chain: 2.52 (dd, J = 16.5, 7.5 Hz, H-1), 3.91 (dd, J = 7.5, 5.5 Hz, H-2), 1.72 (m, H-4), 2.21(m, H-5), 5.13 (m, H-6), 1.66 (d, J = 1.0 Hz, H-8), 1.60 (br,s, H-9), 1.23 (s, H-10); ^{13}C NMR: δ 67.3 (C-6), 41.4 (C-6a), 78.7 (C-11a), 113.3 (C-1a), 133.0 (C-1), 110.4 (C-2), 159.5 (C-3), 103.7 (C-4), 157.7 (C-4a), 117.4 (C-7a), 109.4 (C-7), 141.7 (C-8), 141.5 (C-9), 105.3 (C-10), 151.1 (C-10a); Side chain: 27.2 (C-1), 67.6 (C-2), 80.1 (C-3), 38.3 (C-4), 22.3 (C-5), 125.6 (C-6), 131.7 (C-7), 25.8 (C-8), 17.7 (C-9), 18.6 (C-10).

Lespedezol D_6 : Amorphous powder, λmax nm (log ε): 228 sh (4.14), 282.5 (3.75), 300 (3.86), 335.5 sh (3.30). $[\alpha]_D$ -120.9: (MeOH; c 1.23). CD $[\theta]_{235}$ -42,400, $[\theta]_{289}$ + 10,100 (MeOH; c 0.0126). FAB-MS m/z: 424 $[M]^+$; ^{1}HNMR: δ 3.58 (dd, J = 10.5, 10.5 Hz, H-6), 3.51 (ddd, J = 10.5, 7.0, 4.5 Hz, H-6a), 5.42 (d, J = 7.0 Hz, H-11a) 7.31 (d, J = 8.5, Hz, H-1), 6.54 (dd, J = 8.5, 2.5 Hz, H-2), 6.35 (d, J = 2.5 Hz, H-4), 6.68 (s, H-7); Side chain: 2.61 (dd, J = 17, 8 Hz, H-1), 3.88 (m, H-2), 1.71 (m, H-4), 2.19 (m, H-5), 5.11 (m, H-6), 1.64 (d, J = 1.0 Hz, H-8), 1.57 (br,s, H-9), 1.26 (s, H-10); ^{13}C NMR: δ 67.2 (C-6), 41.4 (C-6a), 78.6 (C-11a), 113.3 (C-1a), 133.0 (C-1), 110.4 (C-2), 159.5 (C-3), 103.9 (C-4), 157.7 (C-4a), 117.4 (C-7a), 109.4 (C-7), 140.7 (C-8), 141.5 (C-9), 105.2 (C-10), 151.2 (C-10a); Side chain: 27.2 (C-1),

67.4 (C-2), 80.0 (C-3), 38.4 (C-4), 22.3 (C-5), 125.6 (C-6), 131.7 (C-7), 25.8 (C-8), 17.7 (C-9), 18.5 (C-10).

Lepedezol Hl : Amorphous powder, λmax nm (log ε): 232 (4.25). 287 (4.24). 332.5 (4.31). FAB-MS *m/z*: 275 $[M+H]^+$; ^{1}H NMR: δ 6.47 (2H, d, J = 2 Hz. H-3), 6.50 (2H, dd, J = 9,2 Hz, H-5), 7.47 (2H, d, J = 9 Hz, H-6); ^{13}C NMR: δ 104.0 (C-3), 110.2 (C-5), 111.5 (C-1), 135.7 (C-6), 167.2 (C-4), 167.4 (C-2), 195.6 (C-α).

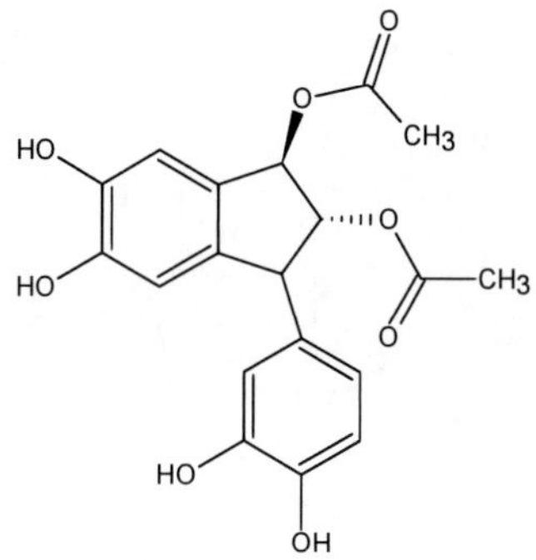

521
Methyl [3-aryl-2-methoxycarbonyl-1-indanyl]acetate

Found in *Helianthus annuus* L. (Compositae), showed antioxidant activity (Kato et al., 1997).

Methyl [3-aryl-2-methoxycarbonyl-1-indanyl]acetate: ^{1}HNMR δ 3.671 (m, H-1), 2.87 (t, J = 9.5 Hz, H-2), 4.26 (brd, J =9.5 Hz, H-3), 6.25 (d, J = 0.9 Hz, H-4), 6.60 (d, J =0.9 Hz, H-7), 2.61 (dd, J = 7.9, 15.9 Hz, H-8a), 2.89 (dd, J =5.7,15.9 Hz, H-8b), 6.57 (d, J = 2.1 Hz, H-2′), 6.69 (d, J = 8.0 Hz, H-5′), 6.50 (dd, J =2.1,8.0 Hz, H-6′); ^{13}CNMR δ 45.1 (C-1), 63.2 (C-2), 55.0 (C-3), 136.9 (C-3a), 112.4 (C-4), 146.1 (C-5), 146.2 (C-6), 110.6 (C-7), 135.4 (C-7a), 39.9 (C-8a), 174.4 (C-9), 176.6 (C-10), 136.2 (C-1′), 116.2 (C-2′), 146.4 (C-3′), 145.3 (C-4′), 116.3 (C-5′), 120.8 (C-6′), 52.2 (9′-OMe), 52.3 (10-OMe).

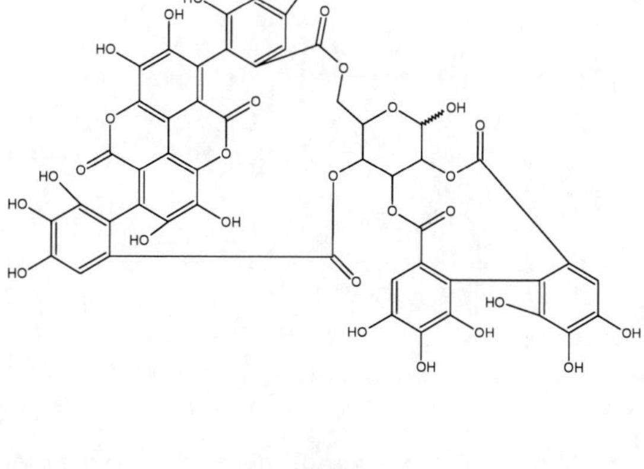

522
Punicalagin

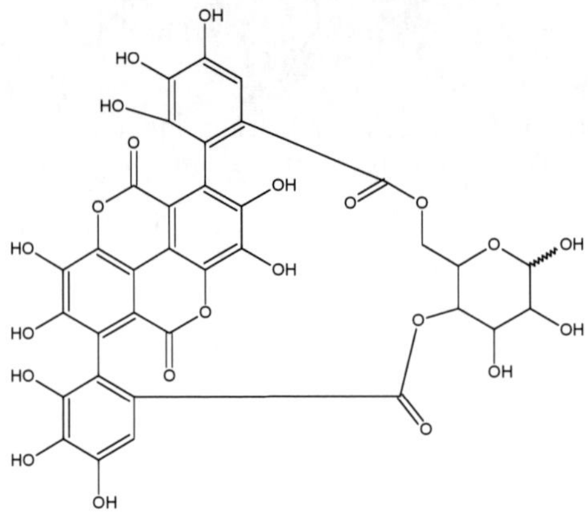

523
Punicalin

Punicalagin and punicalin were isolated from the leaves of *Tenninalia oblongata* L. (Combretacea). Both compounds exerted antioxidative activity and also showed antihepatotoxic activity on acetaminophen-induced toxicity in the rat liver, the changes of several biochemical functions in serum the levels of aspartate aminotransferase (AST) and alanine aminotransferase (AL T) were increased by acetaminophen administration and reduced by punicalagin and punicalin. Histological changes around the hepatic central vein and oxidative damage induced by acetaminophen were also recovered by both compounds (Lin et al., 2001a). The Ethiopian medicinal plant *Conbretum molle* (R. Br. ex G. Don.) Engl and Diels (Combretaceae) showed antiprotozoal activity of the *in vitro* against *Plasmodium falciparum, Trypanosoma brucei rhodesiense, Trypanosoma cruzi* and *Leishmania donovani*. Punicalagin and have IC_{50} values of 1.75 and 1.50 μM against *T. b. rhodesiense* and were relatively less toxic to KB cells (cytotoxic/antiprotozoal ratios of 70 and 48 respectivel), (Asres et al., 2001).

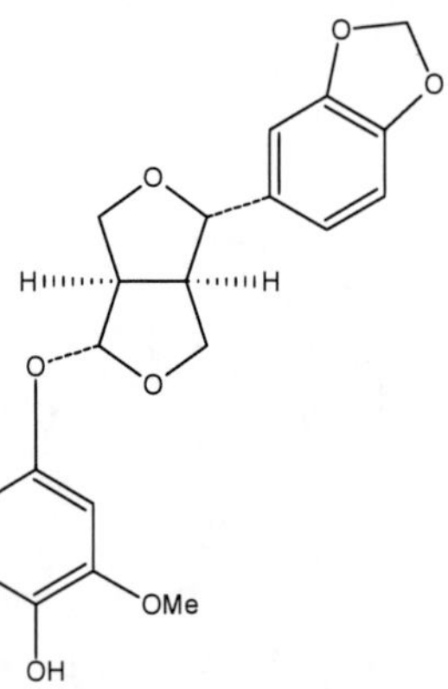

524
Sesamolinol

Isolated from sesame seeds (*Sesamum indicum*) Pedalaceae, showed antioxidant activity (Harbone and Baxter, 1993).

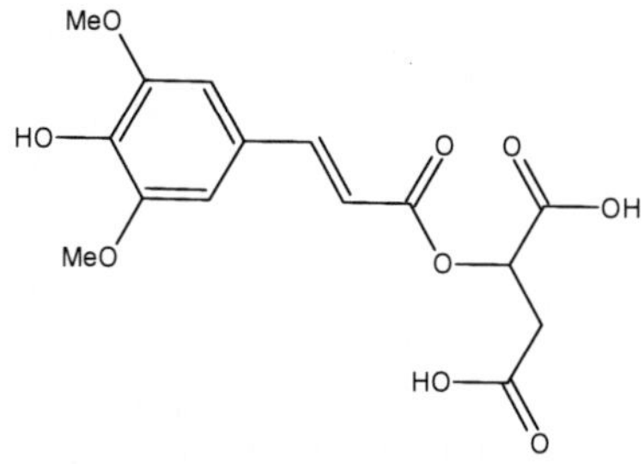

525
trans-Sinapoylmalic acid

Found in *Brassica campestris* L. (Cruciferae), it is known as Atsumi-kabu a variety of red turnip. Showed antioxidant activity (Igarashi, et al., 1990).

trans-Sinapoylmalic acid: A colorless syrup, FAB-MS *m/z* 340 $(M)^+$, 207 $(M -133)^+$. λmax, nm. (MeOH) 224sh, 237, 328; Alkaline hydrolysis afforded malic acid and sinapic acid; ^{1}HNMR (200 MHz. methanol-d_4) δ 2.90-2.97 (2H, m. 3'-H), 388 (6H, s, 2 x OCH_3), 5.50 (1H, dd. 2'-H), 6.44 (1H, d, J = 16Hz. 8-H), 6.90 (2H, s, 2-H and 6-H), 7.66 (1H, d, J = 16Hz. 7-H); ^{13}CNMR (50MHz. methanol- d_4) δ 173.3 (C-1' or C-4′), 173.2 (C-1' or C-4'), 168.4 (C-9), 149.6 (C-3 and C-5), 148.2 (C-7), 139.8 (C-4), 126.6 (C-l), 115.2 (C-8), 107.0 (C-2 and C-6), 70.1 (C-2'), 56.8 (2 x OCH_3), 37.1 (C-3′).

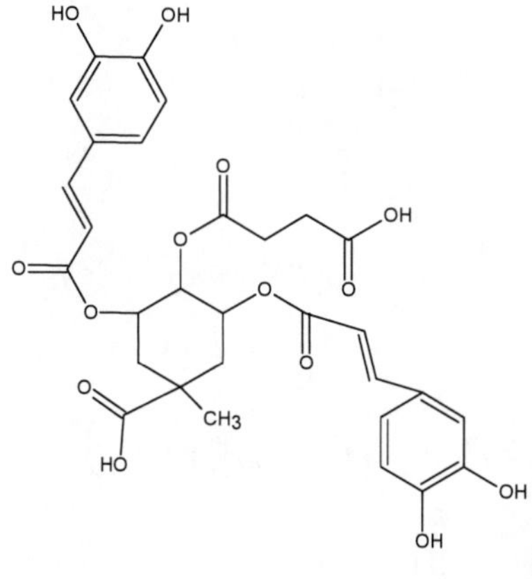

526
4-Succinyl-3,5-dicaffeoylquinic acid

This caffeoylquinic acid derivatives, with antioxidative activity, were isolated from the leaves of garland *Chrysanthemum coronarium* L. (, (Chuda et al., 1996).

4-Succinyl-3,5-dicaffeoylquinic acid: Colorless plates; mp 128-130°C; UV (ethanol): λmax 333 nm (ε 3.16 x 10^4), 247 (1.79 x 10^4), 220 (2.43 x 10^4); $[\alpha]_D$ -204° (methanol, c 1.53); CD (ethanol): λ ext 345 (Δ ε -12.1), 289 (+3.68), 243 (-0.33), 224 (-1.90), 207 (+2.18) nm; IR: νmax 3435, 1714, 1630, 1604, 1516, 1445, 1365, 1277,1178, 1130, 980, 814 cm^{-1};

FAB-MS, *m/z* ; 617 $[M+H]^{+,}$ relative intensity17.0%), 709 ($[M + H + glycerol]^{+}$, 3.8), 455 (2.5), 437 (5.8), 337 (2.5), 163 (100); high-resolution FAB-MS calcd for $C_{29}H_{29}O_{15}$, 617.1506, found 617.1506; ^{1}H NMR: δ 2.274 (dd, J2ax, 2eq = 13.3 Hz, H-2eq), 2.337 (dd, J_{2eq3} = 3.2 Hz, H-2ax), 5:538 (ddd, $J_{2ax,3}$ = 8.3 Hz, H-3), 5.266 (dd, $J_{3,4}$ = 8.0 Hz, H-4), 5.605 (ddd, $J_{4,5}$ = 3,4 Hz, H-5), 2.199 (dd $J_{5,6eq}$ = 6.6 Hz, H-6eq), 2.463 (dd, J $_{5,6ax}$ = 4.0 Hz, H-6ax), 7.139 (d, $J_{6ax,6eq}$ = 14.2 Hz, H-2´), 7.163 (d, $J_{2',6'}$ =2.1, Hz, H-2´´), 6.859 (d, $J_{2'',6''}$ = 2.1 Hz, H-5´), 6.863 (d, $J_{5',6'}$ = 8.2, Hz, H-6´), 7.032 (dd, $J_{5'',6''}$ = 8.2 Hz, H-6´), 7 .050 (dd, $J_{7',8'}$ = 16 Hz, H-6´´), 7,530 (d, $J_{7'',8''}$ =15.9 Hz, H-7´), 7,591 (d, H-7´´), 6,219 (d, H-8´), 6,305 (d, H-8´´), 2,601 (m, 2´´´), 2,601 (m, H-3´´´); ^{13}C NMR: δ 74.1 (C-1), 38.1 (C-2), 68.4 (C-3), 71.8 (C-4), 69.0 (C-5), 36.3 (C-6), 175.5 (C-7´), 127.5 (C-1´), 127.6 (C-1´´), 115.3 (C-2´), 115.2 (C-2´´), 146.2 (C-3´), 146.2 (C-3´´), 148.8 (C-4´), 148.7 (C-4´´), 116.3 (C-5´), 116.3 (C-5´´), 122.7 (C-6´), 122.6 (C-6´´), 146.3 (C-7´), 146.1 (C-7´´), 115.1 (C-8´), 115.6 (C-8´´), 166.3 (C-9´), 166.8 (C-9´´), 172.0 (1´´´), 29.1 (C-2´´´), 29.7 (C-3´´´), 173.5 (C-4´ ´´).

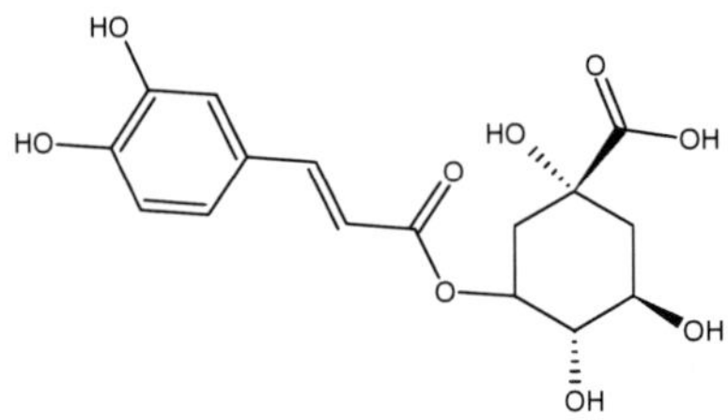

527

Neochlorogenic acid

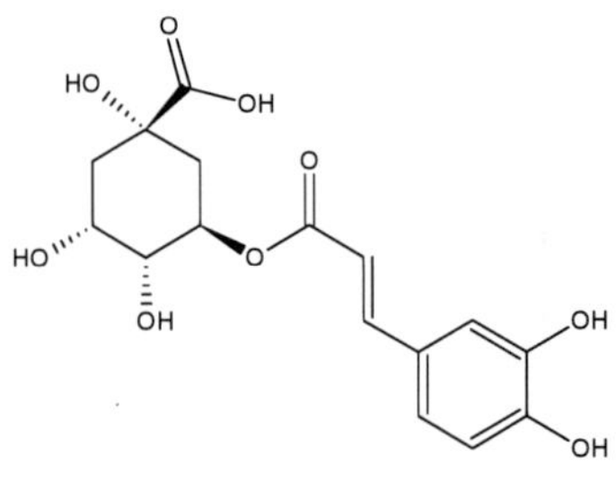

528

Chlorogenic acid

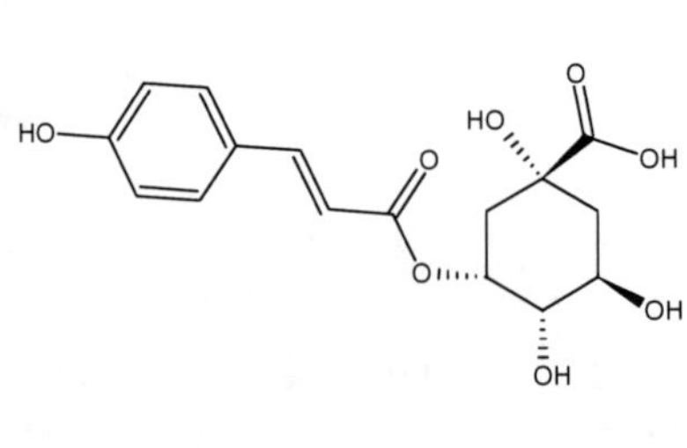

529

3-O-*p*-Coumaroylquinic acid

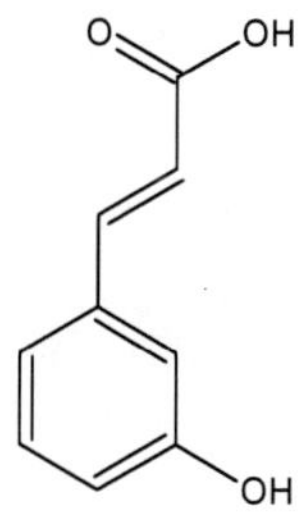

530
m-Coumaric acid

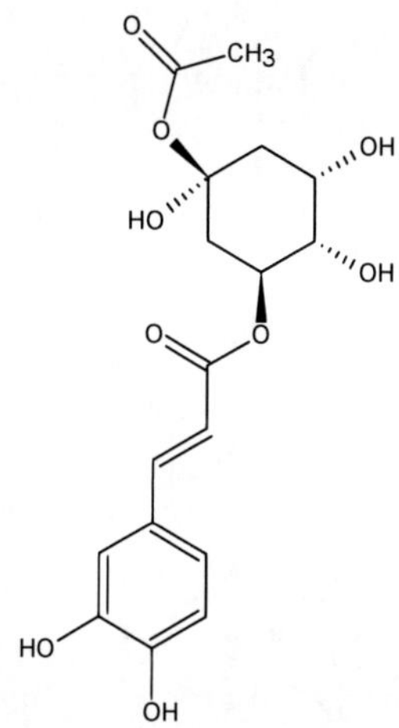

531
Chlorogenic acid methyl ester

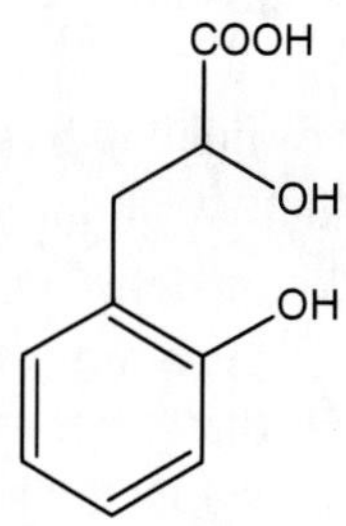

532
2-Hydroxy-3-(O-hydroxyphenyl) propanoic acid

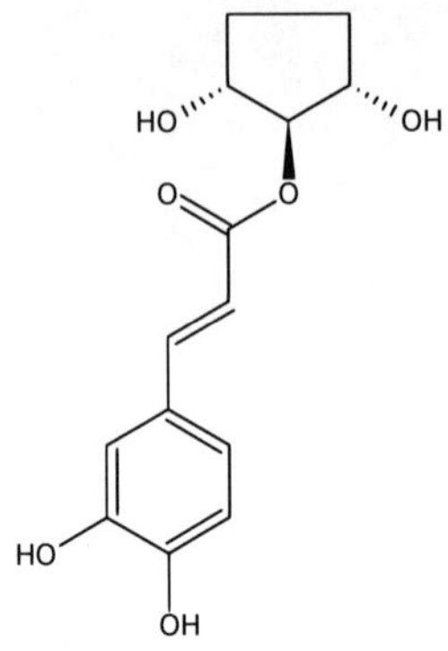

533

1-(3',4'-Dihydroxycinnamoyl)-cyclopenta-2,5-diol

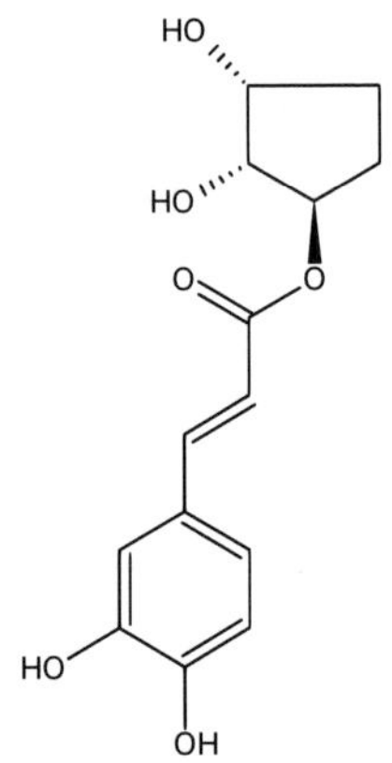

534

1-(3',4'-Dihydroxycinnamoyl)-cyclopenta-2,3-diol

Found in *Prunus cerasus* (Rosaceae) inhibited lipid peroxidation induced by Fe^{2+} at 10 μM concentrations (Donovan et al., 1998, Wang et al., 1999). The mean concentrations of phenolics were 1840 mg/kg, 1397 mg/kg, and 441 mg/L in pitted prunes, extra large prunes with pits, and prune juice, respectively. Hydroxycinnamates, especially neochlorogenic acid, and chlorogenic acid predominated, and these compounds, as well as the prune and prune juice extracts, inhibited the oxidation of LDL. The pitted prune extract inhibited LDL oxidation by 24, 82, and 98% at 5, 10, and 20 μM gallic acid equivalents (GAE). The prune juice extract inhibited LDL oxidation by 3, 62, and 97% at 5, 10, and 20 μM GAE. At a 20-μM concentration, the antioxidant activities of compounds 1-(3',4'-dihydroxycinnamoyl)-cyclopenta-2,5-diol and 1-(3',4'-dihydroxycinnamoyl)-cyclopenta-2,3-diol were comparable to the antioxidant activities of caffeic acid, whereas the first compound showed activity similar to chlorogenic acid. Also, these compounds showed antioxidant activities similar to the commercial antioxidants tert-butylhydroquinone and butylated hydroxytoluene. However, second compound was not active when tested at a 100 μM concentration (Wang et al., 1999).

Chlorogenic acid: ^{1}H NMR (DMSO-d_6): δ 7.45 (lH, d, J = 15.9Hz, H-7'), 7.00(lH, d, J = 2.0 Hz, H-2'), 6.95(lH, dd, J = 8.4, 2.0 Hz, H-6'), 6.75 (lH, d, J = 8.4 Hz, H-5'), 6.18 (lH, d, J

= 15.9 Hz, H-8'), 5.16 (lH, m, H-5), 3.85 (lH, m, H-3), 3.53 (lH, m, H-4), 2.02-1.83 (4H, m, H-2, H-6); ^{13}C NMR (DMSO-d_6): δ 176.1 (COO), 166.1 (C-9'), 148.2 (C-3'), 145.6 (C-4'), 144.4 (C-7'), 125.7 (C-1'), 121,1 (C-6'), 115.8 (C-5'), 115.1 (C-8'), 114.6 (C-2'), 72.9 (C-4), 71.2 (C-l), 71.0 (C-3), 67.3 (C-5), 38.8 (C-2), 35.1 (C-6).

2-Hydroxy-3-(O-hydroxyphenyl) propanoic acid; White solid; IR: (film) Vmax 3316, 1728, 1590, 1406 cm^{-1}; UV λmax (MeOH): 218 (3.04), 253 (3.42), 289 (3.66) nm; CD/ORD measurements gave straight lines indicating that compound was obtained as a racemic mixture; FABMS *m/z* 183 (4) $[M + H]^+$; ^{1}HNMR (CD_3-OD): δ 7.40 (1 H, d, J = 7.32 Hz, H-6'), 7.22 (lH, t, J = 7.57 Hz, 7,32 Hz, H-4'), 7.00 (lH, t, J = 7.57 Hz, 7.32 Hz, H-5'), 6.86 (lH, d, J = 7.57 Hz, H-3'), 4.19 (lH, m, H-2), 2.80 (2H, m, H-3); ^{13}CNMR (DMSO-d_6): δ 178.6 (C-l), 141.6 (C-2'), 133.4 (C-6'), 128.4 (C-4'), 123.8 (C-5'), 121.3 (C-l'), 109.2 (C-3'), 73.4 (C-2), 48.6 (C-3).

l-(3',4'-Dihydroxycinnamoyl)-cyclopenta-2,5-diol; Pale yellow oily compound; IR: (film) Vmax 3351, 2926, 1669, 1599, 1379, 1267, 1076 cm-l; UV λmax (MeOH) 206 (3.99), 215 (4.00), 243 (3.83), 299 (3.86), 325 (3.89) nm; ORD (m deg) 336 (75), 316 (-44), 298 (-40), 260 (35), 240 (-22), and 216 (52) nm; FABMS *m/z* 281 (2) $[M + H]^+$; EIMS *m/z* 180 (93), 163 (100), 145 (20); ^{1}HNMR (DMSO-d_6): δ 7.45 (lH, d, J = 15.9 Hz, H- 7'), 7.01 (lH, d, J = 1.8 Hz, H-2'), 6.96 (lH, dd, J = 8.1 Hz, 1.8 Hz, H-6'), 6.75 (lH, d, J = 8.1 Hz, H-5'), 6.19 (1 H, d, J = 15.9 Hz, H-8'), 5.17 (lH, m, H-l), 3.82 (lH, m, H-2), 3.54 (lH, m, H-5), 1.83 (4 H, m, H-3, H-4); ^{13}CNMR (DMSO-d_6): δ 166.1 (C-9'), 148.2 (C-3'), 145.5(C-4'), 144.4 (C-7'), 125.7 (C-l'), 121.1 (C-6'), 115.8 (C-5'), 115.0 (C-8'), 114.6 (C-2'), 70.9 (C-2, C-5), 67.5 (C-l), 35.2 (C-3, C-4).

l-(3',4'-Dihydroxycinnamoyl)-cyclopenta-2,3-diol; Oily compound; IR (fllm) v max 3372, 1692, 1603, 1277, 1184, 1074 cm^{-1}; λmax (MeOH) 203 (3.95), 215 (3.94), 243 (3.76), 299 (3.81) and 327 (3.90) nm; ORD (m deg) 314 (-58), 288 (-61), and 234 (-61) nm; FABMS *m/z* 281 (2) $[M + H]^+$; EIMS *m/z* 180 (34), 163 (100); ^{1}HNMR (CD_3OD): δ 7.58 (lH, d, J = 15.9 Hz, H-7'), 7.04 (lH, d, J = 1.8 Hz, H-2'), 6.93 (1 H, dd, J = 8.2 Hz, 1.8 Hz, H-6'), 6.76 (lH, d, J = 8.2 Hz, H-5'), 6.30 (lH, d, J = 15.9 Hz, H-8'), 5.35 (lH, m, H-l), 4.14 (lH, m, H-3), 3.64 (lH, dd, J = 8.3 Hz, 3.1 Hz, H-2), 2.15 (2H, m, H-4), 2.15 (lH, m, H-5a), 1.95 (1 H, m, H-5b); ^{13}C NMR (CD_3OD): δ 169.0 (C-9'), 149.4 (C-3'), 146.8 (C-4'), 146.8 (C-7'), 128.0 (C-l'), 122.9 (C-6'), 116.5 (C-5'), 115.8 (C-8'), 115.1 (C-2'), 74.8 (C-2), 73.0 (C-l), 68.3 (C-2), 41.5 (C-5), 36.7 (C-4).

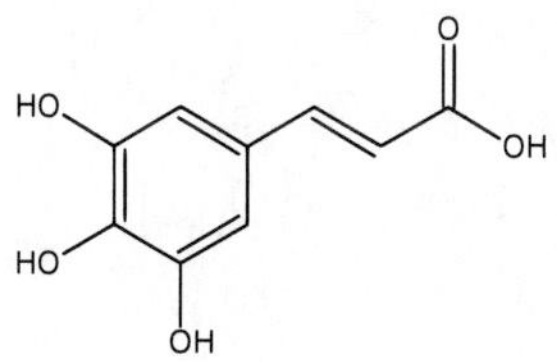

535
Caffeic acid

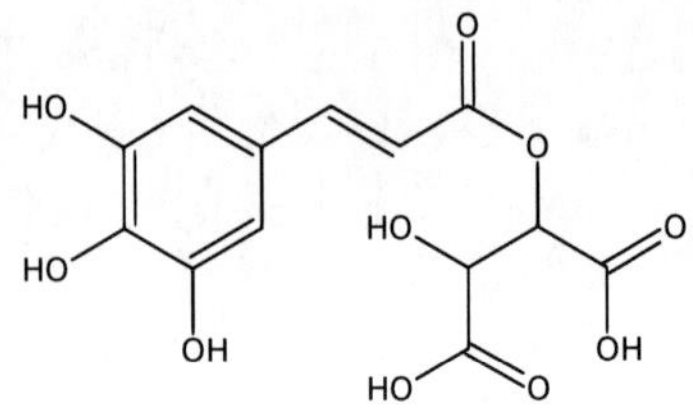

536
Caftaric acid

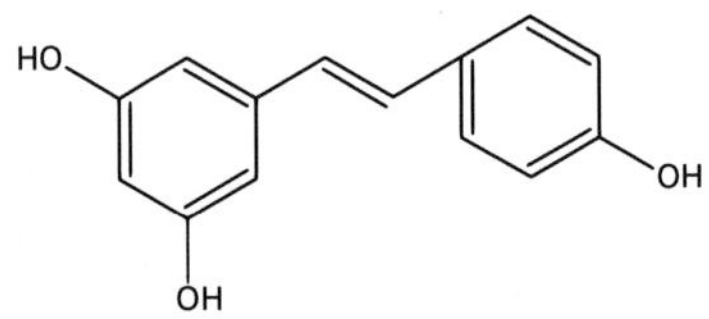

537
trans- Resveratrol

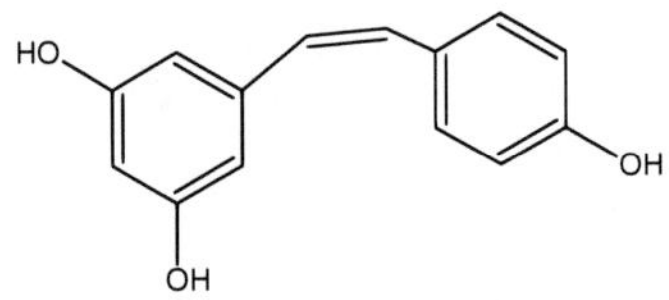

538
cis- Resveratrol

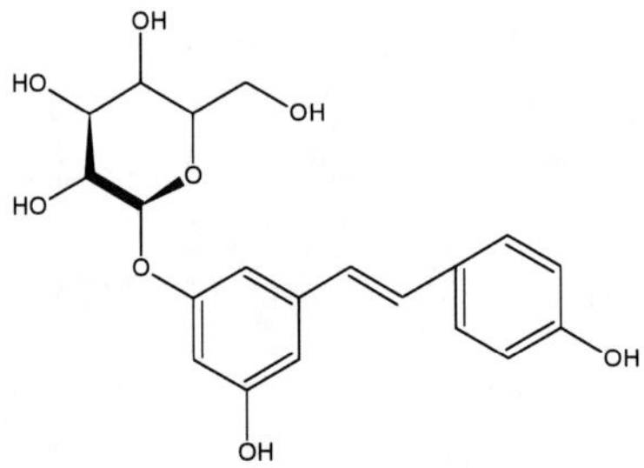

539
trans- Resveratrol-O-glucoside

These phenols compounds were isolated from red wine. The total phenol contents, measured both by HPLC and colorimetrically, correlated very strongly with the antioxidant activity and vasodilation activity (Burns et al., 2000).

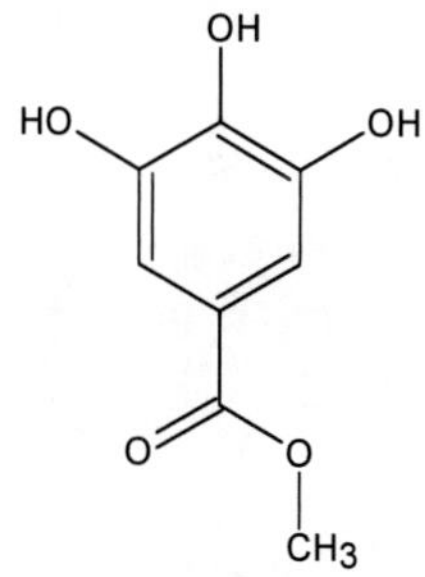

540
Methyl gallate

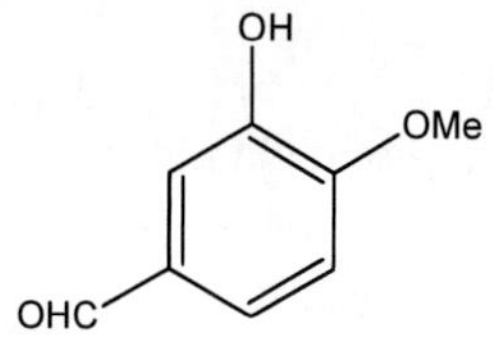

541
Isovanillin

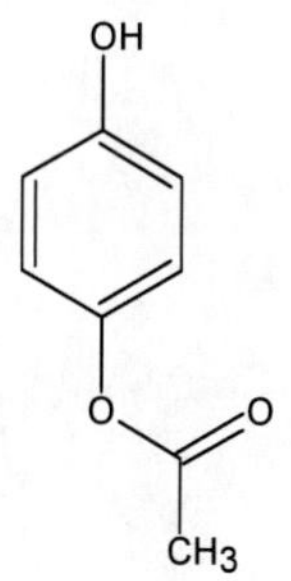

542
4-Hydroxymethylbenzoate

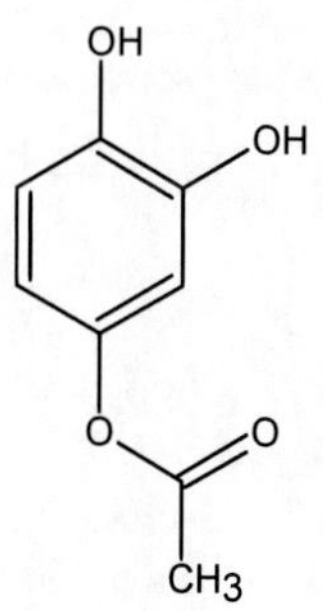

543
3,4-Dihydroxy-methylbenzoate

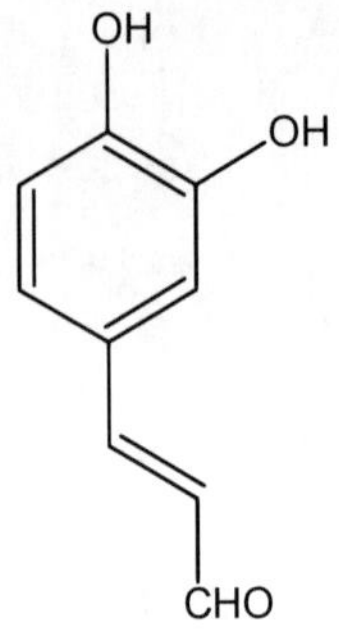

544
Caffeoyl aldehyde

These compounds were isolated from the leaves of *Tachigalia paniculata* (Leguminosae), showed activity in the assays, trolox equivalent antioxidant capacity (TEAC), the coupled oxidation of β-carotene and linoleic acid (autoxidation assay), and the inhibition of xanthine oxidase activity. Isovanillin and caffeoyl aldehyde were moderately actives in the autoxidation test (Cioffi et al., 2002).

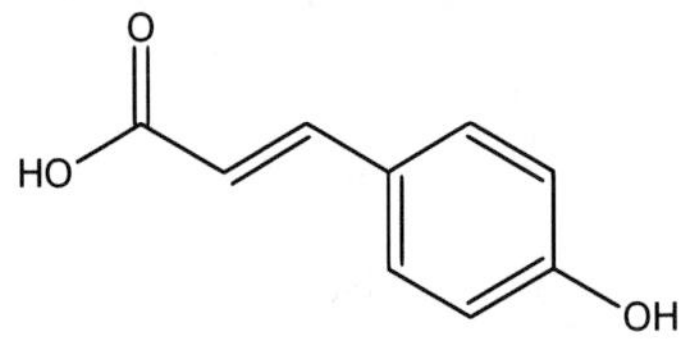

545
p-Coumaric acid

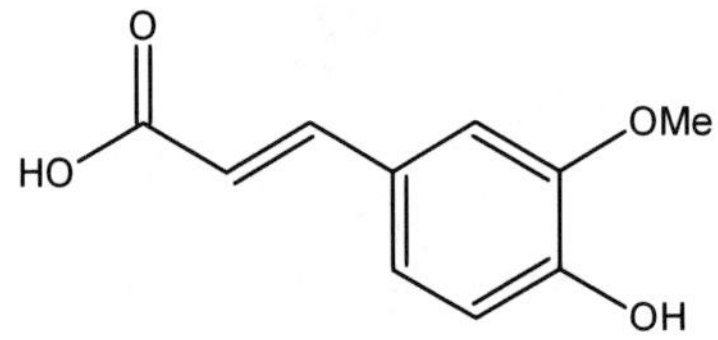

546
Ferulic acid

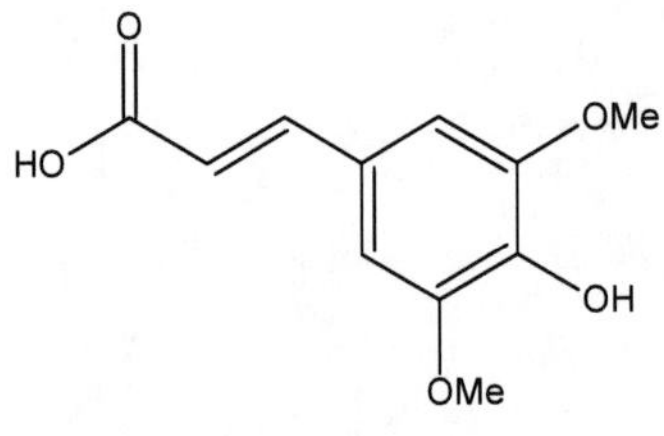

547
Sinapic acid

Four cinnamic acids *p*-coumaric, caffeic, ferulic, and sinapic acid, were isolated from *Avena sativa* L (Graminae). As assessed by the reactivity toward 1,1-diphenyl-2-pilcrylhydrazyl (DPPH), all avenanthramides, showed antioxidant activity and decreased in a similar order as for the corresponding cinnamic acids, that is: sinapic > caffeic > ferulic > p-coumaric acid. (Bratt et al., 2003). Peroxynitrite (ONOO-), formed from a reaction of superoxide and nitric oxide, is one of the most potent cytotoxic species that are known to oxidize cellular constituents including essential proteins, lipids, and DNA. The sinapic acid (3,5-dimethoxy-4-hydroxycinnamic acid), was also isolated from *Brassica juncea* (Cruciferaceae). The sinapic acid is efficiently scavenge native $ONOO^-$ as well as $ONOO^-$ derived from the peroxynitrite donor 3-morpholinosydnohimine hydrochloride (SIN-1). Sinapic acid also showed a significant ability of inhibiting nitration of bovine serum albumin and low-density lipoprotein (LDL) in a dose-dependent manner. Sinapic acid decreased the LDL peroxidation induced by SIN-1-derived $ONOO^-$ (Zou et al., 2002).

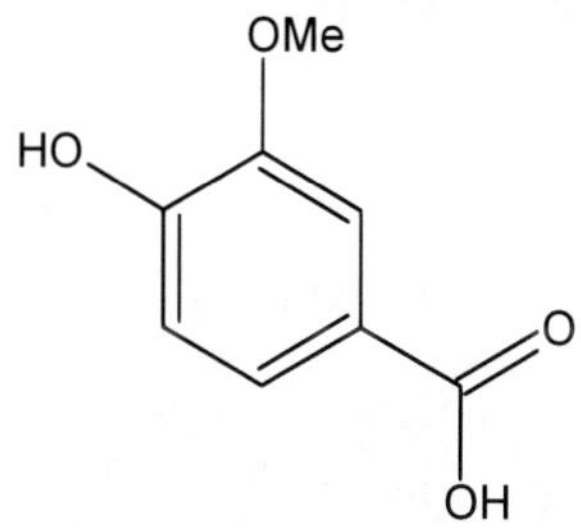

548
Vanillic acid

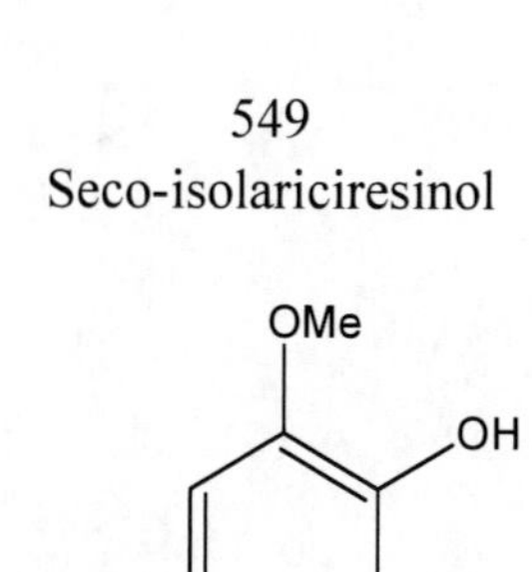

549
Seco-isolariciresinol

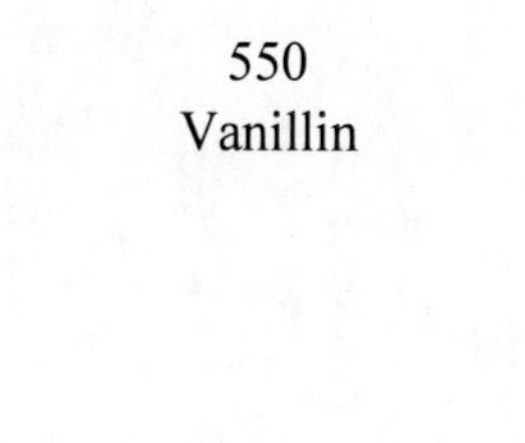

550
Vanillin

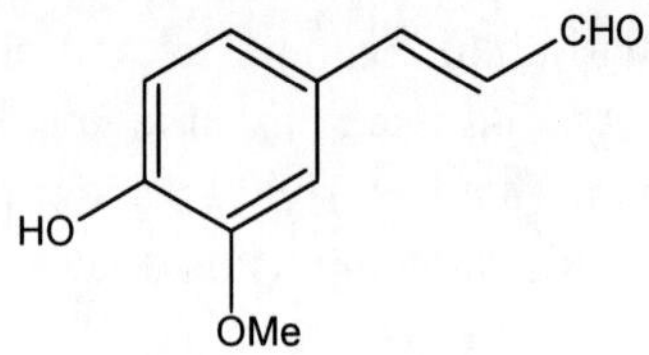

551
Coniferyl aldehyde

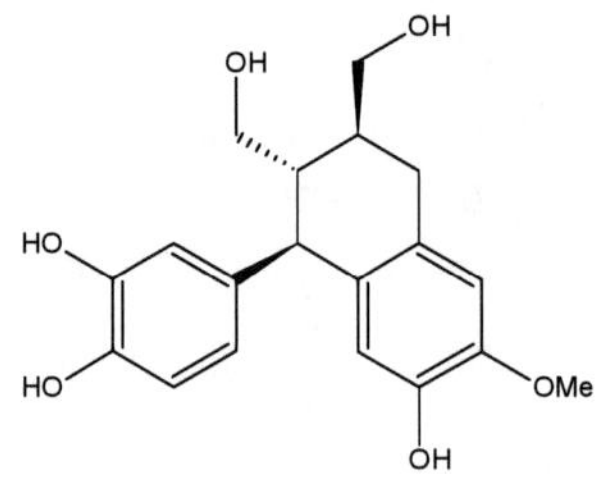

552
Isotaxiresinol

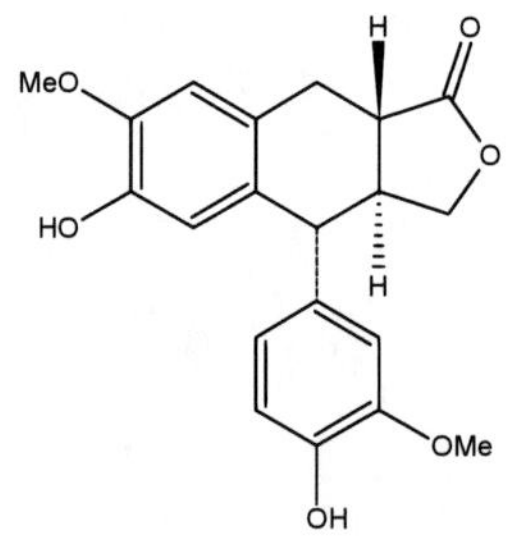

553
α-Conidendrin

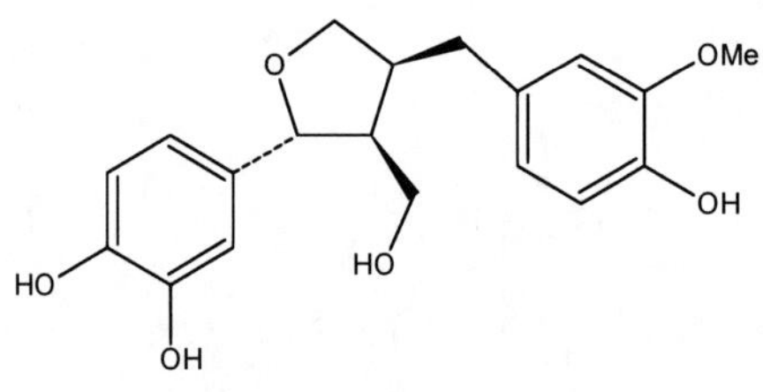

554
Taxiresinol

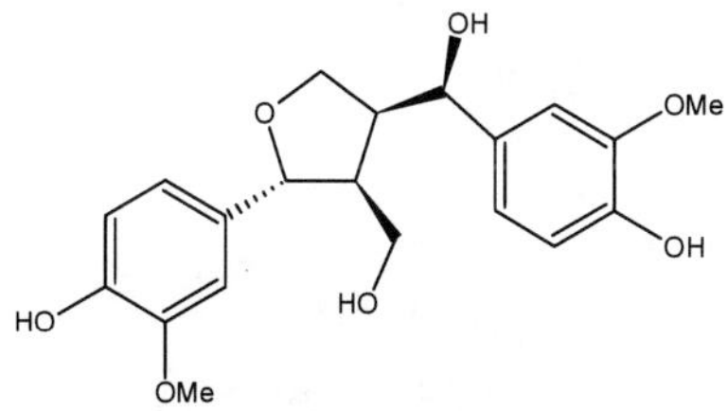

555
(7'R)-7'-Hydroxylariciresinol

Found in *Taxus yunnanensis* (Taxaceae). All compounds showed significant scavenging activity. Taxiresinol possessed the strongest DPPH radical scavenging activity with an IC_{50} value of 18.4 μM. Caffeic acid. a well-known antioxidant used as a positive control, had an IC_{50} value of 25.5 μM, while ascorbic acid had 12.6 μM. The major lignans, are isotaxiresinol, and seco-isolariciresinol also possessed strong scavenging activity toward DPPH radical with IC_{50} values of 21.7 and 28.9 μM, respectively, and other two lignans α-Conidendrin and (7'R)-7'-hydroxylariciresinol showed mild DPPH radical scavenging activities. Among the phenolic compounds, coniferyl aldehyde showed the most potent NO inhibitory activity with an IC_{50} value of 18.0 μM, which is stronger than that of N^G-monomethyl-L-arginine (L-NMMA) used as a positive control having an IC_{50} value of 25.5 μM (Banskota et al., 2003).

Vanillic acid: Yellowish-white powder. El-MS *m/z*: 168 $[C_8H_8O_4]^+$: ^{1}H NMR ($CDCl_3$): δ: 7.54 (1H,d, J = 1.95 Hz, H-2), 7.54 (1H,dd, J = 1.95, 8.79Hz, H-6), 6.82 (1H,d, J = 8.79 Hz. H-5); ^{13}C NMR($CDCl_3$): δ 170.0 (COOH), 125.2 (C-1), 115.8 (C-2), 152.6 (C-3), 148.6 (C-4), 113.7 (C-5), 123.0 (C-6), 56.3 (OCH_3).

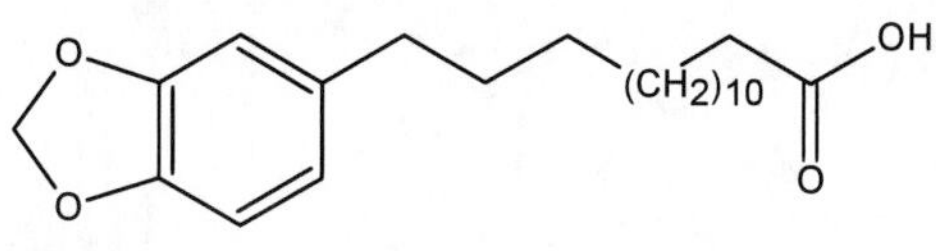

556
15′-(3,4-Methylenedioxyphenyl)- pentadecanoic acid

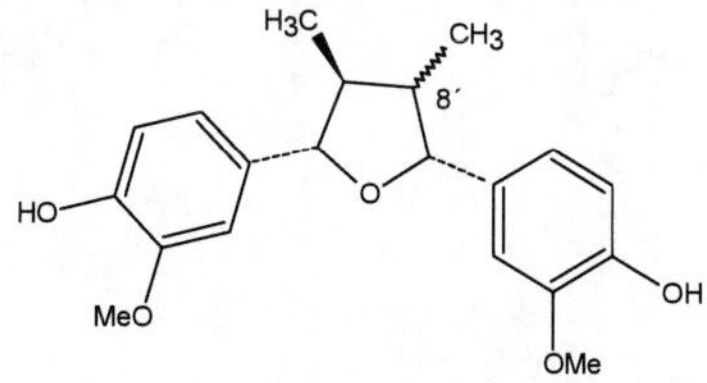

557
8′R Nectandrin B

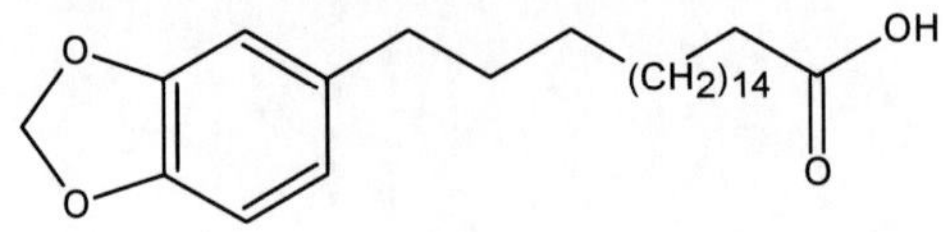

558
19´-(3,4-Methylenedioxyphenyl)-nonadecanoic acid

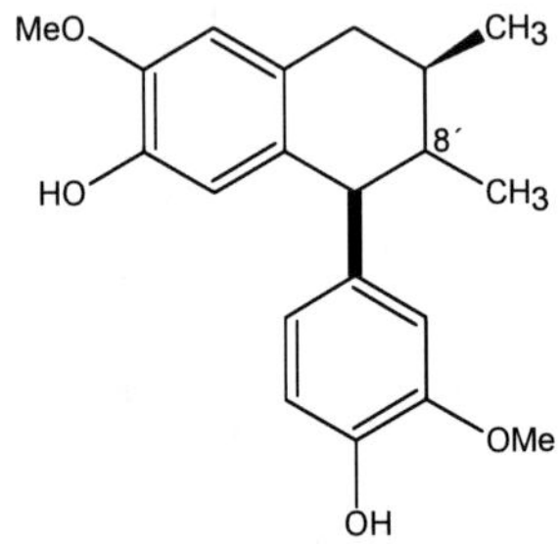

559
8´R Epi-guaiacin

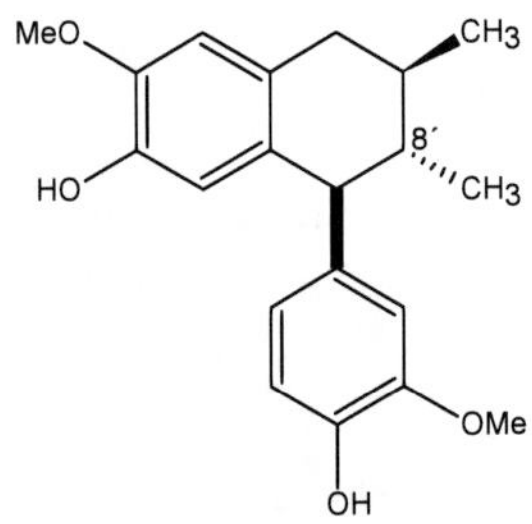

560
8´S Guaiacin

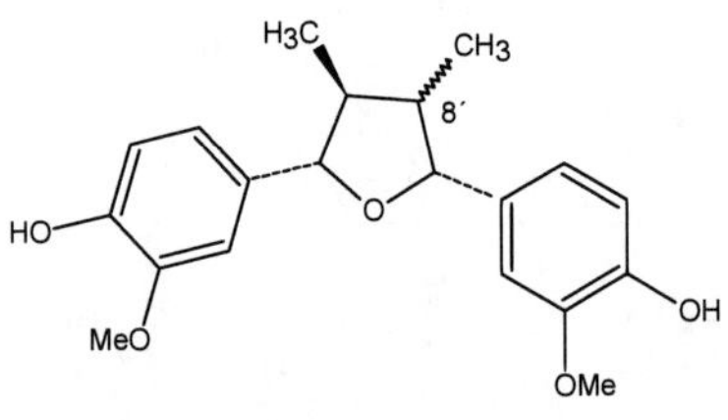

561
8´S Verrucosin

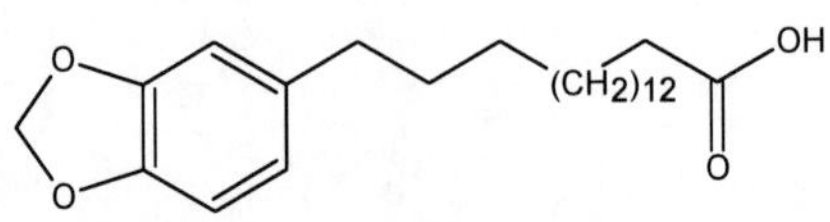

562
17´-(3,4-Methylenedioxyphenyl)-heptadecanoic acid

Lignans were isolated from *Iryanthera juruensis* (Myristicaceae) fruits. The high oxidation potential of the ω-arylalkanoic acids is probably a result of the lack of aromatic hydroxyl groups in these molecules. Lignans *epi*-guaiacin, guaiacin, verrucosin, and nectandrin B showed oxidation potentials in the range 0.8-0.9 V. (Silva et al., 2001).

15′-(3,4-Methylenedioxyphenyl)-pentadecanoic acid: Colorless oil. Found: C, 72.98; H, 9.69. $C_{22}H_{34}O_4$ requires: C, 72.93; H. 9.39%. UV λmax. (MeOH) nm (ε): 284 (7710); IR ν cm^{-1}: 3320, 1721, 1600. 1504; ElMS (70 eV): *m/z* = (rel. int.) 362 $[M]^+$ (10); 135 $[C_7H_5.CH_2O_2]^+$ (100); ESMS: *m/z* = (rel. int.) 385 $[M+Na]^+$ (100), 363 $[M+1]^+$ (45); ^{1}HNMR (300 MHz, $CDCl_3$): δ 1.18 (26H, s, H-4'- H-16'), 1.50 (2H, m, H-3'), 2.28 (2H, t, J = 7.6 Hz, H-2'), 2.45 (2H, t, J = 7.9 Hz, H-17'), 5.85 (2H, s, CH_2O_2), 6.55 (1H, dd, J = 1.6, 7.7 Hz, H-6), 6.61 (1H, d, J = 1.6 Hz, H-2), 6.66 (1H. d, J = 7.7 Hz. H-5); ^{13}C NMR: (75 MHz. $CDCl_3$): δ 179.4 (s, C-1'), 33.9 (t, C-2'), 24.7 (t, C-3'), 28.6-29.0 (t, C-4'- C-12'), 29.0 (t, C-15'), 31.8 (t, C-16'), 35.7 (t, C- 17'), 136.8 (s, C-1), 107.9 (d, C-2), 145.3 (s, C-3), 147.4 (s, C-4'), 108.8 (d, C-5'), 120.9 (d, C-6'), 100.6 (t,. CH_2O_2).

19′-(3,4-Methylenedioxyphenyl)-nonadecanoic acid: Colorless oil. Found: C. 74.39: H. 10.10. $C_{26},H_{42}O_4$ requires: C. 74.64: H. 10.05%. UV λmax. (MeOH) nm (ε): 285 (7710). IR ν cm^{-1}: 3328, 1720, 1605, 1510; ElMS (70 eV): *m/z* = (rel. int.) 135 $[C_7H_5, CH_2O_2]^+$ (29), ESMS *m/z* (rel.int.) 441 $[M+Na]^-$ (100), 419 $[M+1]^-$ (78); ^{1}HNMR (300 MHz, $CDCl_3$): δ 1.18 (32H. s, H-4'- H-20'), 1.50 (2H. m, H-3'), 2.28 (2H, t, J = 7.6 Hz. H-2'), 2.45 (2H, t, J = 7.9 Hz. H-21'), 5.85 (2H, s, CH_2O_2), 6.55 (1H, dd, J = 1.6, 7.7 Hz. H-6), 6.61 (1H, d, J = 1.6 Hz. H-2), 6.66 (1H, d, J = 7.7 Hz. H-5); ^{13}C NMR (75 MHz. $CDCl_3$): δ 179.4 (s, C-1'), 33.9 (t, C-2'), 24.7 (t, C-3'), 28.6-29.0 (t, C- 4'- C-16'), 29.0 (t, C-17'), 31.8 (t, C-18'). 35.7 (t, C-19'), 136.8 (s, C-1), 107.9 (d, C-2), 145.3 (s, C-3), 147.4 (s,C-4), 108.8 (d, C-5), 120.9 (s, C-6), 100.6 (t, CH_2O_2).

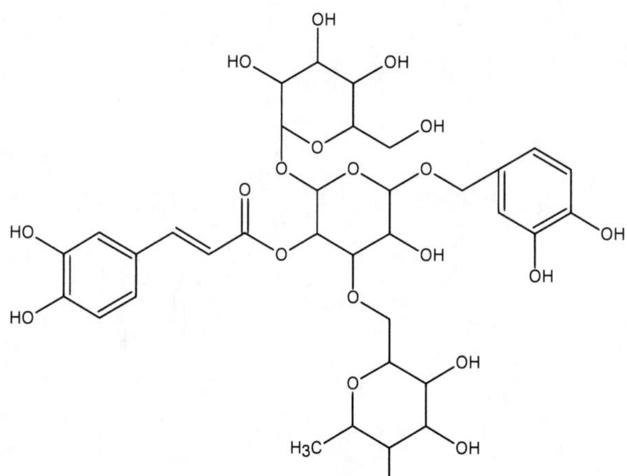

563
Chichoric acid

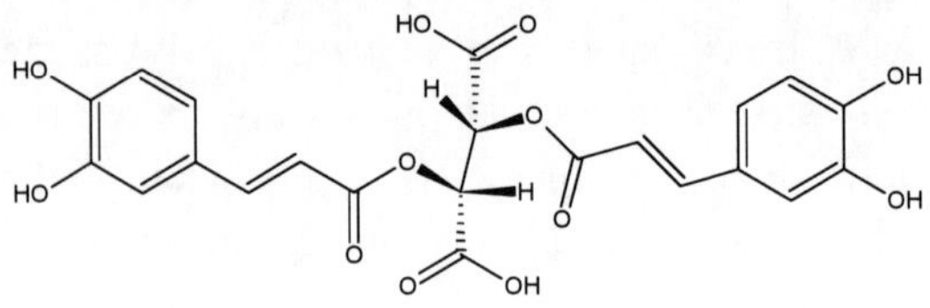

563
Echinacoside

Found in *Echinacea angustifolia, E. pallida* and *E. purpurea* (Compositae). These components of the *Echinacea* root extracts were effective at suppressing the generation of hydroxyl radical generated by the Fenton reaction containing H_2O_2, EDTA, and ascorbic acid in a test system where the generation of ˙OH, and exhibited the greatest DPPH radical scavenging activity (Hu and Kitts, 2000).

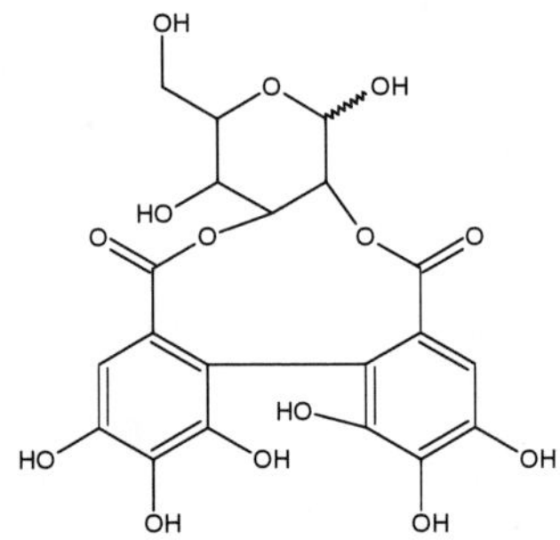

565
2,3-(S)-HHDP-D-glucose

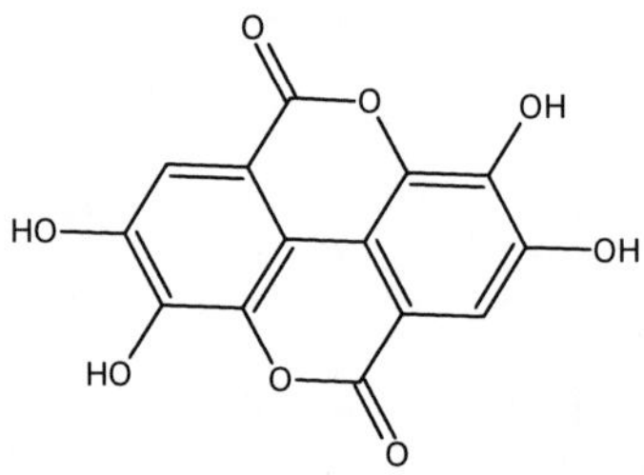

566
Ellagic acid

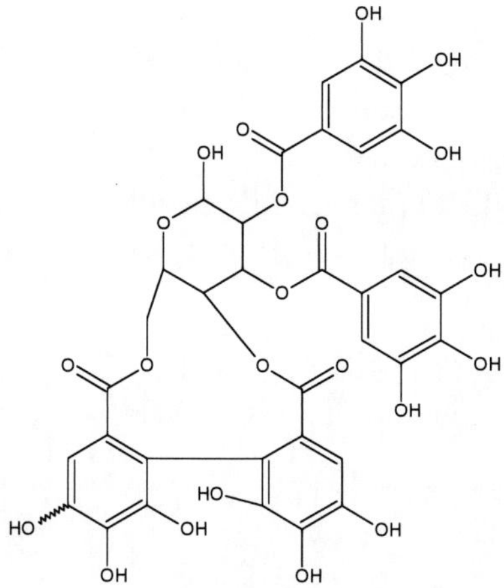

567
1-Desgalloyl eugeniin

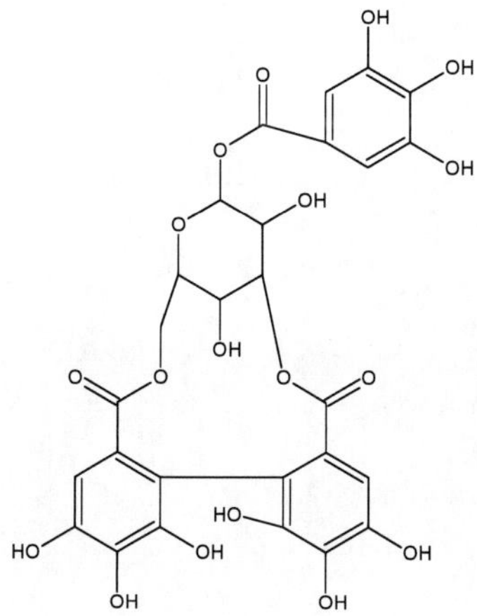

568
Corilagin

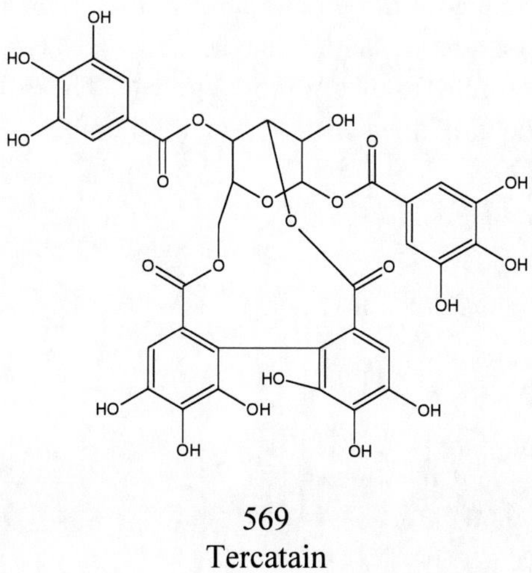

569
Tercatain

The aqueus extract of *Tenninalia catappa* L.(Combretaceous) exhibited superoxide radical scavenger activity and modification of mitomycin C- induced clasto-genicity. The tannin components of *T. catappa* showed ability to prevent lipid peroxidation, superoxide fomation and their free radical scavenging activity. All of these components showed potent antioxidant activity. Punicalagin and punicalin were the most abundant components and had the strongest anti-oxidative effects of this group of tannins (Lin et al., 2001).

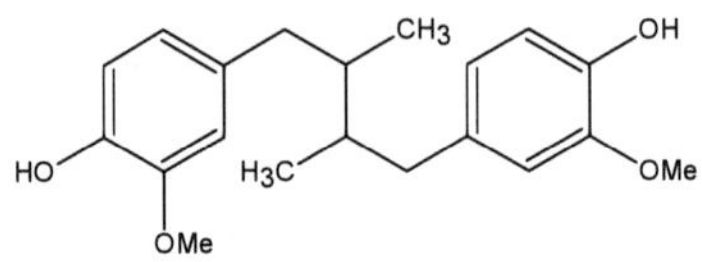

570
Dihydroguayaretic acid

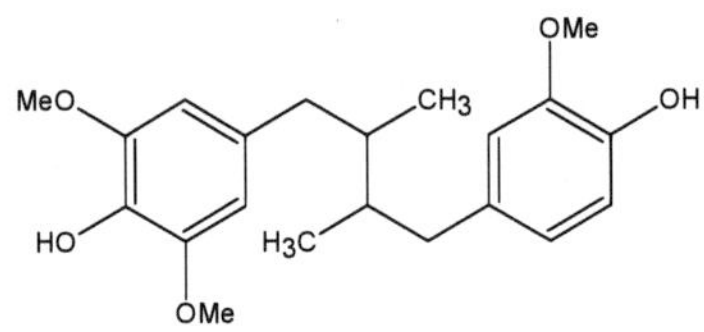

571
Guayacasin

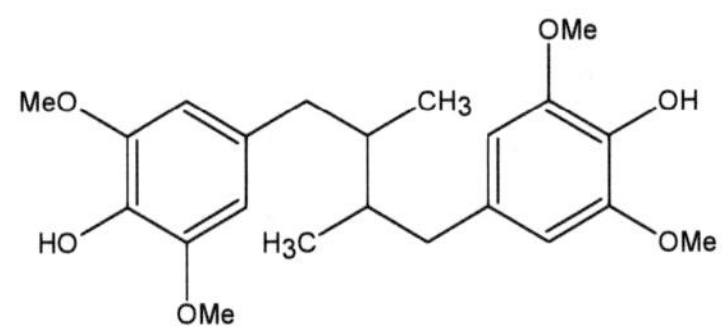

572
Isopregomisin

These lignans were isolated from *Porlieria chilensis* (Zigofilaceae). The structurally similar lignans dihydroguayaretic acid, guayacasin and isopregomisin are powerful antioxidants with activities similar to that of propyl gallate. The differences observed between the different lignans can be explained in terms of the number of nethoxyl groups of each compound (Faure et al., 1990).

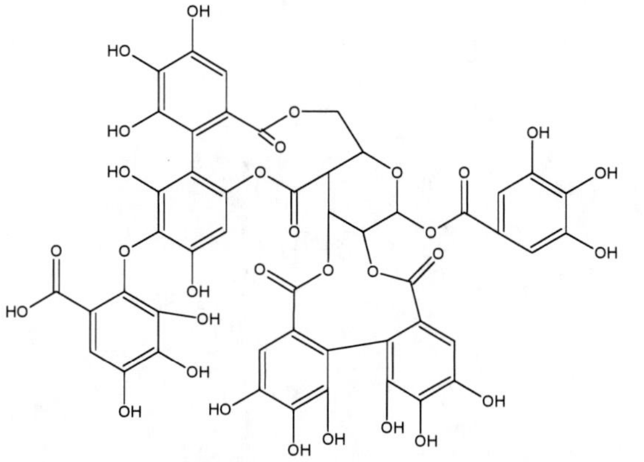

573
Glansrin A

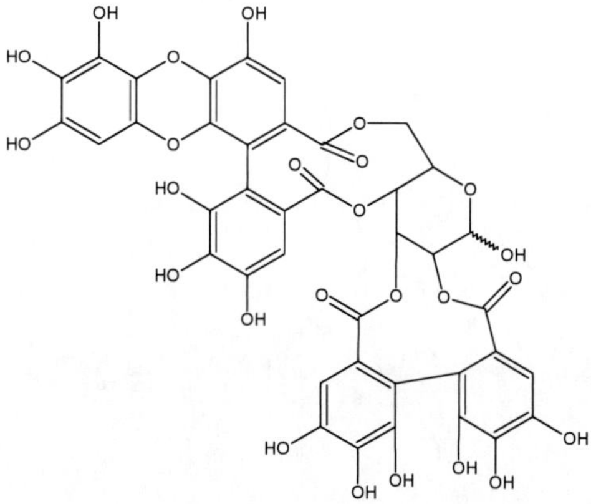

574
Glansrin B

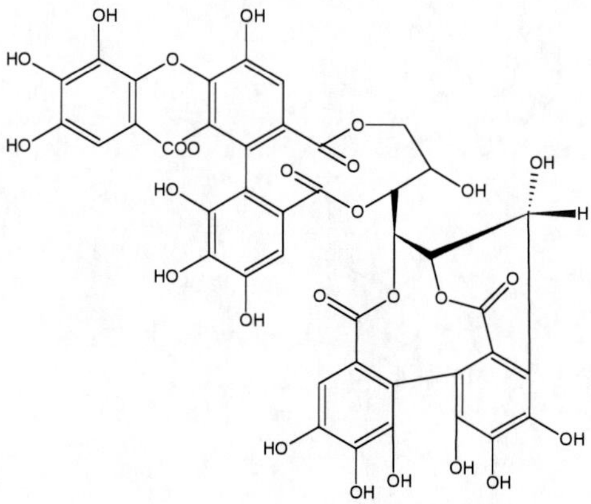

575
Glausrin C

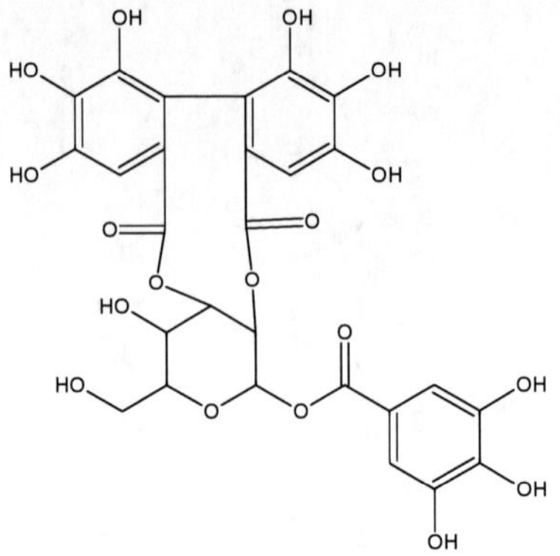

576
Isostrictinin

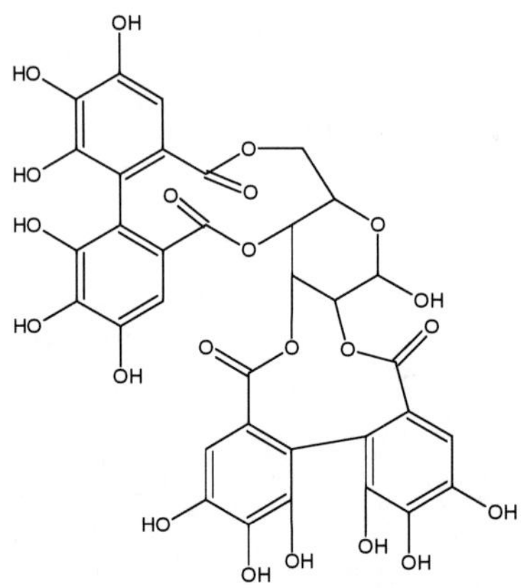

577
Pedunculagin

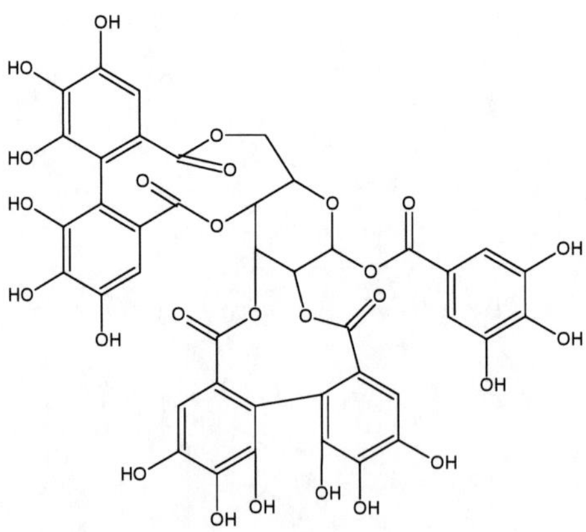

578
Casuarictin

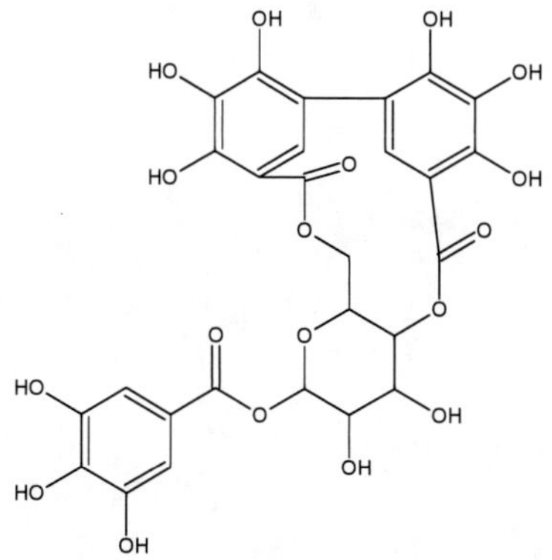

579
Strictinin

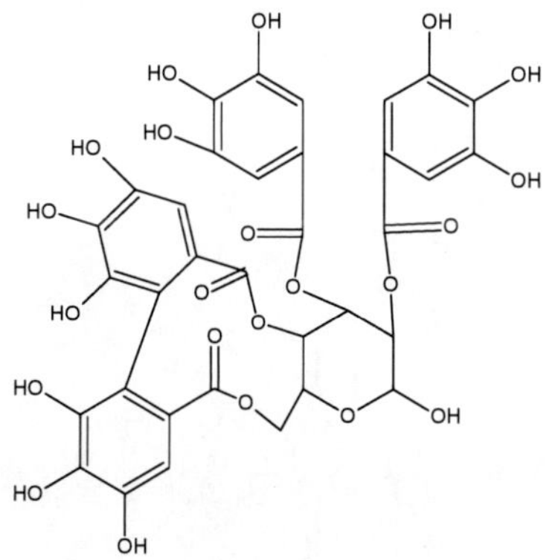

580
Tellimagrandin I

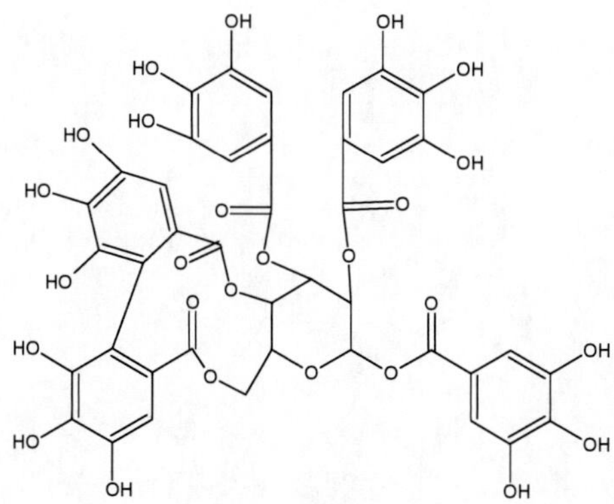

581
Tellimagrandin II

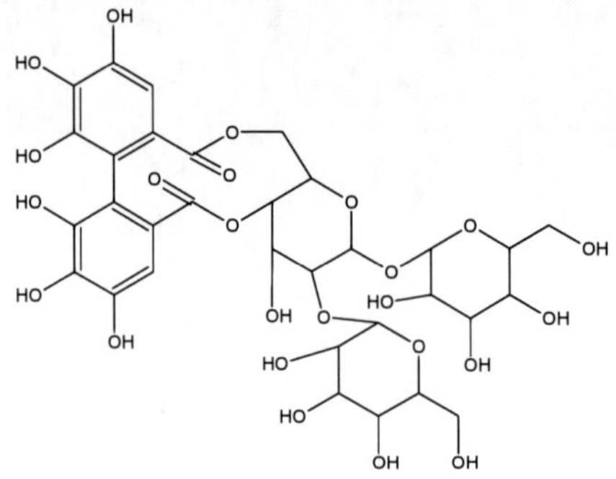

582
1,2-Di-O-galloyl-4,6-HHDP-β-D-glucopyranose

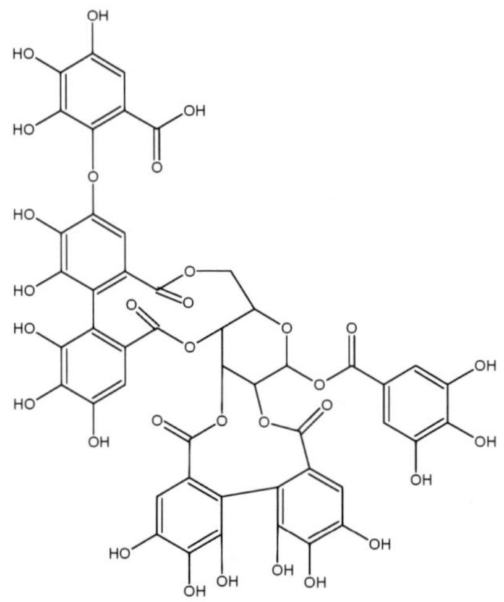

583
Rugosin C

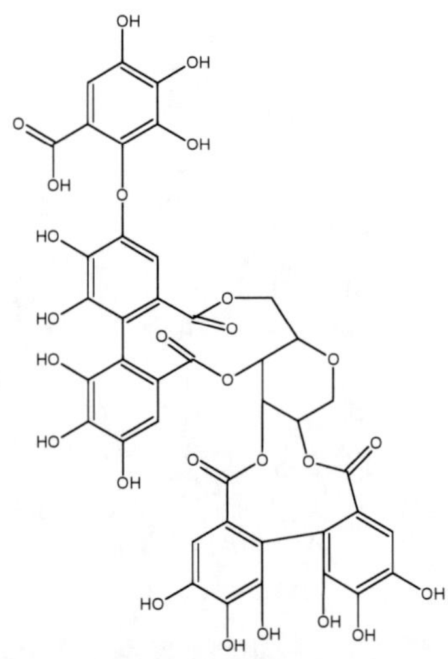

584
Praecoxin A

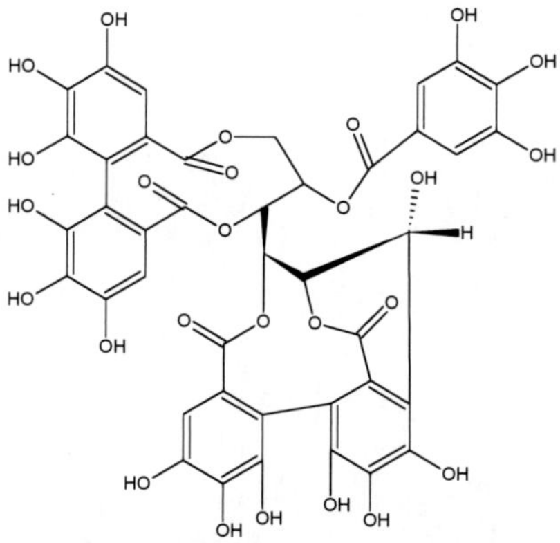

585
Casuarinin

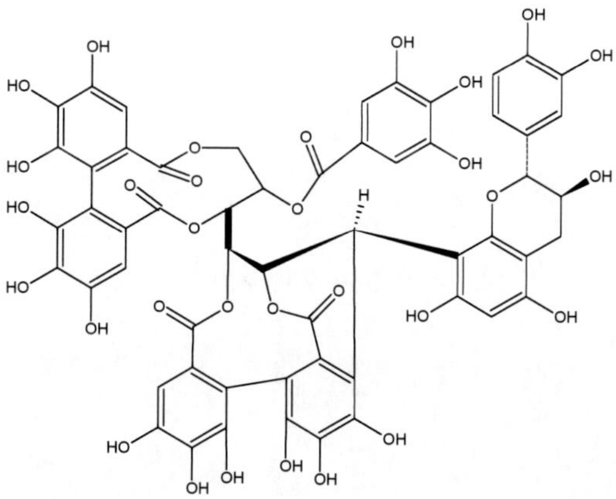

586
Stenophyllanin A

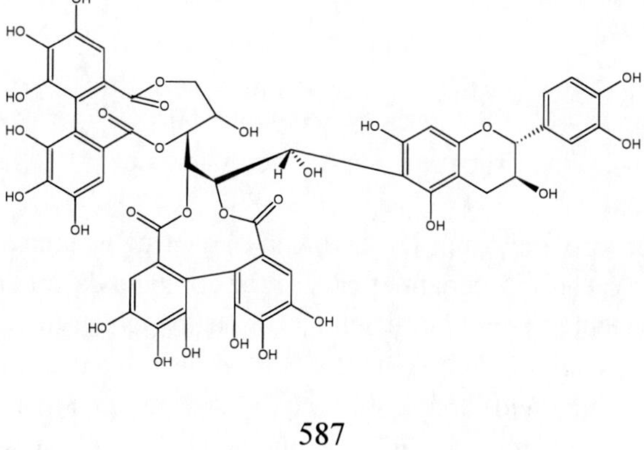

587
Stayuranin

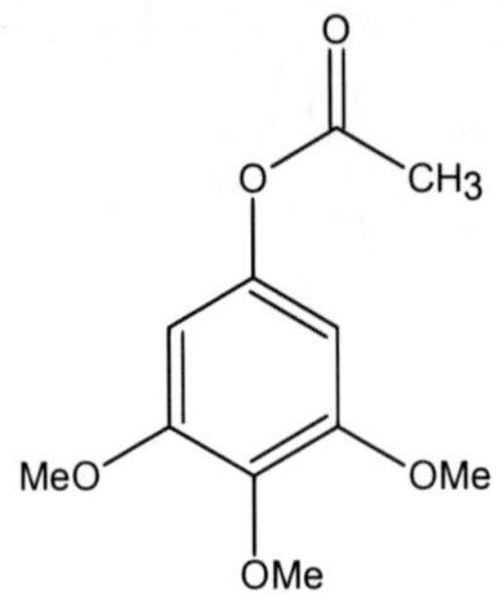

588
A Methyl tri-O-methylgallate

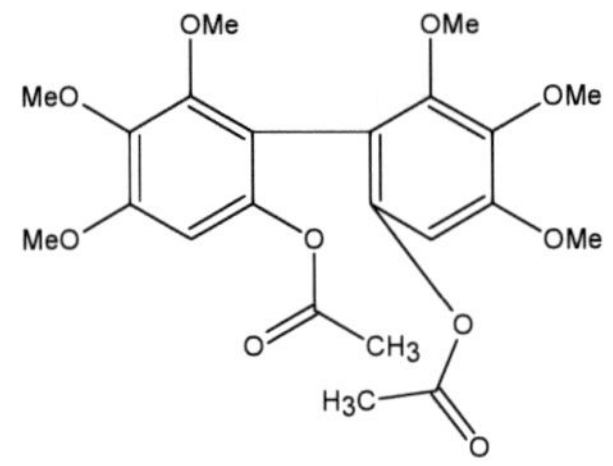

589
Dimethyl hexamethoxydiphenate

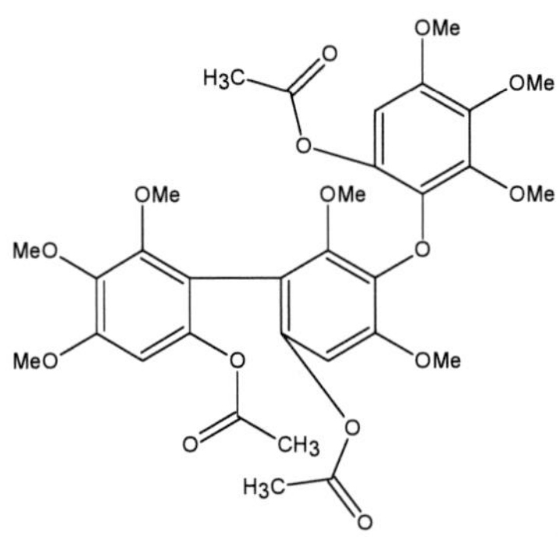

590
Trimethyl octa-O-methylgallate

These tannins were isolated from the n-BuOH extract of walnuts, the seeed of *Juglans regia* L. (Juglandaceae). The antioxidant effects of the compounds obtained were evaluated in superoxide-dismutase (SOD)-like activity and 1, 1-diphenyl-2-picrylhydrazyl (DPPH) radical scavenging activity. Glansrin C and stenophyllanin A exhibited a strong SOD-like effect comparable to L-ascorbic acid and gallic acid used as positive controls. The other polyphenols showed activity with an EC_{50} on the order of 10^{-5} M, with the exception of glansrin A and 2,3-HHDP-D-glucopyranose. Conversely, all of these polyphenols had a more potent reducing ability (E C_{50} value 10^{-7}-10^{-6} M) toward the DPPH radical than L-ascorbic acid and gallic acid. Although the structure-activity relationships are unclear, ellagitannins with a galloyl group plus an HHDP group, or valoneoyl and its isomeric group, tend to exhibit a more potent radical scavenging effect against DPPH than those with only an HHDP group, like 2,3-HHDP-D-glucopyranose and Pedunculagin. This indicates that ellagitannin polyphenols should act as effective antioxidants in walnut (Fukuda et al., 2003).

Glansrin A: Off-white amorphous powder; $[\alpha]_D$ -11° (MeOH; c 0.3); negative FABMS *m/z*: 1103 $[M-H]^-$; (Found: C, 49.6; H, 3.8. $C_{48}H_{32}O_{31}.3H_2O$ requires C, 49.7; H, 3.3%). CD (MeOH) [θ] (nm) + 2.3 x 10^5 (238), -6.9 x 10^4 (263), +6.0 x 10^3 (282); ^{1}H NMR: δ 7.20 (2H, s, galloy1 H-2, 6), 6.93 [lH, s, tergalloy1 (Ter) H-6´´], 6.62 (lH, s, Ter H-3), 6.57 (lH, s, Ter H-3'), 6.48 (lH, s, HHDP H-3), 6.38 (lH, s, HHDP H-3'), glucose protons: δ 6.24 (d, J = 9 Hz, H-1), 5.22 (t, J = 9 Hz, H-2), 5.48 (dd, J = 9, 10 Hz, H-3), 5.24 (t, J = 10 Hz, H-4), 4.54 (dd, J = 6.5, 10 Hz, H-5), 5.37 (dd, J = 6.5, 13 Hz, H-6), 3.92 (d, J = 13 Hz, H-6); ^{13}CNMR: δ Glucose (glc): 92.3 (C-l), 76.0 (C-2), 77.0 (C-3), 69.1 (C-4), 73.4 (C-5), 63.3 (C-6), Galloyl (gall): 119.8 (C-l), 110.2 (C-2, C-6), 146.3 (C-3, C-5), 140.0 (C-4), 165.1 (C-7), Hexahydroxydiphenoyl (HHDP): 115.1 (C-1), 126.0 (C-2), 107.3 (C-3), 145.2 (C-4), 136.6 (C-5), 144.49 (C-6), 168.8 (C-7), 114.2 (C-1'), 126.3 (C-2'), 107.1 (C-3'), 145.2 (C-4'), 136.1 (C-5'), 144.53 (C-6'), 169.4 (C-7'), Tergalloyl (ter): 116.9 (C-1), 131.9 (C-2), 107.7 (C-3), 150.0 (C-4), 137.2 (C-5), 150.0 (C-6), 168.11 (C-7), 117.5 (C-1'), 125.0 (C-2'), 107.9 (C-3'), 145.1 (C-4'), 137.4 (C-5'), 144.9 (C-6'), 168.1 (C-7'), 114.2 (C-1"), 139.0 (C-2"), 139.6 (C-3"), 139.4 (C-4"),142.4 (C-5"), 108.3 (C-6"),172.2 (C-7").

Glansrin B: Light-brown amorphous powder; $[\alpha]_D$ + 90° (MeOH; c 0.1); negative FABMS *m/z*: 905 $[M-H]^-$; (Found: C.50.9; H, 3.9. $C_{40}H_{26}O_{25}.2H_2O$ requires:C, 50.9; H, 3.2%); CD (MeOH) [θ] (nm) + 1.9 x 10^5 (238), -5.6 x 10^4 (266), + 1. 7 x 10^4 (285); ^{1}H NMR: δ 6.73, 6. 71 [1H in total, each s, syzygyl (Syz) H-3'], 6.60, 6.55 (lH in total, each s, Syz H-3), 6.614, 6.618 (lH in total, each s, HHDP H-3), 6.373, 6.368 (lH in total, each s, HHDP H-3'), 5.947,5.941 (lH in total, each s, Syz H-6"), glucose protons δ: 5.46 (d, J = 4 Hz, H-l), 5.08 (dd, J = 4, 10 Hz, H-2), 5.49 (t, J = 10 Hz, H-3), 5.12 (t, J = 10 Hz, H-4), 4.62 (ddd, J = 1.5, 6.5, 10 Hz, H-5), 5.19 (dd, J = 6.5, 13 Hz, H-6), 3.83 (dd, J = 1.5, 13 Hz, H-6), (α-anomer): 5.06 (d, J = 8 Hz, H-1), 4.87 (dd, J = 8, 9 Hz, H-2), 5.25 (dd, J = 9, 10 Hz, H-3), 5.11 (t, J = 10 Hz, H-4), 4.23 (ddd, J = 1, 6.5, 10 Hz, H-5), 5.21 (dd, J = 6.5, 13 Hz, H-6), 3.90 (dd, J = 1, 13 Hz, H-6, β-anomer); ^{13}C NMR: δ 64.1, 64.2 [glucose (Glc) C-6, α and β], 67.3 (Glc C-5, α), 69.7, 70.1 (Glc C-4, α and β), 72.3 (Glc C-5, β), 75.6 (Glc C-2, α), 75.7 (Glc C-3, α), 77.5 (Glc C-3. β), 78.3 (Glc C-2, β), 91.7 (Glc C-1, α), 95.2 (lC) (Syz C-6"), 95.5 (Glc C-1, β), 106.77, 106.83 (Syz C-3), 107.2, 107.30, 107.34, 107.7 (HHDP C-3 and C-3'), 110.4 (1C) (Syz C-3'), 113.9 (1C) (Syz C-I), 114.3. 114.4 (HHDP C-1'), 114.9, 115.0 (HHDP C-l), 115.7 (lC) (Syz C-l'), 124.7 (1C) (Syz C-4"), 125.37, 125.42, 126.5 (lC), 126.6, 126.7 (HHDP C-2, C-2' and Syz C-2), 130.2, 130.3 (Syz C-2'), 130.5 (lC) (Syz C-2"), 133.03, 133.05 (Syz C-5'), 135.3 (lC) (Syz C-3"), 135.5 (lC) (Syz C-5"), 136.0, 136.1 (HHDP C-5'), 136.2 (lC) (Syz C-5), 136.4 (lC) (HHDP C-5). 142.1 (lC) (Syz C-6'), 142.2 (IC) (Syz C-1"), 144.3 (IC), 144,4 (1C) (HHDP C-6 and C-6'), 145.0 (1C) (HHDP C-4'), 145.13, 145.19 (HHDP C-4), 145.27, 145.32 (Syz H-4'), 145.5 (1C) (Syz C-4), 145.6 (1C) (Syz C-6), 167.6, 167.7 (Syz C-7'), 168.1 (1C) (Syz C-7), 168.9, 169.0 (HHDP C-7), 169.5, 169.6 (HHDPC-7').

Glansrin C: Light-brown amorphous powder; $[\alpha]_D$ 3 + 79° (MeOH; c 0.1); negative FABMS *m/z*: 933 $[M-H]^-$; (Found: C, 49.1; H, 3.8. $C_{41}H_{26}O_{26}.4H_2O$ requires:C, 48.9; H,3.4%). CD (MeOH) [θ] (nm): + 1.6 x 10^5 (228), -3.4 x 10^4 (252), +2.8 x 10^4 (287); ^{1}H NMR: δ 6.93 (1H, s, Ter H-6"), 6.89 (1H, s, Ter H-3'), 6.76 (1H, s, Ter H-3), 6.43 (1H, s,

HHDP H-3'), glucose protons δ: 5.65 (d, J = 5 Hz, H-1), 4.68 (dd, J = 2, 5 Hz, H-2), 5.49 (t, J = 2.5 Hz, H-3), 5.12 (dd, J = 3, 8 Hz, H-4), 4.17 (dd, J = 2, 8 Hz, H-5), 4.62 (dd, J = 3, 12 Hz, H-6), 3.92 (d, J = 12 Hz,H-6); ^{13}C NMR: δ Glucose (glc): 67.7 (C-1), 77.0 (C-2), 70.8 (C-3), 77.2 (C-4), 68.3 (C-5), 68.9 (C-6), Hexahydroxydiphenoyl (HHDP): 116.1(C-1), 120.5 (C-2), 116.8 (C-3), 143.68 (C-4), 138.6 (C-5), 143.9 (C-6), 164.9 (C-7), 116.5 (C-1'), 127.7 (C-2'), 105.2 (C-3'), 145.8 (C-4'), 135.0 (C-5'), 145.27 (C-6'), 170.3 (C-7'), Tergalloyl (ter): 114.2 (C-1), 124.7 (C-2), 107.5 (C-3), 145.7 (C-4), 136.3 (C-5), 45.33 (C-6), 168.6 (C-7), 120.3 (C-1'), 133.4 (C-2'), 111.5 (C-3'), 148.4 (C-4'), 142.4 (C-5'), 146.3 (C-6'), 168.3 (C-7'), 112.7 (C-1"), 140.5 (C-2"), 136.6 (C-3"), 143.74 (C-4"), 144.1 (C-5"), 109.8 (C-6"), 163.3 (C-7").

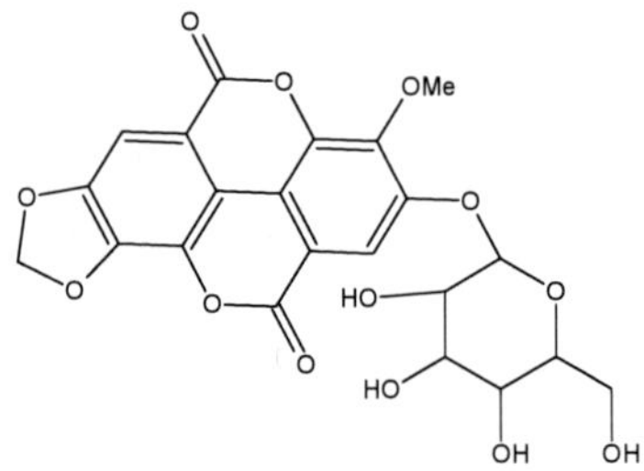

591
3,4-Methylenedioxy-3'-O-methyl-4'-O-glucoside ellagic acid

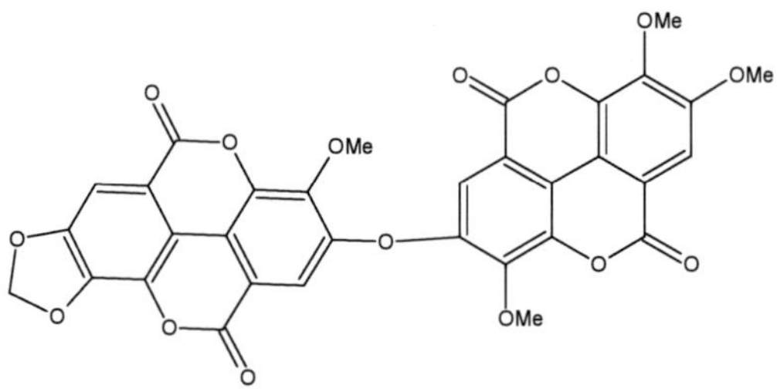

592
Pteleoellagic acid derivative

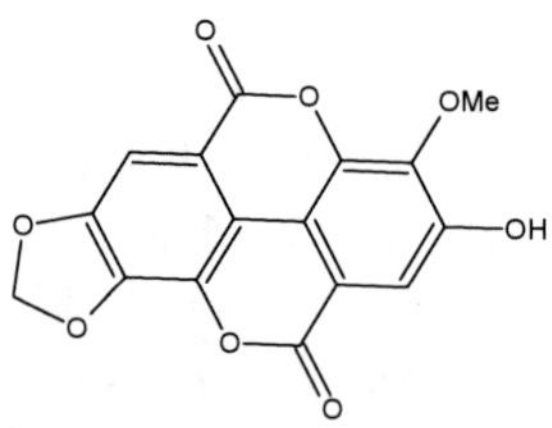

593
3,4-Methylenedioxy-3'-O-methyl ellagic acid

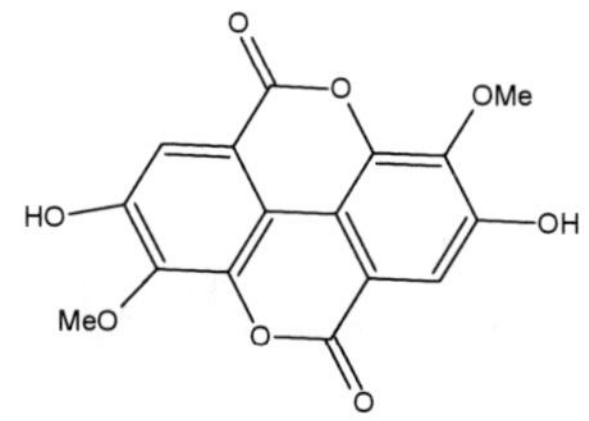

594
3,3'-Di-O-methyl ellagic acid

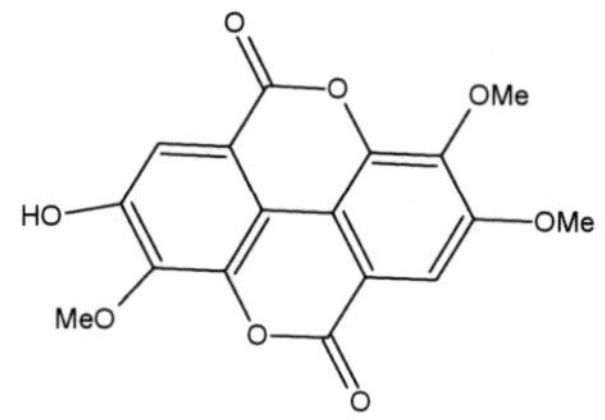

595
3,3',4'-Tri-O-methyl ellagic acid

Found in the stem bark of *Pteleopsis hylodendron* (Combretaceae). 3,4-Methylenedioxy-3'-O-methyl-4'-O-glucoside ellagic acid and 3,3'-di-O-methyl ellagic acid were found to have significant antioxidant activity, and also showed antibacterial activity against different pathogenic bacteria (Rahman et al., 2001).

3,4-Methylenedioxy-3'-O-methyl 4'-O-glucoside ellagic acid: $C_{22}H_{18}O_{13}$; colorless solid; m.p. 230- 232 °C, UV (MeOH): λmax 362, 321, 250, 221. 201 nm; IR: νmax 3395 (OH), 1738, 1711 (lactone carbonyls), 1609 cm^{-1} (aromatic rings). HRFAB MS: *m/z* 491.0740 [M + 1] (calcd. 490.0747) [$C_{22}H_{18}O_{13}$], HREI MS: *m/z* 328.0207 [$C_{16}H_8O_8$, (calcd. 328.0216)], 312.9967 [$C_{16}H_8O_8$ - CH_3]: ^{1}H-NMR (400 MHz, C_5D_5N): δ 6.33 (s, H-3a), 7.74 (s, H-5), 8.44 (s, H-5′), 5.85 (d, J = 7.7 Hz, H-1′′), 4.23 (s, OCH_3); ^{13}C-NMR (100 MHz, C_5D_5N): δ 112.6 (C-1), 132.1 (C-2), 138.9 (C-3), 104.6 (CH_2-C-3a), 151.1 (C-4), 104.0 (CH, C-5), 115.9 (C-6), 158.0 (C-6a), 113.2 (C-1′), 142.0 (C-2'), 142.7 (C-3'), 153.0 (C-4'), 113.5 (CH, C-5'), 114.0 (C-6'), 158.6 (C-6'a), 102.8 (CH, C-1"), 78.3 (CH, C-2"), 79.0 (CH, C-3"), 74.6 (CH, C-4"), 70.9 (CH, C-5"), 62.1 (CH_2, C-6"), 61.7 (CH_3, 3"-OCH_3).

Pteleoellagic acid derivative: $C_{33}H_{18}O_{15}$; yellow solid; m.p.297- 299 °C, UV {DMSO): λ= 366, 282, 258, 248, 224, 200 nm; IR: ν max 3452 (OH), 1742, 1722 (lactone carbonyls), 1619, 1605 cm^{-1} (aromatic rings); HRFAB MS: *m/z* = 655.0649 $[M + 1]^+$ (calcd. 654.0645) [$C_{33}H_{18}O_{15}$], HREI MS: *m/z* 343 [M^+ -$C_{17}H_{11}O_8$], 327 [M^+- $C_{16}H_7O_8$]: ^{1}H-NMR (400 MHz, C_5D_5N): δ 6.29 (s, H-3a), 7.74 (s, H-5), 8.02 (s, H-5′), 7.82 (s, H-5′′), 8.04 (s, H-5′′′), 4.20 (s, 3′-OCH_3), 4.16 (s, 3′′-OCH_3),4.15 (s, 3′′′- OCH_3), (s, 4′′′-OCH_3); ^{13}C-NMR (100 MHz, C_5D_5N): δ 112.5 (C-1), 141.0 (C-2), 139.0 (C-3), 104.4 (CH2, C-3a), 152.2 (C-4), 107.9 (CH, C-5), 116.0 (C-6), 158.1 (C-6a), 112.5 (C-1′), 141.0 (C-2′), 142.5 (C-3′), 154.4 (C-4′), 112.9 (CH, C-5′), 113.6 (C-6′), 158.1 (C-6′a), 112.5 (C-1′′), 141.2 (C-2′′), 142.5 (C-3′′), 154.4 (C-4′′), 112.9 (CH, C-5′′), 113.6 (C-6′′), 159.1 (C-6′′a), 112.5 (C-1′′′), 141.0 (C-2′′′), 141.5

(C-3´´´), 154.2 (C-4´´´), 107.9 (CH, C-5´´´), 113.6 (C-6´´´), 159.1 (C-6´´´a), 61.4 (3-OCH_3), 61.0 (3"-OCH_3), 61.1 (, 3"´-OCH_3), 56.4 (4´´´-OCH_3).

3,4-Methylenedioxy-3'-O-methyl ellagic acid: $C_{16}H_8O_8$; yellow solid: m.p. 298-300 °C; IR: νmax 3505 (OH), 1761,1719 (lactone carbonyls) and 1641, 1619 cm^{-1} (aromatic rings); HREI MS: *m/z* = 328.0228 [$C_{16}H_8O_8$], 313 [M^+-CH_3]; ^{1}H-NMR (400 MHz, C_5D_5N): δ; 6.30 (s,H-3a), 7.74 (s, H-5), 8.04 (s, H-5´), 4.16 (s, 3´-OCH_3); ^{13}C-NMR (100 MHz, C_5D_5N): δ; 112.0 (C-1), 131.1 (C-2), 138.0 (C-3), 104.6 (CH2, C-3a), 150.1 (C-4), 104.4 (CH, C-5), 114.9 (C-6), 158.6 (C-6a), 111.2 (C-1´), 143.0 (C-2´), 142.0 (C-3´), 153.5 (C-4´), 113.5 (CH, C-5´), 114.0 (C-6´), 158.0 (C-6´a).

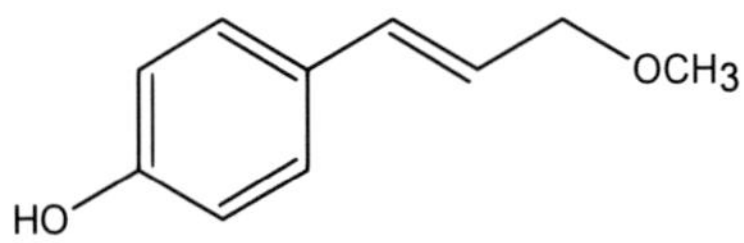

596
(E)-*p*-Coumaryl alcohol γ-O-methyl ether [4-[(1E)-3-methoxy-1-propenyl]phenol

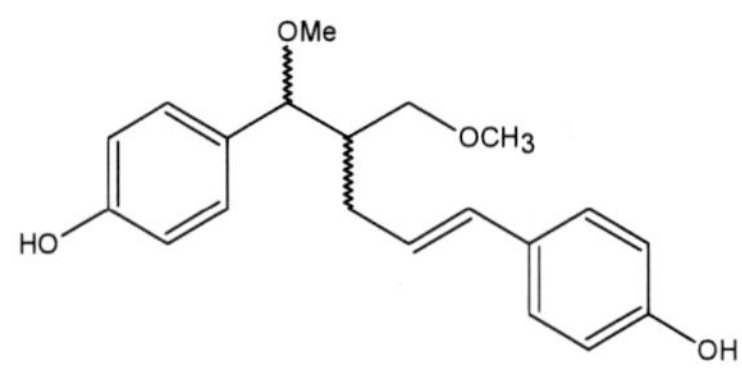

597
(4E)-1,5-Bis(4-hydroxyphenyl)-1-methoxy-2-(methoxymethyl)-4-pentene

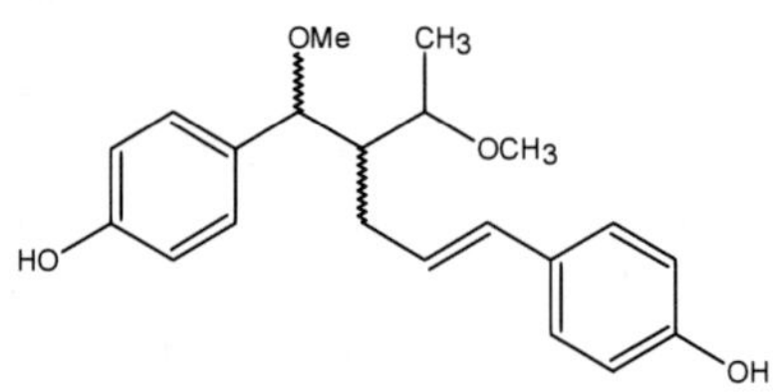

598
(4E)-1,5-Bis(4-hydroxyphenyl)-1-ethoxy-2-(methoxymethyl)-4-pentene

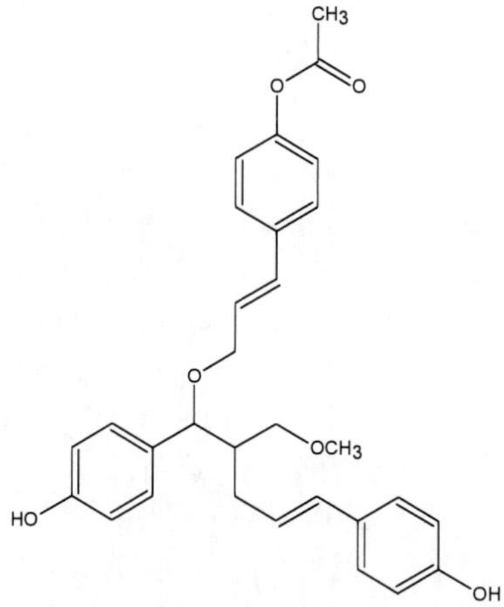

599

(4E)-1,5-Bis(4-hydroxyphenyl)-1-[(2E)-3-(4-acetoxyphenyl)-2-propenoxy]-2-(methoxymethyl)-4-pentene

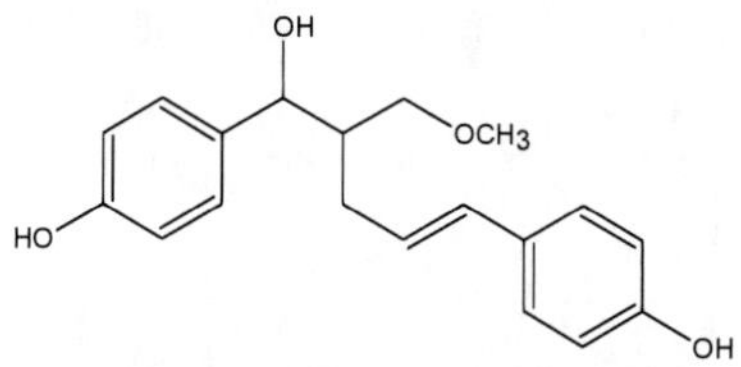

600

(4E)-1,5-Bis(4-hydroxyphenyl)-2-(methoxymethyl)-4-penten-1-o1

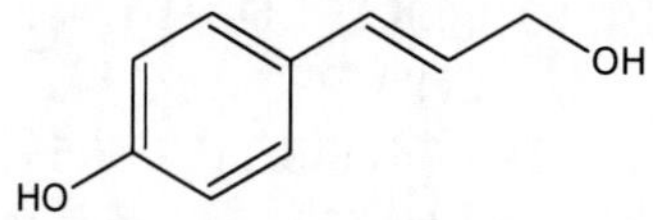

601

(E)-*p*-Coumaryl alcohol [4-[(1E)-3-hydroxy-1-propenyl]phenol

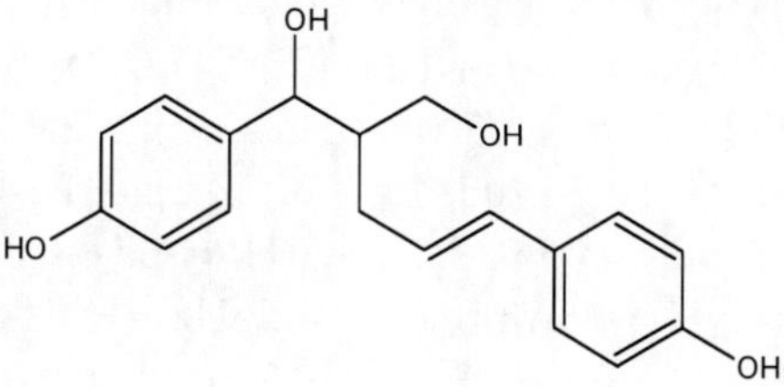

602

(4E)-1,5-Bis(4-hydroxyphenyl)-2-(hydroxymethyl)-4-penten-1-o1

These compounds were isolated from fresh rhizomes of smaller galanga (*Alpinea officinarum* Hance, Zingiberaceae). Differences in antioxidative behavior of the isolated compounds were observed in their inhibition of methyl linoleate autoxidation. The number of hydroxyl group in the phenylpropanoid molecule might influence its antioxidative activity. The most abundant antioxidants were (E)-*p*-coumaryl γ-O-methyl ether and (E)-*p*-coumaryl alcohol. In addition, were isolated five novel henylpropanoids which contain a novel β-γ-linkage in each molecule and dilignols containing β-γ- structure. Smaller galanga is used as a natural food ingredient in Vietnam and southern China. Therefore, consumption of its rhizome might be a good source of dietary antioxidants (Ly et al., 2003).

(E)-*p*-Coumaryl alcohol γ-O-methyl ether [4-[(1E)-3-methoxy-1-propenyl]phenol: White amorphous solid: UV (methanol) λmax 261 nm (4.35): HR-EIMS, *m/z*: 164.0874 (M^+), calcd for $C_{10}H_{12}O_2$, 164.0837: EIMS *m/z*: 164 (M^+, 26%), 132 ($[M–CH_2OH]^+$, 100), 103 (36), and 77 (24); APCIMS (positive), *m/z*: 133.05 ($[M\text{-}OCH_3]^+$, 100%); APCIMS (negative), *m/z*: 162.95 ($[M\text{-}H]^-$, 100%) and 147.45 ($[M\text{-}OH]^-$, 40); ^{1}HNMR (CD_3OD): δ 3.34 (s, 3H, 3-OCH_3), 4,03 (d, J = 6.2 Hz, 2H, H-3), 6,10 (td, J = 6.2, 15.8 Hz, 1H, H-2), 6.52 (d, J = 16,1 Hz, 1H, H-1). 6.74 (d, J = 8.8 Hz, 2H, H-3', H-5'), and 7.25 (d, J = 8.4 Hz, 2H, H-2', H-6'); ^{13}CNMR (CD_3OD): δ 134.2 (C-1), 123.3 (C-2), 74.4 (C-3), 58.0 (C-3-OCH_3), 129.7 (C-1′), 128.8 (C2′,6′), 116.4 (C-3′,5′), 158.5 (C-4′).

(4E)-1,5-Bis(4-hydroxyphenyl).1-methoxy-2-(methoxymethyl)-4-pentene: Pale yellow wax: $[\alpha]_D$ -0,57 (c 0.52, methanol): UV (methanol) λmax (log ε) 261 nm (4,42): HREIMS, m/z: 328,1735 (M^+), calcd for $C_{20}H_{24}O_4$, 328.1675: EIMS, m/z:328 (M^+, 24%), 296 ($[M\text{-}CH_3OH]^+$, 76), 264 ($[M\text{ -}(CH_3OH)_2]^+$, 31), 251 (100), 189 (19),164 (21),137 (94),133 (87), 121 (34),107 (82), and 77 (20): ESIMS (positive), *m/z*: 351.10 ($[M + Na]^+$, 100%); ESIMS (negative), *m/z*: 327.05 ($[M\text{ -}H]^-$, 100%); ^{1}HNMR (CD_3OD): δ 2.02 (m, 2H, H-2, H-3a), 2.09 (m, 1H, H-3b), 3.14 (s, 3H, 1-OCH_3), 3.31 (s. 3H, 6-OCH_3), 3.40 (dd, J = 4,0, 9.2 Hz. IH, H-6a), 3.53 (dd, J=5.1, 9.2Hz, 1H, H-6b), 4.10 (d, J = 7.3Hz, 1H,H-1) 5.87 (td, J = 7.3,15,7 Hz, 1H,H-4), 6.14 (d, J = 15.7 Hz, 1H, H-5), 6.70 (d, J = 8.4 Hz, 2H, H-3", H-5′′), 6.80 (d, J = 8.4 Hz. 2H, H-3', H-5'), 7.10 (d, J = 8.4 Hz, 2H, H-2', H-6'), and 7,12 (d, J = 8.4 Hz, 2H, H-2", H-6′′); ^{13}CNMR (CD_3OD): δ 84.6 (C-1), 56.8 C-1,OCH_3), 46.9 (C-2), 32.3 (C-3), 126.3 (C-4), 132.4 (C-5), 72.7 (C-6), 59.1 (C-6, OCH_3), 132.0 (C-1′), 129.9 (C2′,6′), 116.1 (C-3′,5′), 158.1 (C-4′), 130.9 (C-1′′), 128.1 (C2′′,6′′), 116.2 (C-3′′,5′′), 157.5 (C-4′′).

(4E)-1,5-Bis(4-hydroxyphenyl)-1-ethoxy-2-(methoxymethyl).4-pentene: Yellow wax; $[\alpha]_D$ +2.9 (c 0.17, methanol); UV {methanol) λmax (log ε): 261 nm (4.37); HREIMS, *m/z*: 342.1797 (M^+), calcd for$C_{21}H_{26}O_4$, 342.1831; EIMS, *m/z*: 342 (M^+, 13%), 310 ($[M\text{ -}CH,OH]^+$, 14), 296 ($[M–C_2H_5OH]^+$, (33), 264 (27), 251 (100), 189 (9), 164 (21), 137 (18), 133 (70), 123 (81), 107 (53), and 77 (24); ESIMS (positive), *m/z*: 365.10 ([M^++ Na]+, 100%); ESIMS (negative), *m/z*: 340.95 ($[M\text{-}H]^-$, 100%); ^{1}HNMR (CD_3OD): δ 1.13 (t, J = 7.0, 3H, 1-OCH_2CH_3), 1.98 (m, 2H, H-2, H-3a), 2.07 (m, 1H, H.3b); 3.25 (qd, J = 7.0, 7.3 Hz, 1H, 1-OCH_2CH_3), 3.31 (s, 3H, 6,OCH_3), 3.34 (qd, J = 7,0, 7.3 Hz, 1H, 1, OCH_2CH_3), 3.42 (dd, J = 4.0, 9.5 Hz, 1H, H-6a), 3.55 (dd, J = 5.5, 9.5 Hz, 1H, H-6b), 4,21 (d, J = 7.0 Hz, 1H, H-1), 5.87 (td. J = 7.3, 15.7 Hz. 1H, H-4). 6.14 (d, J = 15.7 Hz, 1H. H-5), 6.68 (d, J = 8.8 Hz. 2H, H-3′′, H-5′′). 6.78 (d, J = 8.8 Hz, 2H, H-3', H-5'). 7.10 (d, J = 8.4 Hz, 2H, H-2', H-6'), and 7.12 (d, J = 8.8 Hz, 2H, H-2′′, H-6′′); ^{13}CNMR (CD_3OD): δ 82.6 (C-1), 65.0 (C-1, OCH_2CH_3), 15.6 (C-1, OCH_2CH_3), 47.0 (C-2), 32.4 (C-3), 126.4 (C-4), 132.4 (C-5), 72.7 (C-6), 59.1 (C-6, OCH_3), 132.8 (C-1′), 129.8 (C2′,6′), 116.0 (C-3′,5′), 158.0 (C-4′), 130.9 (C-1′′), 128.1(C2′′,6′′), 116.2 (C-3′′,5′′), 157.6 (C-4′′).

(4E) - 1,5- Bis (4-hydroxyphenyl)-1-[(2E)-3-(4-acetoxyphenyl)-2-propenoxy]- 2 -(methoxy methyl)-4-pentene: colorless wax; $[\alpha]_D$ +0.58 (c 0.35, methanol); UV (methanol) λmax (log ε): 257 nm (4.67); HREIMS, *m/z*: 488.1809 (M^+). calcd for $C_{30}H_{32}O_6$, 488.2199; EIMS, *m/z*: 488 (M^+, 4%), 338 (6%), 323 (50%), 306 (14), 293 (11), 264 (38), 251 (28), 189 (9), 175 (14), 164 (23), 133 (100),107 (99), and 77 (22); ESIMS (positive), *m/z:* 511.10 $(M + Na]^+$, 100%); ESIMS (negative), *m/z*: 487.05 ($[M-H]^-$, 10%) and 445.00 ($[M-CH_3CO]^-$, 100); ^{1}HNMR (CD_3OD): δ 1.98 (m. 1/5H, H-2), 2.05 (m, 8/5H, H-2, H-3a), 2.11 (m, 4/5H. H-3b), 2.26 (s, 3H, 4'''-$OCOCH_3$), 2.39 (m. 1/5H. H-3a), 2.51 (m, 1/5H, H-3b), 3.03 (dd, J = 5.6. 9.3 Hz, 1/5H, H-6a), 3.18 (s, 3/5 H, 6-0CH_3), 3.26 (dd, J = 4.8, 9.2 Hz, 1/5H, H-6b), 3.31 (s, 12/5 H, 6-OCH_3), 3.44 (dd, J = 4.0, 9.5 Hz, 4/5H, H-6a), 3.59 (dd, J = 5.5, 9.5 Hz, 415H, H-6b). 3.868 and 3.871 (dd. J = 6.6, 12.8 Hz, 1H, H-9'''a), 4.019 and 4.022 (dd, J = 5.5, 12.8 Hz, 1H. H-9'''b), 4.34 (d, J = 7.7 Hz, 1/5H, H-1), 4.37 (d, J = 6.6 Hz. 415H. H-1), 5.90 (td, J = 7.3, 15.7 Hz. 4/5H, H-4), 6.05 (m. 115H. H-4), 6.16 (d, J = 15.7 Hz. 4/5H, H-5), 6.24 (ddd, J = 5.5, 6.6, 16.1 Hz. 1H, H-8´´´). 6.30 (d, J = 16.5 Hz. 1/5H, H-5), 6.51 (d, J = 15.7 Hz, 1H, H-7'''). 6.67 (d, J = 8.8 Hz, 2H, H-3´´, H-5´´), 6.81 (d, J = 8.8 Hz, 2H,. H-3', H-5'), 7.01 (d, J = 8.4 Hz, 2H, H-3''', H-5'''), 7.09 (d, J = 8.4 Hz. 2H, H-2'', H-6´´), 7.14 (d, J = 8.4 Hz, 2H, H-2', H-6'), 7.36 (d, J = 8.4 Hz, 2H, H-2''', H-6'''); ^{13}CNMR (CD_3OD): δ 81.9 (C-1), 47.4 (C-2), 32.5 (C-3), 126.4 (C-4), 132.0 (C-5), 72.7 (C-6), 59.2 (C-6, OCH_3), 132.3 (C-1´), 128.4 (C2´,6´), 116.2 (C-3´,5´), 158.2 (C-4´), 130.9 (C-1´´), 129.9 (C2´´,6´´), 116.1(C3´´,5´´), 157.6 (C-4´´), 136.1 (C-1´´´), 128.2 (C2´´´,6´´´), 122.9 (C3´´´,5´´´), 151.6 (C-4´´´), 171.2 (C -4´´´-$OCOCH_3$), 21.0 (C -4´´´- $OCOCH_3$), 132.5 (C-7´´´), 127.9 (C-8´´´), 70.0 (C-9´´´).

(4E)-1,5-Bis(4-hydroxyphenyl)-2-(methoxymethyl)-4-penten-1-o1: Colorless wax; $[\alpha]_D$ +4.3 (c 0.12. methanol); UV (methanol) λmax (log ε): 261 nm (4.44): HR-EIMS, ml;: 314.1516 (M+). calcd for $C_{19}H_{22}O_4$. 314.1518; EIMS. *m/z*: 314 (M^+, 21%), 296 ($[M - H_2O]^+$, 20), 282 ($[M - CH_3OH]^+$, 34), 264 ($[M -(H_2O + CH_3OH)]^+$, 45), 251 (74), 199 (25), 180 (36),164 (33),160 (35),145 (39).133 (l00), 123 (61),12] (92),107 (97), and 77 (36); ESIMS (positive), *m/z*: 136.85 ($[M + Na]^+$, 100%); ESIMS (negative), *m/z*: 313.00 ($[M-H]^-$, 100%); ^{1}HNMR (CD_3OD): δ 1.92 (m, 1/5H, H-2), 1.99 (m, 8/5H. H-2, H-3a), 2.13 (m, 4/5H, H-3b), 2.25 (m, 1/5H. H-3a), 2.41 (m, 1/5H, H-3b), 3.10 (dd, J = 4.8, 9.2 Hz, 1/5H, H-6a), 3.23 (s, 3/5H, 6-0CH_3), 3.33 (s, 12/5H, 6-OCH_3), 3.34 (dd, J = 5.6, 9.6 Hz. 1/5H. H-6b). 3,43 (dd, J = 4.8, 9.5 Hz. 4/5H, H-6a), 3.56 (dd, J = 4.8, 9.5 Hz. 4/5H, H-6b), 4.60 (d, J = 6.6 Hz, 4/5H, H-1). 4.66 (d, J = 6.6 Hz, 1/5H. H-1), 5.90 (ddd, J = 6.6, 8.1, 15.7 Hz, 4/5H. H-4). 5.99 (ddd, J = 6.8, 8.8, 16.0 Hz, 1/5H. H-4), 6.17 (d. J = 15.7 Hz, 4/5H, H-5), 6.28 (d, J= 15.7 Hz, 1/5H, H-5), 6.68 (d, J = 8.4 Hz, 2H, H-3'', H-5´´), 6.77 (d, J = 8.4 Hz, 2H, H-3', H-5'), 7.13 (d. J = 8.4 Hz, 2H, H2'', H-6´´), and 7.16 (d, J = 8.8 Hz, 2H, H-2', H-6'); ^{13}CNMR (CD_3OD): δ 75.4 (C-1), 47.3 (C-2), 32.5 (C-3), 126.3 (C-4), 132.5 (-5), 73.5 (C-6), 59.2 (C-6, OCH_3), 135.4 (C-1´), 129.1 (C2´,6´), 116.2 (C-3´,5´), 157.8 (C-4´), 130.8 (C-1´´), 128.1 (C2´´,6´´), 115.9 (C3´´,5´´), 157.6 (C-4´´).

(E)-*p*-Coumaryl alcohol [4-[(1E)-3-hydroxy-1-propenyl]phenol: White amorphous solid; UV (methanol) λmax (log ε 260 nm (4.32); HR-EIMS, m/z;: 150.0698 (M^+), calcd for $C_9H_{10}O_2$, 150.0681; EIMS.*m/z*: 150 (M^+, 50%), 132 ($[M-H_2O]^+$, 19}, 107 (97), 103. (23), 94 (43), and 77 (24); APCIMS (negative), *m/z*: 149.00 ($[M-H]^-$, 100%); ^{1}HNMR (CD_3OD): δ 4.18 (dd, J = 1.1, 5.9 Hz. 2H, H-3), 6.16 (td, J = 5.9, 15.8 Hz. 1H, H-2), 6.50 (d, J = 15.8 Hz, 1H, H-1). 6.73 (d, J = 8.4 Hz, 2H, H-3', H-5'), and 7.24 (d, J = 8.8 Hz, 2H, H-2', H-6'); ^{13}CNMR (CD_3OD): δ 131.9 (C-1), 126.6 (C-2), 64.0 C-3), 130.0 (C-1´), 128.7 (C2´,6´), 116.3 (C3´,5´), 158.2 (C-4´).

(4E)-1,5-Bis(4-hydroxyphenyl)-2-(hydroxymethyl)-4-penten-1-o1: Colorless wax: $[\alpha]_D$ -13.2 (c 0.19, methanol): UV (methanol) λmax (log ε): 261 nm (4.37); HR-EIMS, *m/z*: 300.1464 (M^+). calcd for $C_{18}H_{20}O_4$, 300.1362; ElMS. *m/z* :300 (M^+, 5%), 282 ($[M - H_2O]^+$, 15), 264 ($[M -(H_2O)_2]^+$,13), 252 (100),199 (33), 166 (11), 158 (43), 145 (88), 133 (40), 123 (18), 107 (83), and 77 (20); ESIMS (positjve), *m/z*: 322.95 ($[M + Na]^+$, 100%); ESIMS (negative), *m/z*: 298.90 ($[M\text{-}H]^-$, 100%); ^{1}HNMR (CD_3OD): δ 1.91 (m, 1H, H-2), 2.00 (ddd, J = 7.0, 8.1, 15.0 Hz, 1H, H-3a), 2.12 (ddd, J = 7.0, 8.4, 14.0 Hz. 1H. H-3b), 3.65 (dd, J = 5.5, 11.0 Hz, 1H, H-6a), 3.78 (dd, J = 4.8, 11.0 Hz, 1H. H-6b), 4.62 (d, J = 7.3 Hz, 1H, H-1), 5.91 (ddd, J = 7,0, 7.7 15.7 Hz, 1H, H-4), 6.20 (d, J = 15.7 Hz, 1H, H-5), 6.68 (d, J = 8.8 Hz. 2H, H-3", H-5´´), 6.77 (d, J = 8.8 Hz, 2H. H-3', H-5'), 7.13 (d, J = 8.8 Hz, 2H, H-2", H-6), 7.19 (d, J = 8.4 Hz, 2H, H-2', H-6'); ^{13}CNMR (CD_3OD): δ 76.4 (C-1), 49.0 (C-2), 32.4 (C-3), 126.3 C-4), 132.5 (C-5), 63.2 (C-6), 135.7 (C-1´), 129.1 (C2´,6´), 116.0 (C3´,5´), 157.8 (C-4´), 130.9 (C-1´´), 128.2 (C-2´´-6´´), 116.2 (C-5´´-5´´), 157.6 (C-4´´).

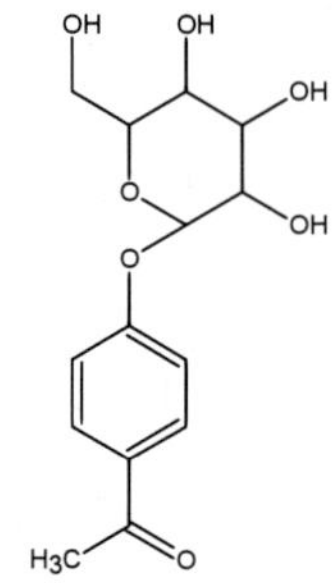

603
Picein

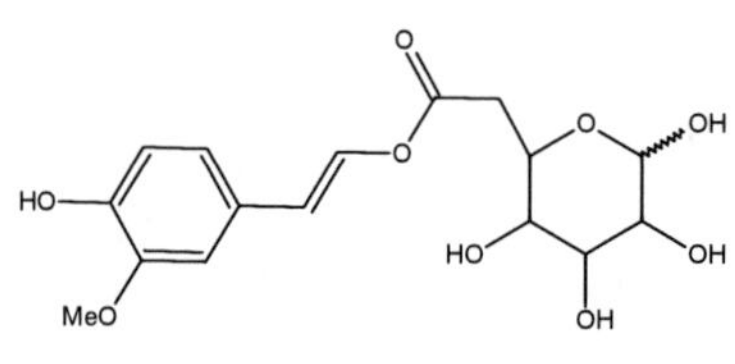

604
6-O-(E)-Feruloyl-(α and β)-glucopyranoside

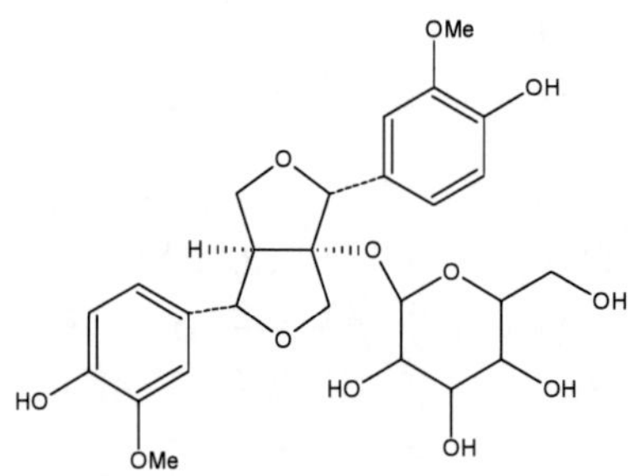

605
(+)-l-Hydroxypinoresinol-l-β-D-glucoside

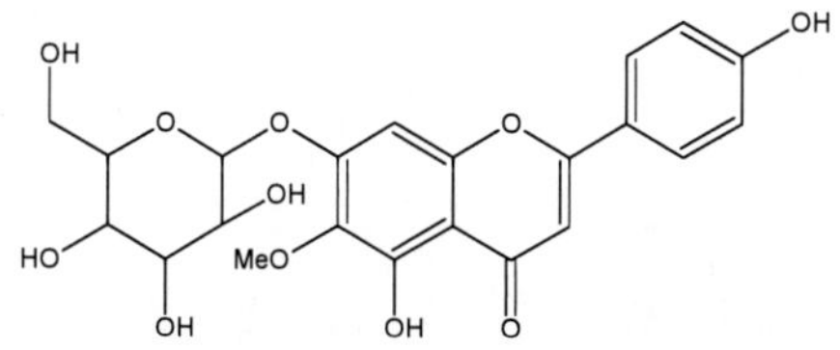

606
Homoplantaginin

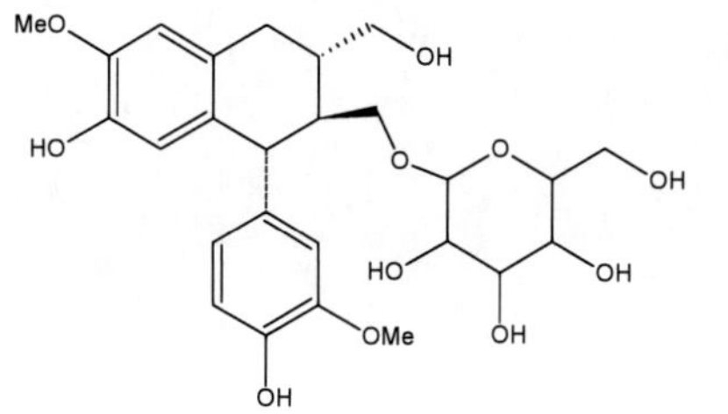

607
(-)-Isolariciresinol 3α-O-β-D-glucopyranoside

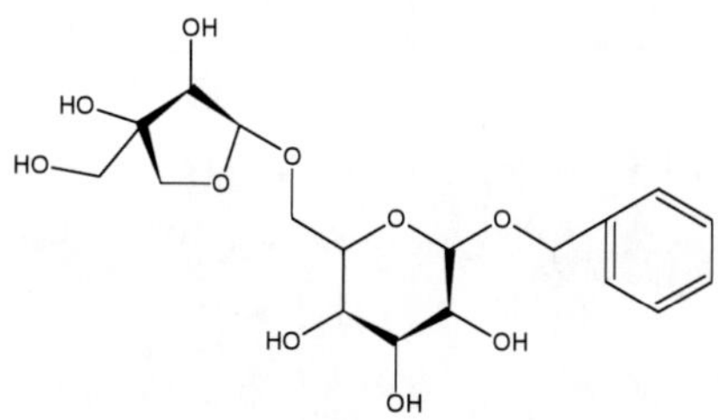

608
Icariside F_2

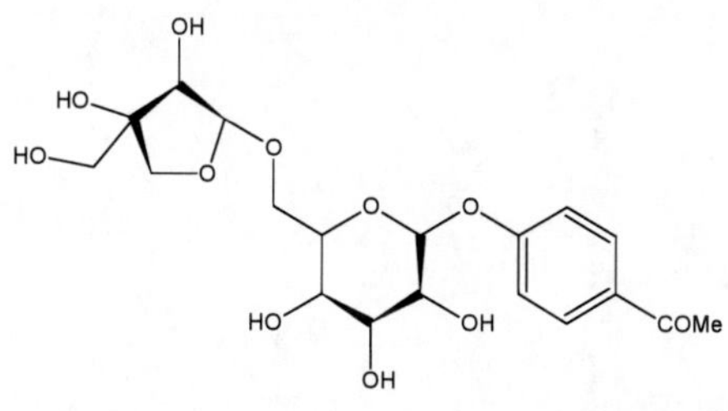

609
4-Hydroxyacetophenone-4-O-β-D-apiofuranosyl-(1→6)-O-β-D-glucopyranoside

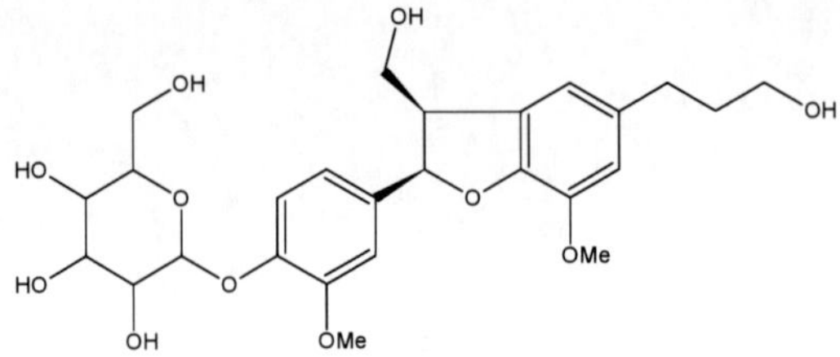

610

2,3-Dihydro-2-(4'-hydroxy-3'-methoxyphenyl)- 3-(hydroxymethyl)-7-methoxy-5-benzofuranpropanol 4'-O-β-glucopyranoside

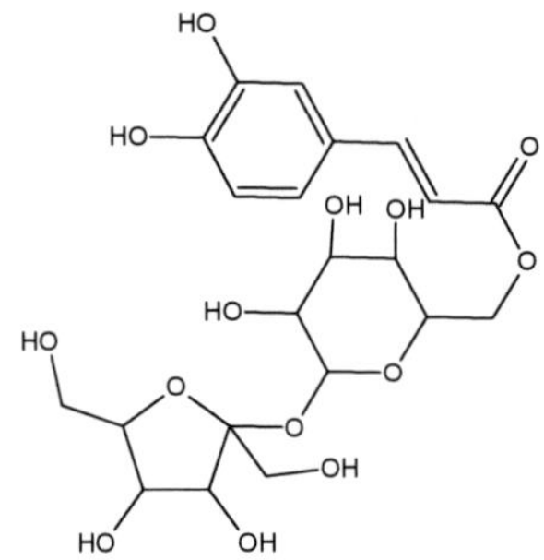

611

6-O-Caffeoyl-β-D-fructofuranosyl-(2→1)- α-D-glucopyranoside

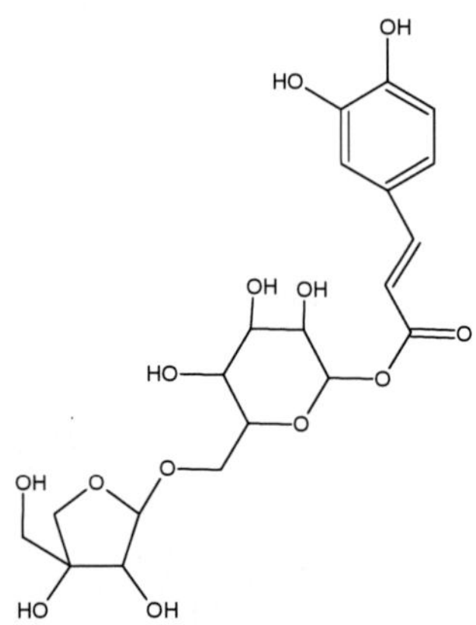

612

1-O-Caffeoyl- β-D-apiofuranosyl, (1→6)-β-D-glucopyranoside

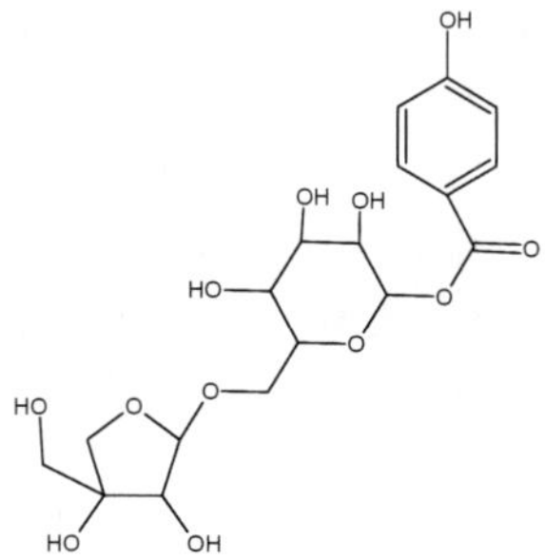

613
1-O-*p*-Hydroxybenzoyl-β-D-apiofuranosyl-(1→6)-β-D-glucopyranoside

Found in Sage (*Salvia officinalis*, Labiatae) these phenolic compounds are widely distributed in nature. It is suggested that their antioxidative activity is related to their conjugated rings and hydroxyl groups, so it is not a surprise that most of phenolic compounds from sage showed some antioxidative activity. Among these compounds, rosmarinic acid and luteolin-7-O-β-glucopyranoside are the most active (Wang et al., 1998). Compounds 6-O-caffeoyl-β-D-fructofuranosyl-(2→1)-α-D-glucopyranoside and 1-O-caffeoyl-β-D-apiofuranosyl, (1→6)-β-D-glucopyranoside were found to be moderately active as antioxidants in the DPPH test and metmyoglobin test (Wang et al., 1999a).

6-O-Caffeoyl-β-D-fructofuranosyl-(2→1)-α-D-glucopyranoside: Colorless crystals (methanol); mp 210-212 ˚C; $[\alpha]_D$ +27.17 °C (c 0.13, MeOH); IR (film) vmax 3363,2927, 1688, 1643, 1616, 1532, 1371, 1066, 992 cm^{-1}; FABMS (positive-ion model) *m/z* 527 $[M + Na]^+$; ^{1}HNMR (400 MHz, DMSO-d_6) δ 7.48 (1H, d, J = 16.4 Hz, H-7), 7.06 (lH, s, H-2), 7.03 (lH, d, J = 8.0 Hz, H-6), 6,76 (lH, d, J =8,0 Hz, H-5), 6.31 (11H, d, J= 16.4 Hz, H-8), 5.21 (lH, d, J = 3.6 Hz, H-1'), 4.34 (lH, d, J = 12.0 Hz, H-6'), 4.12 (lH, dd, J = 12.0, 6.4 Hz, H-6'), 3.99 (lH, t, J = 8.8 Hz, H-5'), 3.89 (lH, t, J = 7.6 Hz, H-3"), 3.79 (lH, m, H-4"), 3.62 (lH, m, H-5"), 3.60 (2H, m, H-6"), 3.50 (lH, m. H-3'), 3.40 (2H, m, H-1"), 3.25 (lH, dd, J = 8.8,4.4 Hz, H-2'}, 3.12 (lH, m, H-4').

1 - O - Caffeoyl - β - D - apiofuranosyl, (1→6) - β - D - glucopyranoside: Colorless crystals (methanol); mp 69-71 °C; $[\alpha]_D$ -21.81 °C (c 0.15, MeOH); IR (film) vmax 3402, 2933, 1709, 1612, 1522, 1269, 1066 cm^{-1}; FABMS (positive-ion model) *m/z* 497 $[M + Na]^+$; ^{1}HNMR (200 MHz CD_3OD): δ 7.64 (lH, d, J =16.0 Hz, H-7}, 7.05 (lH, d, J = 1.8 Hz, H-2}, 6.97 (lH, dd, J=8.2, 1.8 Hz, H-6), 6.76 (lH, d, J=8.2 Hz, H-5), 6.29 (lH, d, J = 16.0 Hz, H-8), 5.53 (lH, d, J = 7.4 Hz, H-1'), 4.96 (lH, d, J = 2.4 Hz, H-1"), 3.94 (lH, d, J = 12.0 Hz, H-6'), 3.92 (lH, d, J = 10.0 Hz, H-4"), 3.89 (lH,d, J = 2.4 Hz, H-2"), 3.73 (lH, d, J = 10 Hz, H-4"), 3.62 (lH, m. H-6'), 3.54 (2H, s, H-5"), 3.47 (lH, m, H-3'), 3.44 (lH, m, H-2'), 3.34 (lH, m, H-5'), 3.29 1H, m, H-4').

1-O-*p*-Hydroxybenzoyl-β-D-apiofuranosyl-(1→6)-β-D-glucopyranoside: Colorless oil; $[\alpha]_D$- 32.75 °C (c 0.04, (c 0.04, MeOH); FABMS (positive-ion model} *m/z* 455 $[M + Na]^+$; ^{1}HNMR (200 MHz CD_3OD): δ 7.96 (2H, d, J = 8.6 Hz, H-2), H-6), 6.85 (2H, d, J = 8.6 Hz, H-3 and H-5), 5.66 (lH, d, J =7.4 Hz, H-1'); 4.97 (lH, d, J = 2.6 Hz, H-")", 3.96 (lH, d, J = 12.0 Hz, H-6), 3.92 (lH, d, J =10.0 Hz, H-4´´), 3.90 (lH, d, J = = 2.2 Hz, H-2"), 3.74 (1H, d, J = 10 Hz, H-4"), 3.60 (lH, m, H-6´), 3.57 (2H, s, H-5"), 3.47 (lH, m, H-3'), 3.44 (lH, m, H-2'), 3.33 (lH, m, H-5'), 3.30 (lH, m, H-4').

Picein, or 4-O-β-D-glucopyranosylacetophenone: Amorphous powder; DCI-MS *m/z*: 316 $[M + NH_4]^-$; ^{1}HNMR (CD_3OD): δ 7.98 (2H, d, J = 8.0 Hz, H-4 and H-8), 7.16 (2H, d, J = 8.0 Hz, H-5 and H-7), 5.04 (lH, d, J = 7.2 Hz, H-1'), 3.90 (lH, m, H-6'), 3.71 (lH, dd, J = 12.0,5.6 Hz, H-6'), 3.31-3.52 (4H, m, H-2', H-3', H-4' and H-5'), 2.56 (3H, s, H-1); ^{13}CNMR (CD_3OD): δ 26.9 (q, C-1), 200.2 (s, C-2), 133.0 (s, C-3), 132.1 (d, C-4), 117.6 (d, C-5), 163.4(s, C-6), 117.6 (d, C-7), 132.1(d, C-8), 101.8 (d, C-1′), 75.1 (d, C-2′), 78.2 (d, C-3′), 71.3 (d, C-4′), 78.7 (d, C-5′), 62.8 (t, C-6′).

6-O-(E)-Feruloyl-(α and β)-glucopyranoside: Amorphous powder; DCI-MS *m/z*: $[M + NH_4]^-$; ^{1}HNMR (DMSO-d_6, 200 MHz): δ 7.57 (d, J = 16.0 Hz, H-7), 7.36 (br s, H-2), 7.15 (br d, J = 8.1 Hz, H-6), 6.81 (d, J = 8.1 Hz, H-5β), 6.70 (d, J = 6.4 Hz, H-5α), 6.51 (lH, d, J = 16.0 Hz, H-8β), 6.50 (d, J = 16.0 Hz, H-8α), 4.20-5.20 (m), 3.84 (s, OCH_3), 3.00-3,80 (m); ^{13}C NMR (DMSO, 75 MHz): δ 166.8 (s, C-9), 149.5 (s, C-3), 148.0 (s, C-4), 145.3 (d, C-7), 125.5 (s, C-1), 123.4 (d, C-6), 115.5 (d, C-5), 114.3 (d, C-8), 111.1 (d, C-2), 97.0 (d,), 76.4 (d), 74.7 (d), 73.6 (d), 70.2 (d), 63.9 (t)] for β-Glc, 92.3 (d), 72.9 (d), 72.2 (d), 70.6 (d), 69.3 (d), 63.9 (t) for α-Glc and 55.7 (q, OCH_3).

(+)-l-Hydroxypinoresinol-l-β-D-glucoside: Amorphous powder; $[\alpha]_D$ -18.2 °C (c 0.11, CD_3OD); DCI-MS *m/z*: 554 $[M + NH_4]^-$; ^{1}HNMR (300 MHz, CD_3OD): δ 6.60-7.11 (6H, m, arom. H), 3.89 (3H, s, OCH_3), 3.85 (3H, s, OCH_3), 3.60-4.90 (m), 2.90-3.55 (m); ^{13}C NMR (77 MHZ, CD_3OD): δ 149.4 (s, C-3"), 148.5 (s, C-3'), 147.5 (s, C-4"), 147.4 (s, C-4'), 133.2 (s, C-1 "), 128.8 (s, C-1 '), 122.4 (d, C-6'), 120.0 (d, C-6"), 116.4 (d, C-5"), 115.4 (d, C-5'), 114,2 (d, C-2'), 111.0 (d, C-2"), 100.1 (d, C-Glc1), 99.2 (s, C-1), 89.9 (d, C-2), 87.0 (d, C-6), 78.4 (d, C-Glc3), 78.0 (d, C-Glc5), 74.9 (d, C-Glc2), 73.5 (t, C-8), 72.2 (t, C-4), 71.3 (d, C-Glc4), 62.6 (t, C-Glc6), 60.2 (d, C-5).

Homoplantaginin. Yellow powder; DCI -MS *m/z*: 463 $[M+ 1]^+$; ^{1}HNMR (DMSO-d_6, 300 MHz): δ 7.94 (2H, d, J = 8.8 Hz, H-2'and H-6'), 7.00 (lH, s, H-8), 6.95 (2H, d, J = 8.8 Hz, H-3'and H-5'), 6.94 (1H, s, H-3), 5.10 (1H, d, J = 7.4 Hz, H-Glc1), 3.10-3.80 (m), 3.75 (3H, s); ^{13}C NMR (DMSO-d_6, 75 MHz): δ 182.1 (s, C-4), 164.2 (s, C-2), 161.3 (s, C-4'), 156.3 (s, C-7), 152.3 (s, C-9), 152.0 (s, C-5), 132.5 (s, C-6), 128.4 (d, C-2' and C-6'), 121.0 (s, C-1'), 115-9 (d, C-3' and C-5'), 105.6 (s, C-10), 102.6 (d, C-3), 100.2 (d, Glc-1), 94.3 (d, C-8), 77.2 (d, Glc-3), 76.7 (d, Glc-5), 73.1 (d, Glc-2), 69.6 (d, Glc-4), 60.6 (t, Glc-6), and 60.2 (q, OCH_3).

(-)-Isolariciresinol 3α-O-β-D-glucopyranoside: Amorphous powder; $[\alpha]_D$ -31.1°C (c 0.22, CD_3OD); DCI-MS *m/z*: 540 $[M + NH_4]^-$; ^{1}HNMR (300 MHz, CD_3OD): δ 6.74 (lH, d, J = 8.0 Hz, H-5'), 6.69 (lH, d, J = 2.0 Hz, H-2'), 6.65 (lH, s, H-8), 6.64 (lH, dd, J = 8.0, 2.0 Hz, H-6'), 6.19 (H, s, H-5), 4.04 (1H, d, J = 7.6 Hz, H-1"), 3.50-3.90 (m), 3.81 (3H, s, OCH_3), 3.78 (3H, OCH_3), 2.70-3.30 (m), 1.97 (2H, m, H-2, H-3); ^{13}C NMR (DMSO-d_6, 75 MHz): δ 148.9 (s, C-3'), 147.2 (s, C-7), 145.9 (s, C-6), 145.2 (s, C-4'), 138.7 (s, C-1 '), 133.7 (s, C-10), 129.2 (s, C-9), 123.5 (d, C-6'), 117.3 (d, C-5), 115.9 (d, C-5'), 113.9 (d, C-2'), 112.3 (d, C-8), 103.8 (d, C-1 "), 78.2 (d, C-5"), 77.8 (d, C-3"), 75.0 (d, C-2"), 71.4 (d, C-4"), 70.7 (t, C-3a), 65.5 (t, C-2a), 62.8 (t, H-6"), 56.4 (q, OCH_3), 56.5 (q, OCH_3), 48.3 (d, C-4), 45.3 (d, C-3), 41.1 (d, C-2), and 33.6 (t, C-1).

Luteolin-7-O-β-glucopyranoside: Yellow powder; FAB-MS m/z: $[M + 1]^+$,449; ^{1}H NMR (DMSO, 200 MHz): δ 7.49 (lH, d, J = 8.0 Hz, H-6'), 7.44 (lH, s, H-2'), 6.92 (lH, d, J = 8.0, H-5'), 6.80 (lH, s, H-3), 6.76 (lH, br s, H-8), 6.46 (lH, br s, H-6), 5.10 (lH, d, J = 6.6 Hz, Glc-1), 3.10-3.80 (m); ^{13}C NMR (DMSO, 50 MHz): δ 182.2 (s, C-4), 164.8 (s, C-7), 163.2 (s, C-2),

161.4 (s, C-5), 157.2 (s, C-9), 150.2 (s, C-4'), 146.0 (s, C-3'), 121.6 (s, C-1 '), 119.4 (d, C-6'), 116.2 (d, C-5'), 113.8 (d, C-2'), 105.6 (s, C-10), 103.4 (d, C-3), 100.1 (d, Glc-1), 99.8 (d, C-6), 95.0 (d, C-8), 77.4 (d, Glc-5), 76.7 (d, Glc-3), 73.4 (d, Glc-2), 69.8 (d, Glc-4), and 60.9 (t, Glc-6).

4 – Hydroxyacetophenone – 4 – O - β - D – apiofuranosyl - (1→6)-O-β-D-glucopyranoside: Amorphous powder; FAB-MS *m/z*: 453 $[M + Na]^+$; FT-IR (KBr): 3411 (OH), 1669 (CO), 1601,1510, 1250 cm^{-1}; ^1H NMR (CD_3OD, 200 MHz): δ 8.01 (2H, dd, J = 8.2, 1.6 Hz, H-4, H-8), 7.15 (2H, dd, J = 8.2, 1.6 Hz), 5.03 (lH, d, J = 7.6 Hz, H-1′), 4.98 (lH, d, J = 2.6 Hz, H-1"), 3.31-4.10 (11H, m), 2.57 (3H, s); 13C NMR δ 26.8 (q, C-1), 199.8 (s, C-2), 132.9 (s, C-3), 132.0 (d, C-4), 117.6 (d, C-5), 163.3 (s, C-6), 117.6(d, C-7), 132.0 (d, C-8), 101.9 (d, C-1′), 75.1 (d, C-2′), 78.3 (d, C-3′), 71.8 (d, C-4′), 77.5 (d, C-5′), 69.1 (t, C-6′), 111.3 (d, C-1′′), 78.2 (d, C-2′′), 80.8 (s, C-3′′), 75.3 (t, C-4′′), 65.7 (t, C-5′′).

Icariside F2 : Amorphous powder; FAB-MS m/z: 425 $[M+ Na]^+$; ^1H NMR (CD_3OD, 200 MHz): δ 7.30-7.70 (5H, m, H-2, 3, 4, 5, 6), 5.06 (lH, d, J = 2.6 Hz, H-1"), 4.93 (lH, d, J = 12.0 Hz, H-7), 4.66 (lH, d, J = 12.0 Hz, H-7), 4.35 (lH, d, J = 7.4 Hz, H-1'), 3.31-4.10 (m, 11H); 13C NMR δ 139.2 (s, C-1), 129.6 (d, C-2), 12.9.6 (d, C-3), 129.0(d, C-4), 129.6 (d, C-5), 129.6 (d, C-6), 72.1(t, C-7), 103.5 (d, C1′), 75.3 (d, C-2′), 78.3 (d, C-3′), 72.1 (d, C-4′), 77.3 (d, C-5′), 69.0 (t, C-6′), 111.3 (d, C-1′′), 78.3 (d, C-2′′), 80.9 (s, C-3′′), 75.4 (t, C-4′′), 65.9 (t, C-5′′).

2,3 - Dihydro – 2 - (4'-hydroxy-3'-methoxyphenyl) -3-(hydroxymethyl)-7-methoxy-5-benzo furanpropanol 4'-O-β-glucopyranoside: Amorphous powder; $[\alpha]_D$ -33.4 °C (c 0.25, CD_3OD); FAB-MS *m/z*: 545 $[M + Na]^+$; ^1H NMR (CD_3OD, 200 MHz): δ 7.14 (lH, d, J = 8.4 Hz, H-5'), 7.04 (lH, d, J = 2.0, H-2'), 6.91(lH, dd, J = 8.4, 2.0 Hz, H-6'), 6.72 (2H, s, H-2 and H-6), 5.56 (lH, d, J = 5.8 Hz, H- 7'), 3.83 (3H, s, OCH_3), 3.81 (3H, OCH_3), 3.2-3.9 (llH, m), 2.63 (2H, m H-7), 1.78 (2H, H-8); ^{13}C NMR (CD_3OD, 50 MHz): δ 151.2 (s, C-4'), 147.9 (s, C-3'), 147.7 (s, C-4), 145.5 (s, C-3), 138.6 (s, C-1'), 137.4 (s, C-5), 129.9 (s, C-1), 119.7 (d, C-6'), 118.2 (d, C-2), 118.2 (d, C-5'), 114.4 (d, C-6), 111.4 (d, C-2'), 103.0 (d, Glc-1), 88.8 (d, C-7'), 78.5.(d, Glc-3), 78.1 (d, Glc-5), 75.1(d, Glc-2), 71.6 (d, Glc-4), 65.3 (t, C-9'), 62.8 (t, Glc-6), 62.6 (t, C-9), 2 x 57.0 (q, OCH_3), 55.9 (d, C-8'), 36.1 (t, C-8), and 33.2 (t, C-7).

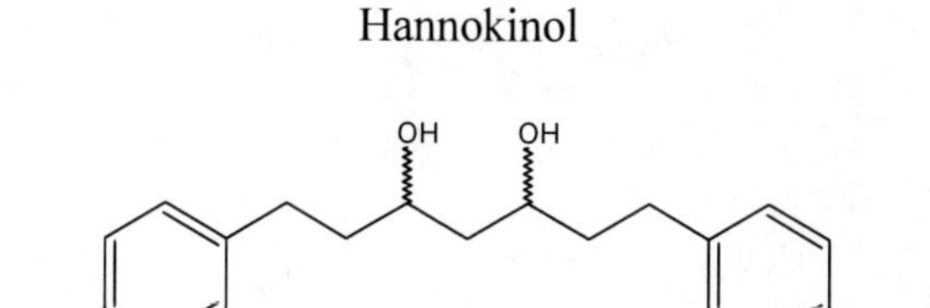

614
Hannokinol

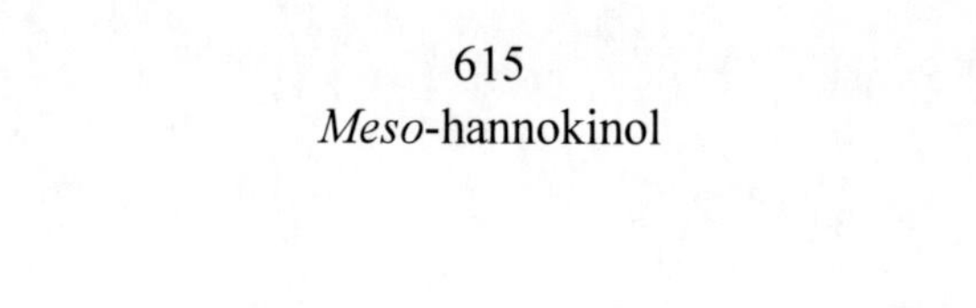

615
Meso-hannokinol

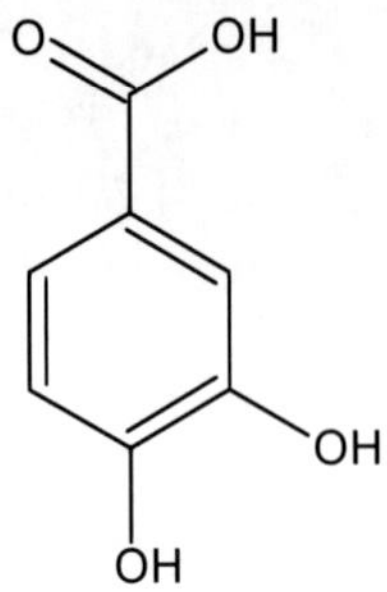

616
Protocatechuic acid

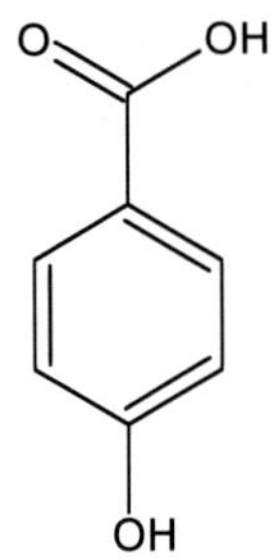

617
p-Hydroxybenzoic acid

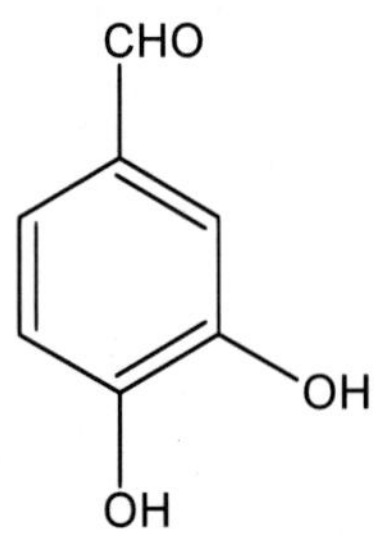

618
Protocatechualdehyde

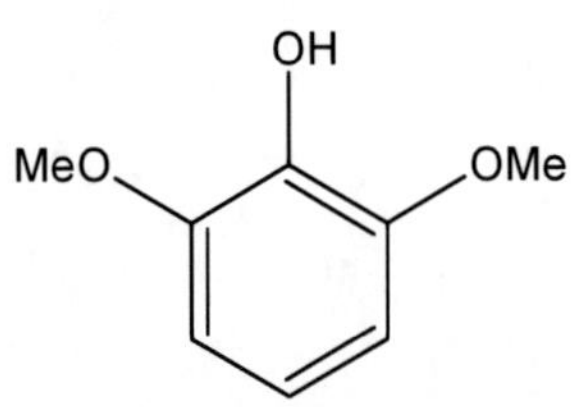

619
2,6-Dimethoxyphenol

Found in the fruits of Amomum tsao-ko. All compounds with the 3',4'-dihydroxy configuration possess marked antioxidant activity. The presence of a second hydroxyl group in the *ortho* or *para* position is known to increase antioxidant activity due to additional resonance stabilization and *o*-quinone or *p*-quinone formation. The presence of phenolic hydroxyl groups increases the antioxidant activity of phenolic acids, while methoxylation of the hydroxyl groups causes a decrease in activity. For protocatechualdehyde and protocatechuic acid, a probable explanation for the difference in their activity is the fact that carboxylic acids are capable of intermolecular hydrogen bonding while aldehydes are not. Because of this bonding, the electron-with drawing capacity of protocatechuic acid is decreased. This makes protocatechuic acid less active toward the DPPH radical. Protocatechualdehyde showed more than 18-fold than antioxidant activity of protocathechuic acid (Martin et al., 2000).

(+)-Hannokinol: White powder. $[\alpha]_D$ + 11.4° (MeOH, c = 0.38). ^{1}H NMR (acetone-d_6): δ 2.55 (2H, ddd, J = 6.84, 9.34, 13.73 Hz,H-Ia,7a), 2.67(2H,ddd, J = 5.62, 9.77, 13.73 Hz, H-Ib, 7b), 1.68 (4H, m, H-2ab, 6ab), 3.88 (2H, m, H-3, 5), 1.57 (2H, t, J = 6.35 Hz. H-4ab), 7.02 (4H, d, J = 8.54 Hz.H-3', 5',3",5"), 6.73 (4H, d, J = 8.54 Hz H-2', 6', 2",6"). ^{13}CNMR

(acetone-d_6): δ 31.8 (C-1, 7), 40.0 (C-2, 6), 68.2 (C-3, 5), 44.8 (C-4), 134.1 (C-1', 1"), 130.0 (C2', 2", 6', 6"), 115.8 (C-3', 3", 5', 5"), 156.1 (C-4', 4").

Meso-hannokinol: Colorless oil. $[\alpha]_D$ 0° (MeOH, c = 0.7). ^{1}H NMR (acetone-d_6): δ 2.56 (2H, td, J = 7.81, 9.34, 13.92 Hz, H-1a,7a), 2.66 (2H, td, J = 7.81, 9.77, 13.92 Hz, H-I b, 7b), 1.68 (4H, q, J = 8.06 H-2ab, 6ab), 3.80 (2H, m, H-3, 5), 1.52 (1H, td, J= 9.52, 1.4, 16 Hz, H-4a), 1.64 (1H, td, J= 2.93, 14.16 Hz, H-4b), 7.02 (4H, d, J = 8.54 Hz, H-3', 5', 3", 5"), 6.73 (4H, d, J = 8.54 Hz, H-2', 6', 2", 6"). ^{13}C NMR (acetone-d_6): δ: 31.4 (C- 1, 7), 41.1 (C-2, 6), 71.6 (C-3, 5), 44.2 (C-4), 134.0 (C-1', 1"),130.0 (C-2', 2", 6', 6"), 115.8 (C-3', 3", 5', 5"), 156.1 (C-4', 4").

2,6-Dimethoxyphenol: Brownish-white powder. El-MS m/z: 154 $[C_7H_{10}O_3]^+$: ^{1}H NMR ($CDCl_3$) δ: 6.80 (1H, dd, J = 8.54, 8.54 Hz, H-4), 6.59 (2H, d, J = 8.54 Hz, H-3, 5), 3.89 (3H, s); ^{13}CNMR($CDCl_3$) δ: 147.2 (C-2, 6), 134.8 (C-1), 119.1 (C-4), 104.8 (C-3, 5), 56.3 (2-OCH_3).

Protocatechualdehyde: White powder. El-MS *m/z*: 138 $[C_7H_{10}O_3]^+$: ^{1}H NMR (acetone-d_6): δ 9.78 (1H, s), 7.36 (1H, d, J= 1.95 Hz, H-2), 7.32(1H, dd, J= 1.95, 7.81 Hz, H-6), 7.00 (1 H, d, J = 7.81 Hz, H-5).

Protocatechuic acid: Brownish white powder. ^{1}H NMR (DMSO-d_6): δ 7.73 (1H, d, J= 1.95 Hz, H-2), 7.19 (1H, dd, J = 1.95, 8.06 Hz, H-6), 6.60 (1H, d, J = 8.06 Hz, H-5); ^{13}CNMR(DMSO-d_6): δ 171.1 (COOH), 129.9 (C-1), 117.6(C-2), 147.7 (C-3), 144.9 (C-4), 114.8 (C-5), 121.0 (C-6).

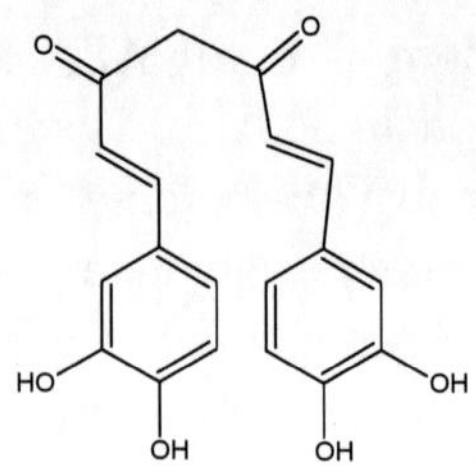

620
Bis-3,4-dihydroxy-cinnamoylmethane

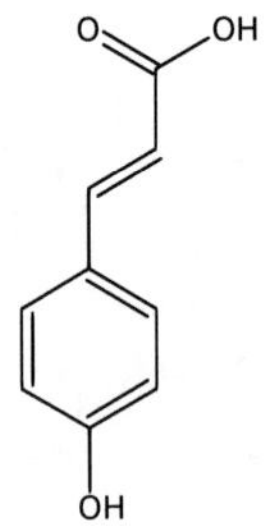

621
p-Hydroxycinnamic acid

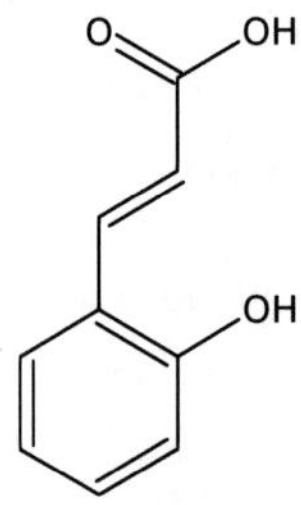

622
O-Hydroxycinnamic acid

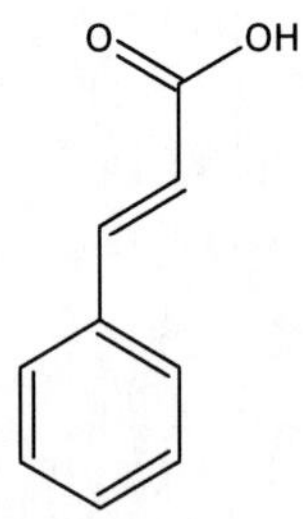

623
Cinnamic acid (*trans*)

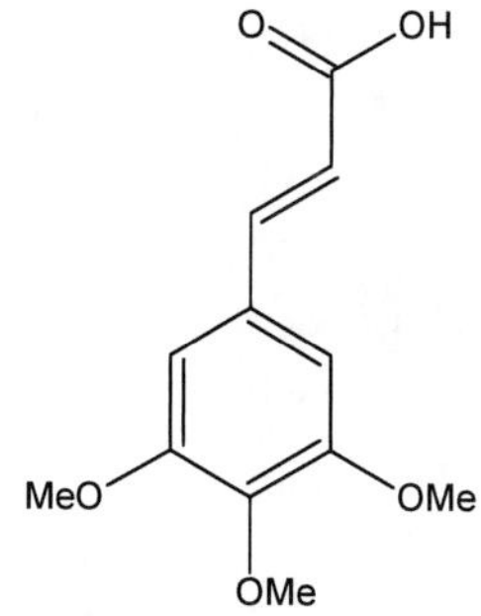

624
3,4,5- Trimethoxycinnamic acid (*trans*)

Demethylated derivatives of curcumin and ferulic acid, viz. bis-3,4-dihydroxy cinnamoylmethane and caffeic acid, are the most potent inhibitors of lipid peroxidation. Complete methylation as in 3,4,5-trimethoxycinnamic acid, also leads to abolition of antioxidant capacity. The hydroxyl group in the benzene ring must be at the *p*-position. Cinnamic acid has no inhibitory effect, on the other hand it potentiates lipid peroxidation. The salient features which emerge from this investigation are: (a) The caffeic acid moiety of curcumin is very potent as an *in vitro* inhibitor of lipid peroxidation. Two caffeic acid molecules when joined together through a methylene bridge lead to bis-demethoxy-curcumin which is the most potent antioxidant studied in this series. (b) Methylation depresses the antioxidant character as is evident from lipid peroxidation in the presence of ferulic acid and caffeic acid, curcumin, bis-3,4-dihydroxycinnamoyl methane, and trimethoxycinnamic acid (Sharma, 1976).

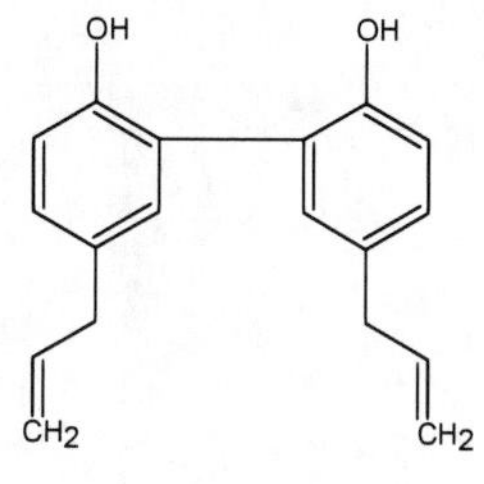

625
Magnolol

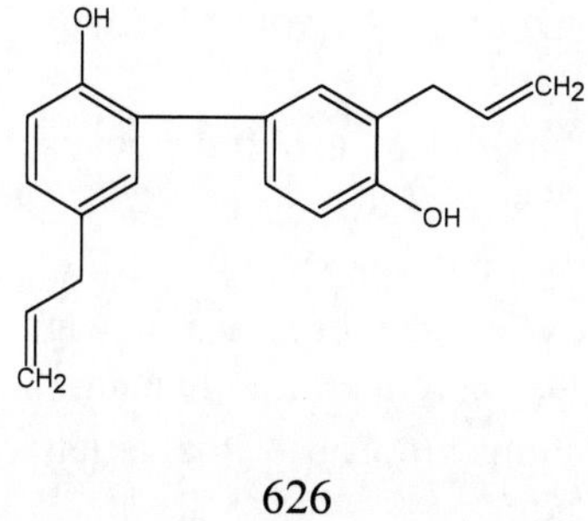

626
Honokiol

Found in *Magnolia officinalis* (Magnolaceae). The antioxidant effects of magnolol and honokiol purified from *M. officinalis* were 1000 times higher than that of α-tocopherol. The IC_{50} values of magnolol and honokiol for inhibition of oxygen consumption (in isolated rat heart mitochondria lipid peroxidation induced with ADP) were 8.0×10^{-8} M and 1.0×10^{-7} M, respectively, while that of α-tocopherol was 1.0×10^{-4} M. Also both exhibited free radical scavenging activities as shown by the diphenyl-*p*-picrylhydrazyl assay, but they were less potent than α-tocopherol (Lo, et al., 1994).

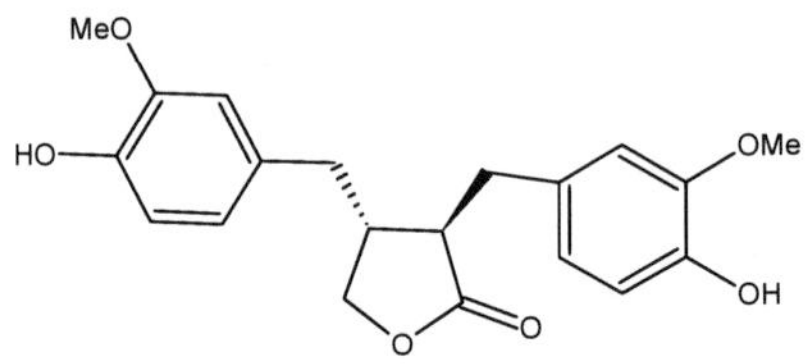

627
Matairesinol

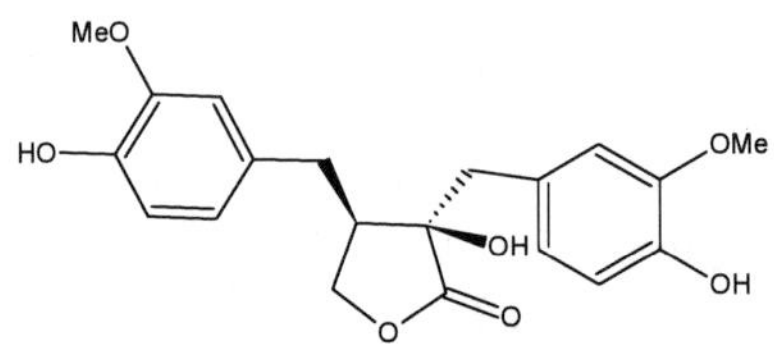

628
Nortrachelogenin

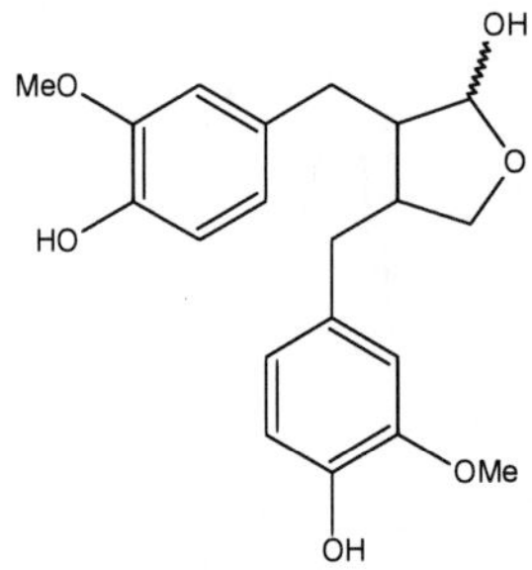

629
Dibenzylbutyrolactollignan

Three lignans with potent antioxidant activity were isolated from in *Cedrus deodara* (Roxb.) Loud (Pinaceae). The presence of two phenolic groups or substituted hydroxyl groups are essential for phosphodiesterase inhibiting activity of lignans and is an important criterion for free radical scavenging activity. The presence in significant amount of free radical scavenging principles in *C. deodara* may therefore explain the frequent use of this medicinal plant in Indian medicinal preparations prescribed in a variety of disease conditions (Tiwari et al., 2001).

Matairesinol: Pale yellow powder, mp 119 °C (EtOH); [M^+] at *m/z* 358 (50), 221 (5), 164 (8), 138 (100), 94 (10), 77 (8); ^{1}HNMR (200 MHz, $CDCl_3$) δ 2.53 (4H, m), 2.95 (2H, br), 3.86 (6H, s), 4.20-4.40 (2H, m), 5.5 (2-OH, bs), 6.40-6.89 (6H, m); ^{13}C NMR (50 MHz, $CDCl_3$) δ 34.48 (C-8), 38.10 (C-8'), 40.90 (C-7), 46.60 (C-7'). 55.70 (2-OMe), 71.30 (OCH_2O), 111.01, 111.53, 114.11, 114.40, 121.21, 121.95, 128.32, 129.45, 129.70, 144.30, 144.43, 146.38 (12 Ar-C), 178.94 (C=O); IR νmax (KBr) cm^{-1} 3560 (-OH), 1765 (lactone carbonyl); $[\alpha]_D$ -37.50 (c 0.5, $CHCl_3$).

(-)-Nortrachelogenin: Amorphous powder; [M^+] at *m/z* 374 (5), 138 (100), 122 (10), 105 (8), 77 (8), 43 (80); ^{1}HNMR (200 MHz, $CDCl_3$) δ 2.40-2.55 (2H, m), 2.65-2.80 (2H, m), 3.10-3.20 (lH, d), 3.85 (3H, s), 3.90 (3H. s), 3.95 (2H, m), 5.60 (2H, d), 6.50-6.80 (6H, m); ^{13}C NMR (50 MHz, $CDCl_3$) δ 31.50 (C-7), 41.90 (C-8), 43.74 (C-7'), 55.94 (d, 2-OMe), 70.26 (OCH_2O), 76.33 (OH), 111.55, 112.81,114.35, 114.56, 116.82, 121.42, 123.12, 126.20, 130.35, 144.27, 144.95, 146.59(12 Ar-C), 178.66 (C=O); IR νmax{KBr)cm^{-1} 3560 (OH),-1760 (lactone carbonyl); $[\alpha]_D$ -30.90 (c 0.44, $CHCl_3$).

Dibenzybutyrolactollignan or 4,4',9-trihydroxy-3,3'-dimethoxy-9,9'-epoxylignan: Amorphous powder; IR νmax (KBr) cm^{-1} 3450, 1765, l712, 1657; $[\alpha]_D$ -41.54 (c 1.00, $CHCl_3$); [M^+] at *m/z* 360; ^{1}HNMR (200 MHz, $CDCl_3$) δ 1.94-2.18 (2H, m), 2.37-2.80 (3H, m), 3.55 (lH, m), 3:90 (6H,bs), 4.15 (2H, m), 5.25 (lH, m), 5.50 (2-OH, bs), 6.42-6.81 (6H, m); ^{13}C NMR (50 MHz, $CDCl_3$) δ 38.35 (C-7'), 39.27 (C-7), 45.76 (C-8'), 52.93 (C-8), 55.73 (OMe), 55.88 (OMe), 72.36 (C-9'), 103.49 (C-9), 111.06 (C-2'), 111.21 (C-2), 114.14 (C-5'), 114.24 (C-5), 121.24 (C-6'), 121.58 (C-6), 131.47 (C-1'), 132.31 (C-1), 143.84 (C-4'), 144.00 (C-4), 146.39 (C-3'), 146.50 (C-3).

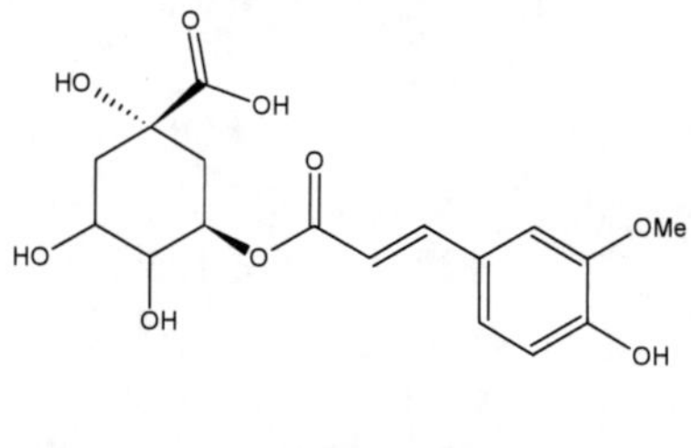

630
3-O-(3′-Methylcaffeoyl)quinic acid

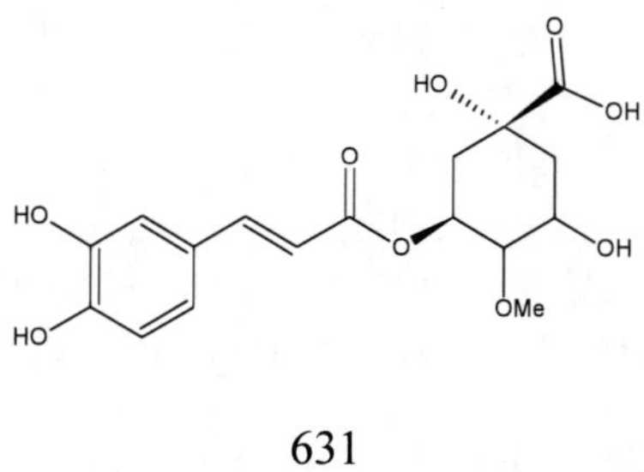

631
5-O-Caffeoyl-4-methylquinic acid

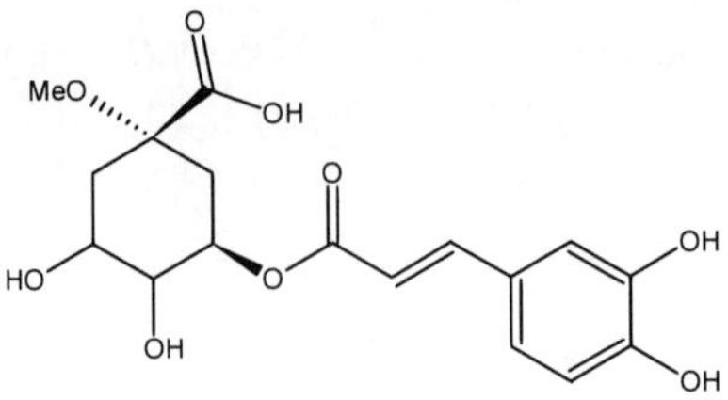

632
3-O-Caffeoyl-1-methylquinic acid

Found in bamboo *Phyllostachys edulis*. In terms of DPPH scavenging activity, compounds 5-O-caffeoyl-4-methylquinic acid and 3-O-caffeoyl-1-methylquinic acid were highly active in the DPPH scavenging and were stronger than chlorogenic acid, caffeic acid, ferulic acid, ascorbic acid, and α-tocopherol. Among the three compounds, compound 3-O-(3′-methylcaffeoyl)quinic acid was the weakest in antioxidant activity, and the DPPH scavenging and lipid peroxidation inhibitory activities (lC_{50} = 16.0 and 29.8 μM) were a little lower than those of chlorogenic and caffeic acids (Kweon et al., 2001).

3-O-(3′-Methylcaffeoyl) quinic acid: Pale yellow powder $[\alpha]_D$ -39.2 (ethanol, c 0.3); mp 195-197 °C; UV λmax in MeOH (nm) 244sh, 299, 328; +NaOH 266sh, 310,374; + $A1C1_3$ 246sh, 300,329; +$A1C1_3$ + HCl 245sh, 299,328; +NaOAc 369; +NaOAc + H_3BO_3 249sh, 305, 349; IR(KBr) vmax 3420, 3250, 1730, 1679, 1523, 1442 cm^{-1}; TLC positive reaction with $FeC1_3$ and bromocresol green; ESI-MS, m/z 369 $[M + H]^+$; FABMS m/z 369 [M + H]+, *m/z* 391 $[M + Na]^+$; HR-FABMS (glycerol matrix), *m/z* 369.1185 $[C_{17}H_{20}O_9 + H]^+$; ^{1}HNMR (CD_3OD, 500 MHz) δ 7.57 (lH, J = 15.9 Hz, H- 7'), 7.04 (lH, d, J = 2.0 Hz, H-2'), 6.93 (lH, dd, J = 8.2 Hz, 2.0 Hz, H-6'), 6.76 (lH, d, J = 8.4 Hz, H-5'), 6.28 (lH, d, J = 15.9 Hz, H-8') 5.34-5.36 (lH, m, H-3), 4.09-4.13 (lH, m, H-5), 3.73 (OCH_3), 3.66 (lH, dd, J = 7.8, 3.3 Hz, H-4), 1.98-2.22 (4H, m, H-2, H-6); ^{13}C NMR (CD_3OD, 125 MHz) δ 176.1 (COO), 168.1 (C-9'), 149.2 (C-3'), 146.4 (C-7'), 146.2 (C-4'), 127.1 (C-6'), 123.7 (C-1 '), 116.8 (C-5'), 115.1 (C-8'), 114.7 (C-2'), 75.8 (C-4), 73.8 (C-1), 72.6 (C-3), 67.3 (C-5), 53.6 (OCH_3), 39.8 (C-2), 36.1 (C-17).

5-O-Caffeoyl-4-methylquinic acid: White powder $[\alpha]_D$ -47.2 (ethanol, c 0.2); mp 200-201 °C; UV λmax in MeOH (nm) 244sh, 296,328; +NaOH 261sh, 308, 372; +$AlCl_3$ 263sh, 306, 351; +$AlC1_3$ + HC1 244sh, 296, 328; +NaOAc 274, 334, 375; +NaOAc + H_3BO_3 252sh, 303, 348; IR (KBr) vmax 3433,2950, 1735, 1680, 1631, 1519, 1445 cm^{-1}; TLC positive reaction with $FeC1_3$ and bromocresol green; ESI-MS, *m/z* 369 $[M + H]^+$; FABMS *m/z* 369 $[M + H]^+$; *m/z* 391 $[M + Na]^+$; HR-FABMS (glycerol matrix), *m/z* 369.1185 $[C_{17}H_{20}O_9$ + H]+; ^{1}HNMR (CD_3OD, 500 MHz) δ 7.61 (lH, 15.9 Hz, H-7'), 7.06 (lH, d, J = 2.0 Hz, H-2'), 6.95 (lH, dd, J = 8.2 Hz, 2.0 Hz, H-6'), 6.78 (lH, d, J = 8.2 Hz, H-5'), 6.35 (lH, d, J = 15.9 Hz, H-8'), 4.23-4.30 (3H, m, H-3, 4, 5), 3.75 (OCH_3), 1.99-2.21 (4H, m, H-2. H-6); ^{13}C NMR (CD_3OD, 125 MHz) δ 176.5 (COO), 168.9 (C-9'), 149.5 (C-3'), 146.9 (C-4'), 146.8 (C-7'), 127.9 (C-6'), 123.0 (C-1'), 116.5 (C-5'), 115.8 (C-8'), 115.1 (C-2'), 74.3 (C-4), 73.9 (C-1), 71.3 (C-3), 70.6 (C-5), 52.9 (OCH_3), 40.8 (C-2), 36.4 (C-17).

3-O-Caffeoyl-1-methylquinic acid: White powder (8.5 mg); $[\alpha]_D$ -16.9 (water, c 0.35); mp 202-203 °C; UV λmax in MeOH (nm) 245sh, 300, 329; +NaOH 266sh, 310, 374; + $AlC1_3$ 355; + $AlC1_3$ + HCl 245sh, 299, 328; +NaOAc 338, 377sh; +NaOAc + H_3BO_3

254,306,351; IR (KBr) νmax 3423, 2957, 1734, 1686, 1631, 1522, 1443 cm^{-1}; TLC positive reaction with $FeCl_3$ and bromocresol green; ESI-MS, *m/z* 369 [M + H]+; FABMS *m/z* 369 $[M + H]^+$; *m/z* 391 $[M+ Na]^+$; HR-FABMS (glycerol matrix), *m/z* 369.1185 $[C_{17}H_{20}O_9+ H]^+$; ^{1}HNMR (CD_3OD, 500 MHz) δ 7.51 (lH, J = 15.9 Hz, H-7'), 7.04 (lH, d, J = 2.0 Hz, H-2'), 6.94 (lH, dd, J = 8.2, 2.0 Hz, H-6'), 6.77 (lH, d, J = 8.4 Hz, H-5'), 6.20 (lH, d, J = 15.9 Hz, H-8'), 5.26-5.30 (lH, m, H-3), 4.12-4.15 (lH, m, H-5), 3.71 (lH, dd, J = 7.5, 3.1 Hz, H-4), 3.70 (OCH_3), 1.93-2.23 (4H, m, H-2, H-6); ^{13}C NMR (CD_3OD, 125 MHz) δ 175.4 (COO), 168.3 (C-9'), 149.7 (C-3'), 147.2 (C-4'), 146.8 (C- 7'), 127.7 (C-6'), 123.0 (C-1'), 116.5 (C-5'), 115.1 (C-2', C-8'), 75.8 (C-4), 72.6 (C-1), 72.1 (C-3), 70.3 (C-5), 53.0 (OCH_3), 38.1 (C-6), 37.8 (C-2).

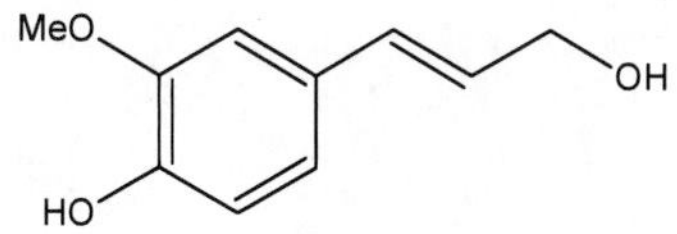

633
Coniferyl alcohol

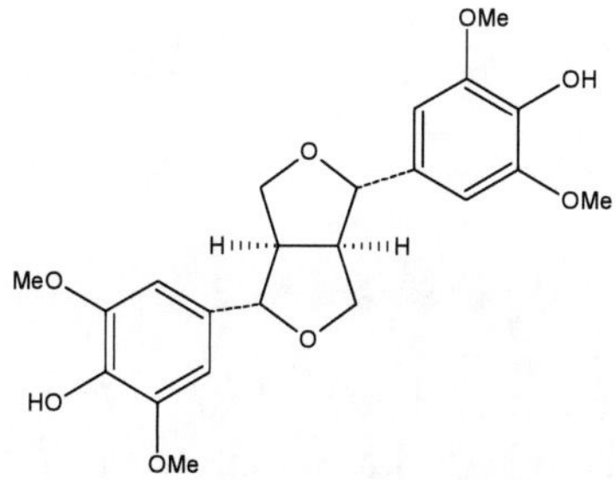

634
Syringaresinol

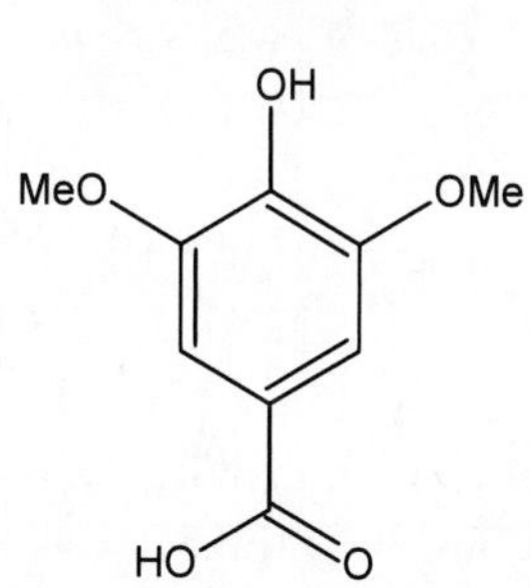

635
Syringic acid

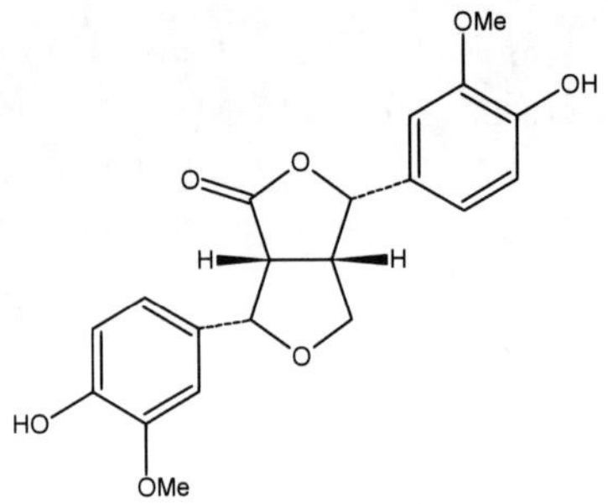

634
4-Ketopinoresinol

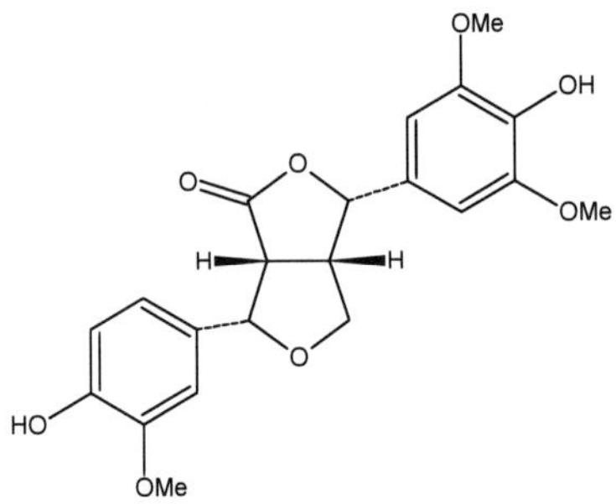

637
Mayuenolide

The isolated compounds from *Coix lachrymal*-jobi var. *ma-yuen* Stapf demonstrated DPPH free radical scavenging activity following the order of syringic acid > syringaresinol > ferulic acid > coniferyl alcohol > mayuenolide > 4-ketopinoresinol. Compounds were isolated from the seed of Coix lachryma-jobi L. var. ma-yuen Stapf (Kuo et al., 2002).

Mayuenolide: Light yellow crystals; mp 215-217 ^{o}C. $[\alpha]_D$ -25.5°. El-MS *m/z* (%) 402 (M^+, 10), 368 (5), 143 (l00), 115 (65). HR-EIMS *m/z*: 402.1321 (calcd. for $C_{21}H_{22}O_8$, 402.1315). UV λmax (MeOH) nm (log ε): 280 (3.1); IR νmax cm^{-1}; (KBr): 3416, 1772, 1619, 1520, 1215; ^{1}HNMR (CD_3OD, 300 MHz): δ 3.18 (m, 1H, H-l), 3.42 (dd, J = 9.1, 3.8 Hz, 1H, H-5), 3.88 (s, 9H, OCH_3 x 3), 4.00 (dd, J = 9.5, 4.5 Hz, 1H, $H\text{-}8_{(a)}$), 4.30 (dd, J = 9.5, 6.8 Hz, 1H, $H\text{-}8_{(eq)}$), 5.29 (d, J = 3.8 Hz, 1H, H-2), 5.32 (d, J = 3.8 Hz, 1H, H-6), 6.47 (s, 2H, H-2', H-6'), 6.83 (lH, d, J = 8.1 Hz, IH, H-5"), 6.86 (dd, J = 8.1, 1.8 Hz, 1H, H-6"), 6.87 (lH, d, J = 1.8 Hz, H-2"); ^{13}C NMR ($CDCl_3$, 300 MHz): δ 50.18 (C-1), 84.71 (C-2), 176.90 (C-4), 53.15 (C-5), 83.35 (C-6), 72.68 (C-8), 56.00 (OCH3), 130.37 (C-l'), 157.3 (C-4′), 116.2 (C-5′), 157.3 (C-6′), 37.8 (C-7′), 55.5 (C-8′), 174.9 (C-9′), 134.96 (C-l"), 145.31 (C-3"), 108.06 (C-2"), 114.38 (C-5"), 117.97 (C-6"), 101.87 (C-2', C-6'), 146.69 (C-3', C-5'), 147.37 (C-4', C-4"),

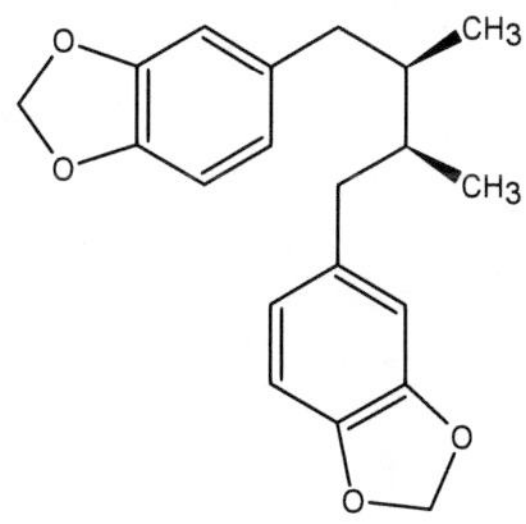

638
Threo-austrobailignan 5

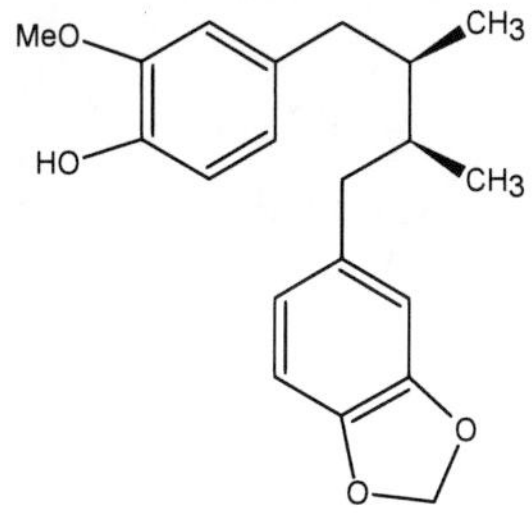

639
Erythro-austrobailignan 6

These lignans together with nectandrin B and meso-dihydroguaiaretic were isolated from the petrol extract of *Myristica argentea* mace (Myristicaceae). Three of them, *erythro*-austrobailignan-6, meso-dihydroguaiaretic acid and nectandrin-B, exert an antiproliferative effect on MCF-7 cells as well as antioxidant activity on the 1,1-diphenyl-2-picrylhydrazyl (DPPH) radical, but not the threo-austrobailignan-5, nectandrin-B also possesses anti-17β-hydroxysteroid dehydrogenase and anti-aromatase activities (Filleur et al., 2001).

Threo-austrobailignan : Colourless oil; $C_{20}H_{22}O_4$; UV (n-hexane) λmax= 236, 288; $[\alpha]_D$ -40 (c 0.01, MeOH); ^{1}HNMR δ 5.58 (d J = 1.4 Hz, H-2), 6.73 (d, J = 7.9 Hz, H-5), 6.54 (dd, J = 7.9, 1.6 Hz, H-6), 2.34 (dd, J = 13.6, 8.3 Hz, H-7), 2.55 (dd, J = 13.6, 6.1 Hz, H-7), 1.73 (m, H-8), 0.81 (d, J = 6.7 Hz, H-9), 6.64 (d, J = 1.5, H-2′), 6.71 (d, J = 7.9 Hz, H-5′), 6.60 (dd, J = 8.0, 1.5 Hz, H-6′), 2.26 (dd, J = 13.4, 9.3 Hz, H-7′), 1.73 (m, H-8′), 0.83 (d, J = 6.7 Hz, H-9′), 5.92 (br s, O-CH_2-O); ^{13}CNMR δ 135.4 (C-1), 109.3 (C-2), 147.4 (C-3), 145.4 (C-4), 107.9 (C-5), 121.6 (C-6), 41.1 (C-7), 38.2 (C-8), 138 (C-9), 135.6 (C-1′), 109.3 (C-2′), 147.5 (C-3′), 145.5 (C-4′), 107.9 (C-5′), 121.8 (C-6′), 39.0 (C-7′), 39.4 (C-8′), 16.1 (C-9′), 100.7 (O-CH_2-O); ElMS, (70 eV): *m/z* (rel. int.) 326 $[M]^+$(13), 136 $[M\text{-}PiC_5H_9]^+$ (32), 135 $[PiCH_2]^+$ (100); Chromatographic behaviour: R_f 0.52 (silica gel 60 F254, n-hexane/EtOAc, 9: 1); R_f 0.45 (silica gel 60F_{254}, cyclo-hexane/toluene, 5: 5); R_t 5.89 min (Merck Lichrosorb Si 60 (7um, 4 x 250 mm), *n*-hexane, 1ml/min].

Erythro-austrobailignan : White crystals; $C_{20}H_{24}O_4$; UV (MeOH): λmax = 227, 282; $[\alpha]_D$ + 8 (c 0.01, MeOH); SE^+-MS (70 eV): *m/z* (rel. int.) = 351 [M+Na]+ (100), 329 [M+H]+ (6), 214 (13), 207 $[M\text{-}Pi]^+$ (37), 205 $[M\text{-}Gu]^+$ (46), 175 (63), 140 (77), 137 $[GuCH2]^+$ (57), 135 $[PiCH2]^+$(53).

Nectandrin B or malabaricanone A: White crystals; $C_{20}H_{24}O_5$; UV (MeOH): λmax = 232,282; $[\alpha]_D$ 0 (c 0.011, MeOH): ^{1}HNMR and ^{13}CNMR values (Purushothaman and Sarada, 1984); El-MS (70eV): *m/z* (%) = 344 $[M]^+$ (18), 192$[GuC_5H_9]^+$ (100), 177 $[GuC_4H_6]^+$ (30), 161 (18),152 $[ArCHO]^+$ (9),151 $[Gu-CO]^+$ (14), 145 (42), 137 $[GuCH_2]^+$ (10), 124 $[GuH]^+$ (12).

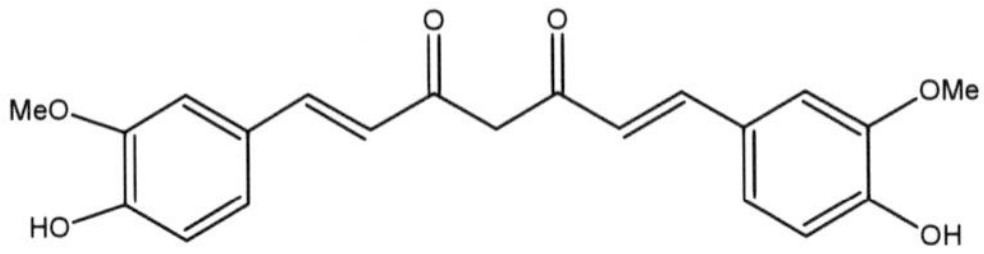

640
Curcumin

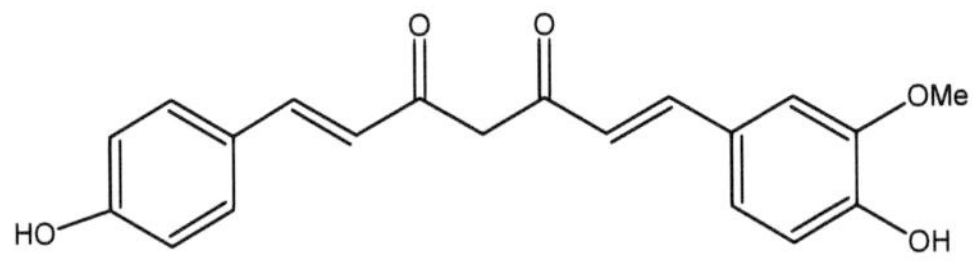

641
Demethoxycurcumin

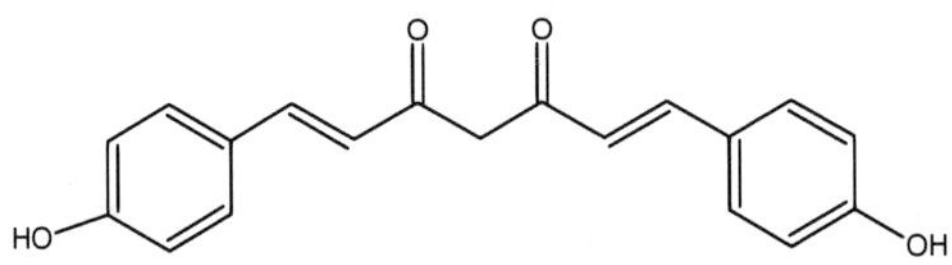

642
Bisdemethoxycurcumin

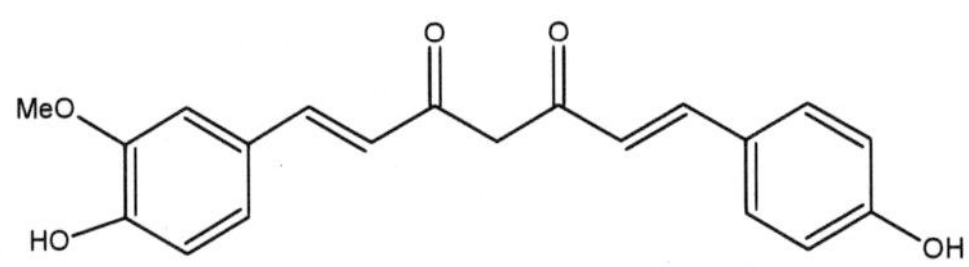

643
(4-Hydroxy-3-methoxycinnamoyl) methane

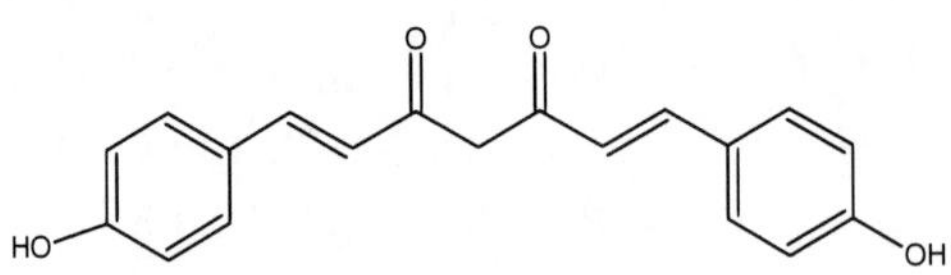

644
Bis-(4-hydroxycinnamoyl)methane

Assay-guided fractionation of the EtOAc soluble fraction of the rhizomes of *Curcuma longa* (Zingiberaceae), furnished three DPPH free radical scavenging diarylheptanoids.

Compounds curcumin, demethoxycurcumin, and bisdemethoxycurcumin showed the DPPH radical scavenging effects with IC_{50} values of 2.8, 39.2, 308.7 μM, respectively L-ascorbic acid and resveratrol as positive controls exhibited IC_{50} values of 22.5 and 25.0 μM, respectively. The Compounds showed significant hepatoprotective effects on tacrine-induced cytotoxicity in human liver-derived Hep G2 cells. The ECso values are 86.9, 70.7, and 50.2 μM respectively. Silybin (IC_{50} = 69.0 μM) and silychristin (IC_{50} = 82.7 μM) were used as positive controls (Song et al., 2001). Compounds showed the peroxynitrite scavenging activities with IC_{50} values or 4.0 ±0.04, 6.4 ±0.30, and 29.7 ±1.29 μM respectively. Penicillamine as positive control exhibited IC_{50} value or 2.38 ±0.34 μM (Kim et al., 2003). The most abundant and most active constituent of the extract as the orange pigment, curcumin. The mechanism of curcumin activity may include metal ion chelation by the central β-diketone group (Osawa et al., 1995).

Curcumin: ^{1}HNMR (400 MHz. acetone-d_6) δ 3.86 (6H. s. 2 x OCH_3). 5.92 (1H, s, H'-4), 6.65 (2H, d, J = 16.0 Hz, H-2, 7). 6.83 (2H, d, J = 8.3 Hz,H-5', 5"), 7 .12 (2H, d, J = 2.0 and 8.3 Hz, H-6', 6"), 7 .28 (2H, d, J = 2.0 Hz, H-2', 2"), 7.55 (2H, d, J = 16.0 Hz, H-1, 6); ^{13}CNMR (100.0 MHz, acetone- d_6) δ 184.96 (C-3. 5), 150.45 (C-4', 4"), 149.21 (C-3', 3"), 141.86 (C-1, 7), 128.64 (C-1', 1"), 124.27 (C-6', 6"), 122.77 (C-2. 6), 116.69 (C-5', 5"), 112.02 (C-2', 2"), 102.13 (C-4), 56.77 (2 x OCH_3).

Demethoxycurcumin: ^{1}HNMR (400 MHz. acetone-d_6) δ 3.86 (3H, s, OCH_3), 5.92 (1H, s, H-4), 6.60 (1H, d, J = 16.0 Hz, H-7), 6.66 (1H, d, J = 16.0 Hz, H-2), 6.83 (1H, d, J = 8.5 Hz, H-5'), 6.85 (2H, d, J = 8.5 Hz, H-3", 5"), 7.11 (1H, dd, J = 1.7, 8.5 Hz, H-6'), 7.28 (1H, d, J = 1.7 Hz, H-2'), 7.51 (2H, d, J = 8.5 Hz, H-2", 6"), 7.55 (1H, d, J = 16.0 Hz, H-6), 7.56 (1H, d, J = 16.0 Hz, H-1); ^{13}CNMR (100.0 MHz, acetone- d_6) δ 185.01 (C-3), 184.92 (C-5), 160.96 (C-4"), 150.47 (C-4'), 149.23, (C-3'), 141.85 (C-1), 141.50 (C-7), 131.42 (C-2", 6"), 128.63(C-1'), 128.16 (C-1"), 124.32 (C-6'), 122.73 (C-2), 122.51 (C-6), 117.25 (C-3", 5"), 116.68 (C-5'), 111.96 (C-2'), 102.17 (C-4), 56.77 (OCH_3).

Bisdemethoxycurcumin: ^{1}HNMR (400 MHz. acetone-d_6) δ 5.92 (1H, s, H-4), 6.61 (2H, d, J = 16.0 Hz, H-2, 6), 6.85 (2H, d, J = 8.5 Hz, H-3', 5'), 6.85 (2H, d, J = 8.5 Hz, H-3",5"), 7.51 (2H, d, J = 8.5 Hz, H-2', 6'), 7.51 (2H, d, J = 8.5 Hz, H-2", 6"), 7 .56 (2H, d, J = 16.0 Hz, H-17); ^{13}CNMR (100.0 MHz, acetone- d_6) δ 184.96 (C-3, 5), 160.91 (C-4', 4"), 141.50 (C-l7), 131.42 (C-2', 6', 2", 6"), 128.16 (C-l', 1"), 122.49 (C-2, 6), 117.24 (C-3', 5',3", 5"), 102.22 (C-4).

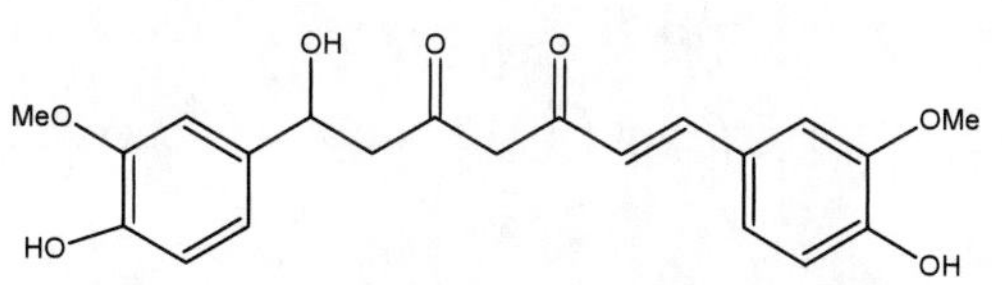

645

1-Hydroxy-l,7-bis(4-hydroxy-3-methoxyphenyl)-(6E)-6-hep-1-tene-3,4-dione

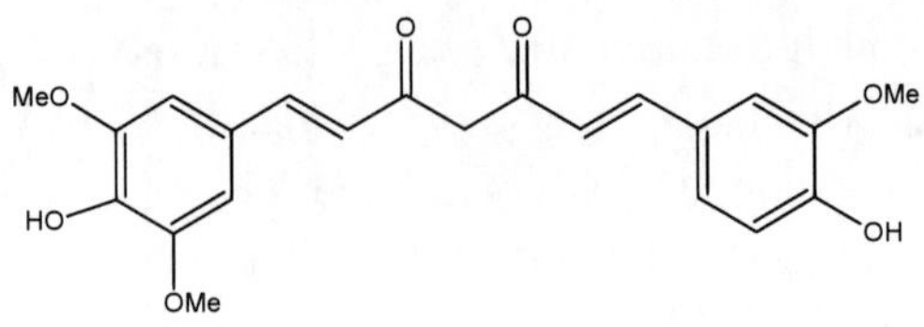

646
1-(4-Hydroxy-3,5-dimethoxyphenyl)-7-(4-hydroxy-3-methoxyphenyl)-(1E,6E)-1,6-heptadiene-3,4-dione

Two curcumins analogue has been isolated from the rhizomes of *Curcuma xanthorrhiza* (Zingiberaceae) along with the known curcuminoids, curcumin, demethoxycurcumin, and bisdemethoxycurcumin. The new compounds showed potent antioxidant activity against autoxidation of linoleic acid in a water-alcohol system (Masuda et al., 1992).

1-Hydroxy-l,7-bis(4-hydroxy-3-methoxyphenyl)-(6E)-6-hep-1-tene-3,4-dione: Pale yellow powder, mp 84-88°; UV λmax 372; El-MS *m/z* 368 $[M-H_2O]^+$; ^{1}HNMR ($COCl_3$): δ 7.30 m (lH, d, J = 15.6 Hz, H-7), 7.03 (lH, dd, J = 7.9, 1.8 Hz, H-6, or H-6"), 7.0 (4H, m, arom.), 6.90 (lH, d, J = 8.4 Hz, H-5′ or H-5"), 6.44 (lH, d, J = 15.8 Hz, H-6), 5.60 (lH, s, H-4), 5.38 (lH, dd, J = 14.0, 3.1 Hz, H-l), 3.95 (3H, s, Me), 3.93 (3H, s, Me), 2.93 (lH, dd, J = 16.6, 14.0 Hz, H-2), 2.63 (lH, dd, J = 16.6, 3.1 Hz, H-2).

1 - (4 – Hydroxy - 3, 5 - dimethoxyphenyl) - 7 -(4-hydroxy-3-methoxy-phenyl)-(1E,6E)-1,6-heptadiene-3,4-dione: Yellow powder, mp 145-146°. UV λmax 429. HREl-MS *m/z* 398.1425 $[M]^+$ (calc. for $C_{22}H_{22}O_7$: 398.1364). ^{1}HNMR ($COCl_3$): δ 7.59 (d, J = 15.8 Hz, H-1), 6.48 (d, J = 15.8 Hz, H-2), 5.79 (s, H-4), 6.46 (d, J = 15.8 Hz, H-6), 7.55 (d, J = 15.8 Hz, H-7), 6.80 (s, H-2′), 6.80 (s, H-6′), 7.05 (d, J = 1.8 Hz, H-2′′), 6.93 (d, J = 8.3 Hz, H-5′′), 7.11 (dd, J = 8.4, 1.8 Hz, H-6′′), 3.94 (s, OMe-3′), 3.94 (s, OMe-5′), 3.955 (s, OMe-3′′), 5.78 (br, s, OH-4′), 5.81 (br, s, OH-4′′),

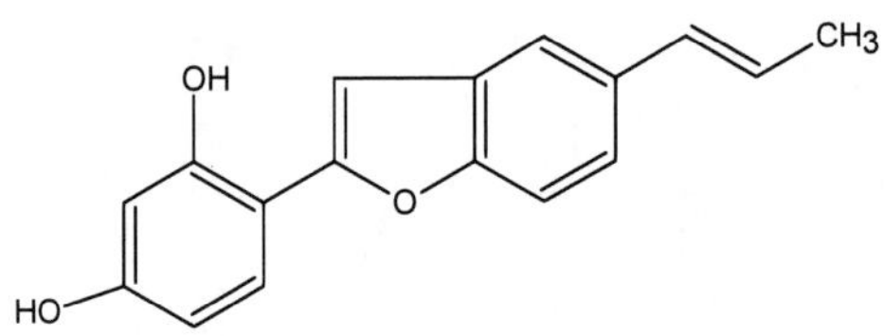

647
2-(2,4-Dihydroxyphenyl)-5-propenyl benzofuran

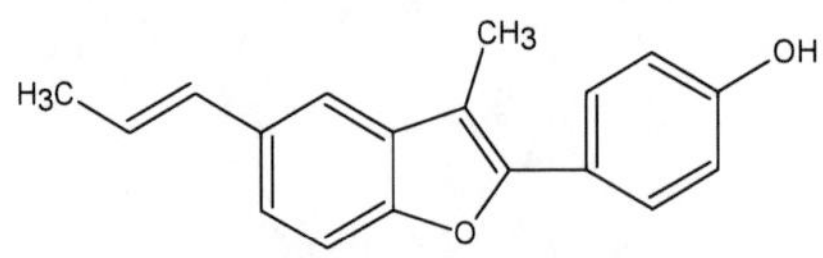

648
Eupomatenoid 6

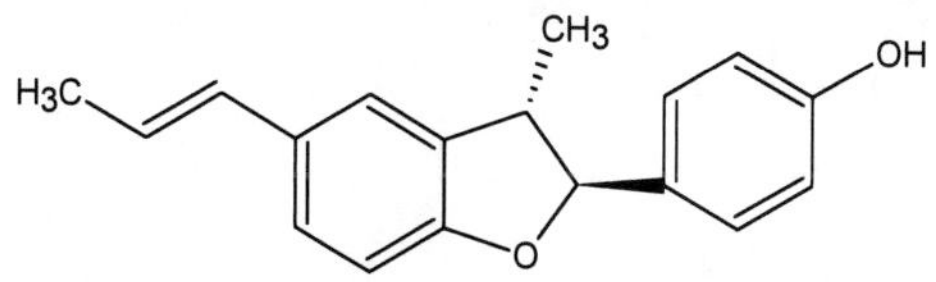

649
Conocarpan

Found in *Krameria triandra* (Krameriaceae) roots. These neolignans possess antilipoperoxidant properties by inhibiting the propagation of the peroxidative cascade induced by hydroxyl radicals in a simple membrane model phosphatidylcholine liposomes. Also showed quench reactive oxygen species (ROS), in complete cell systems (erythrocytes) using the organic hydroperoxide CuOOH (cumene hydroperoxide) and the UVB radiation as free radical generator (Carini et al., 2002).

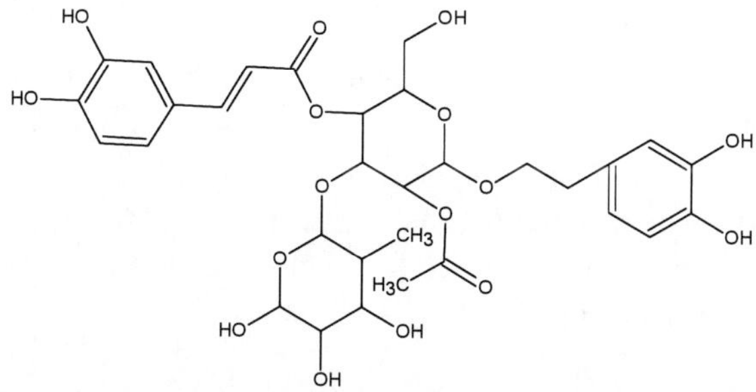

650
2-Acetylacteoside

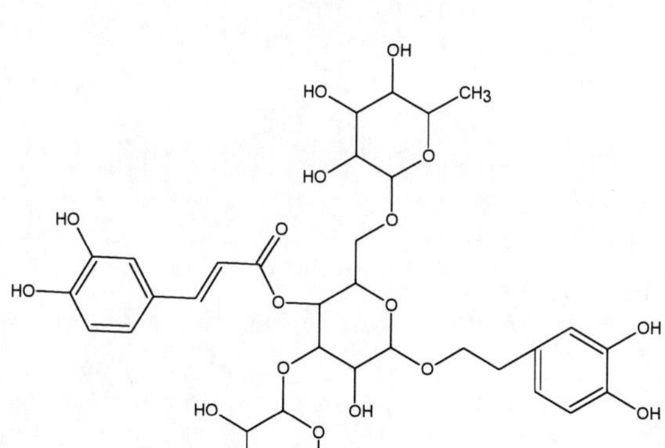

651
Poliumoside

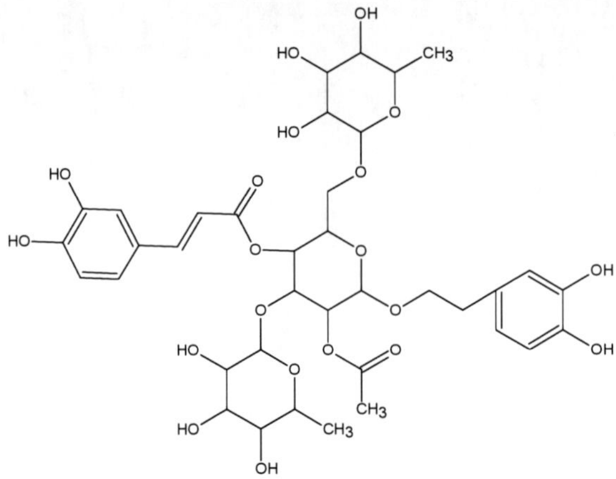

652
Brandioside

Brandisia hancei (Scrophulariaceae) is a medicinal herb in China. From the ethanol extract of this plant were isolated four phenylethanoid glycosides, acteoside, 2'-acetylacteoside, poliumoside and brandioside, were shown to have inhibitory effects on free radical-induced hemolysis of red blood cells and free radical scavenging activities *in vitro*. Brandioside and poliumoside showed stronger antioxidant effect than acteoside and 2'-acetylacteoside and trolox (Dan et al., 2000).

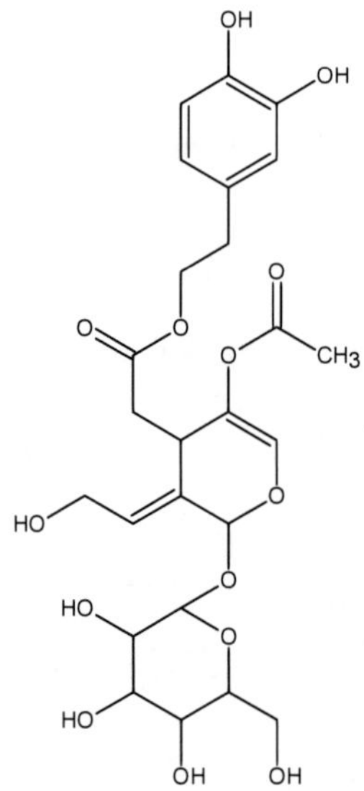

653
10-Hydroxyloleuropein

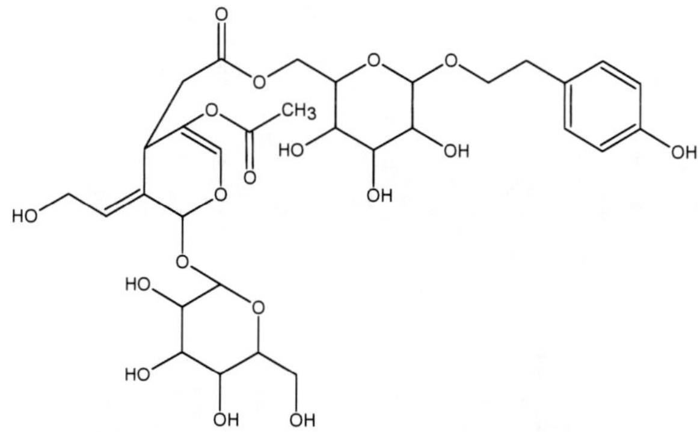

654
Specnuzhenise

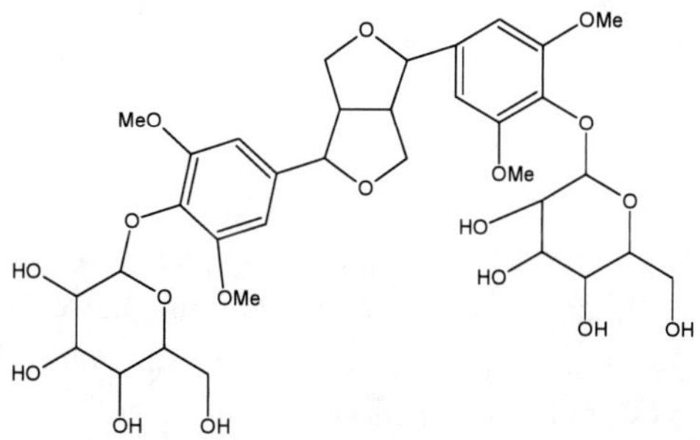

655
Liriodendrin

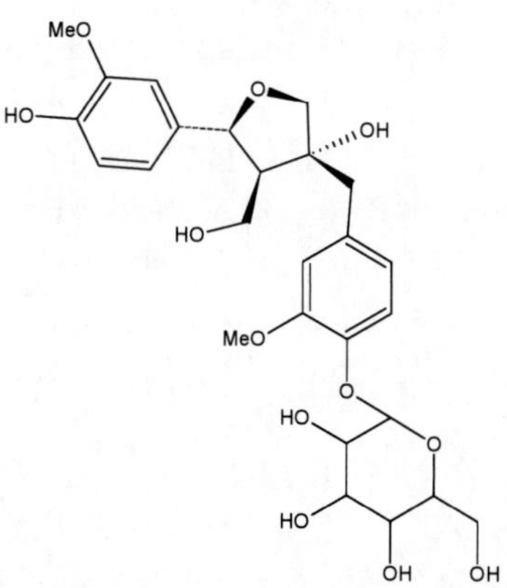

656
8'-α-Hydroxyl-lariciresinol-4'-β-O-D-glucopyranoside

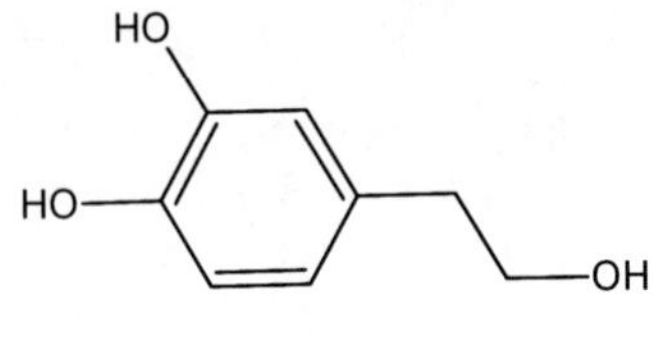

657
3,4-Dihydeoxyphenethyl alcohol

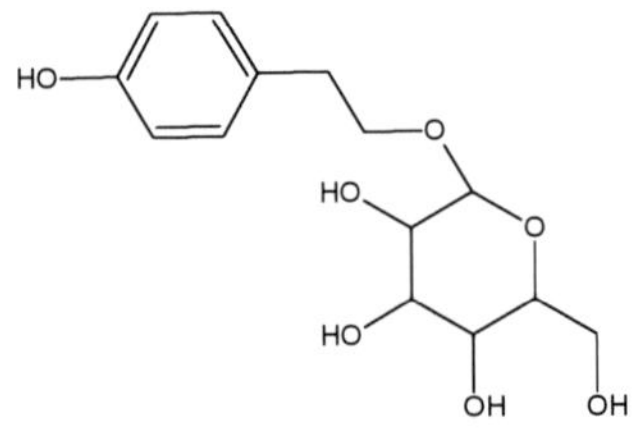

658
p-Hydroxyphenethyl-O-β-O-glucopyranoside

These glucosides were isolated from *Ligustrum sinense* (Oleaceae), This plant grows abundantly in the south of China, being known as "dong-qing", "xiao-la-shu"' and "wen-zi-hua"'. It is used in folk medicine as a remedy to treat hepatitis and inflammation. However, on AAPH screen showed antioxidant activity (Ouyang et al., 2003).

10-Hydroxyoleurapein:^{1}H NMR δ 3.69 (3H, S, OCH_3): 4.84 (1 H, d, J = 7.6 Hz H-1 of Glc), 5.94 (1H, br.s, H-1), 6.13 (1H, t, J = 6.8 Hz, H-8), 6.53 (1H, dd, J = 2.0, 8.0 Hz, H-6'), 6.65 (1H, d, J = 2.0 Hz, H-2'), 6.69 (1H, d, J = 8.0 Hz, H-5'), 7.51 (1H, s, H-3); ^{13}C NMR: δ 173.0 (C-7), 168.5 (C-11), 155.1 (C-3), 146.2 (C-3'), 144.9 (C-4'), 131.1 (C-1'), 130.8 (C-9), 129.4 (C-8), 121.3 (C-6'), 117.1 (C-5'), 116.5 (C-2'), 109.3 (C-4), 94.8 (C-1), 66.8 (C-β), 35.3 (C-α), 101.0 (C_{G-1}, 1-O-Glc), 78.4 (C_G-5), 77.9 (C_G-3), 74.8 (C_G-2), 71.5 (C_G-4), 62.8 (C_G-6).

Specneuzhenide: ^{1}HNMR: δ 7.51 (1 H, s, H-3), 7.04 (2H, d, J = 8.5 Hz, H-2''', H-6'''), 6.68 (2H, d, J = 8.5 Hz, H-3''', H-5'''), 6.08 (1H, t, J = 6.7 Hz, H-8), 5.90 (1H, brs, H-1), 4.80 (1H, d, J = 7.5 Hz H-1 of Glc), 4.32 (1H, d, J = 7.0 Hz H-1 of Gal), 3.67 (3H, s, OCH_3), 1.72 (H-α) (H-α); ^{13}C NMR: δ 173.0 (C-7), 168.7 (C-11), 156.7 (C-4'''), 155.2 (C-3), 103.9 (C-1'''), 130.7 (C-3''', C-5'''), 130.5 (C-9), 125.0 (C-8), 116.2 (C-2''', C-6'''), 109.4 (C-4), 92.2 (C-1), 65.0 (C-β), 36.4 (C-α), 104.4 (C_G-I), 78.3 (C_G-5), 77.9 (C_G-3), 75.1 (C_G-2), 71.5 (C_G-4), 62.7 (C_G-6), 100.9 (C_{Gal}-1), 75.0 (C_{Gal}-5), 4.7 (C_{Gal}-3), 71.0 (C_{Gal}-4), 74.5 (C_{Gal}-2), 72.2 (C_{Gal}-6).

Liriodendrin: ^{1}H NMR: δ 3.74 (3H x 4, s, OCH_3), 4.65 (2H, d, J = 3.8 Hz, H-2, H-6), 4.94 (1H, d, J = 5.2 Hz, H-1 of Glc), 6.64 (2H, s, H-2', H-6'); ^{13}C NMR: δ 53.6(C-1, C-5), 85.0 (C-2, C-6), 71.3 (C-4, C-8), 133.9 (C-1'), 104.3 (C-2', C-6'), 152.6 (C-3', C-5'), 137.2 (C-4'), 102.8 (C_G-1 x 2), 77.2 (C_G-5 x 2), 76.5 (C_G-3 x 2), 74.2 (C_G-2 x 2), 70.0 (C_G-4 x 2), 61.0 (C_G-6 x 2), 56.4 (4 x OCH_3).

8' -α-Hydroxyl-lariciresinol-4'-β-O-D-glucopyranoside: A white crystal, $C_{26}H_{34}O_{12}$, $[\alpha]_D$ -46.67° (c 0.015, C_5H_5N); IR νmax cm^{-1} 3600-3000, 1710, 1600, 1520; FAB-MS *m/z*: 537 $[M-H]^-$, 375 $[M-H-162]^-$; ^{1}H NMR δ 7.21 (d, J = 1.9 Hz, H-2), 6.80 (d, J = 8.0 Hz, H-5), 6.93

(dd, J = 1.9, 8.0 Hz, H-6), 4.80 (d, J = 7.3 Hz, H-7), 2.40 (brs, H-8), 3.80, 3.90 (H-9), 3.76 (3-OCH_3), 7.01 (d, J = 1.8 Hz, H-2′), 7.15 (d, J = 8.2 Hz, H-5′), 7.86 (dd, J = 1.8, 8.2, H-6′), 3.00, 3.09 (d, J = 13.9 Hz, H-7′), 3.70, 3.87 (H-9), 3.76 (3-OCH_3), 4.99 (d, J = 7.2 Hz, H-1′′); ^{13}C NMR: δ 135.5 (C-1), 111.8 (C-2), 149.0 (C-3), 147.3 (C-4), 115.9 (C-5), 120.8 (C- 6), 85.5 (C-7), 62.1 (C-8), 60.7 (C-9), 56.4(3-OCH_3), 134.1 (C-1′), 116.4 (C-2′), 146.8 (C-3′), 150.4 (C-4′), 117.7 (C-5′), 124.0 (C-6′), 40.7 (C-7 ′), 82.4 (C-8′), 78.0 (C-9′), 56.7 (3-OCH_3), 103.1 (C-1′′), 75.0 (C-2′′), 78.0 (C-3′′), 71.5 (C-4 ′′), 78.3 (C-5′′), 62.6 (C-6 ′′).

3,4-Dihydeoxyphenethyl alcohol: ^{1}H NMR: δ 6.68 (1H, d, J=2.0 Hz, H-2), 6.67 (1H, d, J = 8.0 Hz, H-5), 6.54 (1H, dd, J = 2.0, 8.0 Hz, H-6), 3.91 (H-β), 2.75 (H-α), ^{13}C NMR: δ 146.0 (C-4), 144.5 (C-3), 131.5 (C-1), 121.2 (C-6), 117.1 (C-2), 116.3 (C-5), 36.4 (C-α), 72.0 (C-β).

p-Hydroxy-phenethyl-O-β-D-glucopyranoside: ^{1}H NMR: δ 7.04 (2H, d, J = 8.5 Hz, H-2, H-6), 6.69 (2H, d, J = 8.5 Hz, H-3, H.5), 4.29 (1H, d, J = 7.7 Hz H-1 of Glc), 2.80 (H-α); ^{13}C NMR: δ 156.7 (C-4), 130.9 (C-2, C-6), 130.8 (C-1), 116.1 (C-3, C-5), 36.3 (C-α), 72.1 (C-β), 104.3 (C_G-1, β-O-Glc), 78.1 (C_G-5), 77.9 (C_G-3), 75.1 (C_G-2), 71.6 (C_G-4), 62.7 (C_G-6).

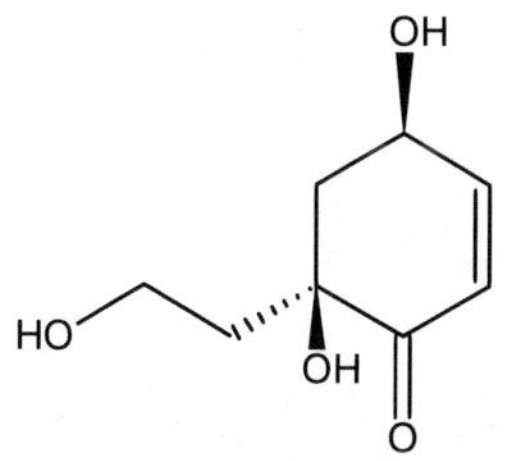

659

cis-2,4-Dihydroxy-2-(2-hydroxyethyl) cyclohex-5-en-l-one

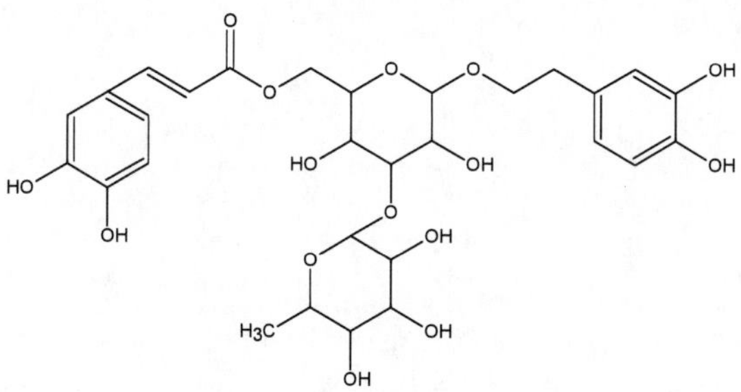

660

Isoacteoside

Found in *Bauhinia tarapotensis* Benth. (Leguminosae). In the DPPH free-radical scavenging test isoacteoside showed the most potent anti-oxidant activity with IC_{50} 1.3 μM, while (-)-Isolariciresinol 3-α-O-β-D-glucopyranoside and (+)-1-hydroxy pinoresinol-O-β-D-glucopyranoside exhibited moderate potencies with IC_{50} values of 10.5 and 20.3 μM, respectively (Braca et al., 2001).

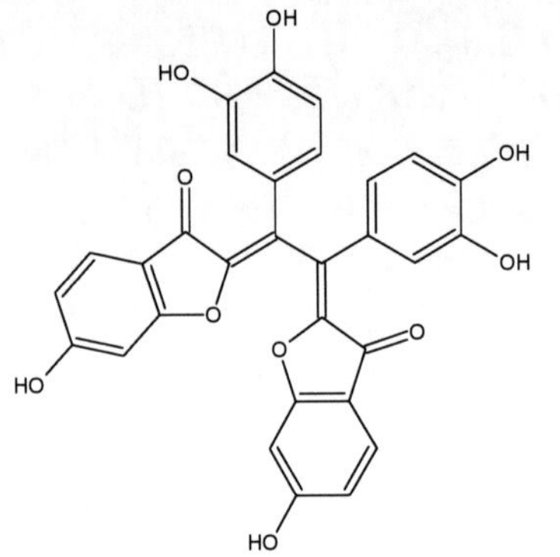

661
Disulfuretin (2,2'-[1,2-bis(3,4-dihydroxyphenyl)-1,2-ethanedylidene]-bis[6-hydroxy-3(2H)-Benzofuranone]

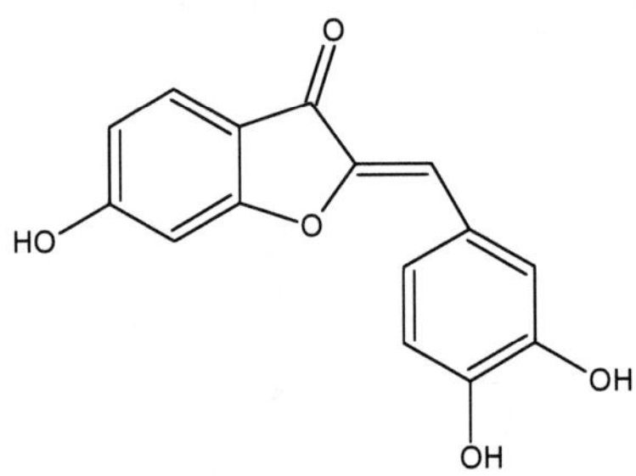

662
Sulfuretin

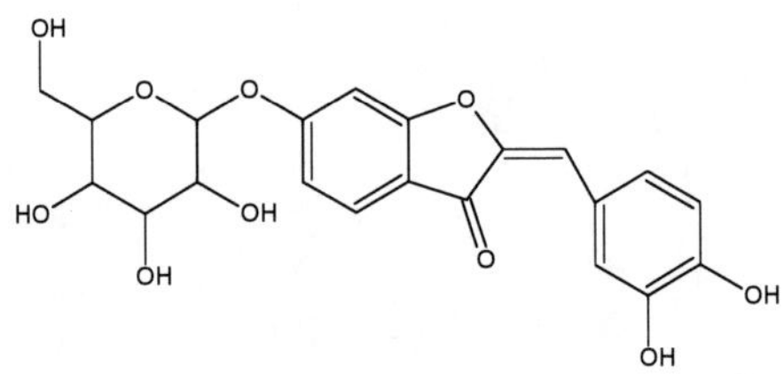

663
Sulfurein

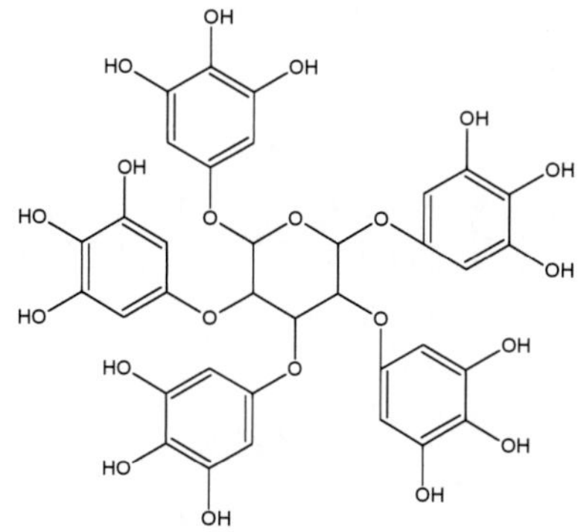

664
Pentagalloyl glucose

These constituents were isolated from EtOAc-soluble of the whole plant *Cotinus coggygri Scop.* (Anacardiaceae). All compounds were found to be potent antioxidants in a 1,1-diphenyl-2-picrylhydrazyl free-radical scavenging assay (Westenburg et al., 2000).

Disulfuretin: Yellow-green powder; UV λmax (EtOH) (log ε) 255 (3.9), 335 (4.2), 408 (4.4) nm; IR (film) νmax 3450, 1890, 1610, 1600 cm^{-1}; ESIMS-MS (positive-ion mode) (35 eV) *m/z* 539 $[M + H]^{+}$(40), 271 (100), 137 (52); Anal. C, 66.86, H, 3.37. Calcd for $C_{30}H_{18}O_{10}$: C, 66.96; H, 3.37; ^{1}HNMR (300 MHz, CD3OD): δ 7.59 (d, J = 8.5 Hz, H-4), 6.48 (d, J = 8.5 Hz, H-5), 6.72 (s, H-7), 7.52 (s, H-2′), 7.11 (d, J = 8.5 Hz, H-5′), 7.30 (dd, J = 8.5 Hz, H-6′), 7.62 (d, J = 8.5 Hz, 4′′), 6.58 (d, J = 8.5 Hz, H-5′′), 6.76 (s, H-7′′), 7.56 (s, H-2′′′), 7.14 (d, J = 8.5 Hz, H-5′′′), 7.38 (dd, J = 8.5 Hz, H-6′′′); ^{13}C NMR (75.4 MHz, CD3OD) δ 142.8 (C-2), 181.0 (C-3), 126.4 (C-4), 110.3 (C-5), 167.0 (C-6), 98.8 (C-7), 172.8 (C-8), 115.2 (C-9), 147.7 (C-10), 126.1 (C-1′), 115.6 (C-2′), 145.2 (C-3′), 146.4 (C-4′), 117.7 (C-5′), 126.1 (C-6′), 143.6 (C-2′′), 182.1 (C-3′′), 126.7 (C-4′′), 110.3 (C-5′′), 168.8 (C-6′′), 99.2 (C-7′′), 174.2 (C-8′′), 115.6 (C-9′′), 147.8 (C-10′′), 126.1 (C-1′′′), 119.8 (C-2′′′), 145.2 (C-3′′′), 146.5 (C-4′′′), 118.1 (C-5′′′), 126.2 (C-6′′′).

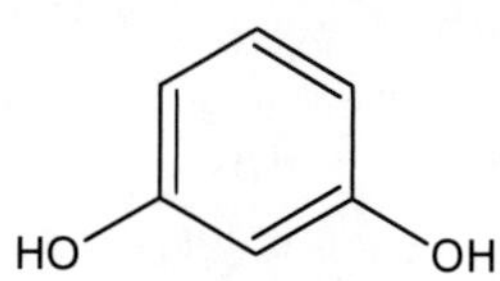

665
Resorcinol

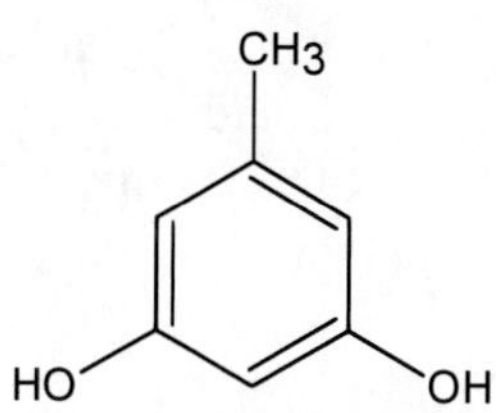

666
Orcinol

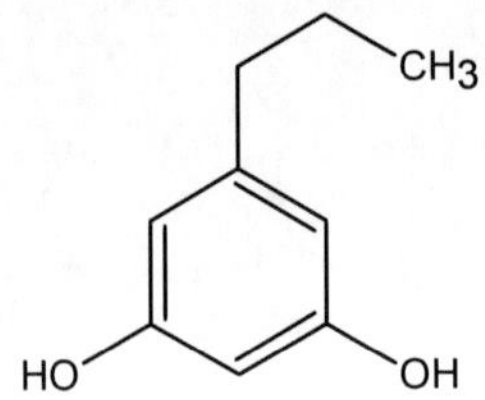

667
Divarin

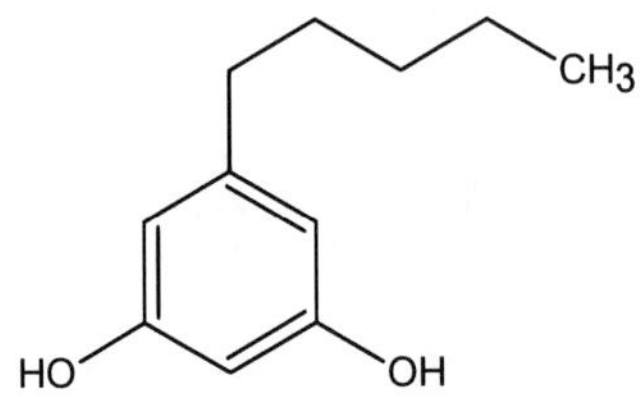

668
Olivetol

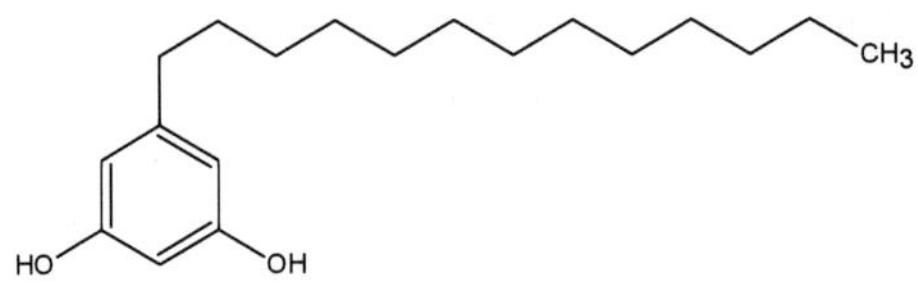

669
Grevillol

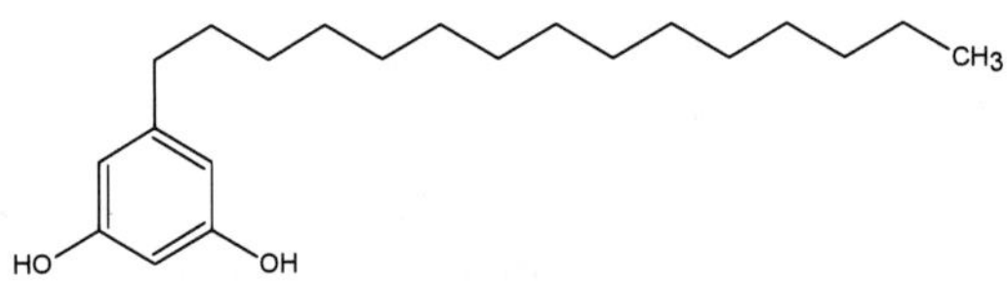

670
Cardol

Alkylresorcinols, resorcinol derivatives with a long hydrocarbon chain at position 5 of the phenolic ring, have been reported to act as antioxidants and protect polyunsaturated fatty acids against peroxidation. Their structures with meta-dihydroxy substitutions do not, however, support active hydrogen donation and scavenging of peroxyl radicals responsible for propagation of lipid oxidation. On DPPH and triacylglycerol oxidation models showed 5-n-pentadecylresorcinol to have very low hydrogen donation and peroxyl radical-scavenging potency (Kamal-Eldin et al., 2000).

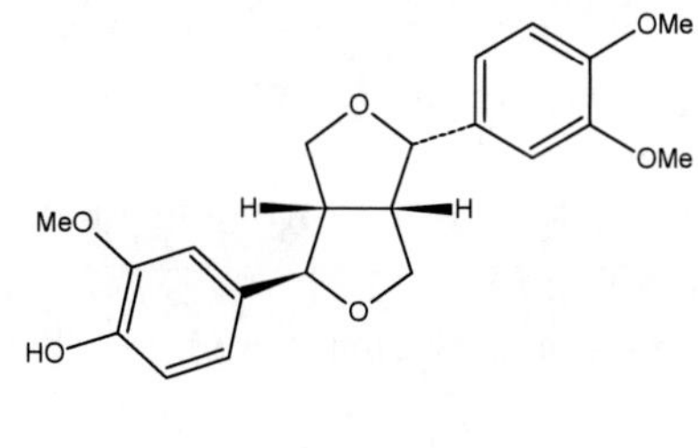

671
(-)-Phylligenin

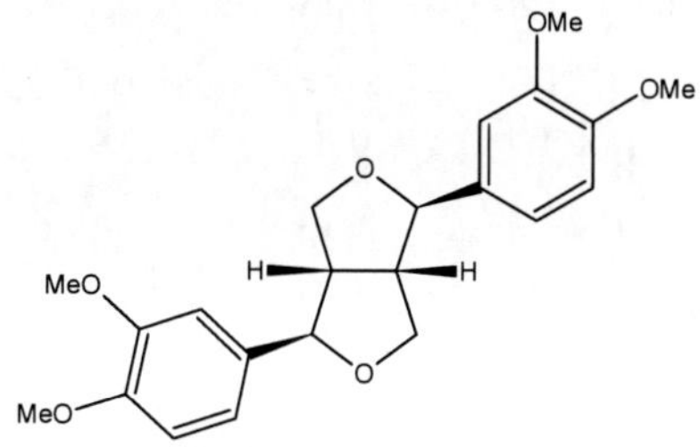

672
(-)-Eudesmin

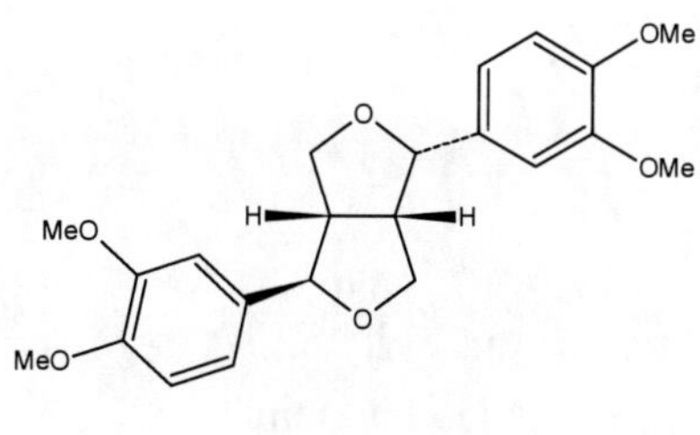

673
(-)-Epieudesmin

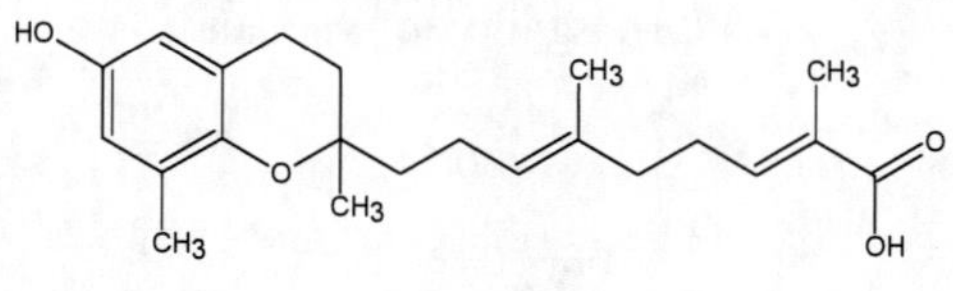

674
Polycerasoidol

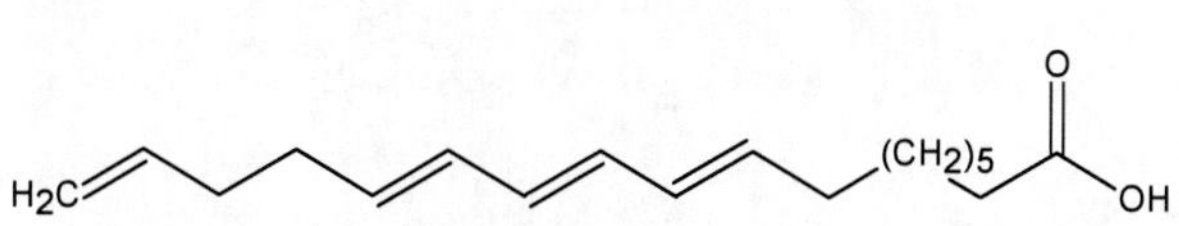

675
Oropheic acid

The CH_2Cl_2 extract of the leaves of *Orophea enneandra* displayed antifungal, antioxidant, and radical scavenging . Showed antioxidant activity against 2,2-diphenyl-l-picrylhydrazyl radical (Cavin et al., 1998a).

Oropheic acid or Octadeca-17-en-9,11,13-triynoic acid: Unstable white amorphous powder: TLC Rf 0.40; UV (MeOH): (log ε) 204 (4.3), 212 (3.6), 253 (2.5), 267 (2.6), 282 (2.6), 310 (2.2) nm; D/CIMS (NH_3, positive ion mode): *m/z* 288 [M + NH_4^-] 100), 271 [M + H^-] (21), 254 (12), 219 (3), 180 (4); LC/'TSP-MS data (NH_4OAc buffer, positive ion mode) $_{m/z}$ 288 [M + NH_4^-] (100), 271 [M + H]$^-$ (3), 231 (12), 182 (2); ^{1}HNMR (CD_3OD, 500 MHz): δ 5.83 (lH, m, H-17). 5.09 (lH, dd, J = 1.5, 17 Hz, H-18 trans), 5.04 (lH, dd, J = 1.5, 10 Hz, H-18 cis), 2.39 (2H, t, J = 7.0 Hz, H-15), 2.30 (2H, t, J = 7.0 Hz, H-8), 2.27 (2H, dd, J = 7.0, 13.9 Hz. H-16), 2.19 (2H, t, J =7.6 Hz, H-2), 1.60 (2H, m, H-3), 1.53 (2H, m, H-7), 1.40 (2H, m, H-6), 1.35 (2H, m, H-5), 1.34 (2H, m. H-4); ^{13}C NMR (CD_3OD, 125 MHz): δ181.41 (C-l), 137.47 (C-17), 116.56 (C-18), 80.42 (C-9), 79.42 (C-14), 66.59 (C-13), 66.17 (C-10), 61.14 (C-11), 60.80 (C-12), 38.10 (C-2), 33.32 (C-16), 30.47 (C-4), 29.94 (C-5), 29.83 (C-6), 29.21 (C- 7), 27.25 (C-3), 19.78 (C-15), 19.75 (C-15).

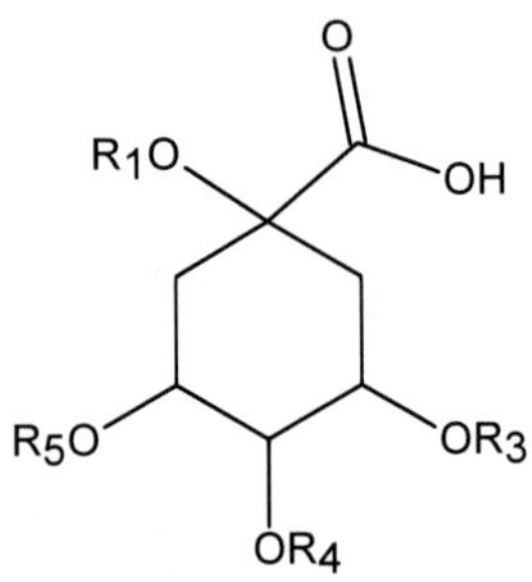

676
R_1 = H , R_5 = H, R_3 = G, R_4 = H
3-O-Galloylquinic acid

677
R_1 = H , R_5 = G, R_3 = G, R_4 = H
3,5 Di-O-galloylquinic acid

678
R_1 = H , R_5 = G, R_3 = G, R_4 = G
4,5 Di-O-galloylquinic acid

679
R_1 = G, R_5= H, R_3 = G, R_4 = H
1,3 Di-O-galloylquinic acid

680
R_1 = G, R_5= H, R_3 = G, R_4 = G
1,3,4 Tri-O-galloylquinic acid

681
R_1 = H, R_5 = G, R_3 = G, R_4 = G
3,4,5 Tri-O-galloylquinic acid

682
R_1 = G, R_5 = G, R_3 = G, R_4 = G
1,3,4,5 Tetra-O-galloylquinic

These tannins were isolated and characterized from different parts of *Guiera senegalensis* J.F. Gmel. (Combretaceae). Interesting results showed that galloylquinic acids (hydrolysable tannins), resulting from a tri- or tetra-substitution of galloyl groups on the quinic acid skeleton, played a crucial role in the inhibitory effect on Fe^{2+} -induced lipid peroxidation in rat liver microsomes and radical scavenger activity in the 1,1-diphenyl-2.picrylhydrazyl (DPPH) test. The effects of all tannins were markedly higher than that of gallic acid (Bouchet et al., 1998).

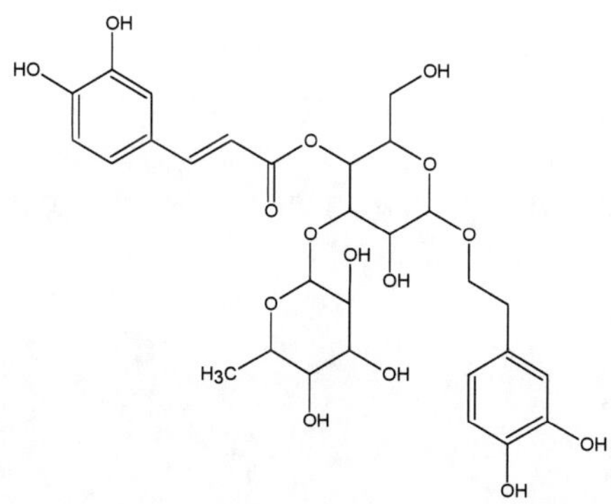

683
Verbascoside

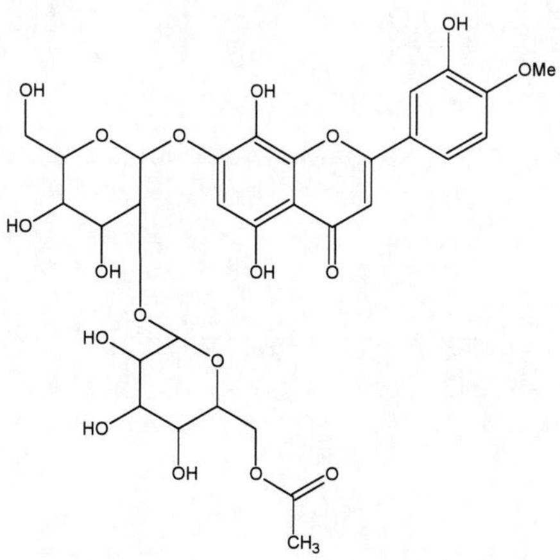

684
Hypolaetin-4'-methylether-7-O-[6‴-O-acetyl-β-D-allopyranosyl-(1‴→ 2")-β-D-glucopyranoside]

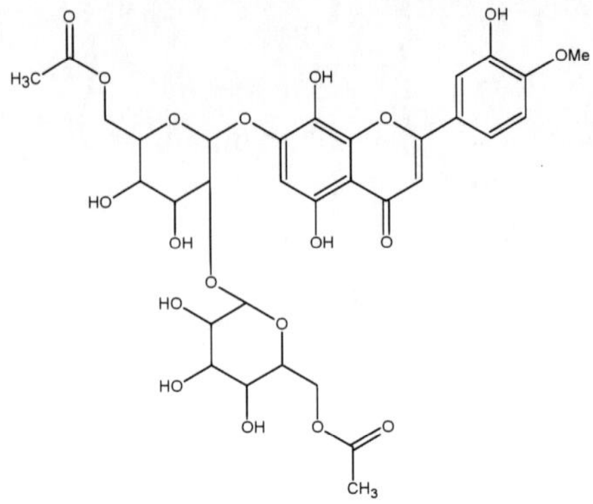

685

Hypolaetin-4'-methylether-7-O-[6′′′-O-acetyl-β-D-allopyranosyl-(1′′′ → 2′′)-β-D-6"-O-acetyl-β-D-glucopyranoside]

The main component from methanol extracts of *Sideritis scardica, S. syriaca* and *S. Montana* (Labiatae) was identified as verbascoside. The flavonoids and phenylpropanoids glycosides have also been found to be effective in rancidity retardation of edible oils, with or without natural tocopherols, and may explain the antioxidant activity exhibited in the HS-GC method. Consequently, the antioxidant activity of *Sideritis* extracts, may explain by the presence of these compounds (Koleva et al., 2003).

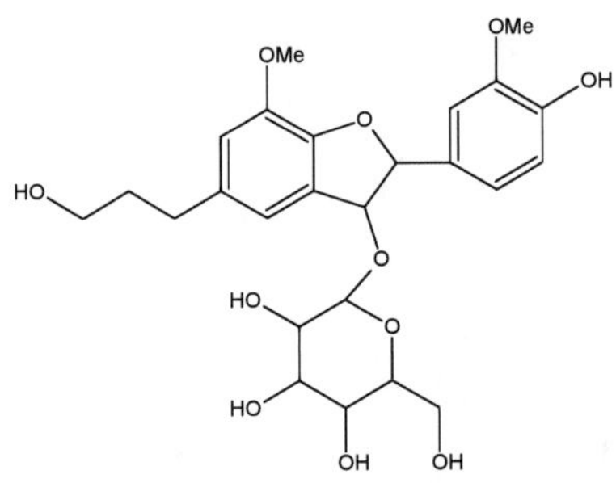

686

Dihydrodehydrodiconiferyl alcohol β-D-glucoside

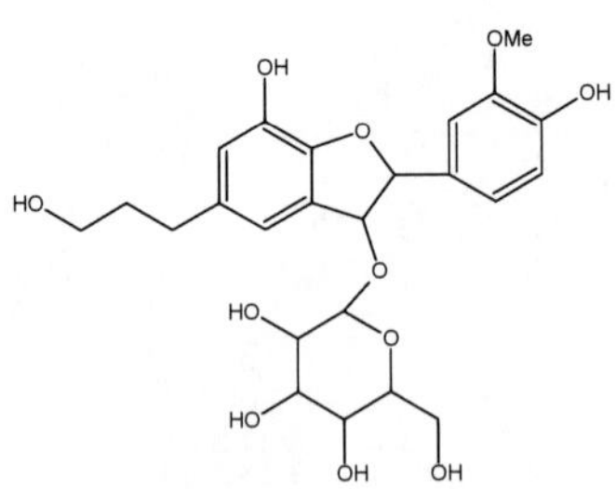

687

2R,2R-2,3-Dihydro-2-(4'-hydroxy-3'-methoxyphenyl)-3-(glucosyloxymethy 1)-7-hydroxy-5-benzofuranpropanol

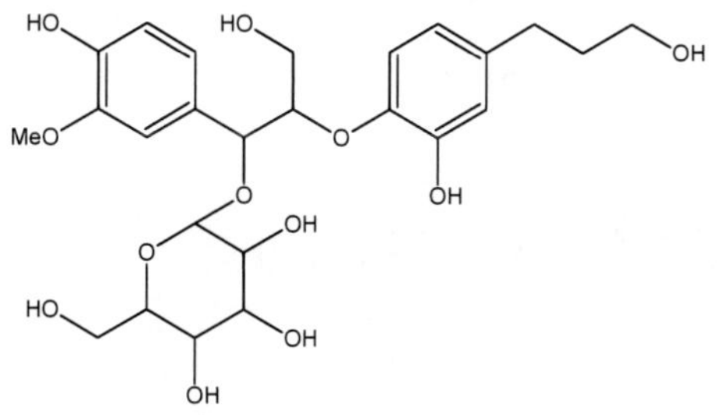

688
1-Glucosyloxy-2-[2-hydroxy-4-(3-hydroxy-propyl)-phenoxy]-1-(4-hydroxyl-3-methoxy-phenyl)-propan-3-ol

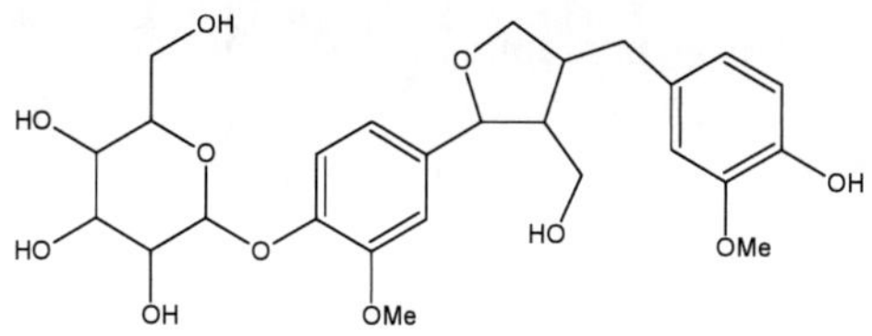

689
Lariciresinol 4-O-β-D-glucoside

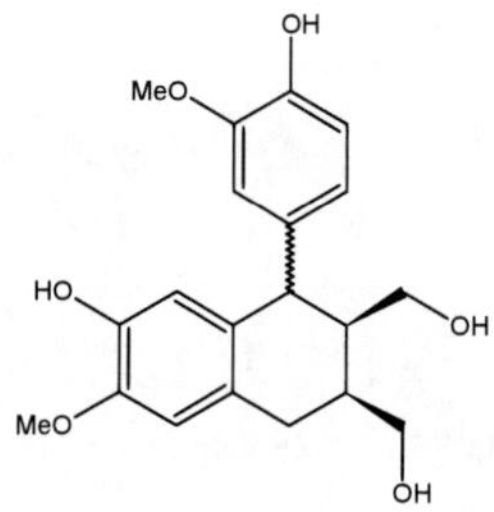

690
(+)-Isolariciresinol

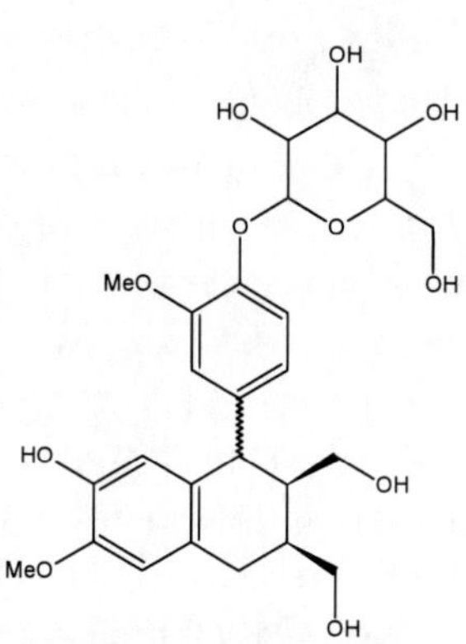

691
(+)-Isolariciresinol 6[a]-O-β-D-glucoside

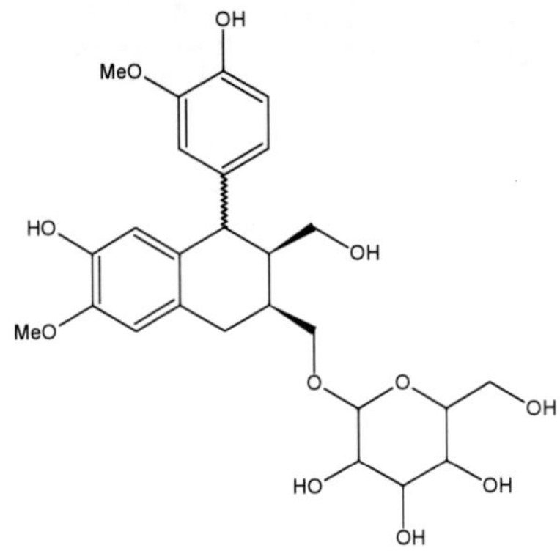

692
Isolariciresinol 6[a]-O-β-D-glucoside

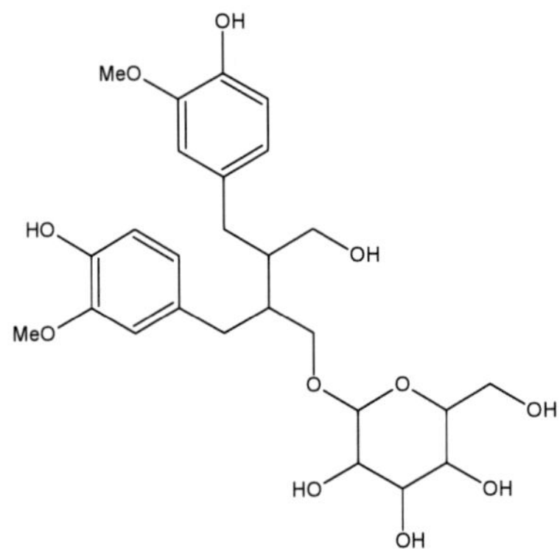

693
Secoisolariciresinol-β-D-glucoside

A German Riesling wine has been fractionated with the aid of countercurrent chromatography and the constituents is reported. These showed antioxidant activity. (Baderschneiderand Winterhalter, 2001).

2*R*, 3*R*, 2,3-Dihydro-2-(4'-hydroxy-3'-methoxypheny1)-3-(glucosyloxymethyl)-7-methoxy-benzofuran-5-propanol (Dihydrodehydrodiconiferyl alcohol β-D-glucoside): UV (MeOH): λmax 282, 213 nm. CD (c =0.05% in MeOH): $[\theta]_{294}$ + 1.7 x 10^3; $[\theta]_{260}$ -0.05 x 10^3; $[\theta]_{243}$ + 4.3 x 10^3; ORD (c = 0.05% in MeOH): $[\phi]_{299}$ +1.6 x 10^3; $[\phi]_{278}$ -0.6 x 10^3; $[\phi]_{249}$ +1.6 x 10^3; ESIMS (positive mode): pseudo molecular ion at *m/z* 545 $[M(522) + Na]^+$, MS/MS of *m/z* 545 at *m/z* 383 $[M + Na–anhydroglucose]^+$, and *m/z* 365 $[M - Na –glucose]^+$; ^{1}HNMR (300 MHz CD_3OD): δ 1.82 (2H, brqi, J = 8 Hz, H9), 2.62 (2H, brt, J = 8 Hz, H8), 3.23 (1H, dd, J = 8.5, 9 Hz, H2"), 3.27-3.40 (3H, m, H3"/H4"/H5"), 3.56 (2 H, t, J = 8 Hz, H10), 3.61-3.71 (2 H, m, H3/H6"a), 3.79-3.90 (2 H, m, H11a/H6"b), 3.83 (3H, s, OCH_3-C3), 3.86 (3 H, s, OCH_3-C3'); 4.10 (1H, dd, J = 9.5, 8 Hz, H11b), 4.35 (1 H, d, J = 7.5 Hz, H1"), 5.58 (1H, d, J = 6 Hz, H2), 6.72 (1H, brs, H4 or H6), 6.76 (1H, d, J = 8.5 Hz, H5'), 6.80 (1H, brs, H6 or H4), 6.86 (1H, dd, J = 8.5, 2 Hz, H6'), 7.00 (1H, d, J = 2 Hz, H2'); ^{13}CNMR (75.5 MHz, CD_3OD): δ 32.9 (C9), 35.8 (C8), 52.9 (C3), 56.5 and 56.8 (2 x OCH_3), 62.3 (C10), 62.8

(C6"), 71.7 (C4"), 72.4 (C11), 75.1 (C2"), 78.0 (C3"), 78.2 (C5"), 89.2 (C2), 104.3 (C1"), 111.0 (C2'), 114.4 (C4 or C6), 116.1 (C5'), 118.3 (C6 or C4), 119.8 (C6'), 129.8 (C5), 134.7 (C1'), 137.0 (C3a), 145.2 (C7), 147.5,147.9 (C7a/C4'), 149.0 (C3').

2*R*, 2*R*-2,3-Dihydro-2-(4'-hydroxy-3'-methoxyphenyl)-3-(glucosyloxymethy1)-7-hydroxyl-5-benzofuranpropanol: UV (MeOH): λmax 282, 207 nm. CD (c = 0.05% in MeOH, d = 1 cm): $[\theta]_{294}$ +2.3 x 10^3; $[\theta]_{263}$ -0.9 x 10^3; $[\theta]_{240}$ +2.1 x 10^3; $[\theta]_{226}$ -1.7 x 10^3; $[\theta]_{210}$ + 10.9 x 10^3; ORD (c = 0.050% in MeOH, d = 1 cm): $[\phi]_{300}$ +2.1 x 10^3; $[\phi]_{282}$ -0.8 x 10^3; $[\phi]_{248}$ +1.2 x 10^3; $[\phi]_{234}$ -1.7 x 10^3; Thermospray-MS: pseudo molecular ion at *m/z* 526 = $[M(508) + NH_4]^+$; *m/z* 329 $[M- glucose + H]^+$; ^{1}HNMR (360 MHz CD_3OD): δ 1.78 (2H, m, H9), 2.55 (2H, m, H8),3.23 (1H, dd, J = 8, 9.5 Hz, H2"), 3.28-3.40 (3H, m, H3"/H4"/H5"), 3.55 (2H,m, H10), 3.62 (1H, m, H3), 3.69 (1H, dd, J = 12/5.5 Hz, H6"a), 3.82 (3H, s, H_3CO-C3'), 3.85 (1H, dd, J = 12.5, 2.5 Hz, H6"b), 3.86 (1H, d, J = 9.5 Hz, H11a), 4.10 (1H, dd, J = 9.5, 8 Hz, H11b), 4.35 (1H, d, J= 8 Hz, H1"), 5.57 (1H, d, J = 6.5 Hz, H2), 6.56 (1H, s, H6), 6.67 (1H, s, H4), 6.75 (1H, dd, J = 8/2 Hz, H6), 6.87 (1 H, d, J = 8 Hz, H5), 7.02 (1 H, d, J = 2 Hz, H2); ^{13}CNMR (90.6 MHz, CD_3OD): δ 32.7 (C9), 35.8 (C8), 53.2 (C3), 56.5 (OCH_3), 62.3 (C10), 62.8 (C6"), 71.7 (C4"), 72.5 (C11), 75.2 (C2"), 78.1 (C3"), 78.2 (C5"), 88.9 (C2) 104.3 (C1"), 110.8 (C2'), 116.0 (C4), 117.0 (C5'), 119.8 (C6), 123.1 (C6'), 129.6 (C5), 135.0 (C1'), 136.8 (C3a), 141.9 (C7), 146.4 (C7a), 147.3 (C4'), 149.0 (C3').

1-Glucosyloxy-2-[2-hydroxy -4-(3-hydroxy-propyl)-phenoxy]-1-(4- hydroxyl-3-methoxy-phenyl)-propan-3-ol: UV (MeOH): λmax 275, 208 nm. CD (0.01% in MeOH, d = 1 cm): $[\theta]_{312}$ -0.4 x 10^3; $[\theta]_{300}$ -0.0 x 10^3; $[\theta]_{271}$ +3.1 x 103; $[\theta]_{246}$ +0.4 x 10^3; $[\theta]_{235}$ + 4.1 x 10^3; $[\theta]_{225}$ +0.7 x 10^3; ESIMS (negative mode): pseudo molecular ion at *m/z* 525 $[M(526) -H^+]^-$;

MS/MS of *m/z* 525 at m/z 507 $[M -H^+ -H_2O]^-$ and *m/z* 345 $[M -H^+ -H2O -anhydroglucose]^-$; ^{1}HNMR (300 MHz CD_3OD): δ 1.77 (2 H, m, H2'''); 2.54 (2H, t, J = 7 Hz, H1'''), 3.24 (1H, dd, J = 819 Hz, H2''''), 3.30-3.45 (3 H, m, H3´´´´/H4''´´´/H5´´´´), 3.53 (2 H, t, J= 7 Hz, H3´´´), 3.65 (lH, dd, J = 12.5, 5 Hz, H6´´´´a), 3.84 (1H, dd, J = 12.5, 2 Hz, H6´´´´b), 3.85 (3 H, s, OCH_3), 3.8-3.9 (2H, m, H3), 4.32 (1H, d, J = 8 Hz, H1´´´´), 4.35 (1H, m, H2), 4.99 (1H, d, J = 5.5 Hz, H1), 6.55 (1H, dd, J = 2, 8 Hz, H5'); 6.67 (1H, d, J = 2 Hz, H3'), 6.74 (1H, d, J = 8 Hz, H5"), 6.88 (1 H, dd, J = 2, 8 Hz, H6"), 6.89 (1H, d, J = 8 Hz, H6'), 7.04 (1 H, d, J = 2 Hz, H2").

1,4-[4-(4-Hydroxy-3-methoxy-benzyl)-3-hydroxymethyltetrahydrofuran-2-yl]-2-methoxy-phenoxy}-β-D-glucopyranose (Lariciresinol 4-O-β-D-glucoside): UV (MeOH) λmax 279, 225, 212 nm. CD (0.01 %in MeOH, d = 1 cm}: $[\theta]_{299}$ +0.4 x 10^3; $[\theta]_{293}$ 0; $[\theta]_{280}$ -1.8 x 10^3; $[\theta]_{259}$ -0.8 x 10^3; $[\theta]_{235}$ -9.5 x 10^3; ESIMS (negative mode): pseudo molecular ion at m/z 521 $[M(522) -H^+]^-$; MS/MS of *m/z* 521 at m/z 359 $[M -H^+$-anhydroglucose$]^-$; ^{1}HNMR (300 MHz CD_3OD): δ 2.35 (1H, qi, J = 6.5 Hz, H8), 2.55 (1 H, dd, J = 11, 13 Hz, H7'a), 2.72 (1H, m, H8'), 2.91 (1H, dd, J = 5.5, 13 Hz, H7'b), 3.66 (1 H, dd, J = 7, 11 Hz, H9a), 3.68 (1 H, dd, J = 2.5, 12 Hz, H6´´a), 3.73 (1H, dd, J = 6.5, 8 Hz, H9'a), 3.83 and 3.86 (2 x 3 H, 2 x s, 2 x OCH_3), 3.86 (2H, m, H9b/H6"b), 4.0 (1H, dd, J = 6.5, 8 Hz, H9'b), 4.82 (1H, d, J = 6.5 Hz, H7), 4.87 (1H, d, J = 7.5 Hz, H1"), 6.66 (1 H, dd, J = 2, 8 Hz, H6'), 6.72 (1H, d, J = 8 Hz, H5'), 6.79{1H, d, J = 2 Hz, H2'), 6.88 (1H, dd, J = 2, 8 Hz, H6), 6.99 (1 H, d, J = 2 Hz, H2), 7.14 (1 H, d, J = 8 Hz, H5).

6*R*, 7*S*, 8*S* – 8 - (4-Hydroxy-3-methoxyphenyl)-6,7-bis-hydroxymethyl-3-methoxy 5,6,7,8-tetrahydro-naphthalen-2-ol [(+)-Isolariciresinol]: UV (MeOH): λmax 284, 212 nm. CD (0.01% in MeOH, d = 1 cm): $[\theta]_{293}$ -8.3 x 10^3; $[\theta]_{277}$ +5.3 x 10^3; $[\theta]_{253}$ +0.3 x 10^3; $[\theta]_{240}$ +

10.2 x 10^3; ORD (0.01% in MeOH, d = 1 cm}: $[\phi]_{320}$; $[\phi]_{299}$ -3.2 x 10^3; $[\phi]_{295}$ 0; $[\phi]_{284}$ +9.4 x 10^3; $[\phi]_{267}$ + 0.3 x 10^3; $[\phi]_{244}$ + 6.2 x 10^3; $[\phi]_{239}$ 0; ESIMS (negative mode): pseudo molecular ion at *m/z* 359 [M(360)-H^+]$^-$; MS/MS of *m/z* 359 at *m/z* 344 [M -H^+-CH_3]$^-$; ^{1}HNMR (300 MHz CD_3OD): δ 1.78 (1H, m, H7), 2.0 (1H, m, H6), 2.78 (2H, brd, J = 8 Hz, H5), 3.40 (1H, dd, J = 4.5, 11 Hz, H7aa), 3.62-3.71 (3 H, m, H7ab/H6a), 3.78 (3H, s, OCH_3-C3'), 3.79 (1H, d, J = 10 Hz, H8), 3.81 (3H, s, OCH_3-C6), 6.18 (1H, brs, H1), 6.61 (1H, dd, J = 2, 8 Hz, H6'), 6.67 (1H, brs, H4), 6.68 (1H, d, J = 2 Hz, H2'), 6.74 (1H, d, J = 8 Hz, H5').

6*R*, 7*S*, 8*S* – 6 – Glucosyloxymethyl – 8 -(4-hydroxy-3-methoxyphenyl)-7-hydroxymethyl-3-methoxy-5,6,7,8-tetrahydro naphthalen-2-ol [(+)-Isolariciresinol 6a-O-β-D-Glucoside]: UV (MeOH): λmax 283, 206 nm. CD (c = 0.01% in MeOH, d = 1 cm): $[\theta]_{293}$ +3.3 x 10^3; $[\theta]_{277}$+ 11.7 x 10^3; $[\theta]_{256}$ -2.4 x 10^3; $[\theta]_{240}$ -29.9 x 10^3; $[\theta]_{224}$ -4.3 x 10^3; $[\theta]_{218}$ -7.4 x 10^3; ESIMS (negative mode}: pseudo molecular ion at *m/z* 21[M(522} -H^+]$^-$; MS/MS of *m/z* 521 at *m/z* 359 [M-H^+-anhydroglucose]$^-$; ^{1}HNMR (400 MHz CD_3OD): δ 1.78 (1H, m, H7), 2.17 (1H, m, H6), 2.84 (1H, brd, H5), 3.20 (1H, dd, J = 8, 9 Hz, H2"), 3.27-3.40 [4H, m, obscured at 3.35 (lH, m, H7aa), (3H, m, H3´´H4´´/H5")], 3.64 (lH, m, H6aa); 3.68 (lH, m, H7ab), 3.72 (1 H, dd, J = 4, 12 Hz, H6´´a), 3.78 (3H, s, OCH_3), 3.80 (3H, s, OCH_3), 3.84 (1 H, brd, J = 10 Hz, H1), 3.87 (1 H, dd, J = 2, 12 Hz, H6"b), 4.02 (1H, dd, J = 6, 10 Hz, H6ab); 4.29 (lH, d, J = 8 Hz, H1"), 6.18 (1H, s, H8), 6.61 (1 H, dd, J = 2, 8 Hz, H6'), 6.64 (1H, s, H5), 6.68 (1H, d, J = 2 Hz, H2'), 6.73 (1H, d, J = 8 Hz, H5'); ^{13}CNMR (90.6 MHz, CD_3OD): δ 33.1 (C5), 37.4 (C6}, 45.3 (C7), 47.9 (C8}, 56.4 (2 x OCH_3), 61.6 (C7a), 62.8 (C6"), 65.2 (C6a), 71.7 (C4"), 73.9 (C6a), 75.2 (C2"), 78.0 (C3"), 78.2 (C5"), 104.6 (C1´´), 112.4 (C4), 113.9 (C2'), 116.0 (C5'), 117.4 (C1), 123.3 (C6'), 129.1 (C9}, 134.2 (C1'), 138.7 (C10), 145.2 (C4'), 145.9 (C2), 147.2 (C3), 149.0 (C3').

6-G1ucosy1oxymethyl-8-(4-hydroxy-3-methoxyphenyl) -7-.hydroxymethyl-3-methoxy-5,6, 7,8-tetrahydronaphthalen-2-ol (Isolariciresinol 6a-O-β-D-glucoside): UV (MeOH): λmax 283,206 nm. CD (c = 0.1% in MeOH, d = 1 cm): $[\theta]_{293}$ +3.3 x 10^3; $[\theta]_{277}$ + 11.7 x 103; $[\theta]_{256}$ -2.4 x 10^3; $[\theta]_{240}$ -29.9 x 10^3; $[\theta]_{224}$ -4.3 x 10^3; $[\theta]_{218}$ -7.4 x 10^3; ESIMS (negative mode): pseudo molecular ion at *m/z* 521 [M(522) -H^+]$^-$; MS/MS of *m/z* 521 at *m/z* 359 [M -H^+ -anhydroglucose]$^-$. ^{1}HNMR (300 MHz CD_3OD): δ 2.11 (1H, m, H7), 2.21 (1H, m, H6), 2.72 (1H, dd, J = 9.5, 17 Hz, H5a), 2.99 (1 H, dd, J = 6, 17 Hz, H5b), 3.16 (1H, dd, J = 8, 9 Hz, H2"), 3.23-3.35 (4 H, m, H7aa/H3"/H4"/H5"), 3.56 (1H, dd, J = 5.5, 9.5 Hz, H6aa), 3.65 (1H, dd, J = 6.5, 11 Hz, H7ab), 3.65 (1H, dd, J = 5.5, 12 Hz, H6"a), 3.75 (3 H, s, OCH_3), 3.83 (3 H, s, OCH_3); 3.84 (1H, dd, J = 2.5, 12 Hz, H6"b), 3.91 (1H, dd, J = 6, 9.5 Hz, H6aa), 4.21 (1 H, d, J = 4.5 Hz, H8), 4.23 (1H, d, J = 8 Hz, H1"), 6.35 (1H, brs, H1), 6.45 (1 H, dd, J = 2, 8 Hz, H6'), 6.64 (1H, d, J = 8 Hz, H5'), 6.69 (lH, brs, H4), 6.70 (1H, d, J = 2 Hz, H2'); ^{13}CNMR (90.6 MHz, D_3OD): δ 33.0-33.2 (C6/C5), 44.7-46.2 (C7/C8), 56.3-56.4 (2 x OCH_3), 62.8-62.9 (C7a/C6"), 71.6 (C4"), 73.4 (C6a), 75.1 (C2"), 77.1 (C3"), 78.2 (C5"), 104.7 (C1"), 112.4 (C4), 115.3-115.5 (C2'/C5'), 117.1 (C1), 124.1 (C6'), 128.6 (C9), 133.0 (C1'), 135.9 (C10), 145.5 (C4 '), 145.8 (C2), 147.8 (C3), 148.3 (C3').

2*R*,3*R*-2,3- Bis-(4- hydroxyl-3-methoxy.benzyl)-butan-1,4-diol [(-)-Secoisolariciresinol]: UV (MeOH): λmax 282, 209 nm. ESIMS (negative mode): pseudo molecular ion at *m/z* 361 [M(362)-H^+]$^-$; MS/MS of *m/z* 361 at *m/z* 346 [M-H^+-CH_3]$^-$; ^{1}HNMR (300 MHz CD_3OD): δ 1.91 (2H, m, H3/H4), 2.56 (2H, dd, J = 7.5, 14 Hz, H7a/H7'a), 2.67 (2H, dd, J = 7,14 Hz, H7b/H7'b), 3.59 (4 H, m,H1/H4), 3.74 (6H, s, 2 x OCH_3); 6.55 (2H, dd, J = 2, 8 Hz, H6'/H6"), 6.59 (2 H, d, J = 2 Hz, H2'/H2"), 6.66 (2 H, d, J = 8 Hz, H5'/H5").

1-Glucosyloxy-2,3-bis-(4-hydroxy-3-methoxy-benzy1)-butan-4-o1 (Secoisolariciresinol β-D-glucoside). UV (MeOH): λmax 282, 210 nm. ESIMS (nnegative mode): pseudo molecular ion at *m/z* 523 [M(524) $-H^+]^-$; MS/MS of *m/z* 523 at *m/z* 361 $[M-H^+$-anhydroglucose$]^-$; ^{1}HNMR (300 MHz CD_3OD): δ 2.01 (1H, m, H3), 2.08 (1H, m, H2), 2.57-2.75 (4H, m, H7'/H7"), 3.21 (1H, dd, J = 8, 9 Hz, H2'''), 3.30-3.40 (3H, m, H3'''/H4'''/H5'''), 3,56 (1H, dd, J = 5.5, 10.5Hz, H1a), 3.57 (1H, dd, J = 6, 11 Hz, H4a), 3.65 (1H, dd, J = 5.5, 11 Hz, H4b), 3.68 (1H, dd, J = 5.5, 12 Hz, H6'''a), 3.76 (6 H, s, 2 x OCH_3), 3.87 (1 H, dd, J = 2, 12 Hz, H6'''b), 3.90 (1H, dd, J = 5.5, 10.5 Hz, H1a); 4.19 (1H, d, J = 8 Hz, H1'''); 6.57 and 6.58 (2H, 2 x dd, J = 2, 8 Hz, H6'/H6"); 6.63 and 6.65 (2H, 2 x d, J = 2 Hz, H2'/H2"); 6.66 and 6.67 (2H, 2 x d, J = 8 Hz, H5'/H5").

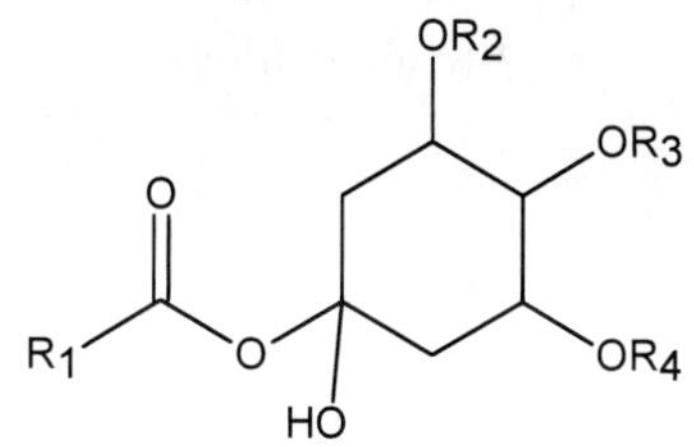

694 R_1=CH_3, R_2=R_3=caffeoyl, R_4=H
Methyl 3,4-di-O-caffeoylquinate

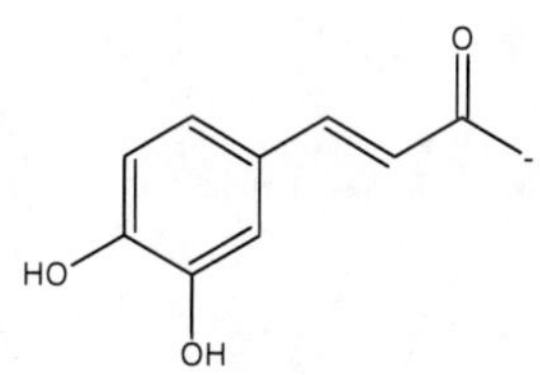

Caffeoyl

695
R_1=H, R_2=R_3=caffeoyl, R_4=H
3,4-Di-O-caffeoylquinic acid

696
R_1=CH_3, R_2=H, R_3= R_4= caffeoyl
Methyl 4,5-di-O-caffeoylquinate

Propolis is a resinous hive product collected by honeybees from various plant sources. It is a popular folk medicine possessing a broad spectrum of biological activities. These dicaffeoylquinic acid possessed more potent hepatoprotective effects than glycyrrhizin at a concentration of 10 μg/mL and methyl 3,4-di-O-caffeoylquinate was the most potent of the three compounds whit antioxidant activity (Banskota et al., 2001).

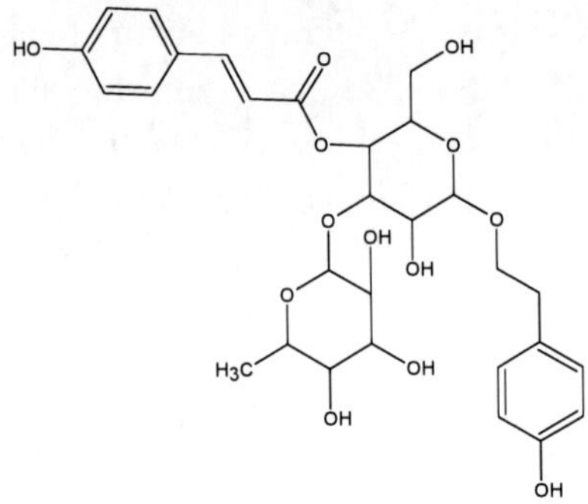

697
Osmanthuside B

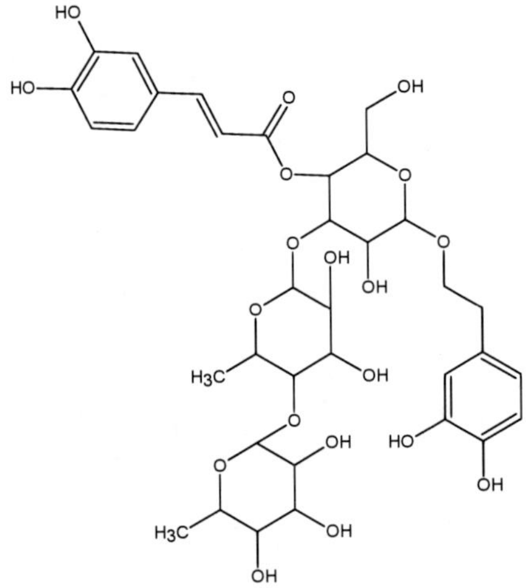

698
Ligupurpuroside A

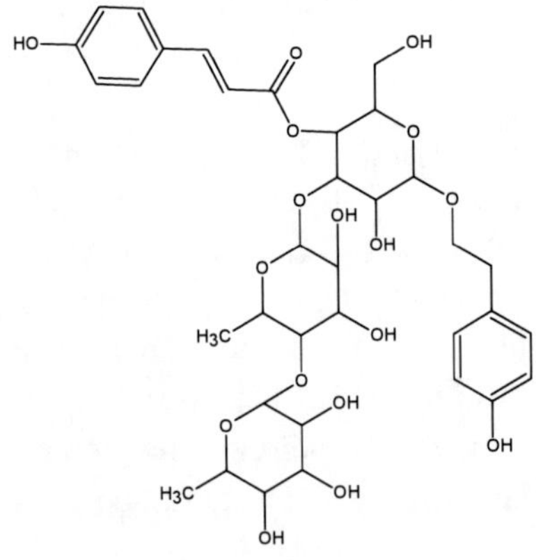

699
trans-Ligupurpuroside B

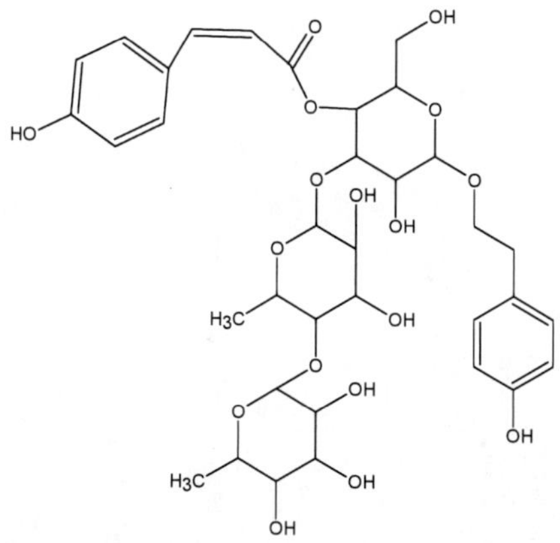

700
cis-Ligupurpuroside B

Tea and kudingcha (bitter tea) are the two most popular beverages consumed in China. It was found that the crude glycoside fraction from the leaves of *Ligustrum purpurascens* (LP) possessed strong protection against oxidation of human low-density lipoprotein (LDL). Acteoside, isoacteoside, and ligupurpuroside A were protective, whereas *cis*-ligupurpuroside B, *trans*-ligupurpuroside B, and osmanthuside B exhibited no protection to human LDL from Cu^{2+} -medicated oxidation. Acteoside, isoacteoside, and ligupurpuroside A were also effective in preventing the peroxyl free radical-induced oxidation of α-tocopherol in human LDL. The antioxidant activities of acteoside, isoacteoside, and ligupurpuroside A were comparable to that observed for a green tea antioxidant, (-)-epicatechin gallate. The inhibitory effect of these three phenylethanoid glycosides on oxidation of human LDL and a-tocopherol was dose-dependent at concentrations of 5-40 μM (Wong et al., 2001).

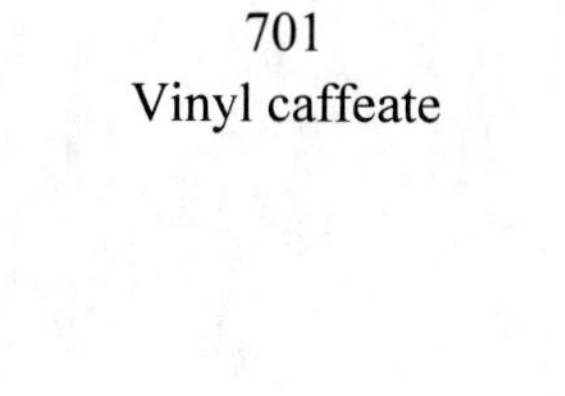

701
Vinyl caffeate

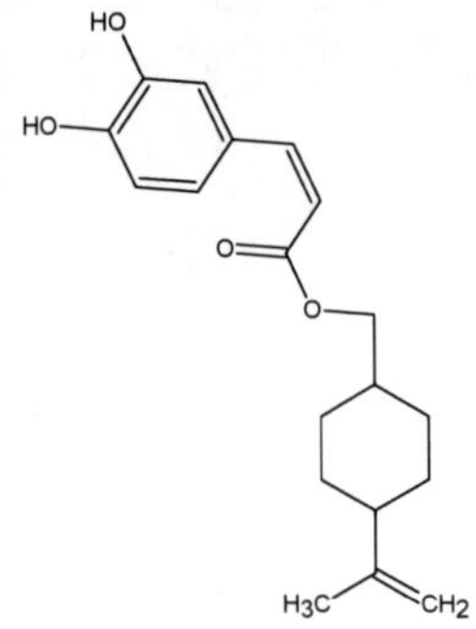

702
trans-p-Menth-8-en-7-yl caffeate

Two antioxidants (vinyl caffeate and *trans-p*-menth-8-en-7-yl caffeate) were isolated from *Perilla frutescens* Brit ton var. crispa (Thunb.). The redox potentials of the l isolated antioxidative compounds were comparable to those of known antioxidants. *trans-p*-menth-8-en-7-yl caffeate was effective to prevent the oxidative degradation of perilla ldehyde in the essential oil of *P. frutescens* (Tada et al., 1996).

Vinyl caffeate: White crystals, mp 133-140°C; HRMS *m/z*: 206.0597 $[M]^+$ (calcd. for $C_{11}H_{10}O_4$: 206.0579); νmax (KBr) cm^{-1}: 3400 (br.), 1710, 1600, 1440, 1290, and 1165; λmax (EtOH) nm (ε): 221 (17.000), 247 (13.000), and 339 (22,000); ^{1}HNMR ($CDCl_3$): δ 7.67 (1H, d, J = 15.9Hz), 7.42 (1H, dd, J= 14.0, 6.4 Hz), 7.12 (1H, d, J = 2.0Hz), 7.05 (1H. dd, J = 8.3, 2.0 Hz), 6.88 (1H, d, J = 8.3 Hz), 6.28 (1H, d, J = 15.8 Hz), 4.97 (1H, dd, J = 14.0, 1.5Hz), 4.63 (1H, dd, J = 6.4, 1.5Hz); ^{13}CNMR ($CDCl_3$): δ164.4, 146.6, 146.5, 143.8, 141.3, 127.5, 123.0, 115.6, 114.5,114.3, 97.8.

trans-p-Menth-8-en-7-yl caffeate: Yellow crystals, mp 155-156°C; HRMS *m/z*: 316.1693 $[M]^+$ (calcd. for $C_{19}H_{24}O_4$: 316.1675); 21.1; νmax (KBr) cm^{-1}: 3500, 3350, 2970, 1685, 1600, 1280, 1180; λmax (EtOH) nm (ε): 217 (15,450), 236 (9,130), 245 (9,730), 322 (16,910); ^{1}HNMR ($CDCl_3$): δ 7.56 (1H, d, J = 15.9 Hz), 7.07 (1H, d, J = 2.0 Hz), 6.98 (1H, dd, J = 8.3, 2.0Hz), 6.81 (1H, d, J = 8.3 Hz), 6.29 (1H, d, J = 15.9 Hz), 4.70 (2H, m), 1.74 (1H, s), 4.02 (2H. d, J = 6.3 Hz), 1.98-1.80 (6H, m), 1.74 (3H, s), 1.40-1.05 (4H, m); ^{13}CNMR ($CDCl_3$): δ 169.4, 151.6, 149.6. 146.8. 146.8, 127.8, 122.9, 116.5, 115.1, 115.1, 108.8, 70.5, 46.7, 38.5, 32.3, 30.8.

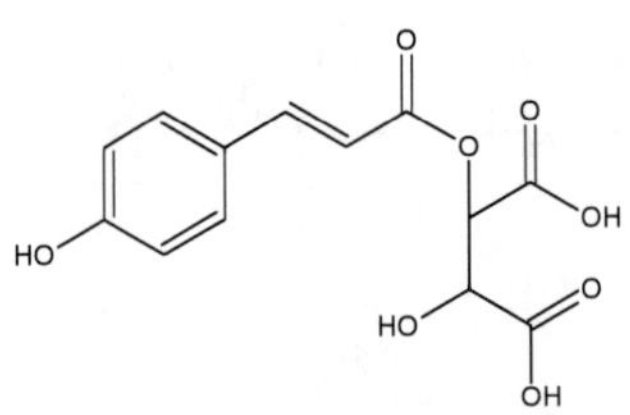

703
(E)-*p*-Coumaroyl tartaric acid

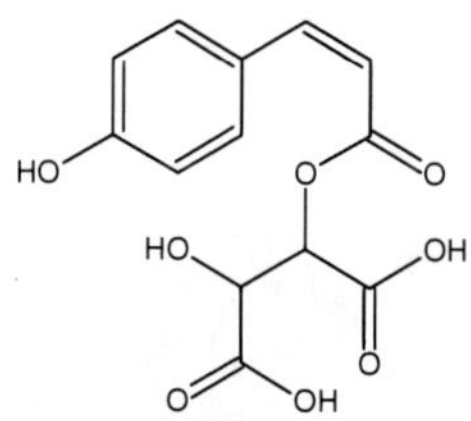

704
(Z)-*p*-Coumaroyl tartaric acid

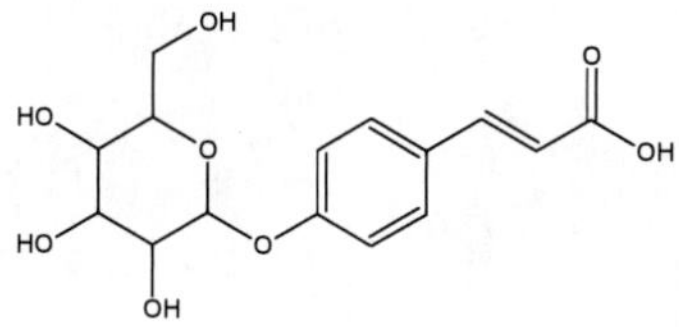

705
4-O-β-D-Glucoside of (E)-*p*-coumaric acid

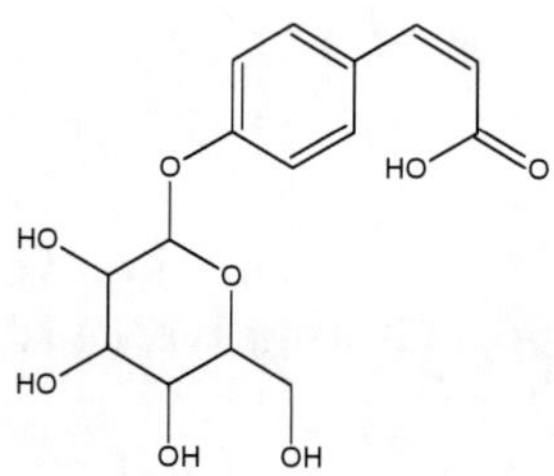

706
4-O-β-D-Glucoside of (Z)-*p*-coumaric acid

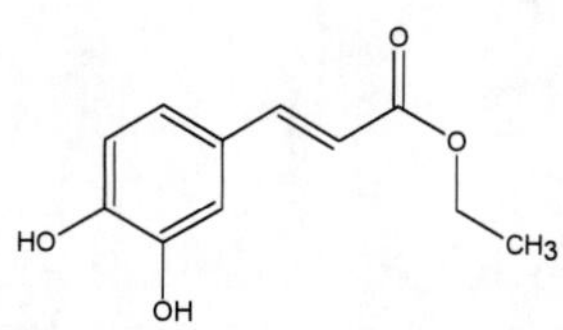

707
€-Ethyl caffeate

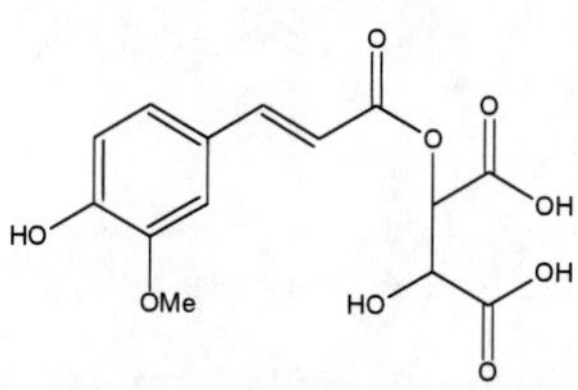

708
€-Feruloyl tartaric acid

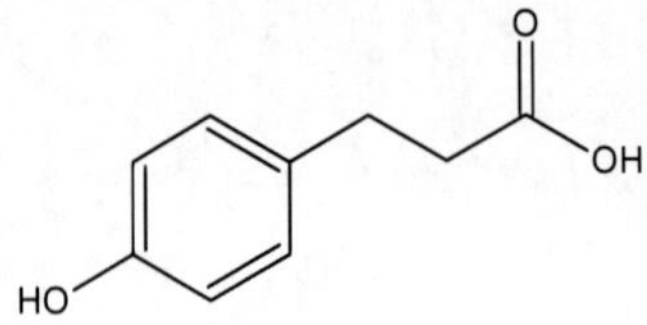

709
p-Hydroxyphenylpropionic acid

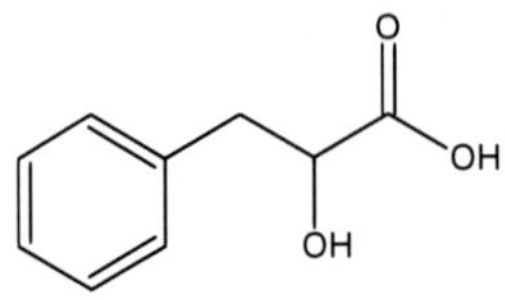

710
2-Hydroxy-3-phenylpropionic acid

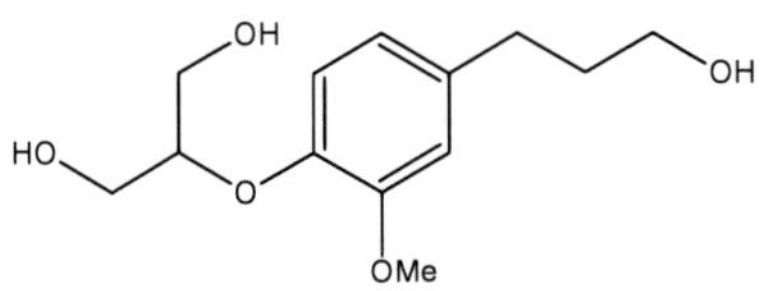

711
2-[4-(3-Hydroxypropyl)-2-methoxyphenoxy]-propan-1,3-diol

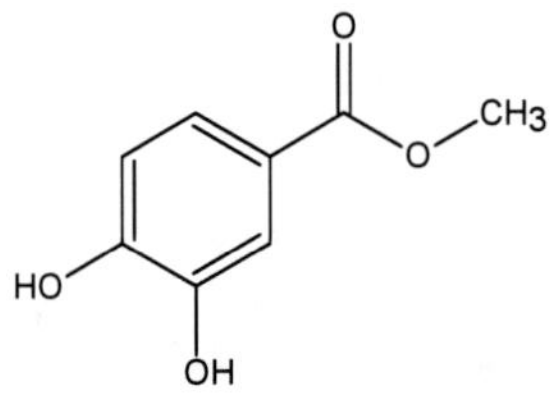

712
Methyl protocatechuic acid

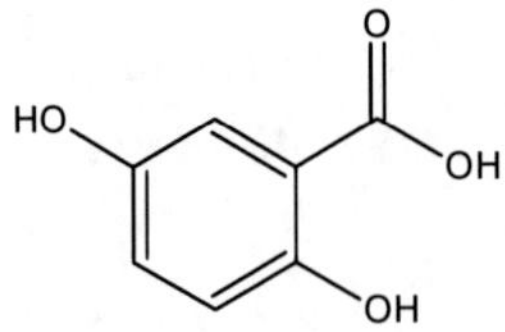

713
Gentisic acid

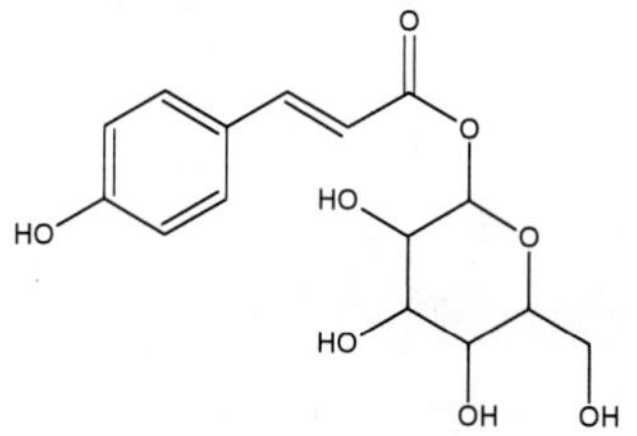

714
(E)-*p*-Coumaroyl glucose ester

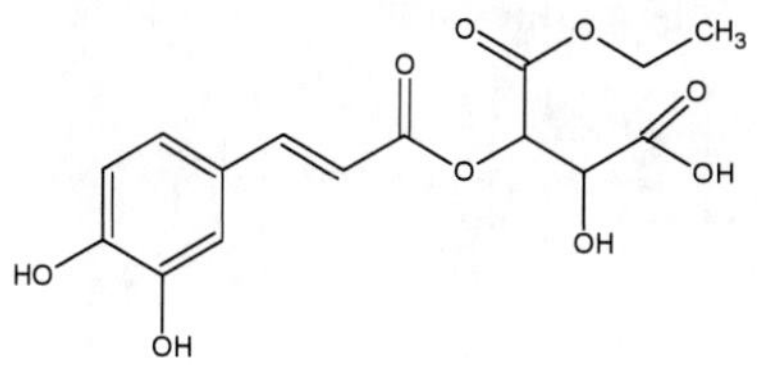

715
(E)-Caffeoyl ethyl tartrate (isomer I)

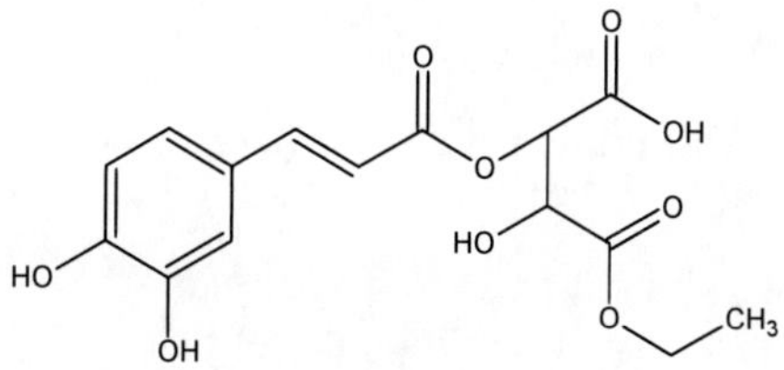

716
(E)-Caffeoyl ethyl tartrate (isomer II)

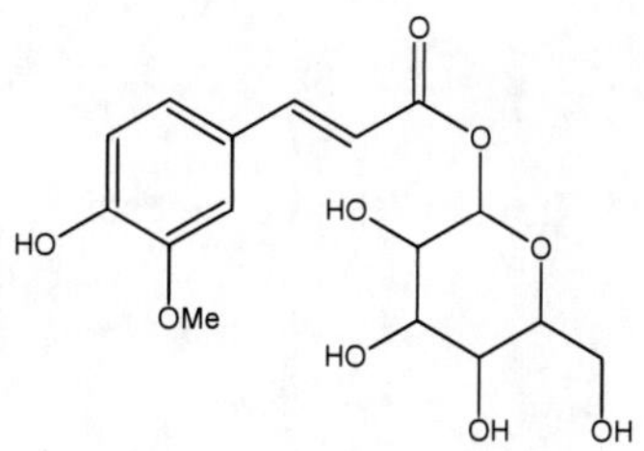

717
(E)-Feruloyl glucose ester

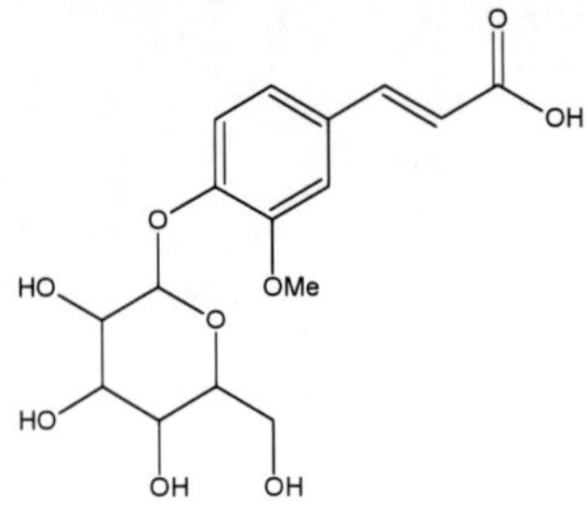

718
4-O-β-D-Glucoside of (E)-ferulic acid

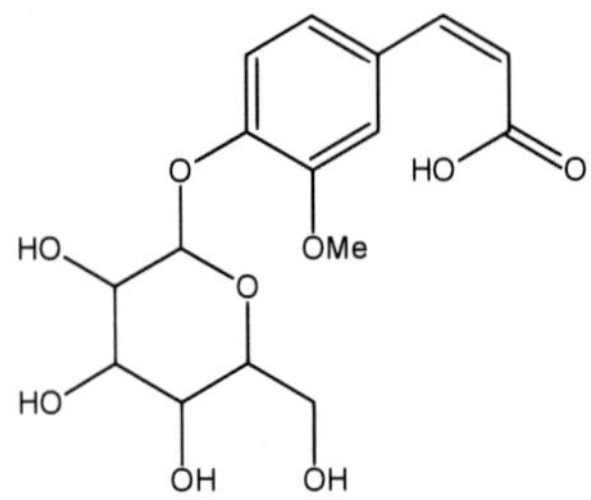

719
(Z)-Ferulic acid 4-O-β-D-glucoside

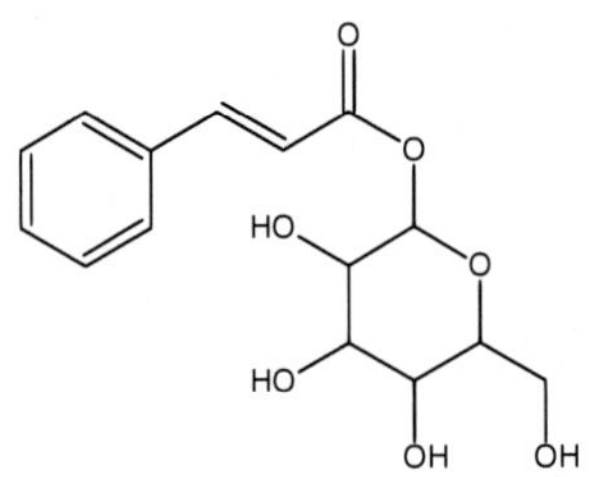

720
Cinnamoyl glucose ester

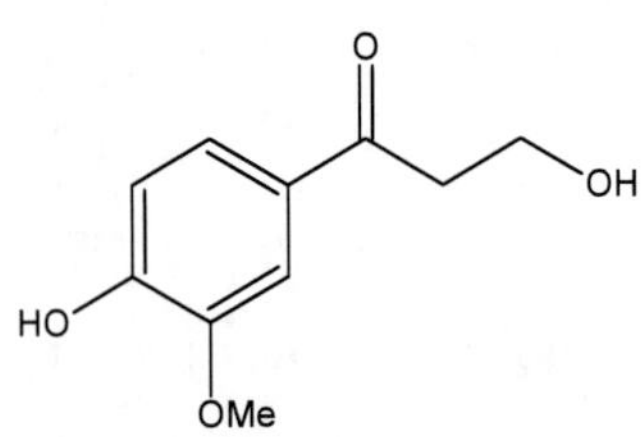

721
3-Hydroxy-1-(4-hydroxy-3-methoxyphenyl)-propan-1-one

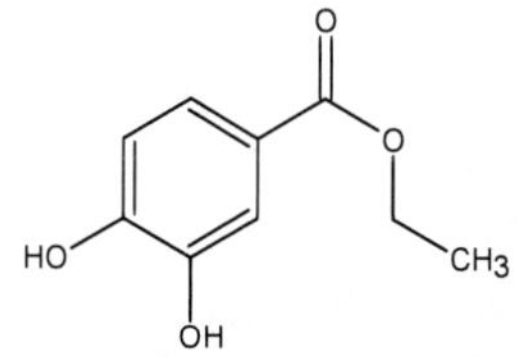

722
Ethyl protocatechuate

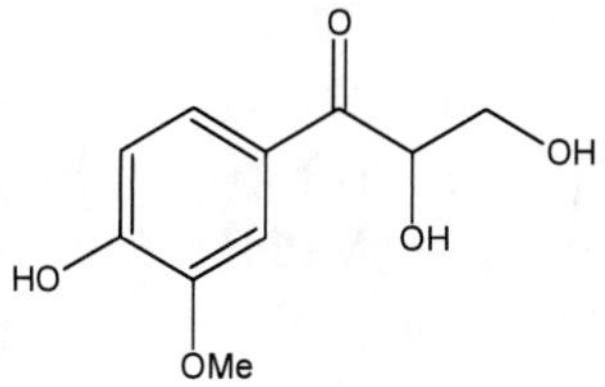

723
2,3-Dihydroxy-1-(4-hydroxy-3-methoxyphenyl)-propan-1-one

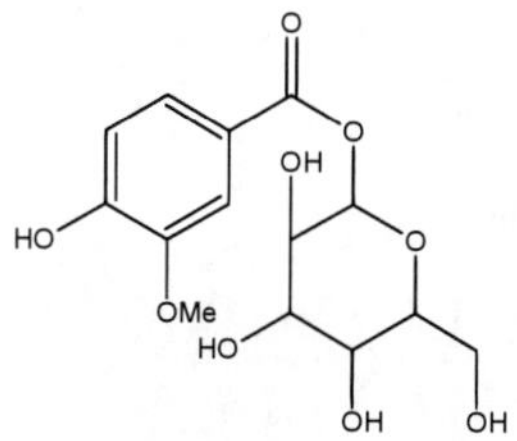

724
Vanilloyl glucose ester

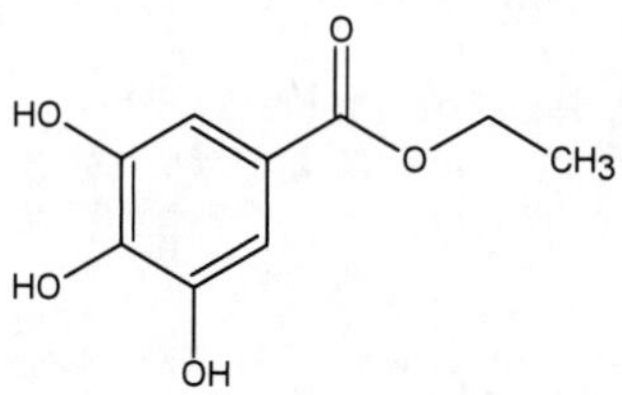

725
Ethyl gallate

A German Riesling wine has been fractionated with the aid of countercurrent chromatography and the constituents are reported together with the determination of their antioxidant activity. The radical scavenging capacity of hydrocinamic acid derivatives does not differ significantly from the other phenpolic antioxidants. The TEAC values were found to be in a range of 0.5 to 1.7 Trolox equivalents. The respective values for the flavonoids quercetin and catechin are 3.8 and 2.7. Surprisingly, in the TEAC test the cinnamic acid derivative with two hydroxy groups, caffeic acid was less active than the monosubstituted p-coumaric acid. In contrast, the antioxidant capacity of the esters of caffeic acid exceeded the activity of *p*-coumaroyl esters. Benzoic acids were somewhat less active in comparison with the homologous derivatives of cinnamic acid, but they provide antioxidant activity as well. Esterification of the carboxylic group of hydroxycinnamic acids slightly decreased the antioxidant activity (Baderschneiderand Winterhalter, 2001).

Glucose Ester of (E)-*p*-coumaric Acid: UV (MeOH): λ max 316,225 nm. Thermospray-MS: pseudo molecular ion at *m/z* 344 $[M(326)+ NH_4]^+$; ^{1}HNMR (250 MHz, CD_3OD) δ 3.37-3.48 (4 H, m, H2"/H3"/H4"/H5"), 3.68 (1 H, dd, J = 12, 4.5 Hz, H6"a), 3.85 (1 H, dd, J = 12, 2 Hz, H6"b), 5.57 (1 H, d, J = 8Hz, H1"), 6.37 (1 H, d, J = 15.5 Hz, H2), 6.81 (2 H, m, H3'/H5'), 7.48 (2 H, m, H2'/H6'), 7.73 (1 H, d, J = 15.5 Hz, H3); ^{13}C NMR (62.9 MHz, CD_3OD) δ 62.4 (C6"), 68.3 (C4"), 74.1 (C2"), 78.1 (C3"), 78.8 (C5"), 95.8 (C1 "), 116.9 (C2/C3'/C5'), 127.0 (C1 '), 131.4 (C2'/C6'), 147.9 (C3), 161.6 (C4'), 167.7 (C1).

Caffeoyl ethyl tartrate, isomer I: UV (MeOH): λ max 330, 205 nm. ESI-MS (negative mode): pseudo molecular ion at *m/z* 339 $[M(340) -H^+]^-$, MS/MS of *m/z* 339 at *m/z* 293 $[M-H^+-C_2H_5OH]^-$ and *m/z* 177 $[M -caffeoyl]^-$; ^{1}HNMR (300 MHz, CD_3OD): δ 1.29 (3 H, t, J = 7 Hz, H´´´) 4.25 (2 H, q, J = 7 Hz, H2'''), 4.43 (1 H, d, J = 2 Hz, H3"), 5.49 (1 H, d, J = 2 Hz, H2"), 6.31 (1 H, d, J = 16 Hz, H2), 6.94 (1 H, d, J = 8 Hz, H5'), 6.97 (1 H, dd, J = 2, 8 Hz, H6'); 7.05 (1 H, d, J = 2 Hz, H2'); 7.61 (1 H, d, J = 16 Hz, H3).

Caffeoyl ethyl tartrate, isomer II: UV (MeOH): λ max 327, 301, 206-nm; EsI-MS (negative model): pseudo molecular ion at *m/z* 339 $[M(340) -H^+]^-$, MS/MS of *m/z* 339 at *m/z* 293 $[M-H^+-C_2H_5OH]^-$ and mlz 177 $[M -caffeoyl]^-$; ^{1}HNMR (300 MHz, CD_3OD): δ 1.22 (3H, t, J = 7 Hz, H1'''), 4.19 (2H, m, H2'''), 5.45 (1H, d, J = 2.5 Hz, H2"), 6.31 (1 H, d, J = 16 Hz, H2); 6.94 (1H,d, J = 8 Hz, H-5), 6.97 (1H, dd, J = 2, 8 Hz, H-6´), 7.05 (1H, d, J = 2 Hz, H-2´), 7.61 (1H, d, J = 16 Hz, H3).

Glucose ester of (E)-ferulic acid: UV (MeOH): λmax 327, 236, 218, 202 nm. Thermospray-MS: pseudo molecular ion at *m/z* 374 $[M(356) + NH4]^+$, 357 $[M + H]^+$. ^{1}HNMR (300 MHz, CD_3OD): δ 3.37-3.48 (4 H, m, H2"/H3"/H4"/H5"), 3.69 (1H, dd, J = 12, 4;5 Hz, H6"a), 3.85 (1H, dd, J = 12, 2 Hz, H6"b), 3.90 (3H, s, OCH_3), 5.58 (1H, d, J = 7.5 Hz, H1 "), 6.40 (1H, d, J = 16 Hz, H2), 6.82 (1H, d, J= 8 Hz, H5'), 7.10 (1H, dd, J = 2, 8 Hz, H6'), 7.20 (1H, d, J = 2 Hz, H2'), 7.72 (1H, d, J = 16 Hz, H3); ^{13}C NMR (62.9 MHz, CD_3OD): δ 56.5 (OCH_3), 62.4 (C6"), 71.2 (C4"), 74.1 (C2"), 78.1 (C3"), 78.8 (C5"), 95.8 (C1"), 111.9 (C2'), 114.8 (C5'), 116.6 (C2), 124.4 (C6'), 127.6 (C1 '), 148.2 (C3), 149.4 (C4'), 151.0 (C3'), 167.7 (C1).

4-O-β-D-Glucoside of (E)-ferulic acid: UV (MeOH): λmax 284, 211 nm. Thermospray-MS: pseudo molecular ion at *m/z* 374 $[M(356) + NH4]^+$, 357 $[M + H]^+$; ^{1}HNMR (300 MHz, CD_3OD): δ.39-3.51 (4H, m, H2"/H3"/H4"/H5"), 3.69 (1H, dd, J = 12, 5 Hz, H6"a), 3.88 (1H, dd, J = 12/2 Hz, H6"b), 3.90 (3H, s, OCH_3), 4.96 (1H, d, J = 7.5 Hz, H1 "), 6.39 (1H, d, J = 16 Hz, H2), 7.13 (1H, dd, J = 2, 8 Hz, H6'), 7.17 (1H, d, J = 8 Hz, H5'), 7.23 (1H, d, J = 2 Hz, H2'), 7.55 (1H, d, J = 16 Hz, H3).

4-O-β-D-Glucoside of (Z)-ferulic acid: UV (MeOH): λmax 267, 207 nm. ESI-MS: pseudo molecular ion at *m/z* 355 $[M(356) -H^+]^-$, MS/MS of *m/z* 355 at m/z 193 $[M-H^+ -anhydroglucose]^-$; ^{1}HNMR (300 MHz, CD_3OD): δ 03.39-3.49 (4H, m, H2"/H3"/H4"/H5"); 3.68 (1 H, dd, J = 12, 5 Hz, H6"a), 3.86 (1 H, dd, J = 12, 2 Hz, H6"b), 3.88 (3H, s, OCH_3), 4.90 (1 H, d, J = 7.5 Hz, H1"), 5.95 (1H, d, J = 13 Hz, H2), 6.45 (1H, d, J = 13 Hz, H3), 7.07 (1H, dd, J = 2, 8 Hz, H6'), 7.10 (1H, d, J = 8 Hz. H5'), 7.55 (1H, d, J = 2 Hz, H2').

Glucose ester of cinnamic acid: Isolated as its tetraacetate: UV (MeOH): λ.max 282, 218 nm. DCI-MS (reactant gas, NH_3): pseudo molecular ion at *m/z* 496 [M(478) + NH4]$^+$. ^{1}HNMR (300 MHz, CD_3OD): δ 2.03-2.09 (4 x 3 H, 4 x s, 4 acetate), 3.90 (1H, ddd, J = 10, 4.5, 2.5 Hz, H5"), 4.14 (1H, dd, J = 12, 2.5 Hz, H6"a), 4.32 (1 H, dd, J = 12, 4.5 Hz, H6"b), 5.18 (1H, dd. J = 10, 9 Hz, H4"), 5.25 (1H, dd, J = 9,8 Hz, 2"), 5.31 (1 H, t, J = 9 Hz, H3"), 5.86 (1H. d, J = 8 Hz, H1"), 6.42 (1H, d, J = 16 Hz, H3), 7.40-7.50 (5 H, m. H2'/H3'/H4'/H5'/H6'), 7.77 (1 H, d, J = 16 Hz, H2).

3-Hydroxy-1-(4-hydroxyl-3-methoxyphenyl)-propan-1-one: UV (MeOH): λmax 276, 228, 206 nm. ESI-MS (negative mode): pseudo molecular ion at *m/z* 195 [M(196)-H$^+$]$^-$, MS/MS of *m/z* 195 at *m/z* 165 [M -H$^+$-CH_2O]$^-$; ^{1}HNMR (300 MHz, CD_3OD): δ 3.16 (2 H, t, J = 6.5 Hz, H2), 3.91 (3H, s, OCH_3), 3.94 (2H, t, J = 6.5 Hz, H3), 6.87(1H, d, J = 8 Hz, H5'), 7.55 (1 H, d, J = 2 Hz, H2'), 7.58 (1 H, dd, J = 2, 8 Hz, H6').

2,3 – Dihydroxy – 1 - (4 – hydroxyl – 3-methoxyphenyl)-propan-1-one: UV (MeOH): λmax 306,279, 231,207 nm. ESIMS (negative mode): pseudo molecular ion at *m/z* 211 [M(212) -H$^+$]$^-$, MS/MS of *m/z* 211 at *m/z* 181 [M -H$^+$ -CH_2O]$^-$. ^{1}HNMR (300 MHz, CD_3OD): δ 3.73 (1H, dd, J = 5.5, 11.5 Hz, H3a), 3.89 (1H, dd, J = 4, 11.5 Hz, H3b), 3.92 (3 H, s, OCH_3), 5.10 (1H, dd, J = 4, 5.5 Hz, H2), 6.88 (1H, d, J = 8.5 Hz, H5'), 7.53 (1H, brs, H2'), 7.58 (1 H, dd, J = 2, 8.5 Hz, H6'); ^{13}C NMR (75.5 MHz, CD_3COCD_3): δ 57.1 (OCH_3), 67.1 (C3), 75.8 (C2), 113.1 (C2'), 116.2 (C5'). 125.4 (C1'/C6'), 149.2 (C3'), 153.8 (C4'), 199.7 (C1).

Ethyl protocatechuate:UV (MeOH): λmax 299, 263, 219, 209 nm. ESIMS (negative mode): pseudo molecular ion at *m/z* 181 [M(182) - H$^+$]$^-$; MS/MS of *m/z* 181 at *m/z* 153 [M-$CH_2CH_3^+$]$^-$; MS/MS of *m/z* 153 at m/z 109 [M- $CH_2CH_3^+$-CO_2]$^-$; ^{1}HNMR (300 MHz, CD_3OD): δ 1.35 (3H, t, J = 7 Hz, H2'), 4.29 (2H, q, J = 7 Hz, H1'), 6.79 (1H, d, J = 8 Hz, H5), 7.41 (1 H, dd, J = 8, 2 Hz, H6), 7.42 (1H, d, J = 2 Hz, H2).

Glucose ester of vanillic acid: UV (MeOH): λmax 294, 265, 222, 205 nm. ESIMS (negative mode): pseudo molecular ion at *m/z* 329 [M(330) -H$^+$]$^-$; MS/MS of *m/z* 329 at *m/z* 167 [M-H$^+$ -anhydroglucose]$^-$; ^{1}HNMR (300 MHz, CD_3OD) δ 3.40-3.50 (4H, m, H2"/H3"/H4"/H5"), 3.70 (1H, dd, J = 12, 5 Hz, H6"a), 3.86 (1H, dd, J = 12, 2 Hz, H6"b), 3.90 (3H, s, OCH_3), 5.68(1H, d, J= 7.5 Hz, H1"), 6.86 (1H, d, J = 8 Hz, H5), 7.62 (1H, dd, J = 8, 2 Hz, H6), 7.65 (1H, d, J = 2 Hz, H2).

Ethyl gallate: UV (MeOH): λmax 281, 219 nm.ESIMS (negative mode): pseudo molecular ion at *m/z* 197 [M(198)-H$^+$]$^-$; MS/MS of *m/z* 197 at *m/z* 169 [M -CH2CH3+]$^-$; MS/MS ofm/z 169 at m/z 125 [M -$CH_2CH_3^+$ -CO_2]$^-$; ^{1}HNMR (300 MHz, CD_3OD): δ 1.35 (3H, t, J = 7 Hz, H2'), 4.27 (2H, q, J = 7 Hz, H1 '), 7.04 (2H, s, H2/H6).

Chapter 7

QUINONES

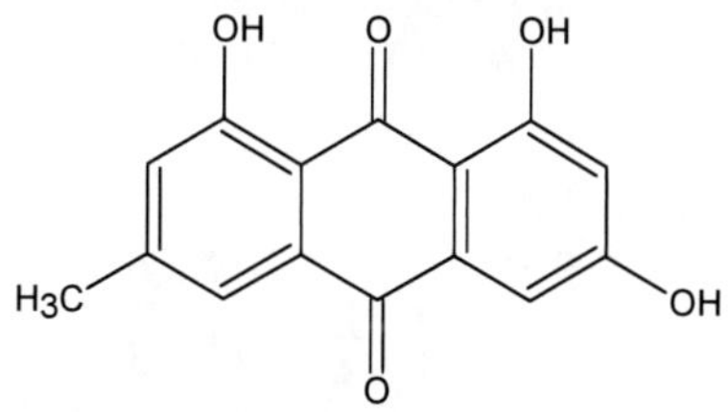

726
Emodin

The anthraquinone emodin containing extract of the root of *Polygonum multiflorum* (Poligonaceae) produced a dose-dependent protection against CCl_4 hepatotoxicity (Chiu et al., 2002).

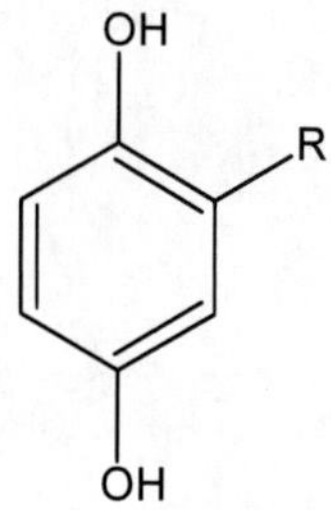

727
R = 10,(Z),13'(E),15'(E)-heptadecatrienyl
l0'(Z),l3'(E),l5'(E)-Heptadecatrienylhydroquinone

728
R = 10,(Z),13'(E)-heptadecadienyl
10'(Z),l3'(E)-Heptadecadienylhydroquinone

729
R = 10′(Z)-heptadecenyl
l0'(Z)-Heptadecenylhydroquinone

Two new antioxidative and cytotoxic hydroquinones, l0'(Z),l3'(E),l5'(E)-heptadecatrienylhydroquinone and 10'(Z),l3'(E)-heptadecadienylhydroquinone, as well as the known l0'(Z)-heptadecenylhydroquinone, were isolated from an EtOH extract of the sap of *Rhus succedanea* L. (Anacardiaceae) These compounds showed antioxidative and cytotoxic activities against five cancer cell lines (Wu et al., 2002).

l0'(Z),l3'(E),l5'(E)-Heptadecatrienylhydroquinone: Pale yellow oil; UV λmax (log ε): 247 (4.02), 277 (3.64), 319 (2.93) nm; IR: νmax 3499, 3012, 2925, 2852, 1594, 1475, 1278, 986, 732 cm^{-1}; HREIMS *m/z* 342.2588 (calcd for $C_{23}H_{34}O_2$, 342.2559); ^{1}HNMR: δ 1.30 (12H, m, H-3'-8'), 1.58 (2H, quintet, J = 7.6 Hz, H-2'), 1.70 (3H, d, J = 6.4 Hz, H-17'), 2.05 (2H, q, J = 6.8 Hz, H-9'), 2.58 (2H, t, J = 7.6 Hz, H.l'), 2.79 (2H, t, J = 6.8 Hz, H-12'), 5.37 (lH, m, H-11'), 5.41 (lH, m, H-10'), 5.49 (lH, m, H-13'), 5.55 (lH, m, H-16'), 5.99 (2H, m, H.14' and -15'), 6.56 (lH, d, J = 2.8 Hz, H-3), 6.58 (lH, d, J = 7.2 Hz, H-6), 6.61 (lH, dd, J = 7.2, 2.8 Hz, H-5); ^{13}C NMR: δ 18.1 (C.17'), 28.0 (C-9'), 30.6-30.8 (C.3'-8'), 31.1 (C-l' and -2'), 31.2 (C-12'), 113.6 (C-5), 120.1 (C-6), 121.9 (C-3), 127.7 (C.16'), 128.1 (C-11'), 130.7 (C-2 and -13'),131.8 (C.10' and 14'),132.9 (C-15'), 144.3 (C-l), 145.9 (C-4); EIMS *m/z* (rel. int) 342 M$^+$ (12), 123 (100).

10'(Z),l3'(E)-Heptadecadienylhydroquinone: Colorless oil; UV λmax (log ε) 242 (3.40), 274 (3.41) nm; IR νmax 3498, 3013, 2925,2853, 1475, 1279,965, 731 cm^{-1}; HREIMS *m/z* 344.2714 (calcd for $C_{23}H_{36}O_2$, 342.2715); EIMS *m/z* (rel. int) 344 M$^+$ (6), 123 (100); ^{1}HNMR δ 0.89 (3H, t, J = 7.6 Hz, H.17'), 1.31 (14H, m, H-3'-8' and H.16'), 1.58 (2H, quintet, J = 7.2 Hz, H-2'), 1.96 (2H, q, J = 6.0 Hz, H.15'), 2.04 (2H, q, J = 6.8 Hz, H-9'), 2.57 (2H, t, J = 7.2 Hz, H.l'), 2.73 (2H, t, J = 6.0 Hz, H-12'), 5.38 (4H, m, H-10', -11', -13', and -14'), 6.55 (lH, d, J = 2.4 Hz, H-3), 6.58 (lH, d, J = 7.2 Hz, H-6), 6.61 (lH, dd, J = 7.2, 2.4 Hz, H-5); ^{13}C NMR δ 14.0 (C-17'), 28.0 (C-9'), 30.3-30.7 (C-3'-8' and C-16'), 31.0 (C-l'), 31.1 (C-2'), 31.3 (C-12'), 35.8 (C-15'), 113.6 (C-5), 120.1 (C-6), 121.9 (C-3), 128.8 (C-11'), 129.8 (C-13'), 130.7 (C-2), 131.3 (C-10'), 131.4 (C.14'), 144.3 (C-l), 145.9 (C.4).

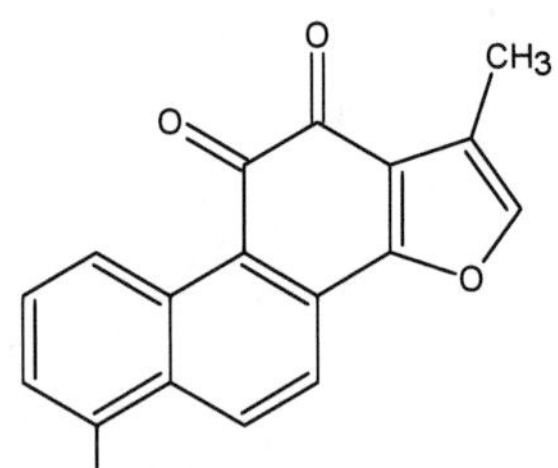

730
Tanshinone-I

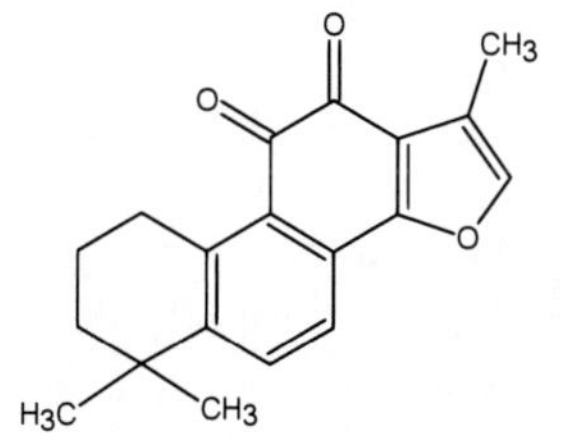

731
Tanshinone IIA

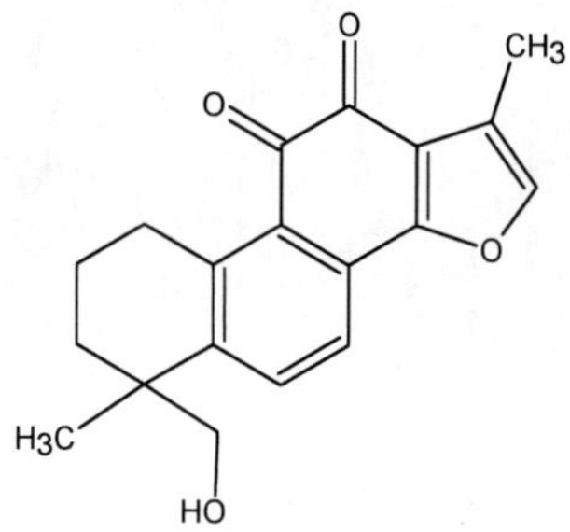

732
Tanshinone IIB

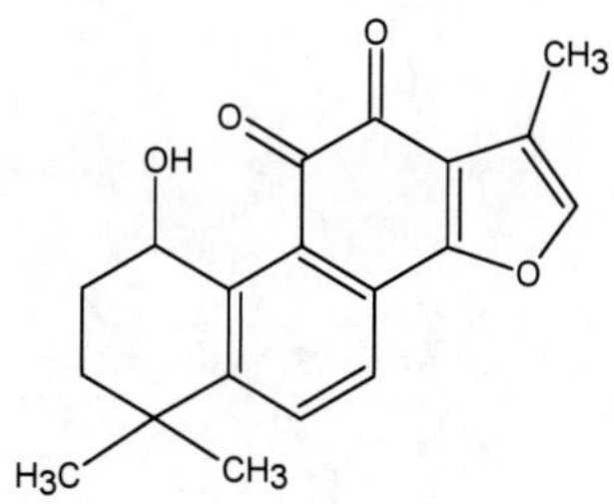

733
Hydroxytanshinone IIA

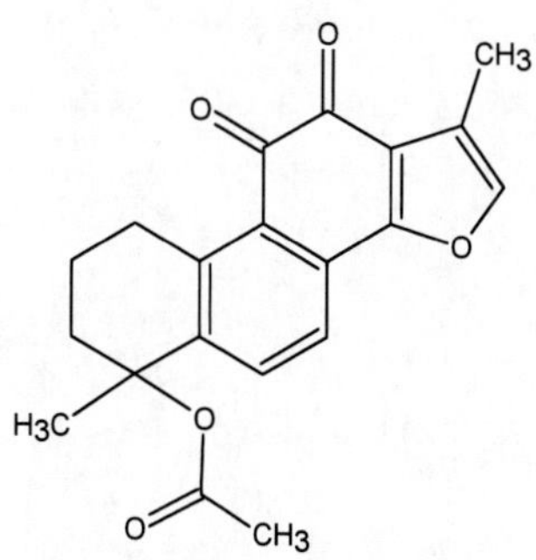

734
Methyl tanshinonate

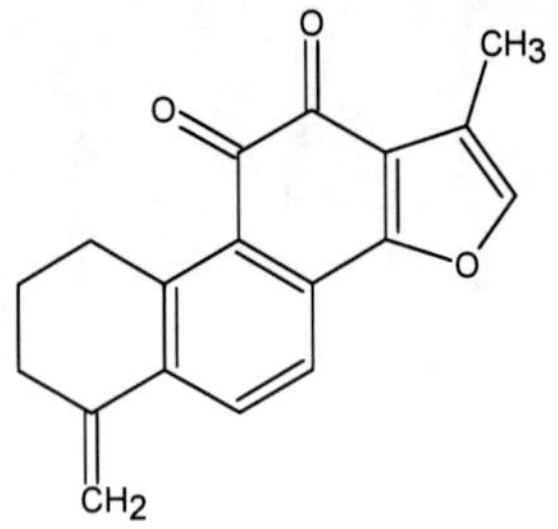

735
Methylene tanshinquinone

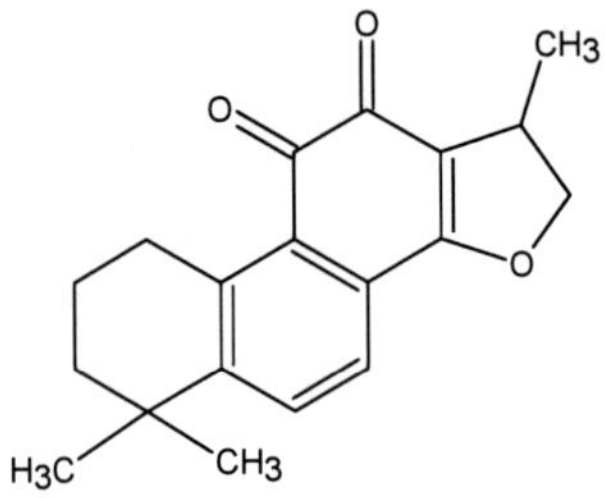

736
Cryptotanshinone

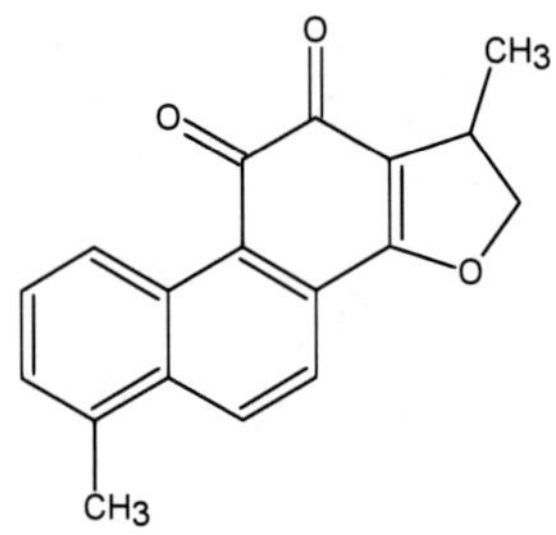

737
Dihydrotanshinone-I

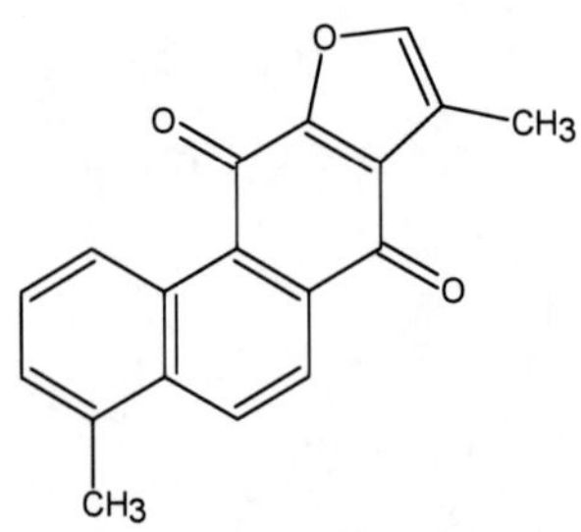

738
Isotanshinone I

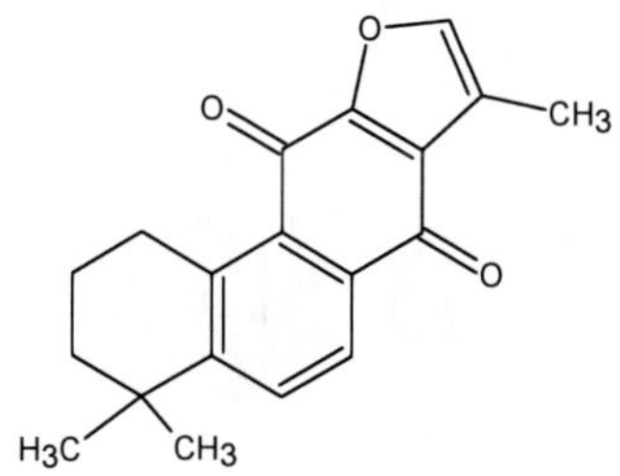

739
Isotanshinone IIA

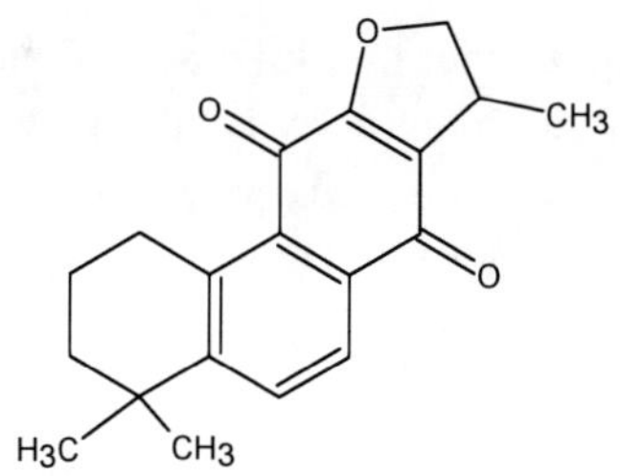

740
Isocrytotanshinone

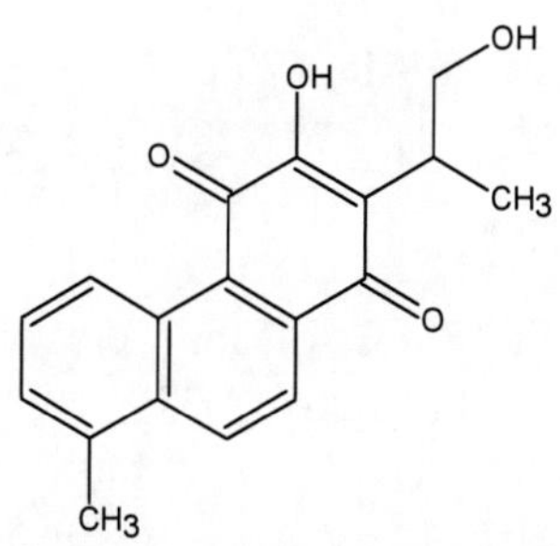

741
Danshenxinkun A

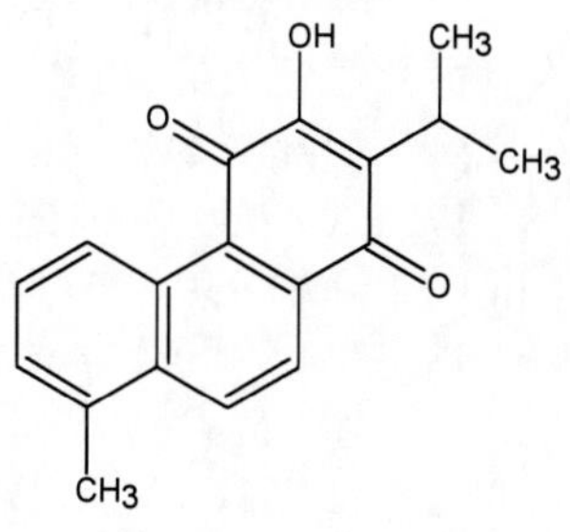

742
Danshenxinkun B

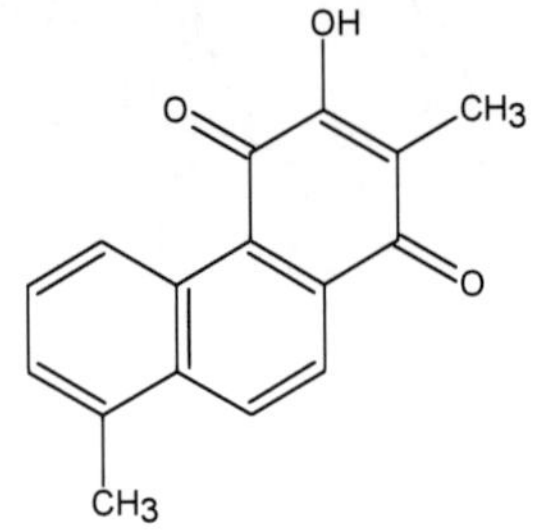

743
Danshenxinkun C

These compounds were isolated from the rhizomes of tanshen (*Salvia miltiorrhiza* Bung, Labiatae). Measurement of the antioxidant activity in lard of these quinones was performed with two methods, the Rancimat method and the Shaal oven method in conjunction with peroxide value determination. Except for tanshinone II_A, all the quinones exhibited antioxidant activities, of which dihydrotanshinone I, tanshinone I, methylenetanshinquinone, and cryptotanshinone showed strong activities comparable to that of BHA and BHT. These compounds may have the potential of being used as natural antioxidants in foods and cosmetics (Zhang et al., 1990).

Danshenxinkun B: mp. 181-183 °C; EIMS, (70 eV): *m/z* (rel. int.) 280 (M^+,100), 265 (9), 247 (10), 237 (12), 219 (8), 209 (7), 189 (9), 169 (21). IR ν cm^{-1}: 3257, 2899, 1653, 1631, 1592, 1404, 1351, 1325, 1220, 1130, 865, 790, 778 .

Tanshinone II_A: mp. 202-204 °C; EIMS, (70 eV): *m/z* (rel. int.) 294 (M^+,100), 279 (52), 261 (87), 251 (73), 233 (19), 222 (14), 178 (20), 165 (33), 152 (16); IR ν cm^{-1}: 2950, 2817,1686,1667, 1580, 1534,1453, 1429, 1403, 1389,1287, 1193,1170, 1152, 961, 917, 838, 707.

Methylenetan shinquinone: 174-176 °C; EIMS, (70 eV): *m/z* (rel. int.) 278 (M^+,100), 263 (20), 250 (12), 249 (33), 235 (80), 222 (8), 207 (5), 194 (15), 179 (25), 178 (34), 165 (25); IR ν cm^{-1}: 2907,1675, 1667,1626, 1572,1534, 996, 959, 840.

Tanshinone I. 232-233 °C; EIMS, (70 eV): *m/z* (rel. int.) 276 (M^+. 45), 261 (2), 248 (100), 222 (4), 219 (4), 191 (24), 189 (17), 176 (9), 165 (7), 152 (4), 139 (3); IR ν cm^{-1}: 1689, 1667, 1595, 1550, 1431, 1333, 1190, 1160. 917. 840, 836. 790. 763. 709

Cryptotanshinone: 174-175 °C; EIMS, (70 eV): *m/z* (rel. int.) 296 (M^+, 64), 281 (10), 268 (32), 253 (100), 235 (14), 178 (5), 171 (18), 152 (10); IR ν cm^{-1}: 2857, 1695, 1647,1626, 1575, 1481, 1441, 1342, 1212, 1176,1156, 948, 919, 704.

Dihydrotanshinone I: 223-225 °C; EIMS, (70 eV): *m/z* (rel. int.) 278 (M^+, 17), 250 (55), 235 (100), 222 (4), 207 (10), 189 (8), 179 (26), 169 (9), 165 (9); IR ν cm^{-1}: 2857, 1684, 1653, 1626, 1563, 1481, 1429, 1337, 1304, 1181, 961, 939, 843, 799.

Tanshinone IIB: 201-204 °C; EIMS, (70 eV): *m/z* (rel. int.) 310 (M^+, 21), 292 (14), 279 (100), 261 (25), 235 (59), 233 (15), 222 (14), 221 (14), 178 (34), 165 (38), 152 (27); IR ν cm^{-1}: 3571, 3472, 2994, 1695, 1667, 1637, 1575, 1531,1379, 1285, 1160, 917, 841, 670.

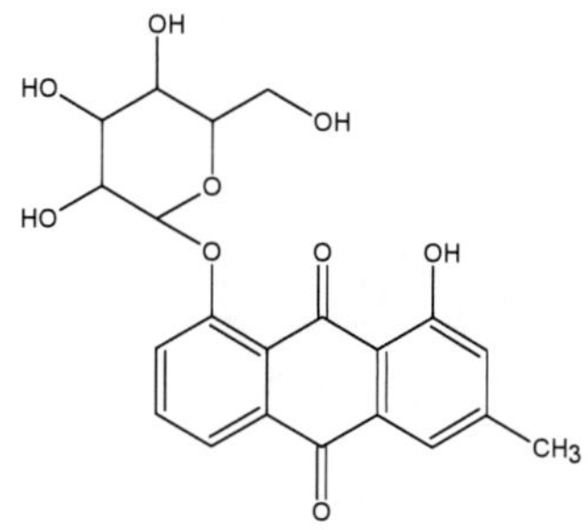

744
Chrysophanol-8-O-β-D-glucopyranoside

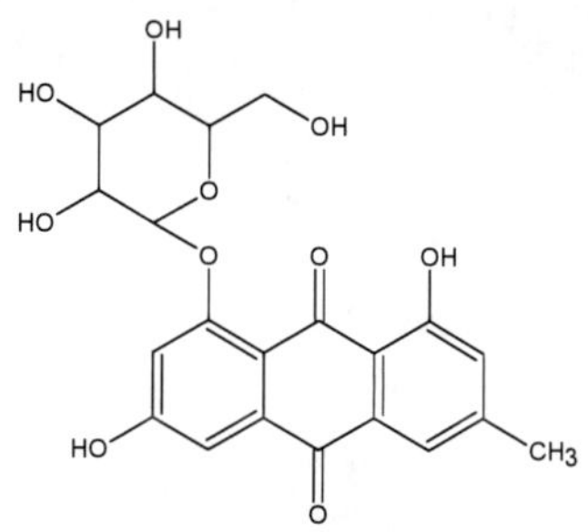

745
Emodin-8-O-β-D-glucopyranoside

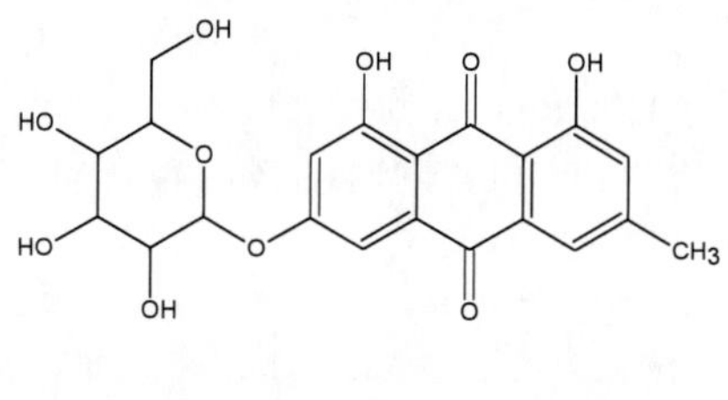

746
Emodin-6-β-D-glucopyranoside

The anthraquinone have been isolated from the roots of *Rumex patiemia* L. Showed radical scavenging properties. Emodin-6-O-β-D-glucopyranoside exhibited a low DPPH radical scavenging (Demirezera et al., 2001).

Emodin-6-O-β-D-glucopyranoside: Amorphous; R_f; 0.08 ($CHCl_3/CH_3OH/H_2O$,80: 20:2); ESI MS *m/z*: 453 (23) $[M+Na]^+$, 413 (29), 365 (5); UV λmax (MeOH) nm: 223, 264, 293 sh, 411: IR νmax (KBr) cm^{-1}: 3432 (OH), 2921, 2850, 1670 (free C=O), 1633 (chelated C=O), 1060 (C-O); ^{1}HNMR (300 MHz; DMSO-d_6): δ 7.17 (bs, H-2), 7.50 (bs, H-4), 7.87 (d, J = 2.1 Hz, H-5), 7.33 (d, J = 2.1 Hz, H-7), 2.42 (s, CH_3), Glucose: 5.07 (d, J = 7.4 Hz, H-1′), 3.10-3.50 (H-2′), 3.10-3.50 (H-3′), 3.10-3.50 (H-4′), 3.10-3.50 (H-5′), 3.55 (dd, J = 3.73 Hz, H-6′); ^{13}C NMR (75.5 MHz; DMSO-d_6): δ 161.71 (C-1), 124.09 (C-2), 147.29 (C-3), 119.55 (C-4), 112.13 (C-5), 159.91 (C-6), 111.94 (C-7), 159.66 (C-8), 186.78 (C-9), 181.95 (C-10),

135.91 (C-11), 115.74 (C-12), 114.57 (C-13), 132.20 (C-14), 21.38 (CH_3), Glucose: 100.67 (C-1′), 73.43 (C-2′), 76.60 (C-3′), 69.71 (C-4′), 77.25 (C-5′), 60.40 (C-6′).

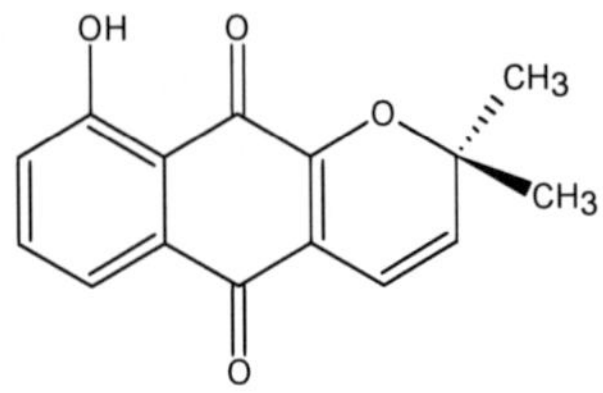

747
α-Caryopterone

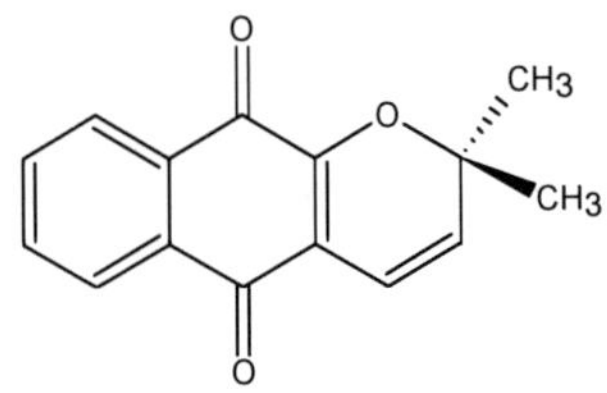

748
Dehydro-α-lapachone

Found in root bark of *Caryopteris x clandonensis* Simmonds (Lamiaceae). The pattern of inhibitory effect of α-caryopterone on the formation of the ESR adduct was similar to that of α-tocopherol. The scavenging effect on the superoxide anion was almost half of α-tocopherol, and dehydro-α-lapachone was not able to scavenge. Thus, the presence of a phenolic hydroxyl group in the potent antioxidant compounds may be important in the observed radical scavenging activity. α-Caryopterone was more efficient than dehydro-α-lapachone. Mesomeric stabilize the quinone radical, could explain the best activity of α-caryopterone. These compounds were tested against the snail *Bulinus truncatus,* an intermediate host snail of a schistosomiasis parasite and showed strong molluscicidal activity with LC_{100}< 4 ppm. In addition, they were found to have potent radical scavenging properties on superoxide radical, (Hannedouche et al., 2002).

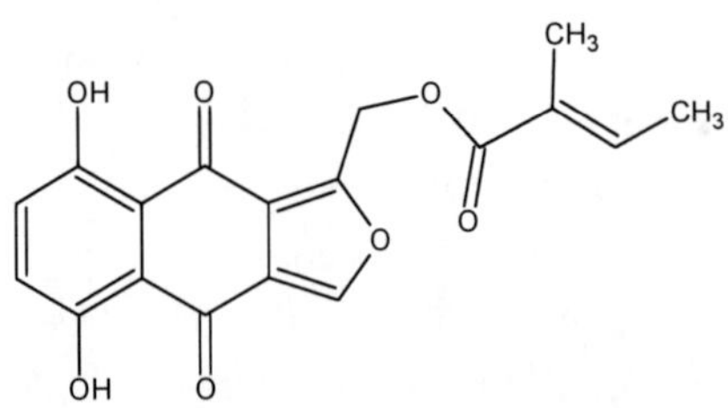

749
5,8-Dihydroxy-l-tigloylmethylnaphtho [2,3-c]furan-4,9-dione

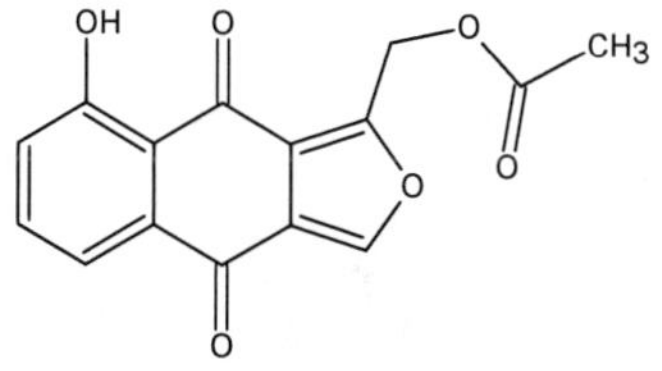

750
l-Acetoxymethyl-8-hydroxynaphtho [2,3-c]furan-4,9-dione

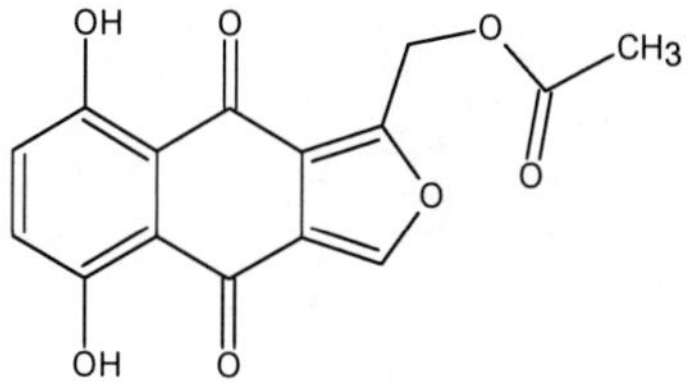

751
l-Acetoxymethyl-5,8-dihydroxynaphtho[2,3-c]furan -4,9-dione

Found in *Bulbine capitata* (Asphodelaceae). All the isofuranonaphthoquinones showed antioxidant and weak antiplasmodial activities (Bezabihl et al., 2001).

5.8-Dihydroxy-l-tigloylmethylnaphtho[2.3-c]furan-4.9-dione: Golden yellow shining solid: m.p.134-136 C; HRMS: *m/z* 342.0740 [M^+] (calcd for $C_{18}H_{14}0_7$. 342.0738); UV ($CHCl_3$): λmax (log ε) 446 (4.00), 295 (3.77), 250 (4.11), 212 nm (3.62); IR (KBr): νmax 1712, 1633, 1601, 1555, 1448, 1342, 1250, 1164 cm^{-1}: EIMS: *m/z* (rel. int.) 342 $[M]^+$ (76), 259 (44), 242 (94), 214 (40), 83 (100). ^{1}HNMR: δ 8.20 (s, H-3), 7.28 (s, H-6), 7.28 (s, H-7), 5.62 (s, H-CH_2O), 6.96 (qq, J = 7.05, 1.4 Hz, H-3′), 1.82 (dq, J = 7.08, 1.11 Hz, H-4′), 1.88 (m, H-5′); ^{13}C NMR: δ 155.9 (C-1), 145.7 (C-3), 184.8 (C-4), 158.6 (C-5), 129.9 (C-6), 129.8 (C-7), 158.6 (C-8), 184.4 (C-9), 119.3 (C-9a), 123.8 (C-3a), 115.0 (C-4a), 114.8 (C-8a), 57.0 (CH_2O), 167.6 (CO), 139.4 (C-2′), 128.2 C-3′), 14.9 (C-4′), 12.5 (C-5′).

l-Acetoxymethyl-8-hydroxynaphtho[2,3-c]furan-4.9-dione: Yellow solid; m.p. 184-186°C; HRMS: *m/z* 286.0503 [M^+] (calcd for $C_{15}H_{10}0_6$. 286.0476); UV ($CHCl_3$): λmax (log ε) 388 (3.67), 305 (3.51), 247 (4.03), 213 nm (3.42); IR (KBr): νmax 1740, 1674, 1635, 1561, 1448, 1325, 1241, 1122 cm^{-1}; EIMS: *m/z* (rel. int.) 286 [M^+] (100),172 (41), 115 (68); ^{1}HNMR: δ 8.17 (s, H-3), 7.81 (dd, J =1.2, 7.5 Hz, H-5), 7.66 (dd, J = 7.5, 8.4 Hz, H-6), 7.28 (dd, J = 1.2, 8.4 Hz, H-7), 5.53 (s, CH_2O), 2.15 (s, CO-CH_3); ^{13}C NMR: δ 158.5 (C-1), 145.6 (C-3), 184.1 (C-4), 155.1 (C-5), 129.8 (C-6), 129.8 (C-7), 155.1 (C-8), 184.6 (C-9), 123.3 (C-9a), 119.4 (C-3a), 114.7 (C-4a), 114.6 (C-8a), 56.6 (CH_2O), 20.8 (CO-CH_3), 170.4 (CO).

l-Acetoxymethyl- 5,8-dihydroxynaphtho[2.3-c] furan-4.9-dione: Orange solid; m.p. 198-200°C; HRMS: *m/z* 302.0432 0503 [M^+] (calcd for $C_{15}H_{10}0_7$. 302.0425); UV ($CHCl_3$): λmax (log ε) 453 (3.96), 447 (3.96), 299 [3.68, 250nm (4.05)]: IR (KBr): νmax 1740,1629,1604, 1157, 1448,1338,1237,1165 cm^{-1}; EIMS: *m/z* (rel. int.) 302 [M^+] (43), 260 (71), 242 (100), 214 (63).^{1}HNMR: δ 8.20 (s, H-3), 7.27 (H-6), 7.27 (H-7), 5.53 (CH_2O), 2.15 (CO-CH_3); ^{13}C NMR: δ 163.6 (C-1), 145.5 (C-3), 178.7 (C-4), 119.9 (C-5), 137.0 (C-6), 124.9 (c-7), 155.1

(C-8), 186.0 (C-9), 123.5 (C-9a), 119.9 (C-3a), 35.6 (C-4a), 117.9 (C-8a), 56.7 (CH_2O), 20.8 ($CO\text{-}CH_3$), 170.4 (CO).

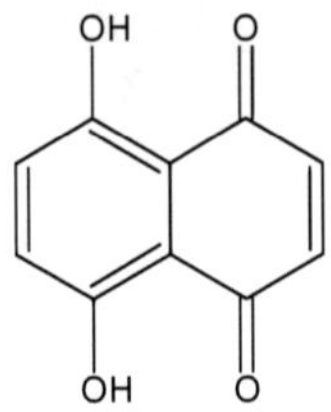

752
Naphthazarin

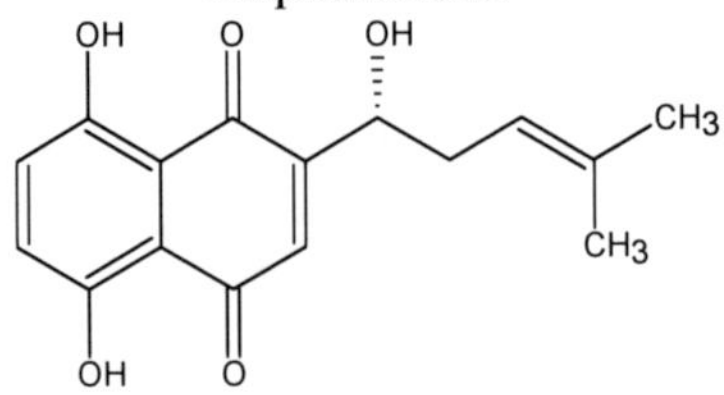

753
Shikonin

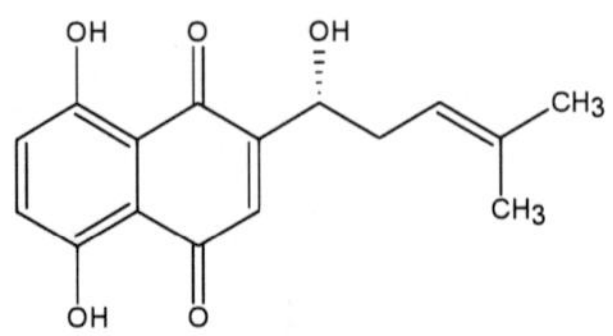

754
Alkannin

Pistacia lentiscus var Chia ev. (Anacardiaceae) which is grown almost exclusively on Chios Island. Greece, gives a resinous exudate. The crude Chios mastic gum (CMG). CMG is basically consumed as chewing gum and also in other culinary- uses, especially in the Greek, Turkish and Arabic kitchen. i.e. in powder form as a food additive, in the form of sugar containing gel as a table-sweetener.Mastic gum mainly consist of triterpenes of the oleanane, euphane and lupine type (amyrin, oleanolic acid, ursolic acid, lupeol, 18-α-glycyrrhetinic acid) and hydroxynaphthoquinones (naphthazarin, shikonin and alkannin) showed 53.5%-78.8% and 27.0%-64.1% LDL protectiye antioxidant activity respectively. The combination effects (68.7%-76.2% LDL protection) of ursolic, oleanolic and ursodeoxycholic acids were almost equal to the effect (75.3%) of the CMG extract in comparable doses (Andrikopoulos et al., 2003).

Chapter 8

TRITERPENOIDS

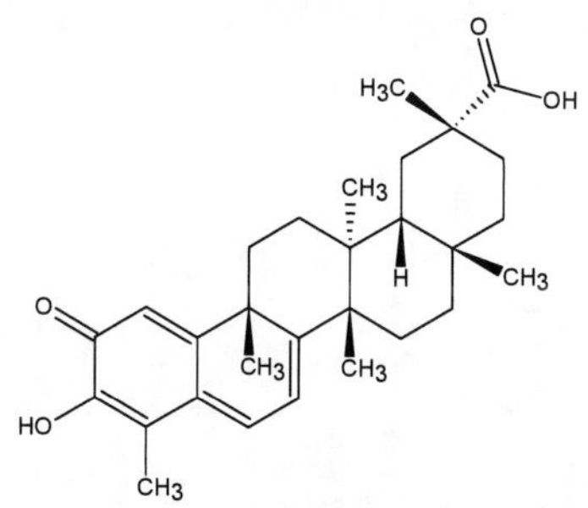

755
Celastrol

Found in *Celastrus scandens* (Celastraceae). The dienone-phenolic triterpene, celastrol showed inhibitory effect, on lipid peroxidation in rat liver mitochondrial membranes induced by ADP and Fe^{2+}.The anti-peroxidative effect of celastrol was very strong: its 50 % inhibitory concentration was 7 μM, and it was about 15 times more effective than a-tocopherol. Celastrol scavenged 1.5 molar equivalents of radicals in homogeneous aqueous ethanolic solution, whereas cysteine and a-tocopherol scavenged one and two molar equivalents of radicals, respectively. The process of anti-peroxidation of celastrol was biphasic, possibly due to stepwise inhibitions of peroxidation in the outer and inner mitochondrial membranes (Sassa et al., 1990).

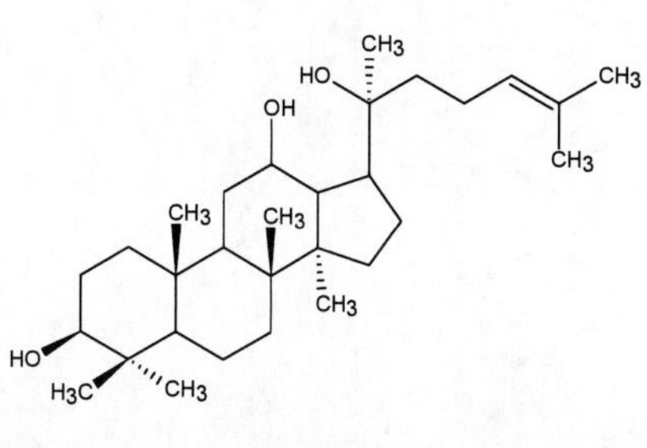

756
Ginsenoside PD

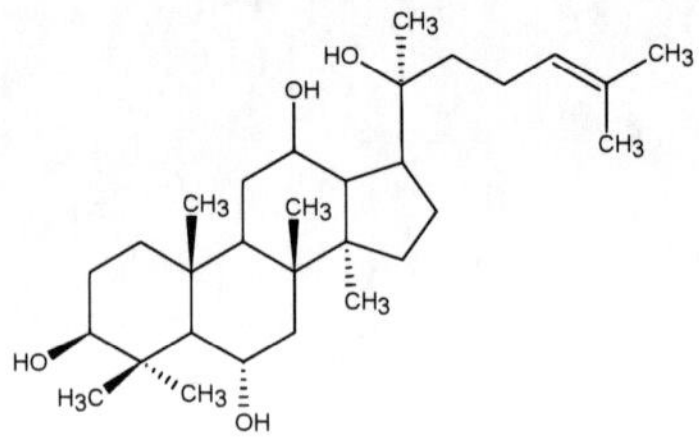

757
Ginsenoside PT

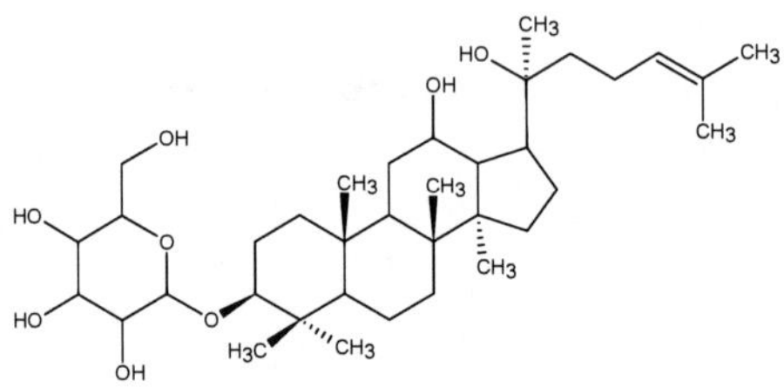

758
Ginsenoside R_{h2}

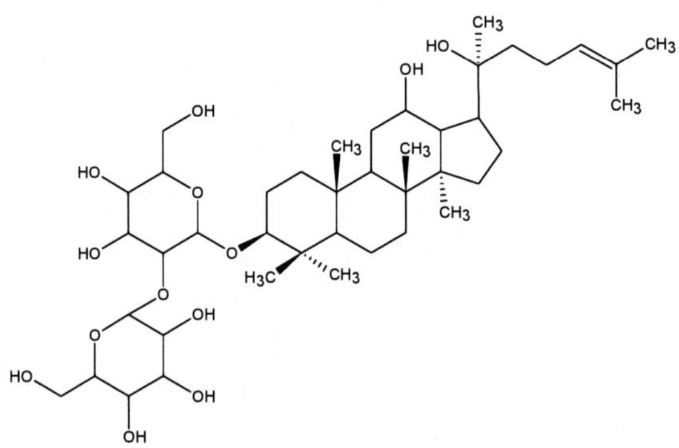

759
Ginsenoside R_{g3}

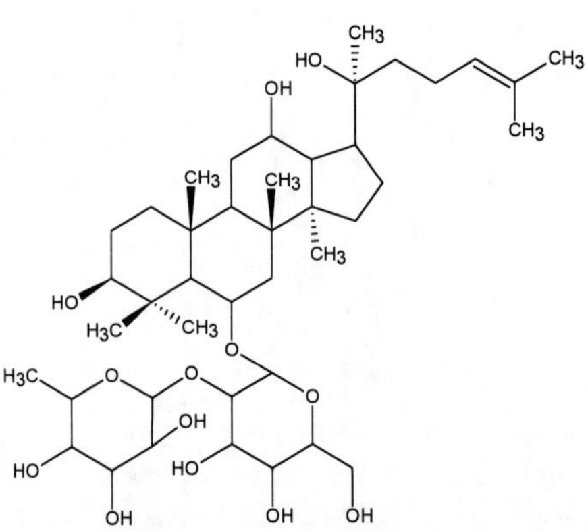

760
Ginsenoside R_{g2}

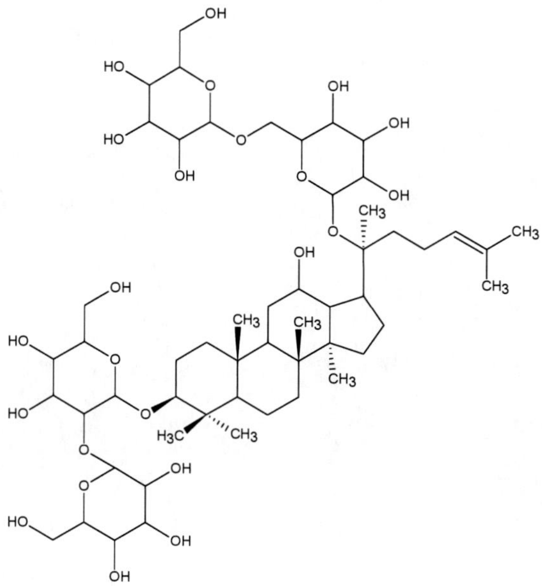

761
Ginsenoside R_{b1}

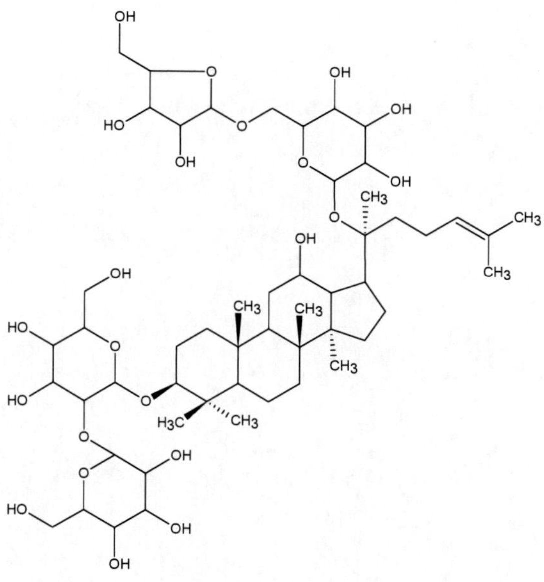

762
Ginsenoside R_c

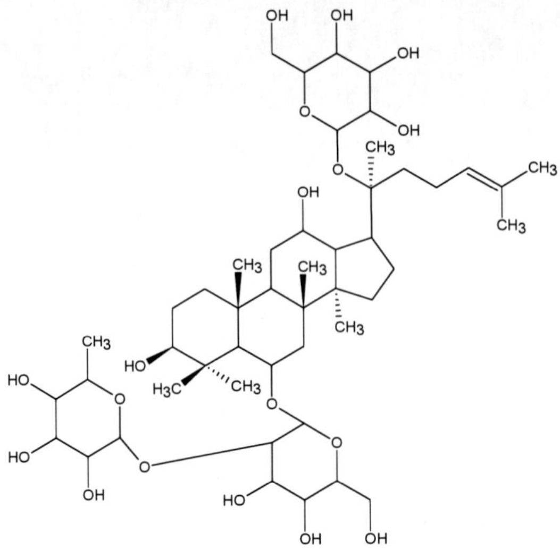

763
Ginsenoside R_e

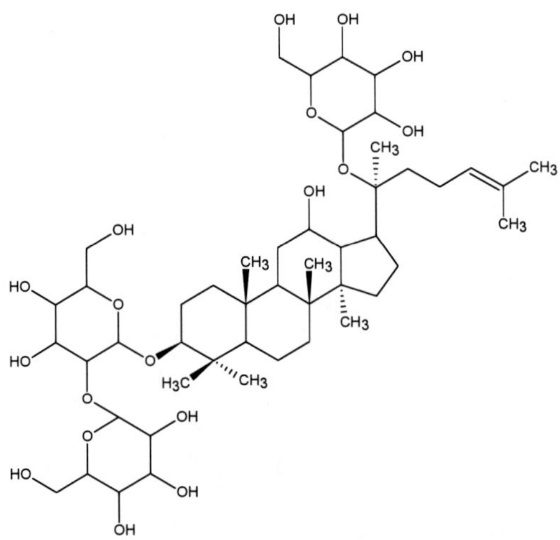

764
Ginsenoside R_d

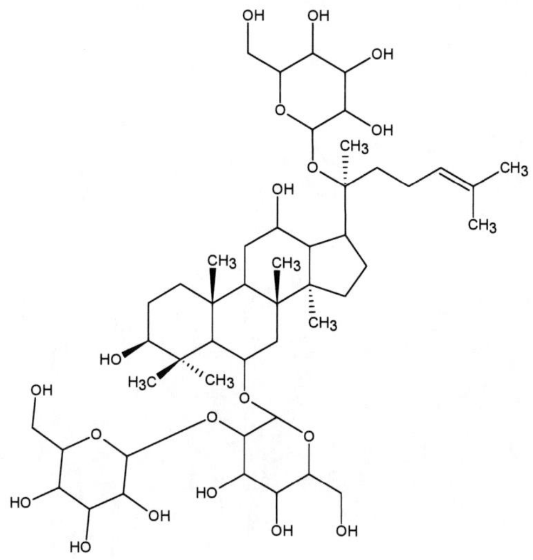

765
Ginsenoside Rl

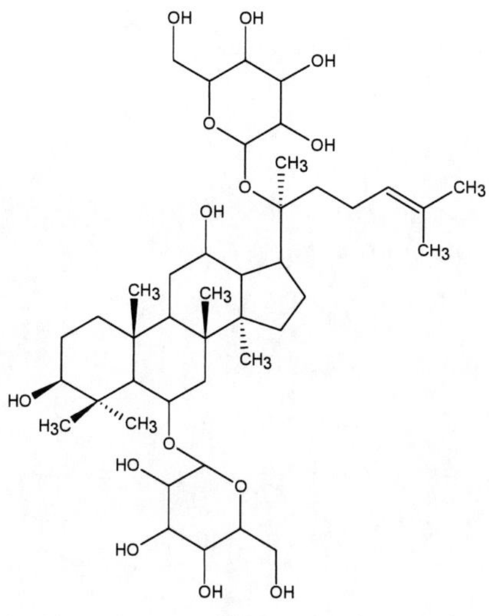

766
Ginsenoside R_{g1}

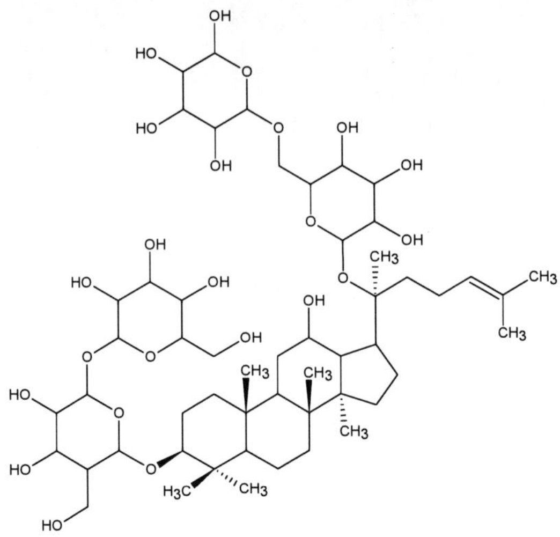

767
Ginsenoside R_{b3}

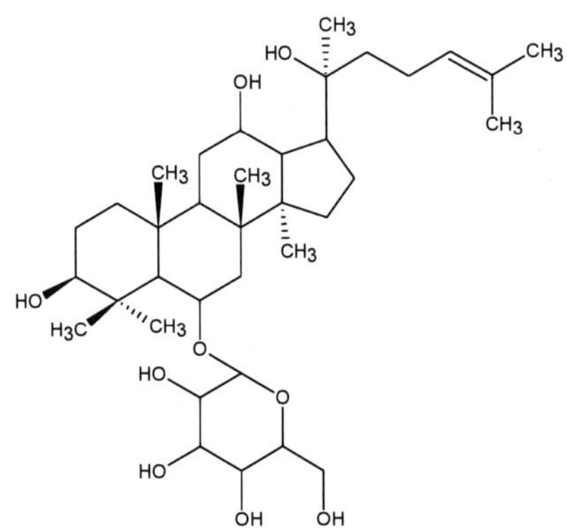

768
Ginsenoside R_{h1}

Ginsenoside, the major active component in *Panax ginseng* C.A. Meyer (Araliaceae), which has been used in traditional Chinese medicine, contains a series of derivatives of the triterpene dammarane being attached by some sugar moieties. Briefly, the central structures of ginsenosides, either protopanaxadiol (PD) or protopanaxatriol (PT) play a prooxidative role in AAPH-induced hemolysis of erythrocytes; in particular, PD generates hemolysis erythrocytes even in the absence of initiator. To the individual ginsenoside, if there are no sugar moieties attached to the 20-position of the triterpene dammarane, the ginsenoside acts as a prooxidant in AAPH-induced hemolysis, that is, Rh2 and Rg2, even in the absence of AAPH, that is, Rg3. A glucose attached to the 6-position instead of the 20-position sugar moieties can also make the ginsenoside an antioxidant, that is, Rh 1. The antioxidants among ginsenosides follow two different mechanisms that can be expressed mathematically by the Boltzmann equation, that is, Rc and Rb 1, and the polynomial equation, that is, Re, Rd, R 1, Rgl, Rb3, and Rhl. The orders of antioxidative ability are Rc >Rbl and Re > Rd > R1 > Rgl > Rb3 >

Rhl, respectively. Sugar moieties attaching to different positions of the triterpene dammarane generate various ginsenosides, which play different roles in AAPH-induced hemolysis, demonstrating there are complicated interactions between sugar moieties and the central structure, the triterpene dammarane (Liu et al., 2003).

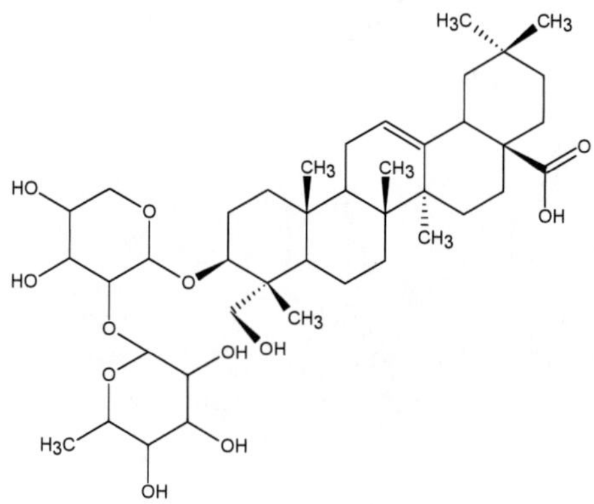

769
Kalopanaxsaponin A

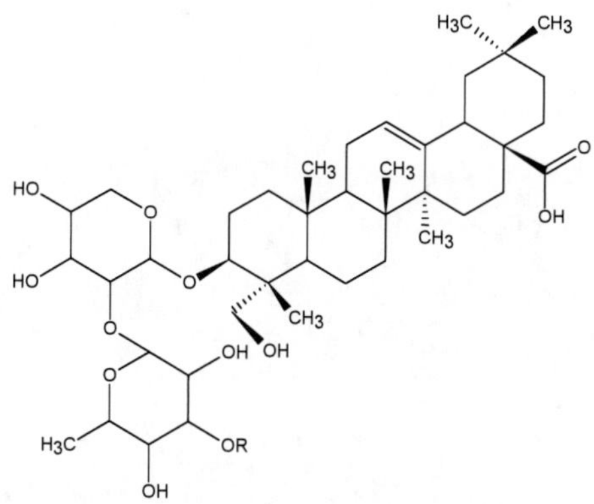

770
R= β-D-xylopyranosyl Kalopanaxsaponin I

The stem bark of *Kalopanax pictus* (Araliaceae), is an anti-rheumatoidal arthritis drug in Oriental medicine. In the rheumatoidal rat, induced by Freund's complete adjuvant (FCA) reagent. Kalopanaxsaponin-A (KPS-A) significantly decreased malondialdehyde formation and the activities of xanthine oxidase and aldehyde oxidase of hepatic non-microsomal systems in FCA reagent-treated rats. In addition, increased activity levels of superoxide dismutase, catalase and glutathione peroxidase were also observed. The effects of KPS-A were more potent than the effects of KPS-I. KPS-A, could reduce rhcumatoidal syndromes through antioxidative mechanisms (Cho et al., 2002).

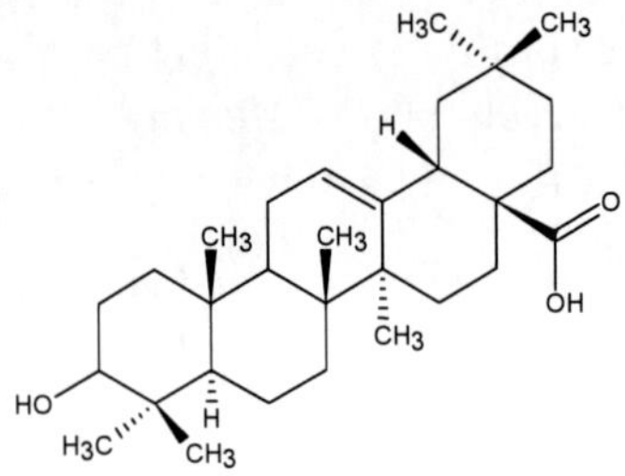

771
Oleanolic acid

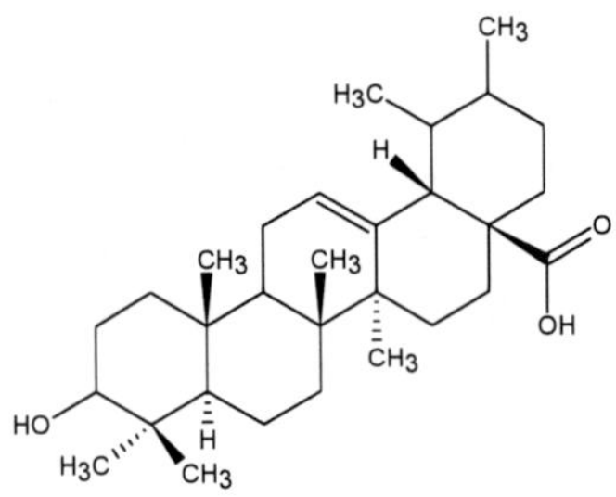

772
Ursolic acid

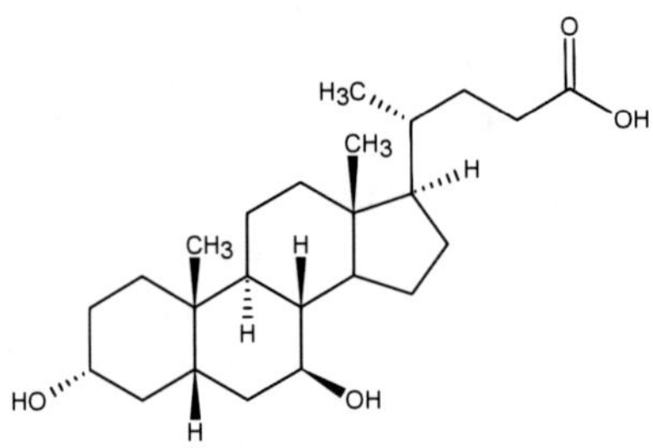

773
Ursodeoxycholic acid

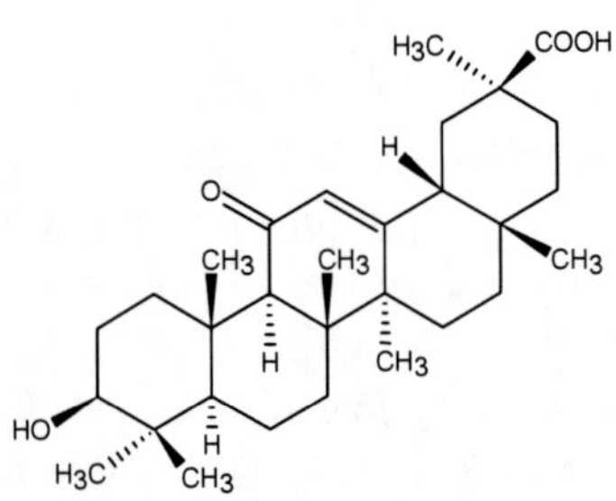

774
18α-Glycyrrheinic acid

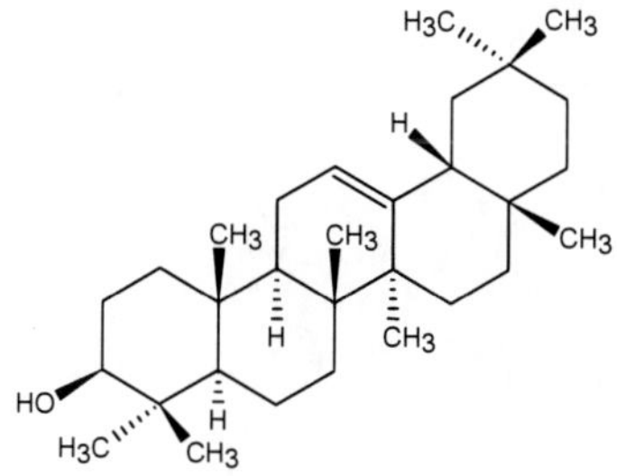

775
Amyrin

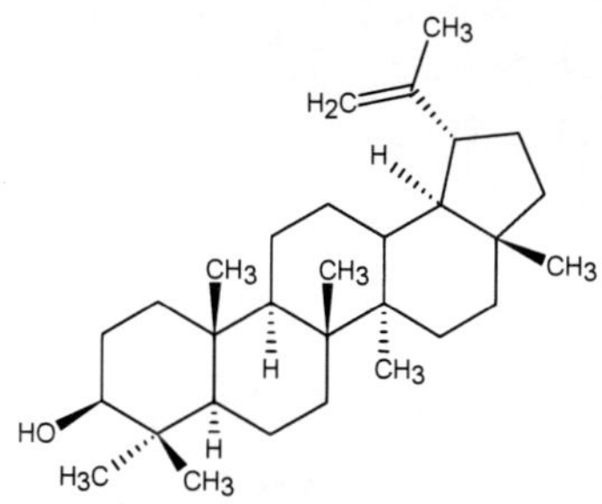

776
Lupeol

Pistacia lentiscus var Chia ev. (Anacardiaceae) which is grown almost exclusively on Chios Island. Greece, gives a resinous exudate. Mastic gum mainly consist of triterpenes of the oleanane, euphane and lupine type (amyrin, oleanolic acid, ursolic acid, lupeol, 18-α-glycyrrhetinic acid) and hydroxynaphthoquinones (naphthazarin, shikonin and alkannin) showed 53.5%-78.8% and 27.0%-64.1% LDL protectiye activity respectiyely. The combination effects (68.7%-76.2% LDL protection) of ursolic, oleanolic and ursodeoxycholic acids were almost equal to the effect (75.3%) of the CMG extract in comparalble doses (Andrikopoulos et al., 2003).

Chapter 9

MISCELLANEOUS

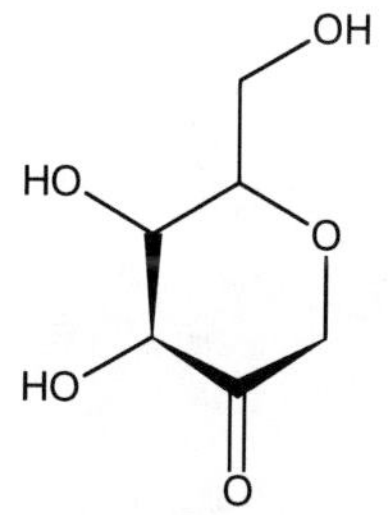

777
1,5-Anhydro-D-fructose

Showed antioxidant effects of 1.5-anhydro-O-fructose (1.5-AF) a unique anhydrohexulose on scavenges DPPH radicals directly in solution and inhibits the formation of hydrogen peroxide and superoxide anion typical reactive oxygen species (ROS). induced by phorbol myristate acetate (PMA) in a dose-dependent manner in THP-l cells (Kazuyo et al., 2002).

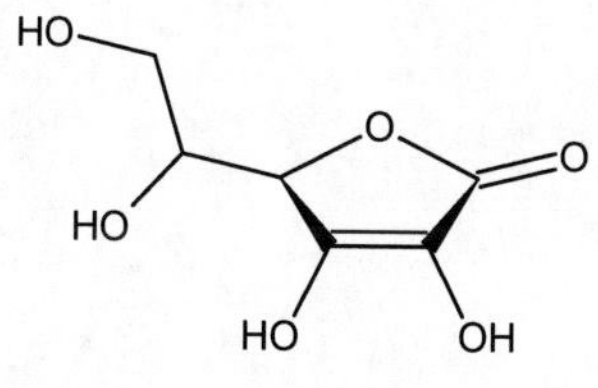

778
Ascorbic acid

Ascorbic acid (vitamin C) has been proposed for a long time as a biological antioxidant. It exists in rather high concentrations in many cellular environments, such as the stroma of chloroplasts, where its level is 2-3 x 10^{-3} M. Ascorbate has been demonstrated in many qualitative studies to possess significant antioxidant activity. 10^{-3} M ascorbate inhibited the photo oxidation of a kaempferol by illuminated spinach chloroplasts. Ascorbate reduces two equivalents of O_2^- (with a second-order rate constant of 3×10^5 M^{-1} sec^{-1}) to produce H_2O_2 and the triketo derivative dehydroascorbic acid. Ascorbate also reacts with $^1O_2$ at a relatively fast rate (about 10^7 M^{-1} sec^{-1}). In a well-controlled *in vitro* kinetic study, vitamin C was shown to inhibit the peroxy radical-initiated oxidation of methyl linoleate in methanol-t-butanol at 37°C. Vitamin C was shown to act as a chain-breaking scavenger for peroxy radicals and also to act as a synergist with vitamin E. The results suggest that vitamin C can donate a hydrogen atom to the vitamin E-derived phenolate radical, thus regenerating its activity (Larson, 1988).

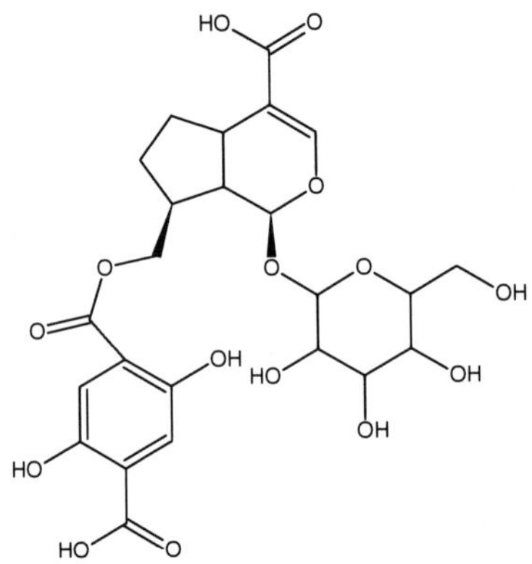

779

Blumeoside A

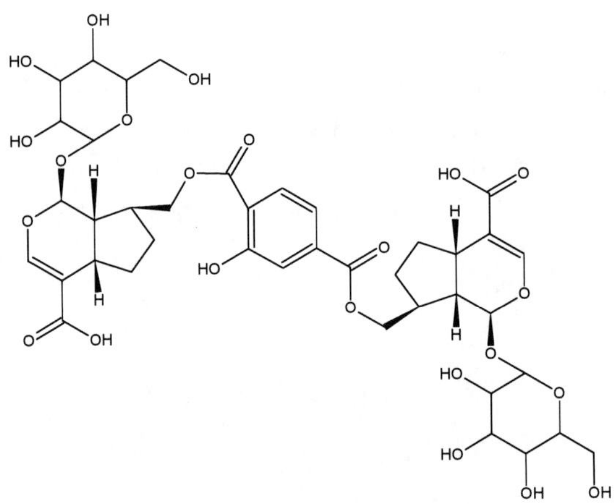

780

Blumeoside B

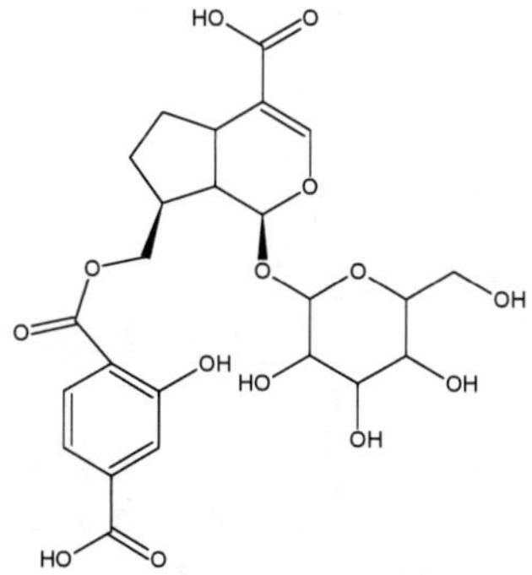

781
Blumeoside C

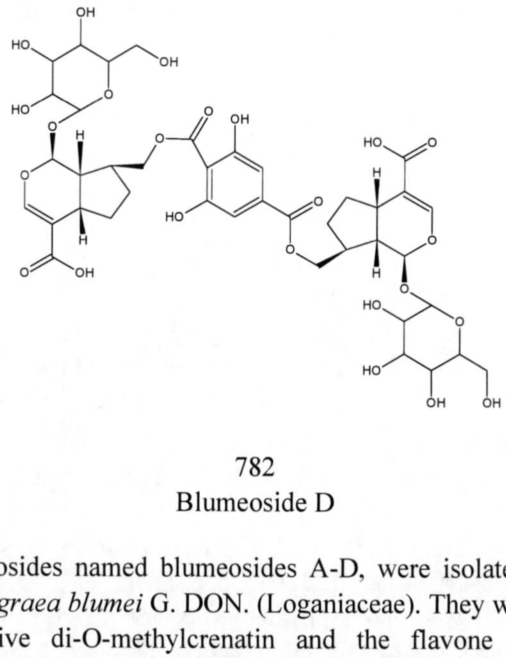

782
Blumeoside D

These iridoid glucosides named blumeosides A-D, were isolated from the methanolic stem-bark extract of *Fagraea blumei* G. DON. (Loganiaceae). They were accompanied by the benzyl-alcohol derivative di-O-methylcrenatin and the flavone C-glucoside swertisin. Blumeosides A and C are 10-O-(2,5-dihydroxyterephthalo)adoxosidic acid and 10-O-(2-hydroxyterephthalo)adoxosidic acid, respectively. In blumeosides B and D, both carboxylic groups of the terephthalic-acid moiety are esterified by adoxosidic-acid units. Blumeosides A-D inhibited bleaching of crocin induced by alkoxyl radicals. Blumeosides A and D also demonstrated scavenging properties towards the 2,2-diphenyl-l-picryl-hydrazyl (DPPH) radical in TLC autographic and spectrophotometric assays (Cuendet et al., 1997).

Blumeoside A: Yellow amorphous powder. M.p.147-149°C; yellowish green fluorescence at 366 nm; HPLC (RP-18, gradient 1): t_R 19.9 min. $[\alpha]_D$-55 (c = 2.3, MeOH); UV (MeOH): 120 (4.42), 362 (3.64); IR (KBr): 3600-3100, 3000-2900, 1660, 1610, 1490, 1440; ES-MS (negative-ion mode): 576.9 ($[M + Na -1H]^-$), 554.9 ($[M-H]^-$), FAB-MS (negative-ion mode): 555 ($[M -H]^-$, $C_{24}H_{27}O_{15}$), 375 ([(adoxo-sidic acid) -H]$^-$); ^{1}HNMR: δ 5.13 (d, J = 6.8 Hz, H-1), 7.38 (d, J = 1.0 Hz, H-3), 2.67 (br, dd, J= 14.7, 6.8 Hz, H-5), 2.11 (m, H-6a), 1.40 (m, H-6b), 1.81 (m, H-7a), 1.40 (m, H-7b), 2.32 (m, H-8), 1.97 (dd, J = 13.7,6.8 Hz, H-9), 4.27 (m, H-10a), 7.30 (s, H-3′), 7.23 (s, H-6′), 4.51 (d, J = 7.8 Hz, H-1′′),

2.96 (t, J = 8.5 Hz, H-2´´), 3.15 (t, J = 8.8 Hz, H-3´´), 3.03 (t, J = 9.3 Hz, H-4´´), 3.13 (m, H-5´´), 3.41 (dd, J = 11.8; 6.5 Hz, H-6´´a), 3.64 (dd, J = 11.8, 2.0 Hz, H-6´´b); ^{13}CNMR): δ 96.35 (C-1), 151.11 (C-3), 110.59 (C-4), 34.78 (C-5), 31.78 (C-6), 27.13 (C-7), 39.00 (C-8), 42.44 (C-9), 67.98 (C-10), 170.09 (C-11), 119.27 (C-1´), 150.56 (C-2´), 117.57 (C-3´), 120.13 (C-4´), 152.03 (C-4´), 117.12 (C-5´), 167.12 (C-6´), 167.84 (C-7´), 99.02 (C-1´´), 73.03 (C-2´´), 76.65 (C-3´´), 69.97 (C-4´´), 77.11 (C-5´´), 61.13 (C-6´´).

Blumeoside B: Cream-colored amorphous powder. M.p. 160-163°C, blue fluorescence at 366 nm; HPLC (RP-18, gradient 1): t_R 226 min. $[\alpha]_D$ -56 (c = 20, MeOH); UV (MeOH): 212 (4.53), 238 (4.36), 317 (3.66); FAB-MS (negative-ion mode): 897 ($[M- H]^-$, $C_{40}H_{49}O_{23}$) ([(blume-oside C) -H]$^-$); ^{1}HNMR: δ 5.16-5.17 (d, J = 7.3 , 7.4 Hz, H-1), 7.50 (br, s, H-3), 2.89 (m, H-5), 2.28 (m, H-6a), 1.46 (m, H-6b), 1.97 (m, H-7a), 1.46 (H-7b), 2.54 (m, H-8), 2.10 (m, H-9), 4.38-4.45 (dd, J = 11.0, 6.3 hz, H-10a), 4.28-4.34 (dd, J = 11.0, 6.8 Hz, H-10b), 7.58 (unresolved, H-3´), 7.59 (unresolved, H-5´), 8.02 (d, J = 8.8 Hz, H-6´), 4.66 (d, J = 7.8, 4.67 Hz, H-1´´), 3.06 (m, H-2´´), 3.35 (m, H-3´´), 3.23 (H-4´"), 3.31 (m, H-5´´), 3.65 (m, H-6´´a), 3.86 (m, H-6´´b); ^{13}CNMR): δ 98.89 (C-1), 153.41 (C-3), 112.07 (C-4), 36.99 (C-5), 33.79 (C-6), 28.82 (C-7), 41.09, (C-8), 44.32 (C-9), 69.78 (C-10), 171.03 (C-11), 119.51 (C-1´), 162.26 (C-2´), 120.97 (C-3´), 137.76 (C-4´), 117.60 (C-5´), 131.71 (C-6´), 170.63 (C-7´), 166.99 (C-8´), 100.97 (C-1´´), 74.60 (C-2´´), 77.90 (C-3´´), 71.48 (C-4´´), 78.36 (C-5´´), 62.80 (C-6´´).

Blumeoside C: Cream-colored amorphous powder.M.p. 171-173°C, blue fluorescence at 366 nm; HPLC (RP-18, gradient 1): t_R 19.2. $[\alpha]_D$ -40 (c = 20, MeOH). UV (MeOH): 215 (4.21), 240 (3.98), 323 (3.36). FAB-MS (negative-ion mode): 539 ($[M-H]^-$, $C_{24}H_{27}O_{14}$; ^{1}HNMR: δ 5.1 (d, J = 7.1 Hz, H-1), 7.42 (s, H-3), 2.77 (m, H-5), 2.16 (m, H-6a), 1.40 (m, H-6b), 1.84 (m, H-7a), 1.40 (m, H-7b), 2.33 (m, H-8), 1.97 (m, H-9), 4.24 (br, d, J = 5.6 Hz, H-10), 7.33 (unresolved, H-3´), 7.33 (unresolved, H-5´), 7.82 (d, J = 8.3 Hz, H-6´), 4.53 (d, J = 7.8 Hz, H-1´´), 2.95-3.80 (part. obscured by H_20 signal, H-C(2") to H-C(5"); ^{13}CNMR: δ 96.40 (C-1), 151.33 (C-3), 110.59 (C-4), 34.93 (C-5), 31.99 (C-6), 27.28 (C-7), 38.99 (C-8), 42.48 (C-9), 67.58 (C-10), 170.46 (C-11), 113.56 (C-1´), 161.96 (C-2´), 117.13 (C-3´), 133.64 (C-4´), 117.65 (C-5´), 130.44 (C-6´), 167.9 (C-7´), 165.54 (C-8´), 99.02 (C-1´´), 73.08 (C-2´´), 76.72 (C-3´´), 69.94 (C-4´´), 77.22 (C-5´´), 61.12 (C-6´´).

Blumeoside D: Yellowamorphous powder. M.p. 156-158°C, blue fluorescence at 366 nm; HPLC (RP-18, gradient 1): t_R 22.9 min $[\alpha]_D$ -36 (c = 4.0, MeOH). UV (MeOH): 222 (4.25), 251 (3.68), 371 (3.43); FAB-MS (negative-ion mode): 913 ($[M-H]^-$, $C_{40}H_{49}O_{24}$), 555 ([(blumeoside A)-H]$^-$); ^{1}HNMR: δ 5.18 (d, J = 7.4 Hz, H-1), 7.48 (s, H-3), 2.88 (m, H-5), 2.28 (m, H-6a), 1.45 (m, H-6b), 1.98 (m, H-7a), 1.45 (m, H-7b), 2,54 (m, H-8), 2.10 (m, H-9), 4.37 (m, H-10), 7.50 (s, H-3´), 7.50 (s, H-6´), 4.68 (d, J = 7.8 Hz, H-1´´), 3.12 (m, H-3´´), 3.15-3.40 (unresolved, H-C(3") to H-C(5"), 3.65 (dd, J = 11.6, 5.4 Hz, H-6´´a), 3.87 (d, J = 11.6 Hz, H-6´´b); ^{13}CNMR: δ 98.83 (C-1), 153.24 (C-3), 112.27 (C-4), 37.04 (C-5), 33.82 (C-6), 28.80 (C-7), 40.92 (C-8), 44.35 (C-9), 69.86 (C-10), 171.24 (C-11), 120.08 (C-1´), 153.84 (C-2´), 118.84 (C-3´), 120.08 (C-4´), 153.84 (C-5´), 118.84 (C-6´), 169.99 (C-7´´), 169.99 (C-8´´), 100.91 (C-1´´), 74.52 (C-2´´), 77.97 (C-3´´), 71.42 (C-4´´), 78.36 (C-5´´), 62.79 (C-6´´).

783

(CS-A-3β) polysaccharide

Three crude polysaccharide fractions. named CS-A. CS-B and CS-C. were prepared from the seeds of *Cuscuta chinensis.* CS-A showed a stimulating effect on concanavalin A or lipopolysaccharide induced mitogenic activity of lymphocytes. An acidic polysaccharide (CS-A-3β) was purified from CS-A by anion exchange and gel filtration chromatography. The polysaccharide showed potent stimulating effects on lymphocyte proliferation and antibody production, but did not significantly influence the serum IgG level. The CS-A-3β had a backbone consisting of α-D-l.4-linked GalpA residues and [β-L-l.2-linked Rhap residues with branches at C-4 of Rhap residues and C-3 of GalpA residues, composed of arabinogalactan and glucobiose. Its had a protective effect against free radical-induced cell toxitity (Bao et al., 2002).

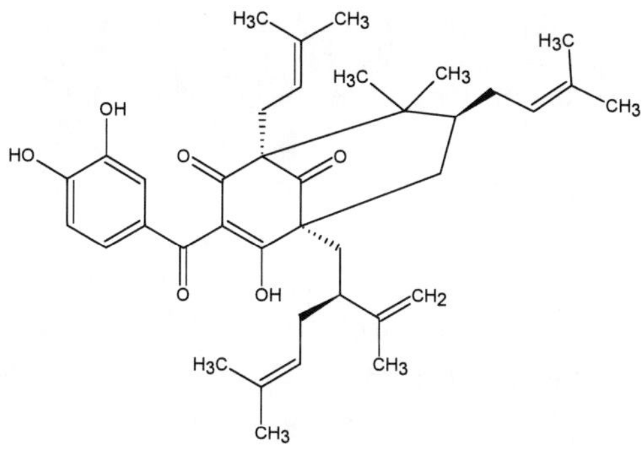

784
Garcinol

Garcinol, a polyisoprenylated benzophenone derivative, was purified from *Garcinia indica* (Clusiaceae) fruit rind. Garcinol exhibited moderate antioxidative activity in the micellar linoleic acid peroxidation system and also exhibited chelating activity at almost the same level as citrate. It also showed nearly 3 times greater DPPH (1,1-diphenyl-2-picrylhydrazyl) free radical scavenging activity than DL-α-tocopherol by weight in aqueous ethanol solution. In a phenazine methosulfate/NADH-nitroblue tetrazolium system, garcinol exhibited superoxide anion scavenging activity and suppressed protein glycation in a bovine serum albumin/fructose system (Yamaguchi, et al., 2000).

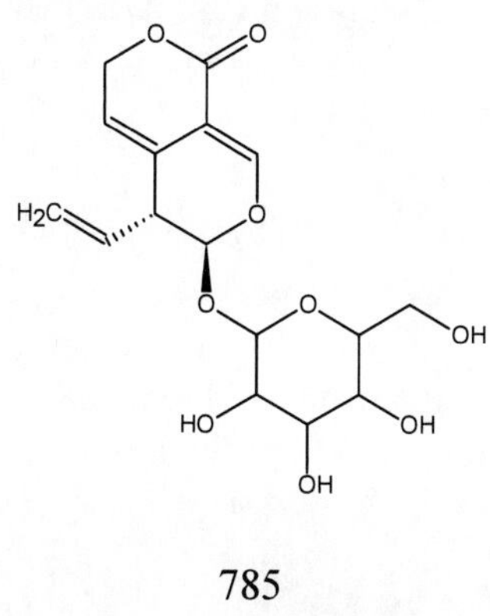

785
Gentiopicroside

Gentiopicroside a secoiridoid glycoside isolated from the methanol extract of the aereal parts of *Centaurium erythraea* showed antibacterial and free radical scavenging activities (Kumarasarnya et al., 2003).

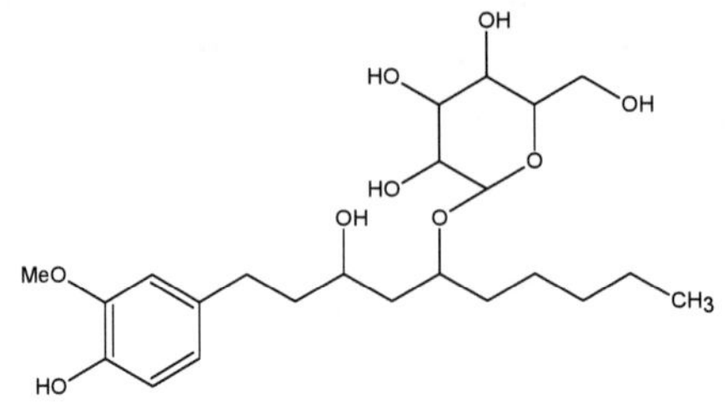

786

5-O-β-D-Glucopyranosyl-3-hydroxy-l-(4-hydroxy-3-methoxyphenyl)decane

Their antioxidative activities were investigated and compared to that of their aglycon, 6-gingerdiol, by a linoleic acid model system and by their DPPH radical-scavenging ability. Found in *Zingiber officinale* Roscoe (Zingeberaceae), (Sekiwa et al., 2000).

5-O-β-D-Glucopyranosyl-3-hydroxy-l-(4-hydroxy-3-methoxyphenyl)decane: Colorless oil; HRFAB-MS m/z 457.2444 [M-H]- (error mmu = +0.7 for C23H38O9); specific rotation [α]D -13.3 (c 0.72. MeOH); lHNMR (CD3OD. 270 MHz): δ 0.90 (3H, t, J = 7.0 Hz, H-10), 1.33 (6H, m, H- 7, 8, 9), 1.50 (2H, m, H-6), 1.59-1.64 (2H, m, H-4), 1.65-1.72 (2H, m, H-2), 2.56 (lH, m, H-1a), 2.68 (lH, m, H-1b), 3.17 (lH, t, J = 8.3 Hz, H-2"), 3.25-3.38 (3H, m, H-3", 4", 5"), 3.68 (lH, dd, J = 5.2, 11.9 Hz, H-6"a), 3.84 (lH, m, H-6"b), 3.83 (3H, s, O-CH3), 3.92 (lH, m, H-3), 3.92 (lH, m, H-5), 4.39 (lH, d, J = 7.6, H-1"), 6.62 (lH, dd, J = 2.1, 6.1, Hz, H-6'), 6.68 (lH, d, J = 7.9 Hz, H-5'), 6.77 (lH, d, J = 1.9 Hz, H-2'); l3CNMR (CD3OD. 270 MHz): δ 32.7 (C-1), 41.3 (C-2), 68.2 (C-3), 42.9 (C-4), 78.8 (C-5), 36.9 (C-6), 26.0 (C-7), 33.1 (C-8), 23.7 (C-9), 14.4 (C-10), 135.4 (C-1′), 113.1 (C-2′), 148.8 (C-3′), 145.4 (C-4′), 116.1 (C-5′), 121.7 (C-6′), 104.4 (C-1′′), 75.5 (C-2′′), 78.2 (C-3′′), 71.6 (C-4′′), 77.8 (C-5′′), 62.8 (C-6′′), 56.3 (OMe).

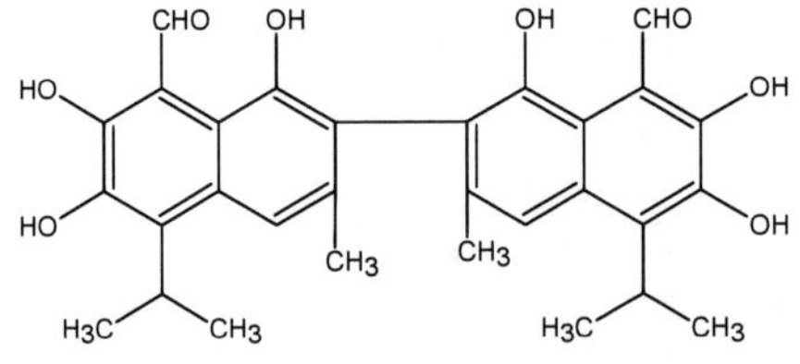

787

Gossypol

Gossypol, was isolated from the fruit of *Gossypium herbaceum* L. (Malvaceae), is powerful inhibitors of iron-induced lipid peroxidation in rat liver microsomes, at low micromolar concentrations (IC50 ≤ 1,5 μM), (Laughton et al., 1989).

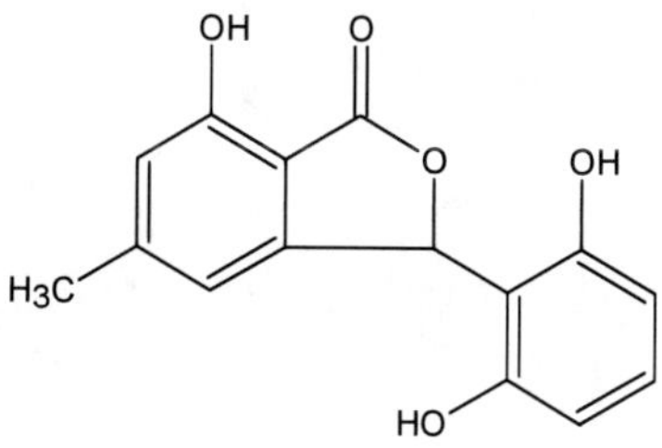

788
Isopestacin

Isopestacin is an isobenzofuranone obtained from the endophytic fungus *Pestalotiopsis microspore*, while a few other isobenzofuranones are known from natural sources. isopestacin is the only one having substituted benzene ring attached at the C-3 position of the furanone ring. This compound possesses antifungal activity and, it also behaves as an antioxidant scavenging both superoxide and hydroxy free radicals (Strobel et al., 2002).

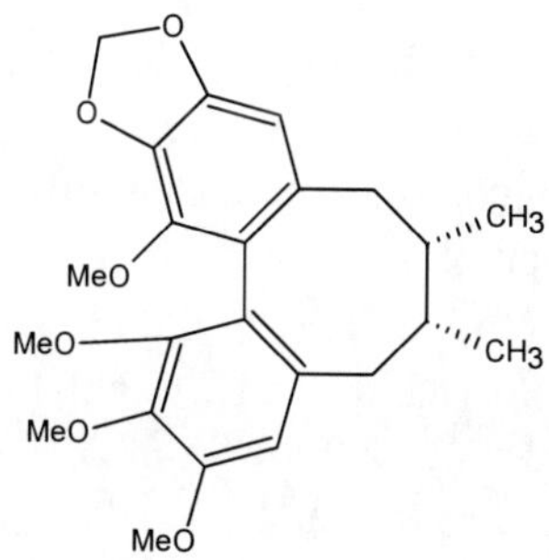

789
Schisandrin B

In vivo antioxidant action of a lignan-enriched extract of the fruit of *Schisandra chinensis* (Schisandraceae) produced a dose-dependent protection against CCl_4 hepatotoxicity. Toxicity was found to correlate well with the degree of enhancement in hepatic mitochondrial glutathione antioxidant status, as evidenced by increases in reduced glutathione level and activities of glutathione reductase. Glutathione peroxidase as well as glutathione S-transferases, in both control and CCl_4-intoxicated mice. α-Tocopherol which did not enhance mitochondrial glutathione antioxidant status, seemed to be less potent in protecting against CCl4 hepatotoxicity (Chiu et al., 2002).

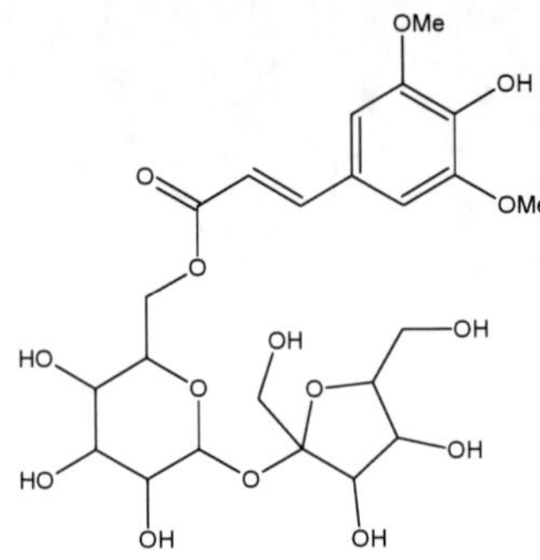

790
6-O-Sinapoyl sucrose

6-O-Sinapoyl sucrose was isolated from *Iberis amara* seeds, showed antioxidative properties. Using the DPPH test, the ester showed lower activity compared with sinapic acid and ascorbic acid, which is considered as a reference compound. On the contrary, in the superoxide scavenging test, the sucrose esterification of sinapic acid increased slightly the antioxidative activity. Therefore, esterification of sinapic acid by a sucrose moiety seems to modulate its antioxidative activity (Fabre et al., 2000).

6-O-Sinapoyl sucrose: Yellow colored amorphous solid; UV max (MeOH) 239, 328 nm; ESI-MS (negative and positive mode): *m/z* 547.2 $[M-H]^-$ and 571.1 $[M+Na]^+$, 587.0 $[M+K]^+$; ^{1}HNMR (400 MHz, CD_3OD): δ 3.29 (lH, m, H-4'), 3.46 (lH, dd, J = 9,7, 3.8 Hz, H-2'), 3.59 (lH, d, J = 12.4 Hz, H-l"a), 3.62 (lH, d, J = 12.4 Hz, H-l"b), 3.72 (lH, t, J = 9.5 Hz, H-3'), 3.79 (lH, m, H-5"), 3.86 (lH, dd, J = 12, 2.9 Hz, H-6"a), 3.87 (6H, s, OMe), 3.88 (lH, m, H-6"b), 4.06 (lH, m, H-4"), 4.08 (lH, m, H-3"), 4.12 (lH, ddd, J= 9.6, 6.4, 1.8 Hz, H-5'), 4.25 (lH, dd, J = 12, 6.4 Hz, H-6'a), 4.5 (lH, dd, J = 12, 1.8 Hz, H-6'b), 6.47 (lH, d, J = 15.9 Hz, H-8), 6.95 (2H, s, H-2 and H-6), 7.63 (lH, d, J = 15.9 Hz, H-7); ^{13}CNMR (100 MHz, CD_3OD): δ 126.6 (C-1), 106.9 (C-2 and C-6), 149.4 (C-3 and C-5), 139.0 (C-4), 147.2 (C-7), 115.7 (C-8), 169.0 (C-9), 93.1 (C-1'), 73.1 (C-2'), 74.6 (C-3'), 71.8 (C-4'), 72.0 (C-5'), 65.1 (C-6'), 64.2 (C-1" and C-6"), 105.1 (C-2"), 79.2 (C-3"), 76.0 (C-4"), 83.8 (C-5"), 56.8 (2 OMe).

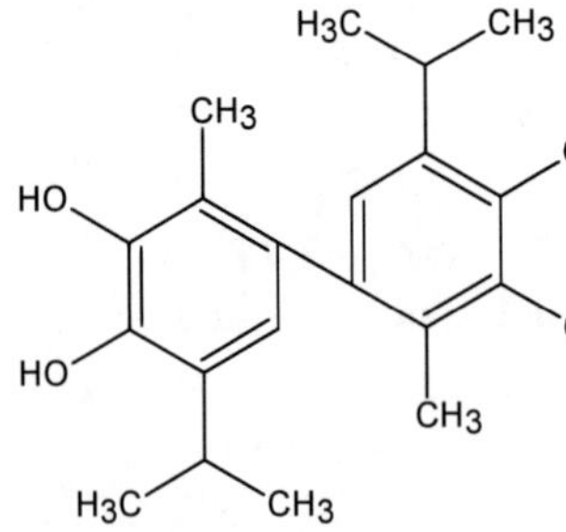

791
3,4,3',4'-Tetrahydroxy-5,5' -diisopropyl-2,2'-dimethylbiphenyl

A biphenyl compound, 3,4,3',4'-tetrahydroxy-*S,S'*-diisopropyl-2,2'-dimethylbiphenyl was isolated as antioxidative components from the leaves of *Thymus vulgaris* (Labiatae) by bioassay-directed fractionation. These compounds inhibited superoxide anion roduction in the xanthine/xanthine oxidase system. Mitochondrial and microsomal lipid peroxidation induced by Fe(I11)-ADP/NADH or Fe(III)-ADP/NADPH were also inhibited by this compounds. It is an extremely potent antioxidant; complete inhibition was observed at 1 μM against both microsomal and mitochondrial peroxidation. Furthermore, also protected red cells against oxidative hemolysis (Haraguchi et al., 1996).

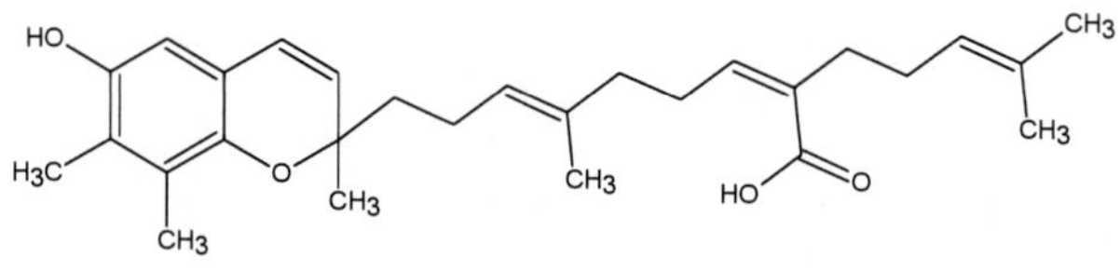

792
2,7,8-Trimethyl-2-[(3E,7Z)-4,12-dimethyl-8-carboxyl-3,7,11-tridecatrienyl]-3-chromen-6-ol

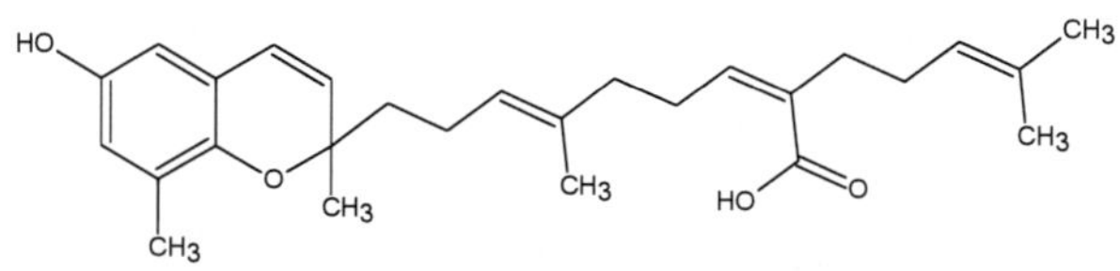

793
Sargachromenol

Two tocotrienols were isolated from *Iryanthera juruensis* (Myristicaceae) fruits. The comparison among oxidation potentials of both tocotrienols and α-tocopherol shows that 2,7,8- trimethyl-2- [(3E,7Z)-4,12-dimethyl-8-carboxyl-3,7,11-tridecatrienyl]-3-chromen-6-ol is the most easily oxidizable compound, Alkyl groups *meta* to the aromatic hydroxyl have little effect on the oxidation ability or phenolic compounds, but inductive effects from either *ortho* or *para* alkyl groups may lower the oxidation potential and make the molecule even easier oxidized than the phenol itself. The lower oxidation potential when compared to α-tocopherol might be due to its C-3 double bond, which is conjugated with the aromatic ring and stabilizes the intermediate cation previous to completion or oxidation steps (Silva et al., 2001).

2,7,8- Trimethyl -2-[(3E,7Z)-4,12-dimethyl-8-carboxyl-3,7,11-tridecatrienyl]-3-chromen-6-ol: Brown oil. Found: C, 76.62; H, 8.76. $C_{28}H_{38}O_4$ requires: C, 76.71; H, 8.68%. UV λmax. (MeOH) nm (ε): 326 (3800). IR ν cm^{-1}: 3408, 2923, 1686, 1634; EIMS (70 e V): *m/z* = (rel. int.) 175 $[189\text{-Me}]^+$; ^{1}HNMR (300 MHz, $CDCl_3$): δ 1.27 (3H, s, H-7), 1.50 (6H, s, H-14', H-16'), 1.59 (3H, s, H-13'), 1.9-2.2 (8H, m, H-1', H-2', H-5', H-10'), 2.04 (3H, s, H-2), 2.05 (3H, s, H-8), 2.2-2.3 (2H, m, H-9'), 2.50 (2H, dd, J = 14.4, 7.3 Hz, H6'), 5.45 (1H, d, J = 9.6 Hz, H-3), 6.14 (IH, d, J = 9.6 Hz, H-4), 6.22 (1H, s, H-5); ^{13}CNMR (75 MHz, $CDCl_3$): δ 11.5 (q, Me-8), 11.9 (q, Me-7), 15.7 (q, C-16'), 17,5 (q, C-14'), 22.6 (t, C- 2'), 25.4 (q, Me-2), 25.8 (q, C-13'), 27.8 (t, C-6'), 28,1 (t, C-10'), 34.4 (t, C-1'), 39.0 (t, C-5'), 40.9 (t, C-9'), 77,7 (s, C-2), 109.9 (d, C-5), 118.7 (s, C-4a), 122.8 (d, C-4), 123.5 (d, C-3'), 123.6 (s, C-8), 125.1 (d, C-11'), 125.1 (s, C-7), 129.6 (d, C-3), 130.6 (s, C-12'), 132.1 (s, C-4'), 134.2 (s, C-8'), 144.7 (s, C-8a), 144.8 (d, C-7'), 146.9 (s, C-6), 171.6 (s, C-15').

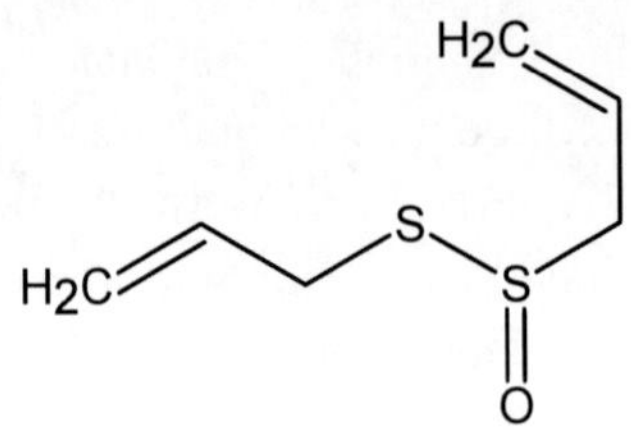

794
Allicin

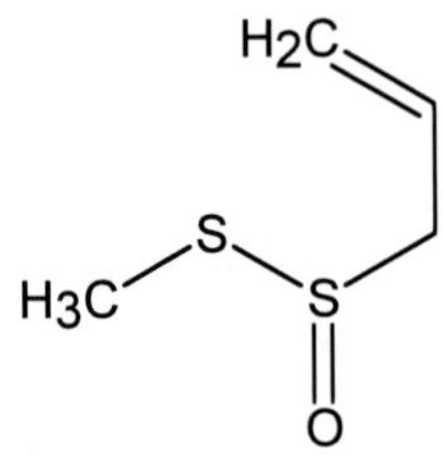

795
Methyl-allyl thiosulfinate

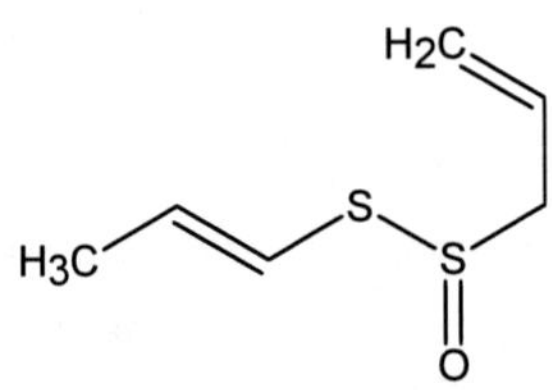

796
1-Propenyl allyl thiosulfinate

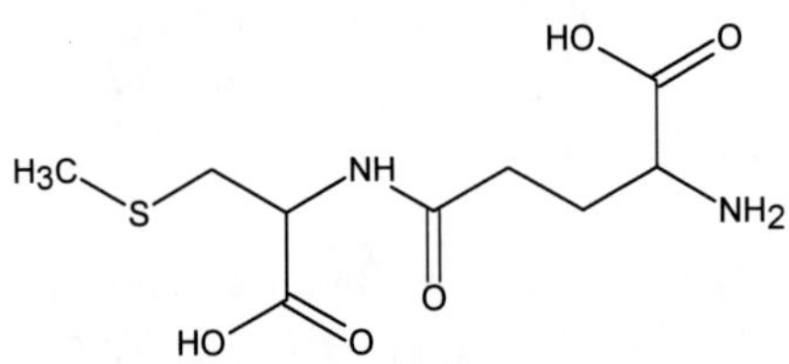

797
L-Glutamyl-S-alkyl-L-cysteine

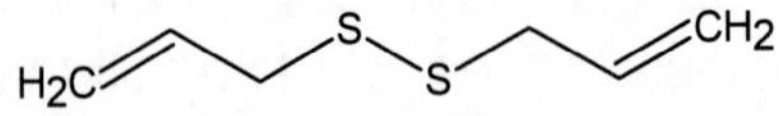

798
Diallyl disulfide

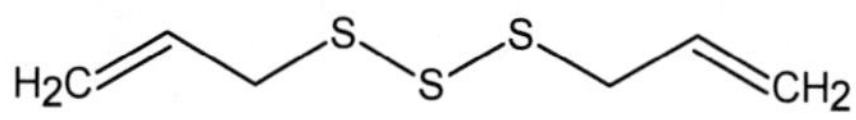

799
Diallyl trisulfide

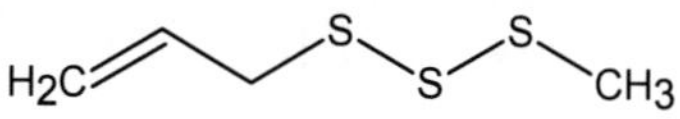

800
Allyl methyl trisulfide

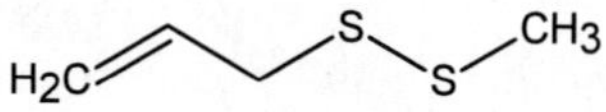

801
Allyl methyl disulfide

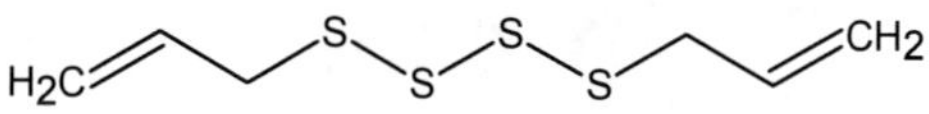

802
Diallyl tetrasulfide

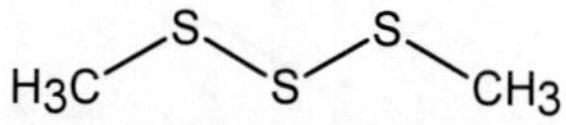

803
Dimethyl trisulfide

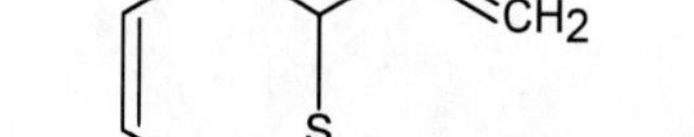

804
Diallyl sulfide

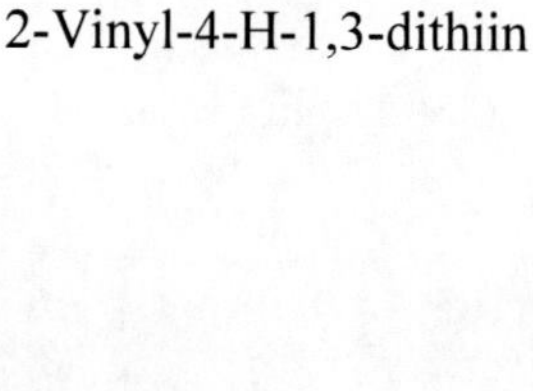

805
2-Vinyl-4-H-1,3-dithiin

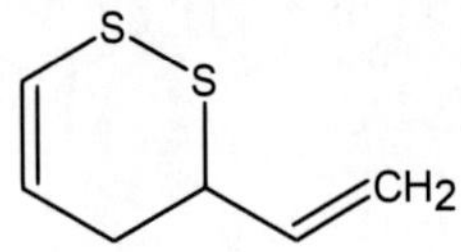

806
3-Vinyl-4-H-1,2- dithiin

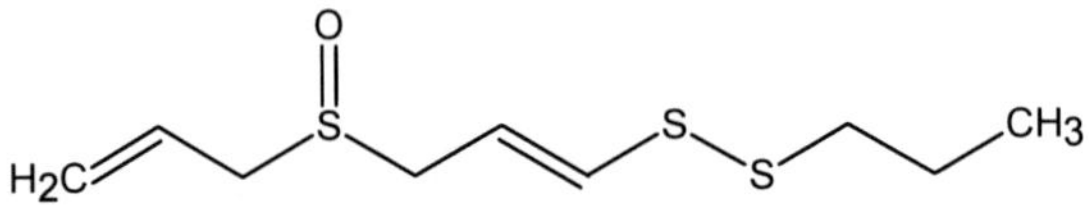

807
E-Ajoene

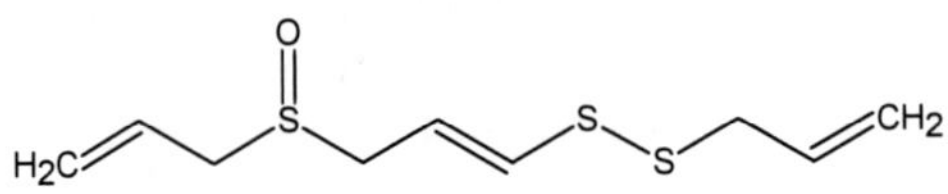

808
Z-Ajoene

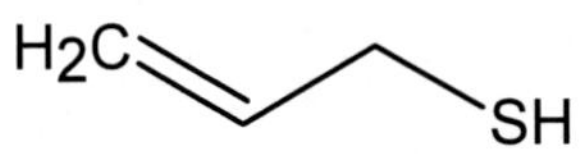

809
Allyl mercaptan

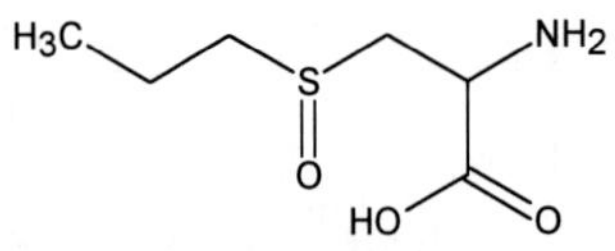

810
Allin

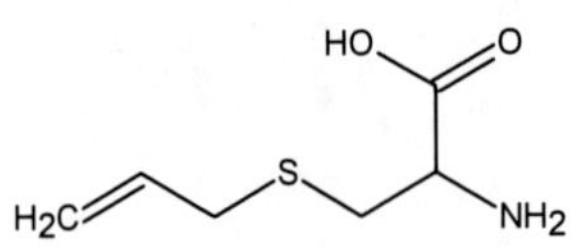

811
S-Allyl cysteine

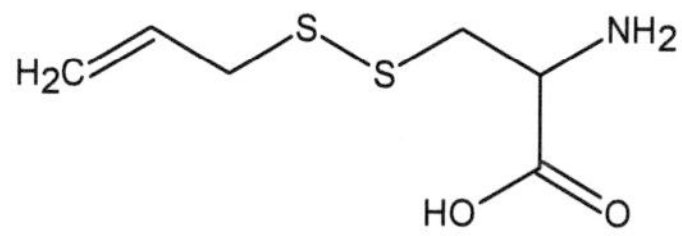

812
S-Allyl mercaptocysteine

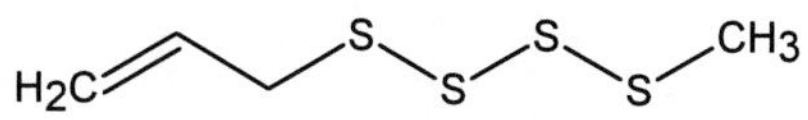

813
Allyl methyl tetrasulfide

Organosulphur were found in *Allium sativum* (Liliaceae). Garlic has played an important dietary and medicinal role throughout the history of mankind. In some Western countries, the sale of garlic preparations ranks with those of leading prescription drugs. The therapeutic efficacy of garlic encompasses a wide variety of ailments, including cardiovascular, cancer, hepatic and microbial infections to name but a few. However, the elucidation of its mechanism for therapeutic action has proved to be more elusive and a unifying theory, which could account for its reported multifarious activities, is yet to emerge. Reactive oxygen species (ROS) seem to be at the core of many disease processes and it is an attractive and convenient hypothesis that garlic might exert its activities through modulatory effects on ROS. A literature search on garlic and its antioxidant potential churned up a surprisingly large amount of data. Various preparations of garlic, mainly aged garlic extract (AGE), have been shown to have promising antioxidant potential. However, the presence of more than one compounds in garlic with apparently opposite biological effects, has added to the complexity of the subject. Raw garlic homogenate has been reported to exert antioxidant potential but higher doses have been shown to be toxic to the heart, liver and kidney (Banerjee et al., 2003). Among the variety of organosulfur compounds, *S*-allylcysteine (SAC) and *S*-allylmercaptocysteine (SAMC), the major organosulfur compounds found in aged garlic extract, showed radical scavenging activity in both chemiluminescence and 1,1-diphenyl-2-picrylhydrazyl (DPPH) assays, indicating that these compounds may play an important role in the antioxidative activity of aged garlic extract (Imail et al ., 1994). Also allicin is active against *Entamoeba histolytica* at 50 μg/ml (100% killing), (Mirelman et al., 1987).

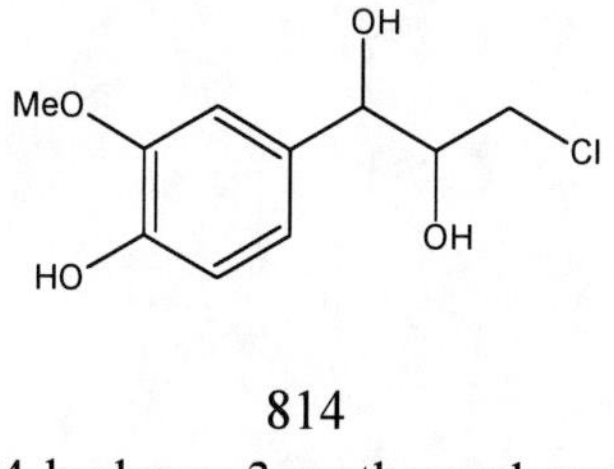

814
Threo-3-chloro-l-(4-hydroxy-3-methoxyphenyl)propane-1,2-diol

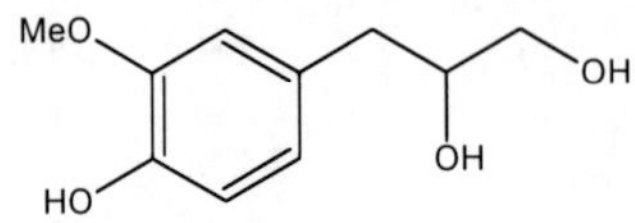

815
3-(4-Hydroxy-3-methoxyphenyl) propane-l ,2-diol

These phenylpropanoid, were isolated from the berries of *Pimentoa dioica* (Myrtaceae). The phenylpropanoids inhibited autoxidation of linoleic acid in a water-alcohol system (Kikuzaki et al., 1999).

Threo-3-chloro-l-(4-hydroxy-3-methoxyphenyl)propane-1,2-diol: Colorless needles, mp 121° (CHC13). $[\alpha]_D$ -2° (EtOH; c 0.52). IR ν cm^{-1}: 3400-3200, 1609, 1522); HR-EIMS 20 eV: found: *m/z* 232.0547 $[M]^+$, $C_{10}H_{13}O_4Cl$ requires: 232.0501, ElMS 20 eV *m/z* (rel. int.): 234 $[M+2]^+$ (0.9, [M+2]/[M] 0.41), 232 [M]+ (2.2), 216 (1), 214 (4), 196 (4), 178 (11), 151 (100), 137 (47); ^{1}HNMR (300 MHz; $(CD_3)_2CO$): δ 5 3.37 (1H, dd, J = 6.5, 11.5 Hz, H-3a), 3.62 (1H, dd, J = 4.5, 11.5Hz, H-3b), 3.81 (1H, m, H-2), 3.84 (3H, s, OCH_3), 4.32 (1H, d, J = 5.0 Hz, 2-OH), 4.45 (1H, d, J = 4.5 Hz, 1-OH), 4.64 (1H, dd, J = 4.5, 5.0 Hz, H-1), 6.79 (1H, d, J = 8.0 Hz, H-5'), 6.85 (1H, dd, J = 2.0, 8.0 Hz, H-6'), 7.04 (lH, d, J = 2.0 Hz, H-2'), 7.55 (IH, s, 4'-OH); ^{13}C NMR ($(CD_3)_2CO$): δ 5 47.4 (C-3), 56.2 (OCH_3), 74.9 (C-1),76.7 (C-2), 111.0 (C-2'), 115.3 (C-5'), 120.2 (C-6'), 134.1 (C-1 '), 146.9 (C-4'), 148.1 (C-3').

3-(4-Hydroxy-3-methoxyphenyl)propane-l ,2-diol: Colorless oil, $[\alpha]_D$ -12° (EtOH; c 0.99). IR ν cm^{-1}: 3600-3200, 1605, 1516, 1274; HR-EIMS 70 eV: found *m/z* 198.0914 $[M]^+$, $C_{10}H_{14}O_4$ requires 198.0892, ElMS 70 eV *m/z* (rel. int.): 198 $[M]^+$, (20), 180 (9), 167 (17), 137 (100); ^{1}HNMR (300 MHz; $(CD_3)_2CO$): δ 2.59 (lH, dd, J = 4.4, 13.2 Hz, H-3a), 2.70 (lH, dd, J = 3.4, 13.2 Hz, H-3b), 3.43 (lH, ddd, J = 5.8, 5.8, 10.8 Hz, H-1a), 3.51 (lH, ddd, J = 4.4, 6.0, 10.8 Hz, H-1b), 3.80 (lH, m, H-2), 3.83 (3H, s, OCH_3), 6.67 (1H, dd, J = 2.0, 8.0 Hz, H-6'), 6.72 (lH, d, J = 8.0 Hz, H-5'), 6.85 (lH, d, J = 2.0 Hz, H-2'), 7.31 (1H, s, 4'-OH); ^{13}C NMR $(CD_3)_2CO$): δ 40.4 (C-3), 56.1 (OCH_3), 66.5 (C-l), 74.2 (C-2), 113.7 (C-2'), 115.5 (C-5'), 122.6 (C-6'), 131.4 (C-1 '), 145.7 (C-4'), 148.0 (C-3').

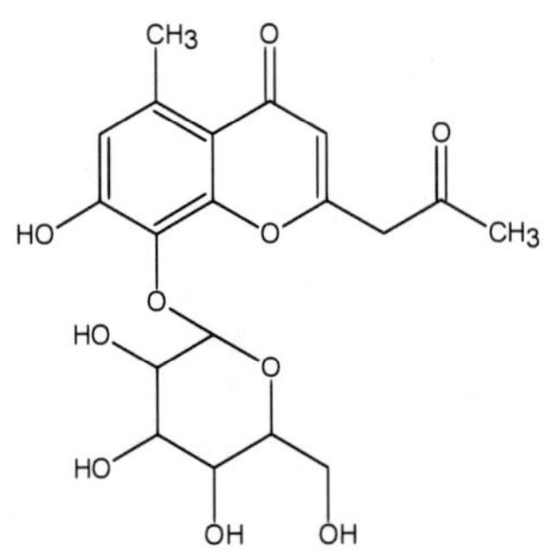

816
Aloesin

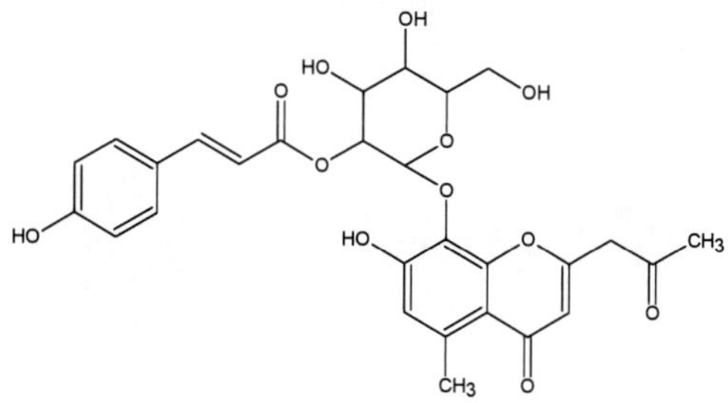

817
Aleresin A

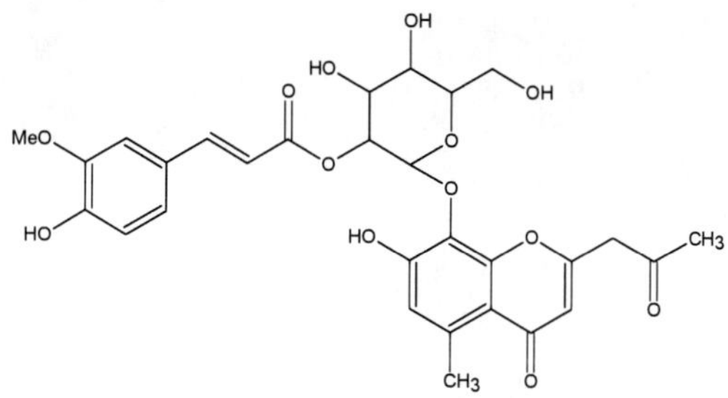

818
Feruloylaloesin

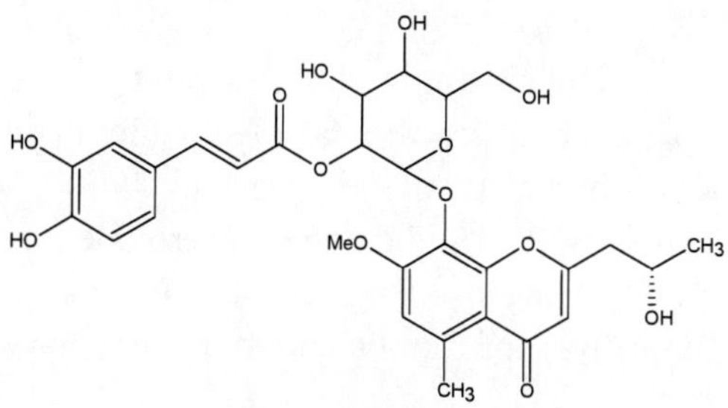

819
Isorabaichromone

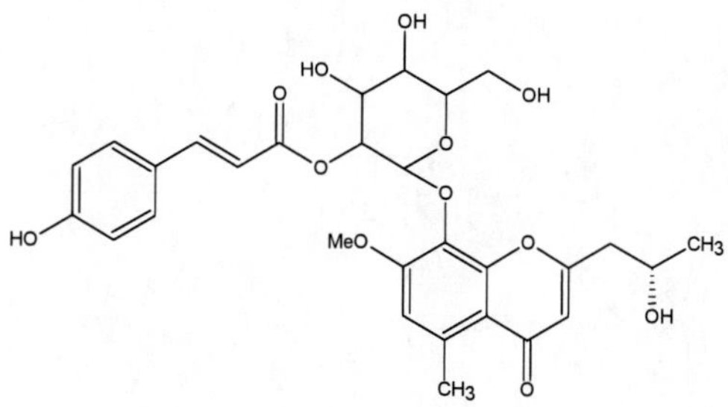

820
Isoaloeresin D

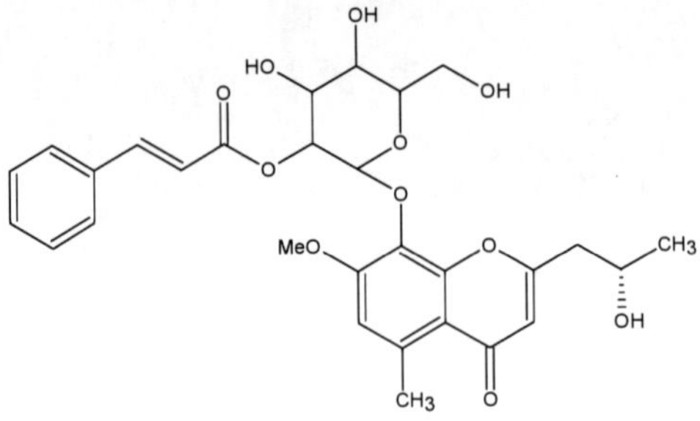

821
Isoaloeresin E

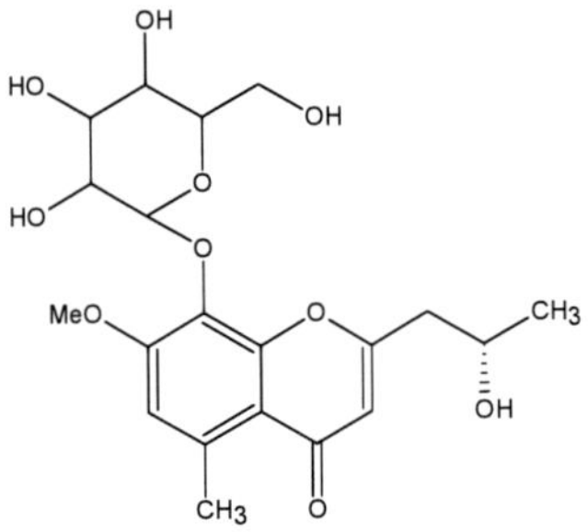

822
7-Methoxy-8-glucosylaloesol

These compounds were isolated from *Aloe vera* (Liliaceae). Among the aloesin derivatives examined, isorabaichromone showed a potent antioxidative activity. As one of the most potent components, isorabaichromone together with feruloylaloesin and *p*-coumaroylaloesin showed potent DPPH radical and superoxide anion scavenging activities. Electron spin resonance (ESR) using the spin trapping method suggested that the potent superoxide anion scavenging activity of isorabaichromone may have been due to its caffeoyl group (Yagi et al., 2002).

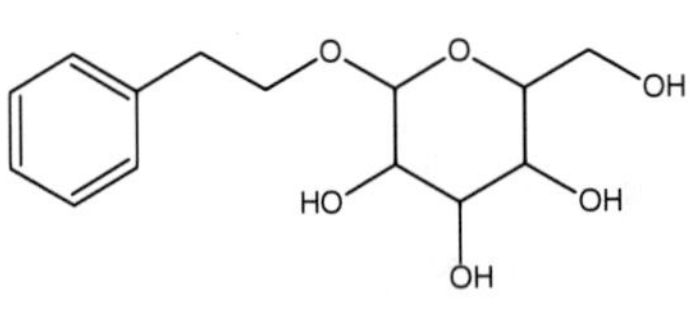

823
Phenethyl β-D-glucopyranoside

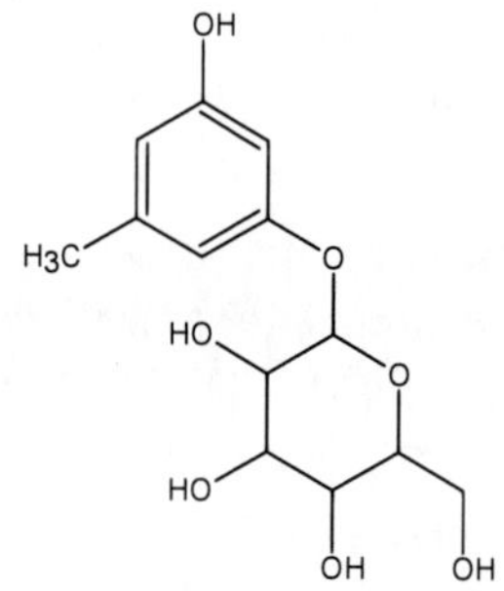

824
Orcinol-β-D-glucopyranoside

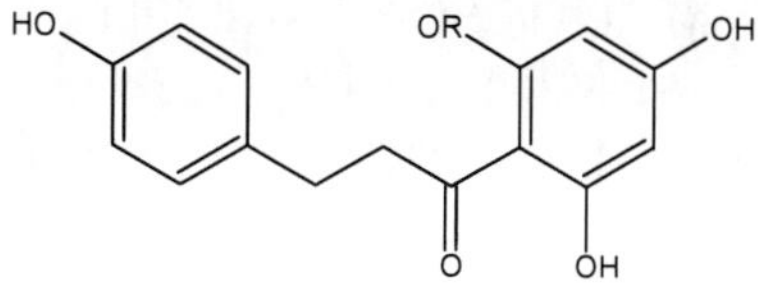

825
R = -β-D-glucopyranoside
Phloretin 2'-O- β-D-glucopyranoside

826
R=-β-D-xylopyranosyl-(1→6)-β-D-glucopyranoside
Phloretin 2'-O-β-D-xylopyranosyl-(1→6)-β-D-glucopyranoside

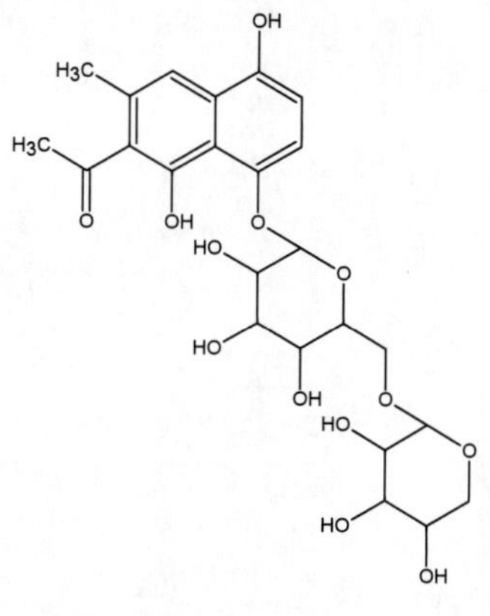

827
Stelladerol-1-(1,5,8-trihydroxy-3-methylnaphthalen-2-yl)ethanone-8-O-β-D- xylopyranosyl-(1-6)-β-O-glucopyranoside)

Daylily (*Hemerocallis* spp. (Hemerocallidaceae) flowers are utilized as an important ingredient in traditional Asian cuisine and are also valued for their reputed medicinal effects. All of these compounds were tested for their antioxidant and cyclooxygenase inhibitory activities. Stelladerol was found to possess strong antioxidant properties, inhibiting lipid oxidation by 94.6% ±1.4 at 10 μM in an *in vitro* assay (Cichewicz and Nair, 2002).

Stelladerol-1-(1,5,8-trihydroxy-3-methylnaphthalen-2-yl)ethanone-8-O-β-D-xylopyranosyl-(1-6)-β-O-glucopyranoside): Yellow amorphous solid; UV (MeOH) λmax (log ε) 223 (4.97), 313 (4.15), 345 (4.18) nm; HRFABMS, *m/z* 527.1756 $[M + H]^+$ (calcd for $C_{24}H_{31}O_{13}$, 527.1765); FABMS, *m/z* 549 $[M + Na]^+$, 527 $[M + H]^+$, 395 $[M\text{-xyl} + 2H]^+$, 233 $[M\text{-xyl-Glc} + 2H]^+$; ^{1}HNMR (600 MHz, DMSO-d_6): δ 9.84 (1H, s, -OH, exchange with D_2O), 9.63 (1H, s, -OH, exchange with D_2O), 7.41 (1H, s, J = 1.5 Hz, H-4), 7.27 (1H, d, J = 8.3 Hz, H-7), 6.76 (1H, d, J = 8.3 Hz, H-6), 4.85 (1H, d, J = 7.5 Hz, H-1'), 4.23 (1H, d, J = 7.5 Hz, H-1"), 4.02 (1H, d, J = 10.5 Hz, H-6'), 3.69 (1H, dd, J = 5.3, 11.3 Hz, H-5"), 3.58

(1H, rn, H-6'), 3.31 (2H, m, H-2', H-5'), 3.29 (1H, rn, H-4"), 3.18 (1H. t, J = 9.0 Hz, H-4'), 3.11 (2H, m, H-3', H-3"), 3.02 (2H, m, H-2", H-5"), 2.51 (3H, s, -$COCH_3$), 2.25 (3H, d, J = 1.5 Hz, -CH_3); ^{13}C NMR (75 MHz, DMSO-d_6): δ 204.9 (s, C=O), 150.2 (s, C-1), 148.2 (s, C-5), 146.8 (s, C-8), 131.3 (s, C-3), 126.2 (s, C-10), 125.5 (s, C-2). 114.0 (s, C-9), 113.9 (d, C-4), 112.1 (d, C-7), 109.0 (d, C-6), 104.2 (d, C-1"), 103.4 (d, C-1'), 76.5 (d, C-3', C-3"), 76.2 (d, C-5'), 73.4 (d, C-2', C-2"), 70.0 (d, C-4'), 69.6 (d, C-4"), 68.6 (t, C-6'), 65.6 (t, C-5"), 32.0 (q, -COCH3), 19.4 (q. -CH_3).

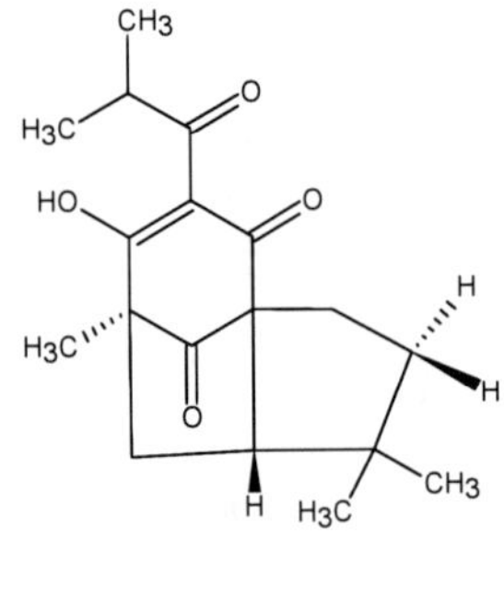

828
Ialibinone E

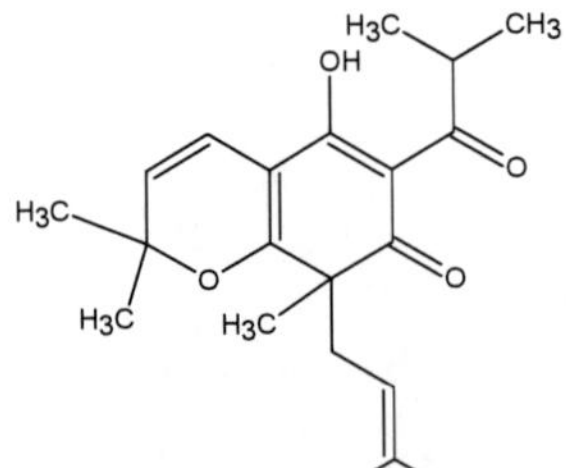

829
Hyperguinone B

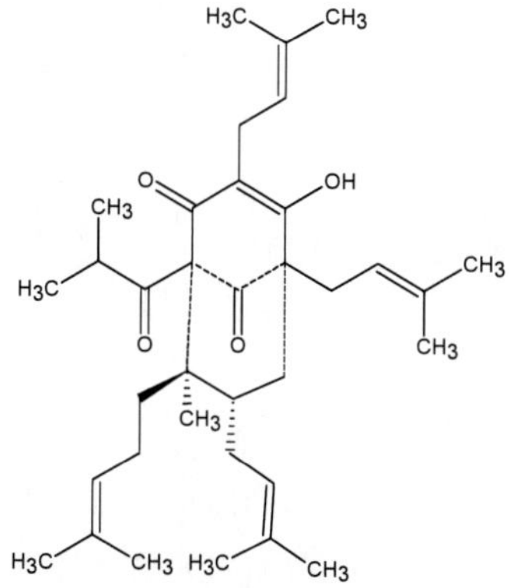

830
Hyperforin

These phloroglucinol derivatives belonging to different carbon skeletons were isolated from *Hypericum papuanum* Ridley (Hipericaceae). Showed a strong reduction of oxygen production by PMNs after stimulation with FMLP for the compounds ialibinone E, hyperguinone B and hyperforin. The IC_{50} values obtained were 2.5 μM, 3.3 μM and 1.8 μM, respectively. Slight modifications of the substituents or variation of the stereochemistry resulted in a significant loss of activity. None of the active compounds showed antioxidative activity after stimulation with OZ. Only ialibinone E, showed potent activity on the production of oxygen in an H_2O_2/horseradish peroxidase system with IC_{50} 1 μM value. Also only ialibinone E was found to be significantly active at lower micromolar concentrations on superoxide-scavenging was not active in a xanthine oxidase assay (urate formation) in concentrations up to 100 μM and its activity is therefore not attributable to the inhibition of this enzyme. None of the active phloroglucinols showed cytotoxic effects against the PMNs (Heilmann et al., 2003).

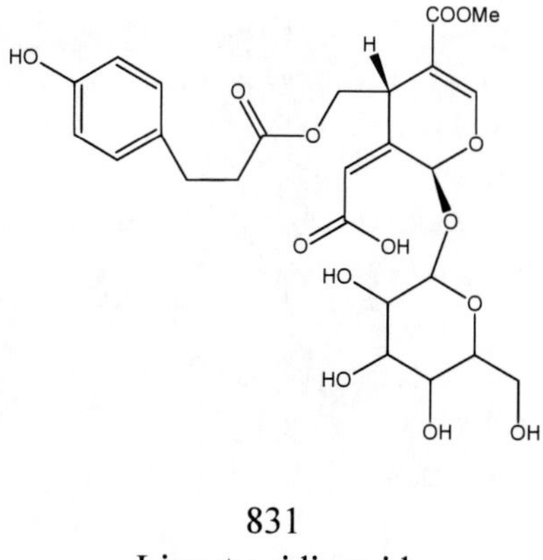

831
Ligustrosidic acid

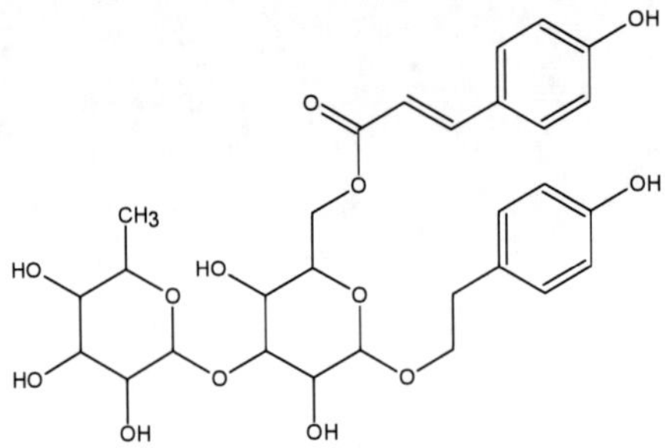

832
Osmanthuside B_6

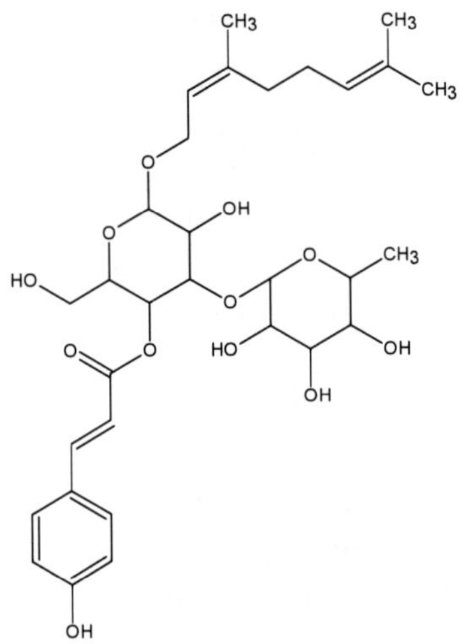

833
Ligurobustoside C

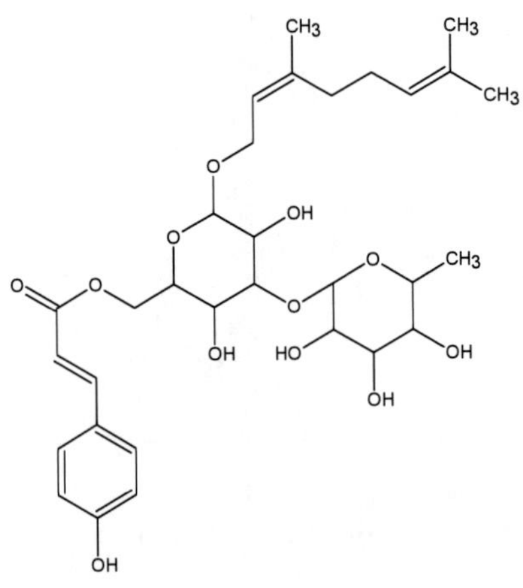

834
Ligurobustoside E

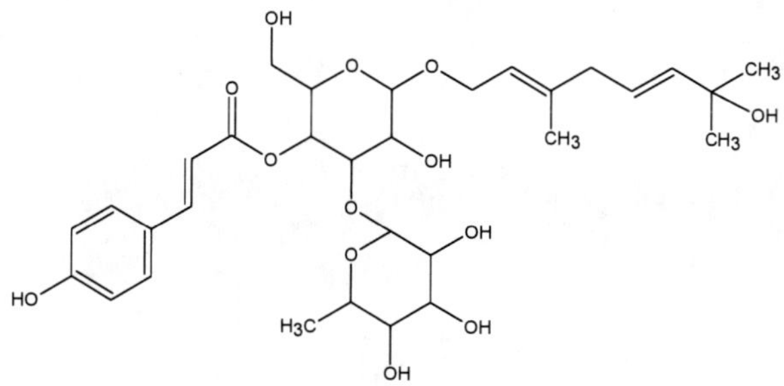

835
Ligurobustoside I

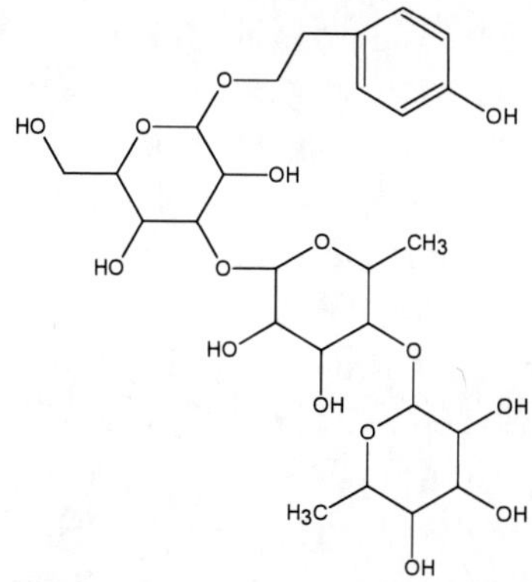

836
Ligurobustoside M

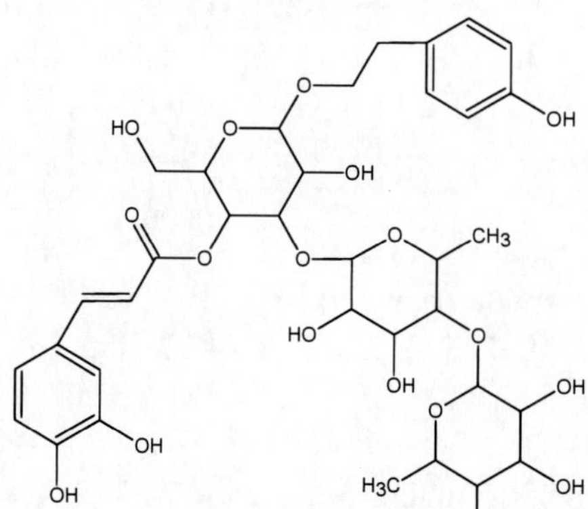

837
Ligurobustoside N

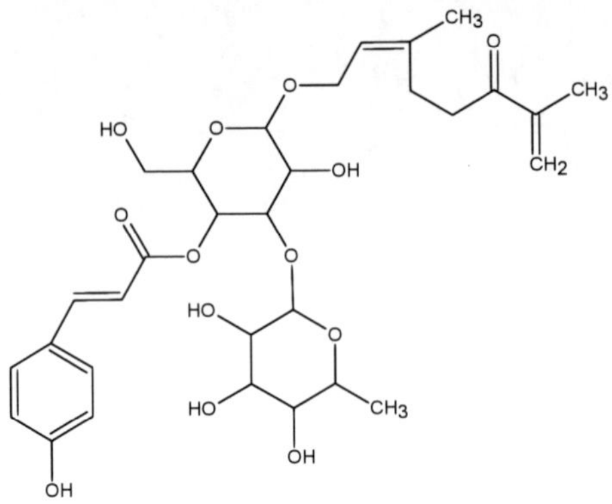

838
Ligurobustoside O

Phytochemical investigation of the ethanol extract of the leaves of *Ligustrum robustum* (Roxb.) (Oleaceae), monitored by a bioassay involving the hemolysis of red blood cells induced by 2,2'-azo-bis(2-amidinopropane) dihydrochloride, led to the isolation of these glycosides, showed stronger antioxidant effects than the standard, trolox (He et al., 2003).

Ligurobustoside M: Amorphous powder, $[\alpha]_D$-67.4° (c 0.03, MeOH); UV λmax (EtOH, log ε): 200 (3.85), 221 (4.10), 283 (3.48),316 (3.35) nm; HRFABMS *m/z* found 593.3149 $[M + H]^+$ ($C_{26}H_{41}O_{15}$ requires 593.3165); ESIMS *m/z* 591 $[M -H]^-$, 445 $[M -rha]^-$, 299 $[M -rha - O-rha]^-$; ^{1}HNMR (300 MHz, CD_3OD): δ aglycone 7.25 (2H, d, J = 8.5 Hz, H_2-2, 6), 6.88 (2H, d, J = 8.5 Hz. H_2-3.5), 3.02 (2H, t, J = 6.5 Hz. H_2-7), 3.77 (lH, m, Ha-8), 4.09 (lH, m, Hb-8), glucosyl 4.47 (lH, d, J = 7.9 Hz, H-l'), 3.58 (lH, t, J = 8.2 Hz, H-2'), 3.58 (lH, t, J = 8.2 Hz, H-3'), 3.50 (1H, t, J= 8.2 Hz, H-4'), 3.52 (m. H-5'), 3.79 (lH, dd, J = 5.5. 11.6 Hz, Ha-6'), 4.04 (lH, dd, J = 3.1, 11.6 Hz, Hb-6'), rhamnosyl 5.39 (lH, d, J = 1.5 Hz, H-1"), 5.35 (lH, d, J = 1.5 Hz, H-1 '''), 3.88 (m, H-2", 2''', 3", 3'''), 3.50 (m, H-4", 4'''), 4.08 (m, H.5", 5'''), 1.46 (3H, d, J = 6.3 Hz, H_3-6"), 1.43 (3H, d, J = 6.3 Hz, H3-6'''); ^{13}C NMR (CD_3COCD_3): δ aglycone 131.0 (C-l), 131.2 (C-2, 6), 116.4 (C-3, 5), 157.1 (C-4), 36.6 (C-7), 72.4 (C-8), glucosyl 104.5 (C-1΄), 76.0 (C-2'), 84.3 (C-3'). 70.7 (C-4'), 78.2 (C-5'), 63.0 (C-6'), rhamnosyl 103.5 (C-1'), 102.7 (C-1'''), 72.4 (C-2"), 72.7 (C-2'''), 73.2 (C-3"), 73.3 (C-3'''), 81.4 (C-4"), 74.2 (C-4'''), 68.8 (C-5"), 70.5 (C-5'''), 18.9 (C-6"), 18.2 (C-6''').

Ligurobustoside N: Amorphous powder, $[\alpha]_D$ -109° (c 0.06, MeOH); UV λmax (EtOH, log ε) 200 (4.1l), 212 (4.00), 226 (3.8l), 318 (4.19) nm; IR Vmax (KBr) cm^{-1}: 3400, 2900,1700, 1618, 1610, 1515, 1450, 1170, 820; HRFABMS *m/z* found 755.3631 $[M+H]^+$ ($C_{35}H_{47}O_{18}$ requires 755.3624); ESIMS *m/z* 753 $[M -H]^-$, 607 $[M -rha]^-$, 591 $[M -caf]^-$, 444 $[M -rha -caf]^-$; ^{1}HNMR (300 MHz, CD_3OD): δ aglycone 7.27 (2H, d, J = 8.5 Hz, H_2-2, 6), 6.89 (2H, d, J = 8.5 Hz, H_2-3, 5), 3.04 (2H, t, J = 6.8 Hz, H_2-7), 3.87 (lH, m, Ha-8), 4.09 (lH, m, Hb-8), glucosyl 4.47 (lH, d, J = 7.9 Hz, H-1'), 5.13 (lH, t, J = 9.5 Hz, H-4'), rhamnosyl 5.39 (lH, d, J = 1.7 Hz, H-1"), 5.25 (lH, d, J = 1.5 Hz, H-1'''), 1.29 (3H, d, J = 6.3 Hz, H3-6"), 1.25 (3H, d, J = 6.3 Hz, H_3-6'''), caffeoyl 7.26 (lH, d, J = 2.0 Hz, H-2''''), 6.99 (lH, d, J = 8.1 Hz, H-5''''), 7.17 (lH, dd, J= 2.0, 8.1 Hz, H-6''''), 7.79 (lH, d, J= 15.9 Hz, H- 7''''), 6.45 (lH. d, J = 15.9 Hz, H-8''''); ^{13}C NMR (CD_3COCD_3): δ aglycone 131.0 (C-l), 131.2 (C-2, 6), 116.6 (C-3, 5), 157.1 (C-4), 36.6 (C-7), 72.6 (C-8), glucosyl 104.5 (C-1'), 76.3 (C-2'), 81.74 (C-3'), 70.6 (C-4'), 76.1 (C-5'), 62.6 (C.6'), rhamnosyl 103.7 (C-1"), 102.9 (C-1'''), 72.7 (C-2"), 74.0 (C-

2'''), 70.8 (C-3''), 72.6 (C-3'''), 81.8 (C-4''). 74.2 (C-4'''), 69.2 (C-5'', 5'''), 19.5 (C-6''), 18.0 (C-6'''), caffeoyl 127.8 (C-1''''), 115.6 (C-2''''), 147.1 (C-3''''), 150.2 (C-4''''), 117.0 (C-5''''), 123.7 (C-6''''), 148.2 (C-7''''), 115.0 (C-8''''), 168.4 (CO).

Ligurobustoside O: Amorphous powder, $[\alpha]_D$ -77.3° (c 0.05, MeOH); UV λmax (EtOH, log ε) 201 (4.32),214 (4.67), 330 (4.15) nm; IR Vmax (KBr) cm^{-1}: 3400,2900, 1700,1595, 1515, 1432, 1170, 825; HRFABMS *m/z* found 623.3343 $[M + H]^+$($C_{31}H_{43}O_{13}$ requires 623.3336); ESIMS m/z 621 $[M -H]^-$, 475 $[M\text{-rha}]^-$, 328 $[M\text{-rha-cou}]^-$; ^{1}HNMR (300 MHz, CD_3OD): δ aglycone 4.46 (2H, d, J = 7.4 Hz, H_2-1), 5.58 (IH, t, J = 7.4 Hz, H_2-2), 2.33 (2H, m, H_2-4), 3.09 (2H, m, H2-5), 6.05 (1H, d, J = 1.5 Hz, Ha-8), 6.28 (lH, d, J = 1.5 Hz, Hb-8), 2.05 (3H, s, H_3-9), 1.91 (3H, s, H_3-10), glucosyl 4.55 (1H, d, J = 7.9 Hz, H-1'), 5.11 (1H, t, J = 9.2 Hz, H-4'), rhamnosyl 5.39 (1H, d, J = 1.3 Hz, H-1''), 1.28 (3H, d, J = 6.0 Hz, H_3-6''), p-coumaroyl 7.66 (2H, d, J = 8.6 Hz, H-2''', 6'''), 7.0 (lH, d, J = 8.6 Hz, H-3''', 5'''), 7.86 (1H, d, J = 15.9.Hz, H-7'''), 6.53 (1H, d, J = 15.9 Hz, H-8'''); ^{13}C NMR (CD_3COCD_3): δ aglycone 66.7 (C-1), 122.2 (C-2), 141.4 (C-3), 37.1 (C-4), 35.5 (C-5), 203.9 (C-6), 147.6 (C-7), 126.3 (C-8), 18.1 (C-9), 17.0 (C-10), glucosyl 103.4 (C-1'), 76.4 (C-2'), 81.9 (C-3'), 70.7 (C-4'), 76.5 (C-5'), 62.7 (C-6'), rhamnosyl 102.9 (C-1''), 72.4 (C-2''), 72.7 (C-3''), 74.1 (C-4''), 70.8 (C-5''), 18.9 (C-6''), p-coumaroyl 127.4 (C-1'''), 131.6 (C-2''', 6'''), 117.2 (C-3''', 5'''), 161.8 (C-4'''), 147.9 (C-7'''), 115.1 (C-8'''), 168.6 (CO).

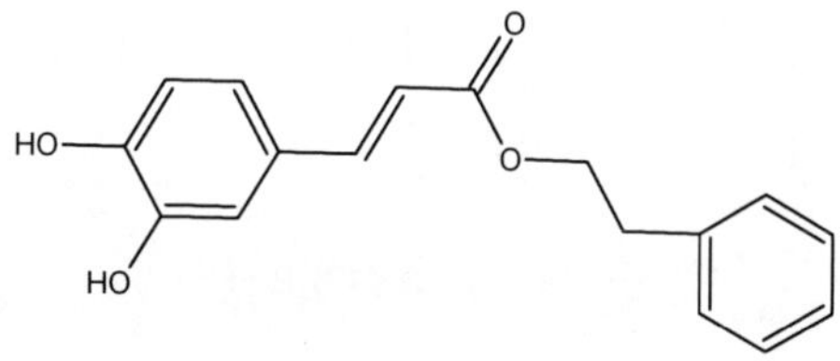

839
Caffeic acid phenethyl ester

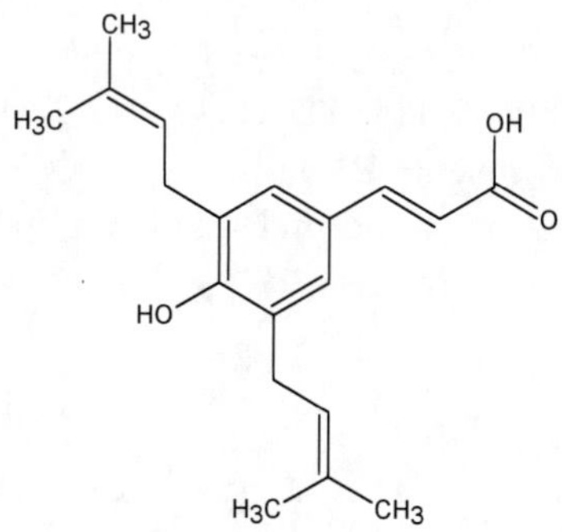

840
Artepillin C

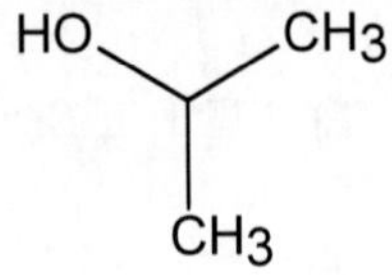

841
Propol

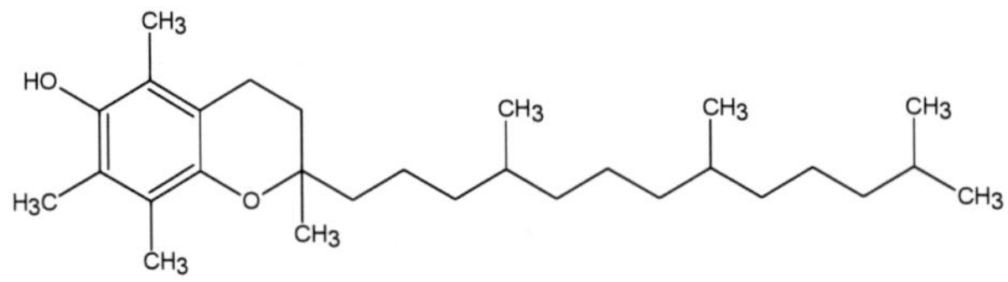

842
α-Tocopherol

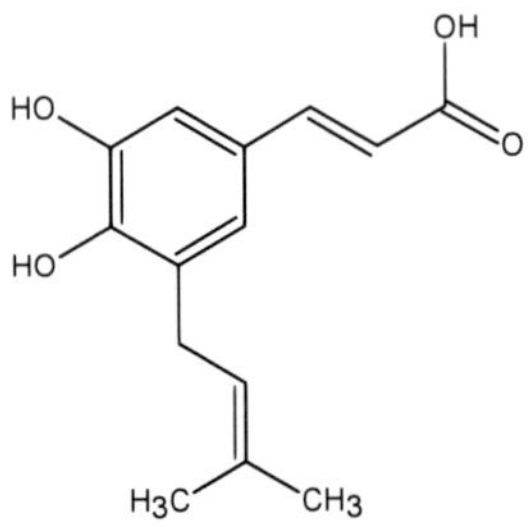

843
3-[3,4-Dihydroxy-5-prenylphenyl)-2-(E)-propenoic acid

These compounds were isolated from Brazilian propolis have activity against peroxidation of linoleic acid in a micelle solution. 3-[3,4-dihydroxy-5-prenylphenyl)-2-(E)-propenoic acid possessed the highest potency (IC_{50} 0.17 μM) among them and was more effective than butylated hydroxy toluene (BHT: IC_{50} 0.36 μM). Artepillin C (IC_{50} 0.44 μM) is the most abundant of the isolated compounds in Brazilian propolis (Banskota, 2001).

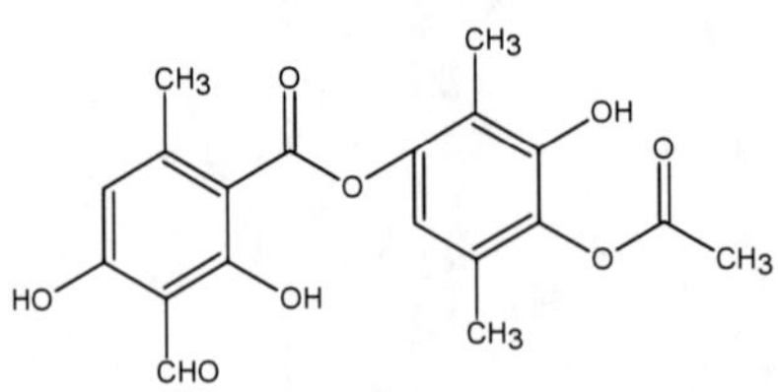

844
Atranorin

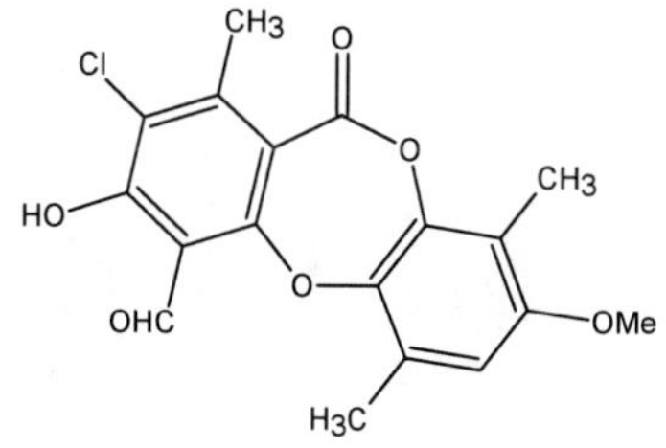

845
Pannarin

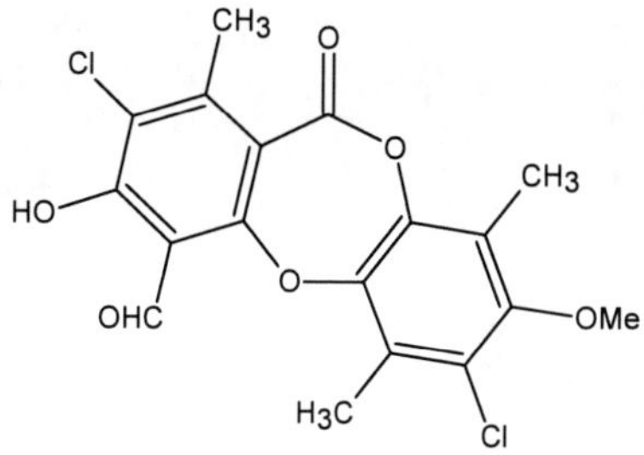

846
1′-Chloropannarin

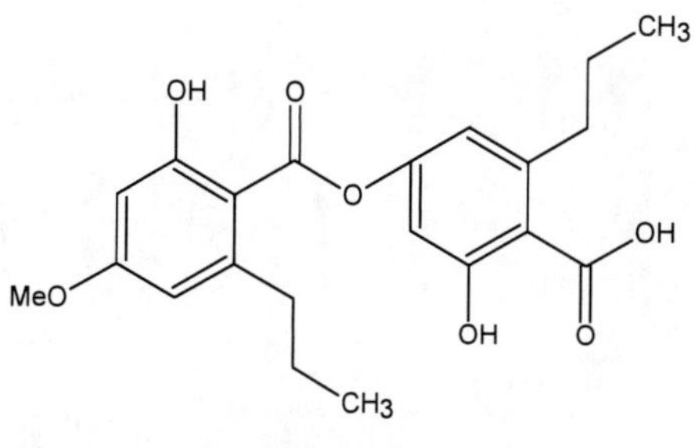

847
Divaricatic acid

The antioxidant activity of lichenic metabolites, depsides and depsidones, showed antioxidant activity. In assay of rat brain homogenate auto-oxidation and β-carotene oxidation. The lichenic metabolites afford a moderate protection in the μM concentration range. The largest effect was measured employing 1′-chloropannarin in the brain homogenate auto-oxidation, where a 66% protection was afforded at 1.7 μM. This protection is very similar to that elicited by addition of the reference antioxidant propylgallate (70% protection at 1.3 μM), (Hidalgo et al., 1994).

ACKNOWLEDGMENTS

The author wishes to thank Escuela Superior de Ingeniería Química e Industrias Extractivas I.P.N. for their support.

References

Ali, B.H. and Blunden, G. 2003. Pharmacological and toxicological properties of *Nigella sativa*. *Phytother. Res.* 17, pp. 299-305.

Andreasen, M.F., Landbo, A., Christensen, L.P., Hansen, A., Meyer, A.S. 2001.

Antioxidant effects of phenolic Rye (*Secale cereale* L.) extracts, monomeric hydroxycinnamates, and ferulic acid dehydrodimers on human low-density lipoproteins.*J. Agric. Food Chem.* 49, pp. 4090-4096.

Andrikopoulos, N.K., Kaliora, A.C., Assimopoulou, A.N., Papapeorgiou, V.P. 2003. Biological activity of some naturally occurring resins, gums, and pigments against in vitro LDL oxidation. *Phytother. Res.* 17, pp. 501-507.

Aquino, R., Morelli, S., Lauro, M. R., Abdo, S., Saija, A., Tomaino.A. 2001. Phenolic constituents and antioxidant activity of an extract of *Anthurium versicolor* leaves. *J. Nat. Prod.* 64, pp. 1019-1023.

Aquino, R., Caceres, A., Morelli, S., Rastrellit, L. 2002. An extract of *Tagetes lucida* and its phenolic constituents as Antioxidants. *J. Nat. Prod.* 65, pp. 1773-1776.

Aruoma, O. I., Halliwell. B., Aeschbach, R., Loligers, I. 1992. Antioxidant and pro-oxidant properties of active rosemary c onstituents: Carnosol and carnosic acid. *Xenobiotica.* 22, pp.257-268.

Asres, K., Bucar, F., Knauder, E., Yardley, V., Kendrick, H., Croft, L. 2001. *In vitro* antiprotozoal activity of extract and compounds from the stem bark of *Combretum molle*. *Phytother. Res.* 15, PP. 613-617.

Azuma, K., Nakayama, M., Koshioka, M., Ippoushi, K., Yamaguchi, Y., Kohata, K., Yamauchi, Y., Ito, H., Higashio, H. 1999. Phenolic antioxidants from the leaves of *Corchorus olitorius* L. *J. Agric. Food. Chem.* 47, pp. 3963-3966.

Baderschneider, B. and Winterhalter, P. 2001. Isolation and characterization of novel benzoates, cinnamates, flavonoids, and lignans from Riesling wine and screening for antioxidant Activity. *J. Agric. Food Chem.* 49, pp. 2788-2798.

Banerjee,S.K., Mukherjee, P.K., and Maulikl, S.K. 2003. Garlic as an Antioxidant: The Good, The Bad and The Ugly. *Phytother. Res.* 17, pp. 97-106.

Banskota, A.H., Tezuka, Y., Nguyen, N. T., Awalel, S., Nobukawa, T., Kadota, S. 2003. DPPH Radical scavenging and nitric oxide inhibitory activities of the constituents from the wood of *Taxus yunnanensis*. *Planta Med.* 69, pp. 500-505.

Bao, Z., Wang, Z., Fang, J. Li, X. 2002. Structural features of an iImmunostimulating and antioxidant acidic polysaccharide from the seeds of *Cuscuta chinensis. Planta Med.* 68, pp. 237-243.

Banskota, A.H., Tezuka, Y., Kadota, S. 2001. Review article : Recent progress in pharmacological research of Propolis. *Phytother. Res.* 15, pp. 561-571.

Barreiros, A.L.B.S., David, J.P., Queirozc, L., David, J. M. 2000. A-type proanthocyanidin antioxidant from *Dioclea lasiophylla. Phytochemistry.* 55, pp. 805-808.

Bergmana, M., Varshavskya, L., Gottliebb, H.E., Grossmana,S. 2001.The antioxidant activity of aqueous spinach extract:chemical identification of active fractions. *Phytochemistry.* 58, pp. 143-152.

Bergman, M., Perelman, A., Dubinsky, Z., Grossman S. 2003. Scavenging of reactive oxygen species by a novel glucurinated flavonoid antioxidant isolated and purified from spinach. *Phytochemistry.* 62, pp. 753-762.

Bezabih, M., Abegaz, B.M., Dufal, K., Croft, K., Skinner-Adams, T., Davis, T.M.E. 2001. Antiplasmodial and Antioxidant isofuranonaphtJ1oquinones from the Roots of *Bulbine capitata.Planta Med.* 67, pp. 340-344.

Bhalla, V.K., Nayak, U.R., Dev. S. 1968.Some flavonoids from *Psoralea corylifolia. Tetrahedron Lett.* 20, pp. 2401-2406.

Biondi, D.M., Rocco, C., Ruberto G. 2003. New dihydrostilbene derivatives from the leaves of *Glycyrrbiza glabra* and Eevaluation of their antioxidant Activity. *J. Nat. Prod.* 66, pp. 477-480.

Bocco, A., Cuvelier, M., Richard, H., Berset, C. 1998. Antioxidant activity and phenolic composition of citrus peel and seed extracts. *J. Agric. Food. Chem.* 46, pp. 2123-2129.

Boverisa, A.D., Galatroa, A., Sambrotta L., Riccoc, R., Gumic, A.A. 2001. Puntaruloa, S. Antioxidant capacity of a 3-deoxyanthocyanidin from soybean. *Phytochemistry.* 58, pp. 1097-1105.

Bouchet, N., Barrier, L., Fauconneau, B. 1998. Radical scavenging activity and antioxidant properties of tannins from *Guiera senegalensis* (Combretaceae). *Phytother. Res.* 12, pp. 159-162.

Braca, A., De Tommasi, N., Di Bari, L., Pizza, C., Politi, P., Morellit, I. 2001. Antioxidant Principles from *Bauhinia tarapotensis . J. Nat. Prod.* 64, pp. 892-895.

Braca, A., Ficob, G., Morellia, I., Simone, F., Tome F., , Tommasi, N. 2003. Antioxidant and free radical scavenging activity "of flavonol glycosides from different *Aconitum* species. *J. Ethnopharmacol.* 86, pp. 63-67.

Bratt, K., Sunnerheim, K., Bryngelsson, S., Fagerlund, A., Engman, L. 2003. Andersson, R.E., Dimerg, L.H. Avenanthramides in Oats (*Avena sativa* L.) and Structure-Antioxidant Activity Relationships. *J. Agric. Food. Chem.* 51, pp. 594-600.

Burda, S and Oleszek,W. 2001. Antioxidant and antiradical activities of flavonoids. *J. Agric. Food. Chem.* 49, pp. 2774-2779.

Burns, J., Gardner, P., O'Neil, J., Crawford, S., Morecroft, I., McPhail, D., Lister, C., Mattthews, D., MacLean, M., Lean, M., Duthie, G.G., Crozier, A. 2000.Relationship among antioxidant activity, vasodilation capacity and phenolic content of red wines. *J. Agric. Food. Chem.* 48, pp. 220-230.

Cakira, A., Mavia, A., Yildirim, A., Duru, M.E., Harmandar, M., Kazazc, C. 2003. Isolation and characterization of antioxidant phenolic compounds from the aerial parts of

Hypericum hyssopifolium L. by activity-guided fractionation. *J. Ethnopharm.* 87, PP. 73-83.

Camara, B. and Moneger, R. 1978. Free and sterified carotenoids in green and red fruits of *Capsicum annuum*. *Phytochemistry.* 17, pp. 91-93.

Carini, M., Aldini, G., Orioli, M. 2002. Roberto Maffei FacinoAntioxidant and photoprotective activity of a lipophilic extract containing neolignans from *Krameria triandra* Roots. *Planta Med.* 68, pp. 193-197.

Cartron, E., Carbonneau, M., Fouret, G., Descomps, B., Claude L. Leger, C.L. 2001. Specific antioxidant activity of caffeoyl derivatives and other natural phenolic compounds: LDL protection against oxidation and decrease in the proinflammatory lysophosphatidylcholine production. *J. Nat. Prod.* 64, pp. 480-486.

Cavin, A., Hostettmannl, K., Dyatmyko, W., Potteratl, O. 1998. Antioxidant and lipophilic Constituents of *Tinospora crispa. Planta Med.* 64, pp. 393-396.

Cavin, A., Potterat, O., Wolfender, J., Hostettmann,K., Dyatmykot, W. 1998a. Use of on-flow LC/^{1}H NMR for the study of an antioxidant fraction from *Orophea enneandra* and isolation of a polyacetylene, lignans, and a tocopherol Derivative. *J. Nat. Prod.* 61, PP. 1497-1501.

Che, Z., Wong, I.Y.F., Leung, W.S., He, Z., Huang, Y. 2002. Characterization of Antioxidants Present in Bitter Tea (*Ligustrum pedunculare*). *J. Agric. Food Chem.* 50, pp. 7530-7535.

Chiu, P.Y., Mak, D.H.F., Poon, M. K.T., Ko, K.M. 2002. *In vivo* antioxidant action of a lignan-enriched extract of *Schisandra* Fruit and an anthraquinone-containing extract of *Polygon* root in comparison with Schisandrin B and Emodin. *Planta Med.* 68, pp. 951-956.

Cho, J., Huh, K., Kim, S., Lee, K.,Lee, H., Park, H. 2002. Kalopanaxsaponin A from *Kalopanax pictus*, a potent antioxidant in the rheumatoidal rat treated with Freund's complete adjuvant reagent. *J. Ethnopharmacol.* 79, pp. 113-118.

Choi, J.S., Chung, Y.H., Kang,S.S., Jung,M.J., Kim, J.W., No, J.K., Jung, R.A. 2002. The structure-activity relationship of flavonoids as scavengers of peroxynitrite. *Phytother. Res.* 16, pp. 232-235.

Chuda, Y., Ono, H., Ohnishi-Kameyama, M., Nagata, T., Tsushida T. 1996. Structural identification of two antioxidant quinic acid derivatives from garland (*Chrysanthemum coronarium* L.). *J. Agric. Food Chem.* 44, pp. 2037-2039.

Chung , H.S., and Wooj, W.S. 2001. A Quinolone Alkaloid with Antioxidant Activity from the Aleurone Layer of Anthocyanin-Pigmented Rice. *J. Nat. Prod.* 64, pp. 1579-1580.

Cichewicz, R.H., Nair, M.G. 2002. Isolation and characterization of stelladerol, a new antioxidant naphthalene glycoside, and other antioxidant glycosides from Edible Daylily (*Hemerocallis*) Flowers. *J. Agric. Food. Chem.* 50, pp. 87-91.

Cioffi, G., D'Auria, M., Braca, A., Mendez, J., Castillo, C., Morelli, I., De Simone, F De Tommasit, N. 2002. Antioxidant and Free-Radical Scavenging Activity of Constituents of the Leaves *Tachigalia paniculata. J. Nat. Prod.* 65, pp. 1526-1529.

Cioffi, G., Escobar, L.M., Braca, A., De Tommasit, N. 2003. Antioxidant chalcone glycosides and flavanones from *Maclura* (*Chlorophora*) *Tinctoria. J. Nat. Prod.* 66, pp. 1061-1064.

Cuendet, M., Hostettmann, K., Potterat, O., Dyatmiko, W. 1997. Iridoid glucosides with free radical scavenging properties from *Fagraea blumei. Helvetica Chim Acta.* 80, pp. 1144-1152.

Dan, Z., Man, H.K., Xu, L.H., Li,P., But, P. 2000. Antioxidant activity of phenylethanoid glycosides from *Brandisia hancei*. *J. Ethnopharm.* 71, pp. 483-486.

Demirezera, L.O., Kuruiiziim-Uza, A., Bergereb, I., Schieweb, H.J., Zeeckb, A. 2001. The structures of antioxidant and cytotoxic.agents from natural source: anthraquinones and tannins from roots of *Rumex patientia. Phytochemistry,* 58, pp. 1213-1217.

Donovan, J.L., Meyer,A.S., Waterhouse, L. 1998. Phenolic composition and antioxidant activity of prunes and prune juice (*Prunus domestica*). *J. Agric. Food Chem.* 46, pp. 1247-1252.

Dorman, H.J., Kosar, M., Kahlos, K., Holm, Y., Hiltunes, R. 2003. Phenolic diterpenes, flavones, and rosmarinic acid distribution during the development of leaves, flowers, stems, and roots of *Rosmarinus officinalis*. Antioxidant Activity. *J. Agric. Food Chem.* 51, pp. 4247-4253.

Dorman, H.J., Kosar, M., Kahlos, K., Holm, Y., Hiltunes, R. 2003. Antioxidant properties and composition of aqueous extracts from *Mentha* species, hybrids, varieties, and cultivars. *J. Agric. Food Chem.* 51, pp.4563-4569.

Dufalla, K.G., Ngadjuib, B.T., Simeonc, K.F., Abegaz B.M., Croft, K.D.2003. Antioxidant activity of prenylated flavonoids from the West African medicinal plant *Dorstenia mannii*. *J. Ethnopharmacol.* 87, pp. 67-72.

Dziedzic, S.Z., Hudson, B.J.F., Barnes, G., 1985. Polyhydroxydihydrochalcones as Antioxidants for Lard. *J. Agric. Food Chem.* 33, pp. 244-246.

Edge, R., McGarvey, D.J., Truscott, T.G. 1997. New Trends in Photobiology . The carotenoids as anti oxidants a review. *J. Photochem. Photobiol*, 41, pp. 189-200.

Escuder, B., Torres, R., Lissi, E., Labbe, C., Faini, F. 2002. Antioxidant capacity of abietanes from *Sphacele salviae.Nat. Prod. Lett.* 16, pp. 277-281.

Fabre, N., Urizzi, P., Souchard, J.P., Frechard, A., Claparols, C., Fouraste, I., Moulisa, C. 2000. An antioxidant sinapic acid ester isolated from *Iberis amara*. *Fitoterapia.* 71, pp. 425-428.

Faure, M., Lissi, E., Torres, R., Videla, L. Antioxidant activities of lignans and flavonoids. *Phytochemistry.* 29, pp. 3773-3775.

Filleur, F., Le Bail, J. C., Duroux, J. L. Simon, A., Chulia, A.J. 2001. Antiproliferative, anti-Aromatase, anti-17f3-HSD and antioxidant activities of lignans isolated from *Myristica argentea*. *Planta Med,* 67, pp. 700-704.

Fraga CG, Martino VS. Ferraro GE, Goussio JD, Boveris A. 1987. Flavonoids as anti-oxidants evaluated by *in vitro* and *in situ* liver, chemiluminiscence. *Biochem Pharmacol* 36, pp. 717-720.

Frankel, E. N., Huang, S., Aeschbach, R., Prior, E. 1996. Antioxidant activity of a rosemary extract and its constituents, carnosic acid, carnosol, and rosmarinic acid, in bulk oil and oil-in-water emulsion. I. *Agric. Food Chem.*44, pp. 131-135.

Fukuda, T., Ito, H., Yoshida, T. 2003. Antioxidative polyphenols from walnuts (*Juglans regia* L.). *Phytochemistry.* 63, pp. 795-801.

Garg,A., Garg, S., Zaneveld, L.J.D., and Sinpla, A. K. 2001. Chemistry and Pharmacology of The Citrus Bioflavonoid Hesperidin. *Phytother. Res.* 15, pp. 655-669.

Gao, Z., Huang, K., Yang, X., Xu, H. 1999.Free radicals scavenging and antioxidant activities of flavonoids extracted from radix of *Scutellaria baicalensis Georgi. Biochim. Et Biophy. Acta* 1472, pp. 643-650.

Gunatilaka, A.A.L., Kingston, D.G.I., 1998. DNA-damaging natural products with potential anticancer activity. In: Rahman, A.-U. (Ed.), Studies in Natural Products Chemistry, Elsevier. 20, pp. 457-505.

Hannedouche, S., Souchard, J.P., Jacquemond-Collet, Moulis C.M. 2002. Molluscicidal and radical scavenging activity of quinones from the root bark of *Caryopteris x clandonensis. Fitoterapia,* 73, pp. 520-522.

Haraguchi, H., Hashimoto, K., and Yagif, A. 1992. Antioxidative Substances in Leaves of *Polygonum hydropiper. J. Agric. Food Chem.* 40, pp.1349-1351.

Haraguchi, H., Ishikawa H., Kubo, I. 1997. Antioxidative Action of Diterpenoids from *Podocarpus nagi. Planta Med.* 63, pp. 213-215.

Haraguchi, H., Inoue,J., Tamura, T., Mizutani, K. 2002. Antioxidative components of *Psoralea corylifolia* (Leguminosae). *Phytother. Res.* 16, pp. 539-544.

Harbone,J.B., and Baxter, H. 1993. Phytochemical Dictionary. A handbook of bioactive compounds from plants. Taylor and Francis, London.

Hatano, T., Miyatakea, H., Natsumeb, M., Osakabeb, N., Takizawab, T., Itoa, H., Yoshidaa, T. 2002. Proanthocyanidin glycosides and related polyphenols from cacao liquor and their antioxidant effects. *Phytochemistry.* 59, pp. 749-758.

He, Y and Shahidi, S. 1997.Antioxidant activity of green tea and its catechins in a fish meat model system. *J. Agric. Food Chem.* 45, pp.4262-4266.

He, Z., Lau, K., But, P.P., Jiang, R., Dong, H., Ma, S., Fung, K., Ye, W., Sun, H. 2003. Antioxidative glycosides from the leaves of *Ligustrum robustum. J. Nat. Prod.* 66, pp. 851-854.

Heilmann, J, Merfortl, I., and Weiss, M. 1995. Radical Scavenger Activity of Different 3',4'-Dihydroxyflavonols and 1,5 -Dicaffeoylquinic Acid Studied by Inhibition of Chemiluminescence. *Planta Med.* 61, pp. 435-438.

Heilmann, J., Winkelmann, K., Sticher, O. 2003. Studies on the antioxidative activity of phloroglucinol derivatives isolated from *Hypericum Species. Planta Med.* 69, pp. 202-206.

Hidalgo, M.E., Fernandez, E., Quilhot, W., Lissi, E. 1994. Antioxidant activity of depsides and depsidones. *Phytochemistry.* 37, pp. 1585-1587.

Hiroyuki Haraguchi, H., Saito, T., Ishikawa, H., Date, H., Kataoka, S., Tamura, Y., Mizutani, K.1996. Antiperoxidative components in *Thymus vulgaris. Planta Med.* 62, pp. 217-221.

Hisamoto, M., Kikuzaki, H., Ohigashi, H, Nakatani, N. 2003 Antioxidant Compounds from the Leaves of *Peucedanum japonicum* Thunb. *J. Agric. Food Chem.* 51, pp.5255-5261.

Hisatomi, E., Matsui, M., Kobayashi, A., Kubota, K. 2000. Antioxidative activity in the pericarp and seed of Japanese pepper (*Xanthoxylum piperitum* DC). *J. Agric. Food Chem.* 48, pp. 4924-4928

Ho, K.Y., Huang, J.S., Tsai, C.C., Lin, T.C., Hsu, Y.F., Lin, C.C. 1999. Antioxidant activity of tannin components from *Vaccinium vitis-idaea* L.*J. Pharm. Pharmacol.* 51, pp. 1075-1078.

Hou, W., Chen, Y., Hsien-Jung Chen, Lin,Y., Yang, L., and Lee, M. 2001. Antioxidant activities of trypsin inhibitor, a 33 KDa root storage protein of sweet potato (*Ipomoea batatas* (L.) Lam cv. Tainong 57). *J. Agric. Food Chem.* 49, pp.2978-2981.

Houlihan, C., Ho, C., Chang, S. 1985. The structure of rosmariquinone a new antioxidant isolated from *Rosmarinus officinalis* L. *J. Am.Oil Chem.* 62, pp. 96-98

How, W., Hsu, F., Lee, M. 2002. Yam (dioscorea batatas) tuber mucilage exhibited antioxidant activities *in vitro. Planta Med.* 68, pp. 1072-1076.

Hu, C. and Kitts, D. D. 2000. Studies on the antioxidant activity of *Echinacea* root extract. *J. Agric. Food Chem.* 48, pp.1466-1472.

Huang, S. and Frankel E.N. 1997. Antioxidant activity of tea catechins in different lipid systems *J. Agric. Food Chem.* 45, pp. 3033-3038.

Hwang,B.Y., Kim, H.S., Lee, J.H., Hong, Y.S., Ro, J. S., Lee, K.S., Lee, J.J.2001. Antioxidant benzoylated flavan-3-o1 glycoside from *Celastrus orbiculatus. J. Nat. Prod.* 64, pp. 82-84.

Ibañez, E., Kubatova, A., Señorans, F.J., Cavero, S., Reglero, G., Hawthorne, S.B. 2003. Subcritical water extraction of antioxidant compounds from Rosemary Plants. *J. Agric. Food Chem.* 51, pp. 375-382.

Igarashi, K., Itoh, M., Harada, T. 1990. Major antioxidative substances in leaves of Atsumi-kabu (Red Turnip, *Brassica campestris* L.). *Agric. Biol. Chem.* 54, pp. 1053-1055.

Igile G.O., Oleszek, W., Jurzysta, M., Burda, S., Fafunso, M., Fasanmade, A.A. 1994. Flavonoids from *Vernonia amygdalina* and their antioxidant activities. *J. Agric. Food Chem.*42, pp. 2445-2448.

Imai1, J., Ide, N., Nagae, S., Moriguchi, T., Matsuura, H., Itakura, Y.1994. Antioxidant and radical scavenging effect of aged Garlic extract and its constituents. *Planta Med.* 60, pp. 417-420.

Ioset, J., Marston, R., Cupta, M.P., Hostettmann, K. 2001. A methylflavan with free radical scavenging properties from *Pancratium littoral. Fitoterapia.* 72, pp. 35-39.

Kahkonen, M.P., Heinonen, M. 2003. Antioxidant activity of anthocyanins and their aglycons. *J. Agric. Food. Chem.* 51, pp. 628-633.

Kamal-Eldin, A., Pouru, A., Eliasson, C., Aman, P. 2000. Alkylresorcinols as antioxidants: hydrogen donation and peroxyl radical-scavenging effects. *J. Science Food Agric.* 81, pp. 353-356.

Kato, T., Takahashi, W., Susuki, Y. 1997. Isolation and synthesis of new antioxidants from sunflower seeds. *Nat. Prod. Lett.* 9, pp. 161-165.

Kazuyo, K., Sarker, K.P., Maruyama, I., Hizukuri, S. 2002. Antioxidant effects of 1 ,5-anhydro-D-fructose, a new natural sugar, *in vitro. Planta Med.* 68, pp. 16-19.

Kikuzaki, H., Hara, S., Kawai, Y., Nakatani, N. 1999. Antioxidative phenylpropanoids from berries of *Pimenta dioica. Phytochemistry,* 52; pp. 1307-1312.

Kim, J., Kim, A. R, Chun, H.Y., Han, S.Y., Kim, B.S., Choi, J.S. 2003. *In vitro* peroxynitrite scavenging activity of diarylheptanoids from *Curcuma longa. Phytother. Res.* 17, pp. 481-484.

Kim, J.H., Lee, H. J., Yeon, S. C., Choi, D. H., Lee, S. S., Kang, J. K., Chae, C. H. 2004. Antioxidative effects of 7-hydroxy-3-methoxy-cadalene extracted from *Zelkova serrata* on 4-(methylinitros amino)-1-(3- pyridyl)-l-butanone-induced oxidative stress in A/J Mice. *Phytother. Res.* 18, pp. 425-427.

Kitta, K., Hagiwara, Y., Shibamoto, T. 1992. Atioxidative activity of an isoflavonoid, 2"-O-Glycosylisovitexin isolated from Green Barley Leaves. *J. Agric. Food. Chem.* 40, pp. 1843-1845.

Koleva, I.I., Linseen, J.P.H., Van Beek, T.A., Evstatieva, L.N., Kortenska, V., Handjieva, N. 2003. Antioxidant activity screening of extracts from *Sideritis* species (Labiatae) grown in Bulgaria. *J. Sci. Food Agric.* 83, pp. 809-819.

Kolhir, V. K., Bykov, V.A., Baginskaja, A.I., Sokolov, S. Y., Glazova, N.G., Leskova, T. E and Sakolich, G. S. 1996. Antioxidant activity of a dihydroquercetin isolated from *Larix gmelinii* (Rupr.) Rupr. Wood. *Phytother. Res.* 10, pp.478-482.

Kolhir,V. K., Bykov,V.A., Teselkin, Y.O., Babenkova, I.V., Tjukavkina, I. A., Rulenko,I.A., Kolesnik, Y.A., Eichholz, a.A. 1998. Use of a new antioxidant diquertin as an adjuvant in the therapy of patients with acute Pneumonia. *Phytother. Res.* 12, pp. 606-608.

Kosina, P., Kien, V., Gebhardt, R., Grambal, F., Ulrichov, J., Walterov, D. 2002. Antioxidant properties of silybin glycosides. *Phytother. Res.* 16, pp. S33-S39.

Kostova, I. 2001. *Fraxinus ornus* L. *Fitoterapia,* 72, pp. 471-480.

Kubo, I., Chaudhuri, S.K., Kubo, Y., Sanchez, Y., Ogura, T., Saito, T., Ishikawa, H., Haraguchi, H. Cytotoxic~nd Antioxidative Sesquiterpenoids from *Heterotheca inuloides. Planta Med.* 62, pp. 427-430.

Kumarasarnya, Y., Nahar L., Sarker, S.D. 2003.Bioactivity of gentiopicroside from the aerial parts of *Centaurium erythraea. Fitoterapia.* 74, pp. 151-154

Kuo, Y., Chien, S., Kuo, C. 2002. Antioxidative 7 -oxodehydropodocarpane- type trinorditerpenes from the bark of *Taiwania cryptomerioides. Planta Med.* 7, pp. 1020-1023.

Kuo, C., Chiang, W., Liu, G., Chien, Y., Chang, J., Lee, C., Lo, J., Huang, S., Shih, M., Kuo, Y. 2002. 2,2'-Diphenyl.1-picrylhydrazyl radical-scavenging active components from adlay (*Coix lachryma-jobi* L. Var. ma-yuen Stapf) Hulls. *J. Agric. Food Chem.* 50, pp.7548-7555.

Kweon, M., Hwang, H., Sung, H. 2001.Identification and Antioxidant Activity of Novel Chlorogenic Acid Derivatives from Bamboo (*Phyllostachys edulis*}. *J. Agric. Food Chem.* 49, pp.4646-4655.

Larson A. R. 1988. Review article number 30. The antioxidants of higher plants. *Phytochemistry.* 27, pp. 969-978.

Laughton, M.J., Halliwell, B., Evans, P.J., Hoult, J.R.1989. Antioxidant and pro-oxidant actions of the phenolics quercetin, gossypol and myricetin. Effects on lipid peroxidation, hydroxyl radical generation and bleomycin-dependent damage DNA. *Bioch. Pharmacol.* 38, pp. 2859-2865.

Lea, C. H. and Swoboda, P. A. T.1956, On the antioxidant activity of the flavonols, gossypetin and quercetagetin. *Chem. Ind.* pp. 1426-1428.

Limasset B, Le Doucen C, Dore J.C, Ojasoo T, Damon M, Crastes de Paulet A. 1993. Effects of flavonoids on the release of reactive oxygen species by stimulated human neutrophils. *Biochem Pharmacol.* 46, pp. 1257-1271.

Lin, C., Hsu, Y., Lin, T. 2001. Antioxidant and Free Radical Scavenging Effects of the Tannins of *Terminalia catappa* L. *Anticancer Res.* 21, pp. 237-244.

Lin, C., Hsu, Y., Lin, T., Hsu, H. 2001a. Antioxidant and hepatoprotective effects of punicalagin and punicalin on acetaminophen-induced liver damage in rats. *Phytother. Res.* 15, pp. 206-212.

Liu, Z., Luo, X., Liu, G., Chen, Y., Wang, Z., Sun, Y. 2003. *In vitro* study of the relationship between the structure of ginsenoside and its antioxidative or prooxidative activity in free radical induced hemolysis of human erythrocytes. *J. Agric. Food. Chem.* 51, pp. 2555-2558.

Lo, Y., Teng, C., Chen, C., Chen, C., Hong, C. 1994.Magnolol and honokiol isolated from *Magnolia officinalis* protect rat heart mitochondria against lipid peroxidation *Bioch. Pharmacol.* 47, pp. 549-553.

Luo, X., Basile, M.J.,Kennelly, E.J. 2002. Polyphenolic Antioxidants from the Fruits of *Chrysophyllum cainito* L. (Star Apple). *J. Agric. Food Chem.* 50, pp.1379-1382.

Ly, T.M., Shimoyamada, M., Kato, K., Yamamuchi, R. 2003. Isolation and characterization of some antioxidative compounds from the rhizomes of smaller galanga (*Alpinia officinarum* Hance). *J. Agric. Food Chem.* 51, pp.4924-4929.

Manguro, L.O., Ogweno, J., Kraus, W. 1996. A flavonol glycoside from *Myrsine Africana* leaves. *Phytochemistry.* 43, pp. 1107-1109.

Markham, K.R., and Mitchell, K. A. 2003. The mis-identification of the major antioxidant flavonoids in young Barley (*Hordeum vulgare*) Leaves. Z. Naturforsch. 58c, pp. 53-56.

Martin, T.S., Kikuzaki, H., Hisamoto, H., Nakatani, N. 2000. Constituents of amomum tsao-ko and their radical scavenging and antioxidant activities. *J. Am. Chem. Soc.* 77, PP. 667-673.

Martindale, A. 1989. The extra Pharmacopoeia. 28th/29th ed. Pharmaceutical Press, London. pp. 234-237.

Martinez-Valverde,I., Periago,M.J., Provan, G., Chesson, A. 2002. Phenolic compounds, lycopene and antioxidant activity in commercial varieties of tomato (*Lycopersicum esculentum*). *J. Science Food Agric.* 82, pp. 323-330.

Mathiesen, L., Malterud, K. E., Sund, R. B. 1995. Antioxidant Activity of Fruit Exudate and C- Methylated Dihydrochalcones from *Myrica gale. Planta Med.* 61, pp. 515-518.

Mathisen, E., Diallo, D., Andersen, O. M., Malterud, K. E. 2002. Antioxidants from the Bark of *Burkea Africana*, an African Medicinal Plant. *Phytother. Res.* 16, pp. 148-153.

Matsufuji,H., Nakamura, H., Chino, M., and Takeda M. 1998. Antioxidant Activity of Capsanthin and the Fatty Acid Esters in Paprika (*Capsicum annuum*). *J. Agric. Food Chem.* 46, pp.3468-3472.

Mirelman, D., Monheit, D., Varon, S. 1987. Inhibition of growth of Entamoeba histolytica by allicin, the active principle of garlic extract (*Allium sativum*). *J. Infect. Dis.* 156, pp. 493-500.

Miura, K., Kikuzaki, H., Nakatani, N. 2002. Antioxidant activity of chemical components from sage (*Salvia officinalis* L.) and thyme (*Thymus vulgaris* L.) Measured by the oil stability index method. *J. Agric. Food Chem.* 50, pp.1855-1851.

Miyase,T., Sanoa, M., Nakaia, H., Muraokaa, M., Nakazawaa, M., Suzukia, M., Yoshinob, K., Nishiharab, Y., Tanai J. 1999. Antioxidants from *Lespedeza homoloba* (I). *Phytochemistry.* 52, pp. 303.310.

Miyase,T., Sanoa, M., Yoshinob, K., Nonaka, K. 1999a. Antioxidants from *Lespedeza homoloba* (II). *Phytochemistry.* 52, pp.311-319.

Miyase, T., Sano, M., ,Nakai. H., Muraoka. M., Nakazawa. M., Suzuki, M., Yoshino, K., Nishihara. Y., Tanai, T., 1999. Anlioxidanls from *Lespedeza homoloba* (II). *Phylochemistry.* 52, pp. 303-310.

Modak, B., Torres, R., Lissi, E., Monache, F.D. 2003. Antioxidant capacity of flavonoids and a new arylphenol of the resinous exudates from *Heliotropium sinuatum. Nat. Prod. Res.* 17, pp. 403-407.

Montoro, P., Carbone,V., De Simone, F., Pizza, C., and De Tommasi, N. 2001. Studies on the constituents of *Cyclanthera pedata* fruits: Isolation and structure elucidation of new flavonoid glycosides and their antioxidant activity. *J. Agric. Food Chem.* 49, pp.5156-5160.

Nobuji, N., and Kikuzaki, H. 1987. A New Antioxidative Glucoside Isolated fr~ Oregano (*Origanum vulgare* L.). *Agric. Food Chem.* 10, pp. 2727-2732.

Nakatani, N., Kayano, S., Kikuzaki, H., Sumino, K., Katagiri, K., and Mitani, T. 2000. Identification, quantitative determination, and antioxidative activities of chlorogenic acid isomers in Prune (*Prunus domestica* L.). *J. Agric. Food Chem.* 48, pp. 5512-5516.

Nakatani, N., Ikeka, K. 1987. Structure of lignans from Papuan Mace (*Myristica argentea* Warb.). *Chemistry Express.* 2, pp. 627 -630.

Noreen, Y., El-Seedi,H., Perera, P., Bohlin, L. 1998a. Two new isoflavones from *Ceiba pentandra* and their effect on cyclooxygenase-catalyzed prostaglandin biosynthesis. *J. Nat. Prod.* 61, pp. 8-12.

Nukata. M., Yamamoto. I., Sasaki. K. 1996. Antioxidative substances from several mushrooms. *Nippon Nogeikagaku Kaishi.* 70, pp. 58-61.

Nukataa, M., Hashimotob, T., Yamamotoa, I., Iwasakib, N., Tanakab, M., Sakawab,Y. 2002. Neogrifolin derivatives possessing anti-oxidative activity from the mushroom *Albatrellus ovinus*. *Phytochemistry*. 59, pp. 731-737.

Ochi, T., Takaishi, Y., Kogure, K. and Yamauti, I. 2003. Antioxidant Activity of a New Capsaicin Derivative from *Capsicum annuum*. *J. Nat. Prod.* 66, pp. 1094-1096.

Osakabe. N., Natsume, M.. Adachi, T., Yamagishi, M., Hirano. R.Takizawa. T., Ilakura, H., Kondo. K. 2000. Effects or cacao liquor poly phenols on the susceptibility or low-density lipoprotein to oxidation in hypercholesterolemic rabbits. *J. Atheroscl.Thromb.* 7, pp. 164-168.

Osawa, T., Katsuzaki, H., Hagiwara, Y., Hagiwara, H., Shibamoto, T. 1992. A novel antioxidant isolated from young Green Barley Leaves. *J. agric. Food Chem.* 40, pp. 1135-1138.

Osawa, T., Sugiyama, Y., Inayoshi, M., Kawakishi, S. 1995. Antioxidative activity of tetrahydrocurcuminoids. *Biosci. Biotech. Biochem.* 59, pp. 1609-1612.

Ouyang, M., He, Z., Wu, C. 2003. Anti-oxidative activity of glycosides from *Ligustrum sinense*. *Natural. Prod. Res.* 17, pp. 381-387.

Paiva-Martins, F and Gordon, M.H. 2001.Isolation and characterization of the antioxidant component 3,4- Dihydroxyphenylethyl 4- formyl-3- formylmethyl-4- hexenoate from Olive (Olea europaea) Leaves. . *J. Agric. Food Chem.* 49, pp. 4214-4219.

Palozza, P and Krinsky, N.I. 1992. Antioxidant effects of carotenoids *in vivo* and *in vitro*: An Overview. Methods in Enzymology, Academic Press., 213, pp. 403-419.

Peng, Z.F., Strackb, D., Baumertb, A., Subramaniama, R., Goha, N.K., Chiaa, T. F., Tana, S.N., Chiaa,L.S. 2003. Antioxidant flavonoids from leaves of *Polygonum hydropiper* L. *Phytochemistry.* 62, pp. 219-228.

Purushothaman KK. Sarada A. 1984. Structure of malabaricanol. A lignan from the aril of *Myristica malabarica* Lam. *Indian Journal of Chemistry*; 23 B, pp. 46-48.

Pham, T.Q., Cormier,F., Farnworth, E., Tong, V., and Van Calsteren M. 2000. Antioxidant properties of crocin from *Gardenia jasminoides* Ellis and study of the reactions of crocin with linoleic acid and crocin with oxygen. *J. Agric. Food Chem.* 48, pp. 1455-1461.

Rahman, A.U., Ngounou, F.N., Choudhary1, M.I., Malik, S., Makhmoor, T., Alam, M.N. 2001. New antioxidant and antimicrobial ellagic acid derivatives from *Pteleopsis hylodendron. Planta Med.* 67, pp. 335-339.

Rezk, B. M., Haenen, G. R.M.M., Vijgh, W J.F., Basta, A. 2002. The antioxidant activity of phloretin: the disclosure of a new antioxidant pharmacophore in flavonoids. *Bioch. Biophysical Res.* 295, pp. 9-13.

Rios, J.L., Mañez, S., Paya, M., Alcaraz, M.J. 1992. Antioxidant activity of flavonoids from *Sideritis javalambrensis.Phytochemistry,* 31, pp. 1947-1950.

Sakushimao, A., Maoka, T., Ohnoo, K., Coskunc, M., Guvenc, A., Erdurak, C.S., , Ozkanc, A.M., Seki, K., Ohkurao, K. 2000. Major Antioxidative Substances in *Boreava orientalis* (Cruciferae). *Nat. Prod. Lett.* 14, pp. 441-446.

Sanbongi, C., Osakabe, N., Natsume,M., Takizawa, T., Gomi, S., Osawa, T. 1998. Antioxidative polyphenols isolated from *Theobroma cacao. J. Agric. Food Chem.* 46, pp. 454-457.

Sassa, H., Takaishi, Y., Terada H. 1990.The triterpene celastrol as a very potent inhibitor of lipid peroxidation in mitochondria. *Biochem. Biophys. Comm.* 172, pp. 890-897.

Sekiwa, Y., Kubota,K., Kobayashi, A. 2000. Isolation of novel glucosides related to gingerdiol from Ginger and their antioxidative activities. *J. Agric. Food. Chem.* 48, Pp. 373-377.

Sharma, O.P. 1976. Antioxidant activity of curcumin and related compounds. *Biochemical Pharmacology*. 25, pp. 1811-1812.

Silva, D.H.S., Davino,S.C., Barros, S.B.D., Yoshida, M.1999.Dihydrochalcones and flavonolignans from *Iryanthera lancifolia. J. Nat. Prod.* 62, pp. 1475-1478.

Silva, D.H., Pereira, F.C., Zanoni, M.V.B., Yoshida, M. 2001. Lipophyllic antioxidants from *Iryanthera juruensis* fruits. *Phytochemistry.* 57, pp. 4437-442.

Song, E., Cho, H., Kim, J., Kim, S., An, N., Kim, J., Lee, S., Kim, Y. 2001. Diarylheptanoids with free radical scavenging and hepatoprotective activity *in vitro* from *Curcuma longa. Planta Med.* 67, pp. 876-877.

Strobel, G., Ford, E., Worapong, J., Harper, J. K., Arif, A.M., Grant, D.M., Fung,P.C.W., Chau, R.M.W. 2002. Isopestacin, an isobenzofuranone from *Pestalotiopsis microspora*, possessing antifungal and antioxidant activities. *Phytochemistry.* 60, pp. 179-183.

Tachibana,Y., Kikuzaki, H., Lajis, N., Hj. Nakatani, N. 2001. Antioxidative Activity of Carbazoles from *Murraya koenigii* Leaves. *J. Agric. Food Chem.* 49, pp.5589-5594.

Tada, M., Matsumoto, R., Yamaguchi, H., Chiba, K. 1996. Novel antioxidants isolated from *Perilla frutescens* Brit ton var. crispa (Thunb.). *Biosci. Biotech. Biochem.*60, pp. 1093-1095.

Tagashira, M., and Ohtake, Y. 1998. A New Antioxidative 1 ,3-Benzodioxole from *Melissa officinalis.Planta Med.* 64, pp. 555-558.

Tiwari, A.K., Srinivas, P.V., Kumar, S.P., Rao, J.M. 2001. Free radical scavenging active components from *Cedrus deodara. J. Agric. Food Chem.* 49, pp.4642-4645.

Tambe, Y., Tsujiuchr, H., Honda, G., Ikeshiro, Y., Tanaka. S. 1996. Gastric cytoprotection of the non-steroidal anti-inflammatory sesquiterpene, *p*-caryophyllene. *Planta Med.* 62, pp. 469-470.

Tamura, H. and Yamagami, A. 1994. Antioxidative activity of monoacylated anthocyanins isolated from Muscat Bailey A Grape. *J. Agric. Food. Chem.* 42, pp. 1612-1615.

Tapia, A., Feresin G.E., Bustos, D., Astudillo L., Theoduloz, C., Schmeda-Hirschmann G. 2000. Biologically active alkaloids and a free radical scavenger from *Prosopis* species. *J. Ethnopharmacol.* 71, pp. 241-246.

Tiew, P., Ioset, J.R., Kokpol, U., Chavasirr, W., and Hostettmann K. 2003. Antifungal, antioxidant and larvicidal activities of compounds isolated from the heartwood of *Mansonia gagei. Phytother. Res.* 17, pp. 190-193.

Torel, J., Cillard, J., Cillard, P. Antioxidant activity of flavonoids and reactivity with peroxy radical . *Phytochemistry.* 25, pp. 383-385.

Torres, J.L., and Bobett, R. 2001. New Flavanol Derivatives from Grape (*Vitis vinifera*) Byproducts. Antioxidant Aminoethylthio- Flavan-3-ol Conjugates from a Polymeric Waste Fraction U sed as a Source of Flavanols. *J. Agric. Food Chem.* 49, pp. 4627-4634.

Traykovaa, M., Traykovb, T., Hadjimitovab, V., Krikorianc, K., Bojadgievaa N. 2003. Antioxidant properties of galantamine Hydrobromide. *Z. Naturforsch.* 58c, pp. 361-365.

Tsai, J., Chou, C., Chen, C., Chiou, W. 2002. Antioxidant activity of piperlactam S: Prevention of copper-induced LDL peroxidation and amelioration of free radical-induced oxidative stress of endothelial cells. *Planta Med.* 69, pp. 3-8.

Vedavanam, K., Srijayanta, S., O' Reilly, J., Raman, A., Wisemanl, H. 1999.Antioxidant Action and Potential Antidiabetic Properties of an Isoflavonoid-containing Soyabean Phytochemical Extract (SPE). *Phytother. Res.* 13, pp. 601-608.

Wagner, H., Chari, V., Sonnenbichler.1976. ^{13}CNMR-spektren naturlich vorkkommender flavonoide. *Tetrahedron Lett.* 21, pp. 1799-1802.

Wang, M., Li, J., Rangarajan, M., Shao, Y., LaVoie, E.J., Huang, T., Ho, C. 1998. Antioxidative Phenolic Compounds from Sage (*Salvia officinalis*). *J. Agric. Food Chem.* 46, pp. 4869-4873.

Wang, H., Nait, M, G., Strasburg, G.M., Booren, A.M. Gray, J.I. 1999a. Antioxidant Polyphenols from Tart Cherries (*Prunus cerasus*). *J. Agric. Food Chem.* 47, pp. 840-844.

Wang, H., Nair, M.G., Strasburg, G.M., Booren, A.M., Gray, J. I. 1999b. Novel Antioxidant compounds from tart cherries (*Prunus cerasus*). *J. Nat. Prod.* . 62, pp. 86-88.

Wang, H., Nair; M.G., Strasburg, G.M., Chang, Y., Booren, A.M., Gray, J.I., DeWitt, D.L. 1999c. Antioxidant and antiinflammatory activities of anthocyanins and their aglycon, cyanidin, from tart cherries. *J. Nat. Prod.* 62, pp. 294-296.

Wang, M., Shao, Y., Li, J., Zhu, N., Rangarajan, M., LaVoie, E., Ho, C.1999a. Antioxidative phenolic glycosides from Sage (*Salvia officinalis*). *J. Nat. Prod.* 62, pp. 454-456.

Wang, S., Kuo, Y., Chang, H., Kang, P., Tsay, H., Lin, K., Yang, N., Shyur, L. 2002. Profiling and characterization antioxidant activities in *Anoectochilus formosanus* Hayata. *J. Agric. Food Chem.* 50, pp. 1859-1865.

Watanabe, M., Ohshita, Y., Tsushida, T. 1997.Antioxidant compounds from Buckwheat (*Fagopyrum esculentum* Moench) Hulls. *J. Agric. Food Chem.* 45, pp. 1039-1044.

Watanabe, W. 1998.Catechins as antioxidants from Buckwheat (*Fagopyrum esculentum* Moench) Groats. *J. Agric. Food Chem.* 46, pp. 839-845.

Watanabe, M. 1999. Antioxidative phenolic compounds from Japanese Barnyard Mollet {*Echinochloa utilis*) grains. *J. Agric. Food Chem.* 47, pp. 4500-4505.

Westenburg, H.E., Lee, K., Lee, S.K., Fong, H.H.S., van Breemen, R.B., Pezzuto, J.M., Kinghorn, D. A. 2000. Activity-Guided isolation of antioxidative constituents of *Cotinus coggygria. J. Nat. prod.* 63, pp. 1696-1698.

Wong, I.Y.F., He, Z., Huang, Y., Chen, Z. 2001. Antioxidative activities of phenylethanoid glycosides from *Ligustrum purpurascens .J. Agric. Food Chem.* 49, pp. 3113-3119.

Wu, P., Lin,S., Huang, C., Robin Y., Chiou, Y. 2002. Antioxidative and cytotoxic compounds extracted from the Sap of *Rhus succedanea. J. Nat. Prod.* 65, pp. 1719-1721.

Yagi, A. Kabash, A. Okamura, N., Haraguchi, H., Moustafa, S.M., Khalifa, T. I. 2002. Antioxidant, free radical scavenging and anti-inflammatory effects of aloesin derivatives in *Aloe vera. Planta Med.* 68, pp. 957-960.

Yagi, A., Kabashl, A., . Mizunol, K., Moustafa, S. M., Khalifa, T.I., Tsuji, H.2003. Radical scavenging glycoprotein inhibiting Cyclooxygenase-2 and thromboxane A_2 Synthase from *Aloe vera* Gel. *Planta Med.* 69, pp. 269-271.

Yamaguchi, F., Ariga, T., Yoshimura,Y., Nakazawa, H. 2000. Antioxidative and anti-glycation activity of garcinol from *Garcinia indica* fruit Rind. *J. Agric. Food Chem.* 48, pp. 180-185.

Yan, X., Murphy, B.T., Hammond, G.B., Vinson, J.A., Neto, C.C. 2002. Antioxidant Activities and Antitumor Screening of Extracts from Cranberry Fruit (*Vaccinium macrocarpon*). *J. Agric. Food Chem.* 50, pp. 5844-5849.

Yesilada E., Tsuchiya K, Takaishi Y., Kawazoe K. 2000. Isolation and characterization of free radical scavenging flavonoid glycosides from the flowers of *Spartium junceum* by activity-guided fractionation. *J. Ethnopharmacol.* 73, pp. 471-478.

Ying, B. and Kubo, I. Complete ^{1}H and ^{13}CNMR assignments of totarol and its derivatives.1991. *Phytochemistry.* 30, pp. 1951-1955.

Yokozawa, T., Dong , E., Wu, Z., Shimizu, M. 1997. Antioxidative activity of flavones and flavonols *in vitro*. *Phytother. Res.* 11. pp. 446-449.

Yoshimasa, N., Watanabe, S., Miyake, N., Kohno, H., Osawa, T.2003. Dihydrochalcones: Evaluation as novel radical scavenging antioxidants. *J. Agric. Food Chem.* 51, pp.3309-3312.

Yun, B., Lee, I., Ryoo, I., and Yoo, I. 2001. Coumarins with Monoamine Oxidase Inhibitory Activity and Antioxidative Coumarino-lignans from *Hibiscus syriacus*. *J. Nat. Prod.* 64, pp.1238-1240.

Zhang, K., Bao, Y., Wu, P., Rosen, R.T., Ho, C. 1990. Antioxidative components of Tanshen (*Salvia miltiorrhiza* Bung). *J. Agric. Food Chem.* 38, pp. 1194-1197.

Zhang, H.L., Nagatsu, A., Watanabe, T., Sakakibara, J., Okuyama, H. 1997. Antioxidative compounds isolated from safflower (*Carthamus tinctorius* L.) Oil Cake. *Chem. Pharm. Bull.* 45, pp. 1910-1914.

Zhao WH, Gin GW, Xu RS. 1999. Constituents from the roots of *Acanthopanax setchuenensis*. *Fitoterapia.* 70, pp. 529-531.

Zheng, W., and Wang, S. Y. 2001. Antioxidant activity and phenolic compounds in selected herbs. *J. Agric. Food Chem.* 49, pp.5165-5170.

Zou, Y., Kim, A.R., Kim, J.E., Cho, J.S., Chung, H.Y. 2002. Peroxynitrite scavenging activity of sinapic acid (3,5.dimethoxy-4-hydroxycinnamic acid) isolated from *Brassica juncea*. *J. agric. Food. Chem.* 50, pp. 5884-5890.

INDEX

A

B

C

D

E

F

M

N

O

P

Q

R

S

T

U

V

W

Z